and relates c-pgm to a-pgm; however, it also uses information about <u>why</u> a-pgm is correct. Basically, one expresses the verification conditions of a-pgm in terms of the relation a-constraint, parameterized over the set E. One then relates the functions c-f to a-pgm via the mappings CA and CE and one proves correctness of c-pgm by working through a checklist of abstract verification conditions.

3. AN ABSTRACT MARKING ALGORITHM.

The general purpose of marking algorithm is to "mark" all cells in some multi-linked structure that are currently in use. The unmarked cells can then be collected and used by the system. In other words, given a finite static set of memory nodes N, a set of "immediately accessible" nodes S_0 (where $S_0 \subseteq N$), and a static relation $R_0 \subseteq N \times N$ (corresponding to "direct reachability" with respect to a list structure imposed on N), the purpose of a marking algorithm is to "mark" all nodes in S_0 as well as all nodes reachable from S_0 by one or more applications of the relation R_0.

Abstractly we have:

a-vec $\triangleq$ <N, S_0, R_0, marked, S, R>, where
N is a fixed set of nodes;
S_0 is a fixed subset of N;
R_0 is a fixed relation on N
marked, S are dynamically varying subsets of N;
and R is a dynamically varying relation on N.

The abstract specifications are given by

a-in-spec(a-vec) $\triangleq$ (marked = S_0) <u>and</u> (S = S_0) <u>and</u> (R = R_0)
a-out-spec(a-vec) $\triangleq$ marked = R_0*(S_0), where * denotes closure
a-inv(a-vec) $\triangleq$ (S $\subseteq$ marked $\subseteq$ R_0*(S_0) $\subseteq$ (marked $\cup$ R*(S))) <u>and</u> (R $\subseteq$ R_0)
a-term-cond(a-vec) $\triangleq$ (S = { }) <u>or</u> (R = { })

Based on the above, we can now write the following abstract marking algorithm (a-pgm).

<u>a-pgm</u>:

<u>while</u> S $\neq$ { } <u>and</u> R $\neq$ { } <u>do</u> a-vec: = a-f(a-vec) <u>od</u>

The program a-pgm will be partially correct provided we can show that a-inv(a-vec) really is invariant under a-f. This, of course, is impossible without some knowledge of a-f; however, the constrained functional mapping approach (cf. Section 2) can be used in this situation.

Let the entity E = <A, B, C, D> be as follows:

A, B, C are subsets of N; and
D $\subseteq$ N x N.

Notationally, let D_2 = range(D), and let a-f(a-vec) be denoted by

a-vec' = <N, S_0, R_0, marked', S', R'>.

Let a-constraint(a-vec, E. A-vec') be defined as the conjunction of the following conditions:

1. $S' = (S \cup A) - B$
2. $marked' = marked \cup C$
3. $R' = R - D$
4. $A \subseteq marked \cup C$
5. $C \subseteq R_0{*}(S_0)$
6. $B \subseteq marked \quad C$
7. $R(B) \subseteq (S \cup A) - B$
8. $D_2 \subseteq marked \cup C \cup ((S \cup A) - B)$
9. $R(D_2) \subseteq ((S \cup A) - B) \cup (R - D)((S \cup A) - B)$.

The intuition behind these sets is as follows: At each stage of the algorithm we can reach all accessible unmarked cells by starting with the set S and applying relation R. At each step we can do one (or more) of the following: (1.) mark some cells (C), (2.) add cells (A) to S, (3.) remove unnecessary cells (B) from S, or (4.) remove unnecessary pairs (D) from the relation R.

The following theorem can be proved in a straightforward manner.

THEOREM 3.1: Any function a-f satisfying a-constraint preserves the invariance of a-inv.

As a result of Theorem 3.1, the bulk of the work needed to show correctness of a concrete marking program involves producing mappings CE, where CE(c-vec) = <A, B, C, D>, and proving that the concrete function c-f manipulates c-vec in such a manner that a-constraint is satisfied at the abstract level. In other words, we want to prove Step 2(iv) of Level 3 in the previous section.

4. INTERLUDE: CODED STRUCTURAL GRAPHS.

Concrete implementations of marking algorithms generally operate on some multilinked data structure, such as a heap. It is therefore desireable to have a model for such a data structure as well as an assertion language to describe the semantics of operations on the data structure. The theory of coded structural graphs provides both a model and an assertion language for data structures of this type and the interested reader is referred to [12,13] for a detailed description of the theory. In this section we shall simply state the main ideas and indicate how they pertain to marking algorithms.

CODED STRUCTURAL GRAPHS

A *coded structural graph space* (*CSG-space*) consists of a triple G = <nodeset nameset, codeset> of sets. A *coded structural graph* on G is a function.

$$\gamma : nodeset \times nameset \rightarrow (codeset \times nodeset) \quad \{\lambda\}$$

Example: In the Schorr-Waite Algorithm we will have

nodeset = N

nameset = {ALINK, BLINK}

Texts in Theoretical Computer Science
An EATCS Series

G. Păun G. Rozenberg A. Salomaa

DNA Computing

New Computing Paradigms

With 74 Figures

Authors

Prof. Dr. Gheorghe Păun
Institute of Mathematics
of Romanian Academy
PO Box 1-764
RO-70700 Bucharest, Romania
and
Research Group on Natural Computing
Department of Computer Science
and Artificial Intelligence
Sevilla University
Avenida Reina Mercedes s/n
41012 Sevilla, Spain
gpaun@us.es

Prof. Dr. Grzegorz Rozenberg
Leiden Institute of Advanced Computer Science
University of Leiden
Niels Bohrweg 1
2333 CA Leiden, The Netherlands
rozenber@liacs.nl

Prof. Dr. Arto Salomaa
Turku Centre of Computer Science
Lemminkäisenkatu 14 A
20520 Turku, Finland
asalomaa@utu.fi

Series Editors

Prof. Dr. Wilfried Brauer
Institut für Informatik der TUM
Boltzmannstr. 3
85748 Garching, Germany
Brauer@informatik.tu-muenchen.de

Prof. Dr. Grzegorz Rozenberg

Prof. Dr. Arto Salomaa

Library of Congress Cataloging-in-Publication Data
Păun, Gheorghe, 1950-
DNA computing : new computing paradigms / G. Păun, G. Rozenberg,
A. Salomaa.
p. cm. -- (Texts in theorectical computer science)
Includes bibliographical references and index.

1. Molecular computers. I. Rozenberg, Grzegorz. II. Salomaa,
Arto. III. Titel. IV. Series.
QA76.887.P38 1998
511.3--dc21 98-18927
CIP

ACM Computing Classification (1998): F.1.2, F.4.1, F.4.3, I.1.3

ISBN 978-3-642-08388-4

First edition 1998, 2nd corr. printing 2005

Springer is a part of Springer Science+Business Media
springeronline.com

Cover design: KünkelLopka, Heidelberg

Preface

The authors are much indebted to many friends and collaborators whose contributions to automata and language theory approach to DNA computing can be recognized in the present book. The bibliography specifies their names and we shall not repeat them here. Some of them have also read previous versions of various chapters, suggesting modifications which have improved the readability of the text. Many thanks are due in this respect to Tom Head, Hendrik Jan Hoogeboom, Vincenzo Manca, Alexandru Mateescu, Victor Mitrana, Andrei Păun, and Nikè van Vugt. In particular, we are grateful to our biologist friends Hans Kusters and Paul Savelkoul for many illuminating discussions. Anu Heinimäki drew the pictures in the Introduction. The expert assistance and timely cooperation of Springer-Verlag, notably Dr. Hans Wössner, is gratefully acknowledged.

Gheorghe Păun, Grzegorz Rozenberg, Arto Salomaa
Leiden, July 1998

Contents

Introduction: DNA Computing in a Nutshell

From silicon to carbon. From microchips to DNA molecules. This is the basic idea in DNA computing. Information-processing capabilities of organic molecules can be used in computers to replace digital switching primitives.

There are obvious limits to miniaturization with current computer technologies. For a drastic innovation, it was suggested already a long time ago that the basic components should go to the molecular level. The result would be much smaller than anything we can make with present technology. *Quantum computing* and *DNA computing* are two recent manifestations of this suggestion. This work is about the latter.

Figure 1

Computers have a long history. Mechanical contrivances designed to facilitate computations have existed for ages. While the earliest-known instrument of calculation of any importance is the abacus, the present-day electronic computers depicted in Fig. 1 have gained such a dominant position in our society that most of our activities would have to be abandoned but for their help. Yet present-day computers have many drawbacks. Because of numerous

Figure 2

intractable problems, it seems that the computer in Fig. 1 is not the end of the long road of development.

Figure 3

When the road continues, we might see a DNA computer. In the one shown in Fig. 2, all operations with the test tubes have to be carried out by the user.

A more advanced model is depicted in Fig. 3, where some robotics or electronic computing is combined with DNA computing, and the majority of the operations with the test tubes is carried out automatically, without the intervention of the user.

A famous forerunner of present-day computers, Charles Babbage, set out around 1810–1820 to build an automatic computer, a "Difference Engine," as well as a more ambitious computing machine, an "Analytical Engine." The failure to construct either of the machines was due mainly to the lack of sufficiently accurate machine tools, and of mechanical and electrical devices that became available only during the 20th century. Perhaps we face today a similar situation with respect to DNA computers. Biochemical techniques are not yet sufficiently sophisticated or accurate. In particular, the techniques have not yet been adequately developed towards the specific needs of DNA computing. It is most likely that the waiting period here will be much shorter than in Babbage's case.

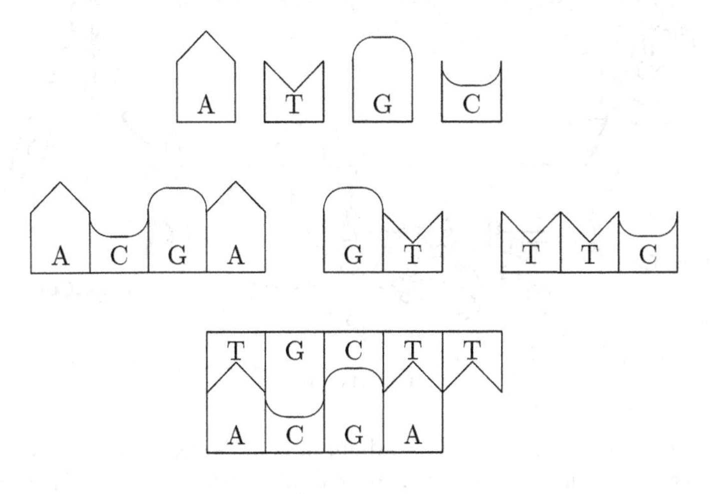

Figure 4

The high hopes for the future of DNA computing are based on two fundamental features:

(i) The massive *parallelism* of DNA strands,

(ii) Watson–Crick *complementarity*.

We now describe the two features briefly.

(i) Most of the celebrated computationally intractable problems can be solved by an exhaustive search through all possible solutions. However, the insurmountable difficulty lies in the fact that such a search is too vast to be carried out using present technology. On the other hand, the density of information stored in DNA strands and the ease of constructing many copies of them might render such exhaustive searches possible. A typical example would be the cryptanalysis of a ciphertext: all possible keys can be tried out simultaneously.

(ii) Watson–Crick complementarity is a feature provided "for free" by the nature. When bonding takes place (under ideal conditions) between two DNA strands, we know that the bases opposite each other are complementary. When we know one member of a bond, we know also the other; there is no need to check it in any way. This results in a powerful tool for computing because, as we will see latter, complementarity brings the universal twin-shuffle language to the computing scene. By encoding information in different fashions on the DNA strands subjected to bonding, we are able to make far-reaching conclusions based on the fact that bonding has taken place.

Figure 5

Let us elaborate further the paradigm of complementarity. DNA consists of polymer chains, usually referred to as *DNA strands.* A chain is composed of *nucleotides*, and nucleotides may differ only in their *bases.* There are four *bases*: A (adenine), G (guanine), C (cytosine), and T (thymine). The familiar double helix of DNA arises by the bonding of two separate strands. The phenomenon known as *Watson–Crick complementarity* comes into picture in the formation of such *double strands.* Bonding happens by the pairwise attraction of bases: A always bonds with T, and G with C. Complementarity and the formation of double strands is presented schematically in Fig. 4. (Important details such as the orientation of the strands are omitted in this nutshell exposition.)

Figures 5 and 6 illustrate the importance of complementarity, in particular, how different things would be if complementarity were not provided for us by nature. In Fig. 5, the users of a DNA computer face the hopeless task of finding matches from huge piles of single strands. If the situation of Fig. 5 were the actual reality, the prospects for DNA computing would be very bleak, and perhaps also the theory presented below in Part II of this book would seem rather uninviting. But the situation of Fig. 5 is not the actual reality. The user can readily enjoy, as shown in Fig. 6, the result after the matching strands have found each other.

The paradigm of complementarity, or some generalization or modification thereof, will be present throughout the mathematical theory discussed in Part II of this book. Part I is a general introduction to DNA computing including an introduction (Chap. 1) to basic concepts of molecular biology needed in this book. It also discusses some prospects for laboratory realizations. For instance, the error rate of operations with DNA strands can make a really dramatic difference. Thus, the ultimate success of DNA computing depends heavily on the development of proper laboratory techniques.

Figure 6

There are many reasons to investigate "DNA computing" other than the solution of computationally hard problems by using DNA strands as a support for computation. On the one hand, it is important to try to *understand* how nature "computes" (remember that just by manipulating DNA the extraordinary sophistication and performance of life are obtained). On the other hand, as we shall see in the following chapters, "computing by DNA" leads to *computing paradigms* which are rather different from those customary in present-day computer science: new data structures, new types of operations on these new data structures or on classic ones (strings, languages), new computability models. Even if building DNA computers will prove to be unrealistic (error prone, for instance), an alternative could be the implementation of the new computing paradigms in silicon frameworks.

One can go further with these speculations: classic theoretical computer science is grounded on rewriting operations; this is true for most automata and language theory models. As we shall see, nature manipulates the DNA molecules in a computing manner by using operations of a quite different type: cut and paste, adjoining, insertion, deletion, etc. We shall prove that by using such operations we can build computing models which are equivalent in power with Turing machines. Thus, the computability theories can be reconstructed in this new framework. Whether or not this has practical significance for computer science applications is a premature question.

Part I

Background and Motivation

Chapter 1

DNA: Its Structure and Processing

The term "genetic engineering" is a very broad generic term used to cover all kinds of manipulations of genetic material. For the purpose of this book this term describes the *in vitro* (hence outside living cell) manipulation of DNA and related molecules. These manipulations may be used to perform various kinds of computations.

In this chapter we present the basic structure of the DNA molecule, and then the "tool box" of available techniques for manipulating DNA that are applicable in DNA computing.

1.1 The Structure of DNA

DNA is the molecule that plays the central role in DNA computing, and hence in this book. In the biochemical world of large and small molecules, polymers and monomers, DNA is a polymer which is strung together from monomers called *deoxyribonucleotides*. DNA is a crucial molecule in living cells (*in vivo*) and it has a fascinating structure which supports two most important functions of DNA: coding for the production of proteins, and self-replication so that an exact copy is passed to the offspring cells.

Let's look into the structure of a *DNA* (*DeoxyriboNucleic Acid*) *molecule*, to the extent needed for this book. As said above, the monomers used for the construction of DNA are deoxyribonucleotides, where each deoxyribonucleotide consists of three components: a *sugar*, a *phosphate group*, and a *nitrogenous base*. The name of the sugar used here is *deoxyribose* which explains the prefix "deoxyribo" used above. To simplify our terminology, we will use the simpler term "nucleotide" rather than "deoxyribonucleotide".

This (deoxyribose) sugar has five carbon atoms – for the sake of reference there is a fixed numbering of them. Since the base also has carbons, to avoid confusion the carbons of the sugar are numbered from $1'$ to $5'$ (rather than from 1 to 5). The phosphate group is attached to the $5'$ carbon, and the base is attached to the $1'$ carbon. Within the sugar structure there is a hydroxyl group (OH) attached to the $3'$ carbon.

Different nucleotides differ only by their bases, which come in two sorts: *purines* and *pyrimidines*. There are two purines: *adenine* and *guanine*, abbreviated A and G, and two pyrimidines: *cytosine* and *thymine*, abbreviated C and T, that are present in nucleotides. Since nucleotides differ only by their bases, they are simply referred to as A, G, C, or T *nucleotides*, depending on the sort of base they have.

The structure of a nucleotide is depicted (in a very simplified way) in Fig. 1.1, where B is one of the four possible bases (A, T, C, G), P is the phosphate group, and the rest (the "stick") is the sugar base (with its carbons enumerated $1'$ through $5'$).

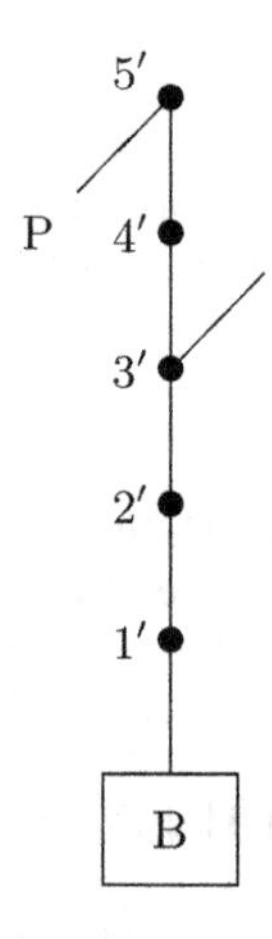

Figure 1.1: A schematic representation of a nucleotide

For readers who have more affinity with chemistry, Fig. 1.2 gives the standard (but still simplified) picture of the chemical structure of a nucleotide.

RNA (*RiboNucleic Acid*) is another polymer that is of crucial importance for living cells. Its structure is quite close to that of DNA. It is strung together from monomers called *ribonucleotides*. A ribonucleotide differs from a (deoxyribo)nucleotide in two ways.

(1) It contains the *ribose sugar* which differs from the deoxyribose sugar in that it has the hydroxyl (OH) group, rather than the hydrogen (H), attached to the $2'$ carbon.

(2) The thymine base is replaced in a ribonucleotide by the *uracil* base, denoted U. Hence the four possible bases are A, U, C, and G.

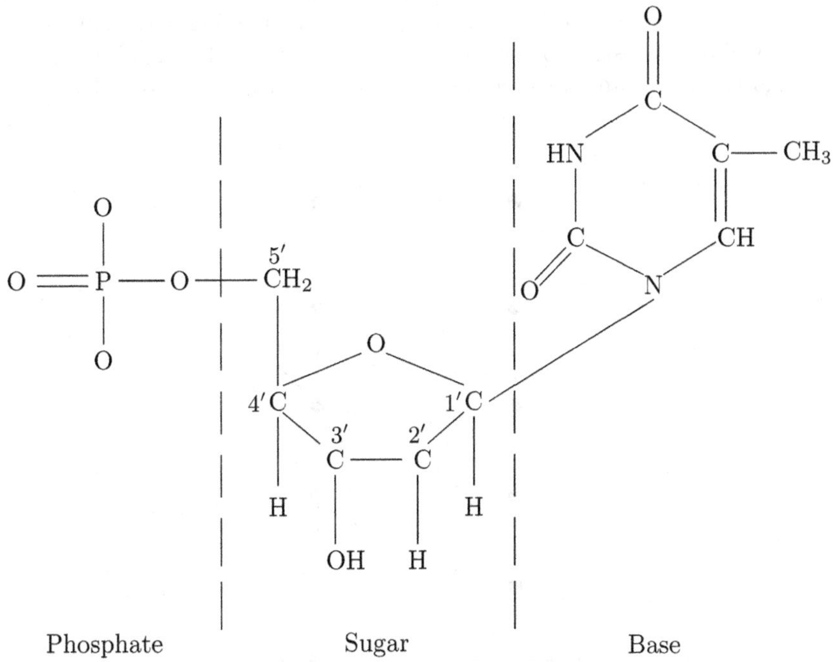

Phosphate Sugar Base

Figure 1.2: The chemical structure of a nucleotide with thymine base

It may be interesting for the reader to know that the ribonucleotide with the adenine base (and with a triple phosphate group – this is just another technical detail that we omit in our description) is called the ATP molecule, which is the main source of energy in living cells.

Nucleotides can link together in two different ways.

(1) The $5'$-phosphate group of one nucleotide is joined with the $3'$-hydroxyl group of the other forming a *phosphodiester bond*, which is a strong (*covalent*) bond – this is illustrated in Fig. 1.3.

Note that the resulting molecule has the $5'$-phosphate group of one nucleotide, and the $3'$-OH group of the other nucleotide available for bonding. This gives the molecule the *directionality*; we can talk about the $5'-3'$ direction, or the $3'-5'$ direction. This directionality is crucial for understanding the functionality and the processing of DNA (it is also crucial for the use of words in modeling such polymers as we do in this book).

(2) The base of one nucleotide interacts with the base of the other to form a *hydrogen bond*, which is a weak bond. This bonding is the subject of the following restriction on the base pairing: A and T can pair together, and C and G can pair together – no other pairings are possible.

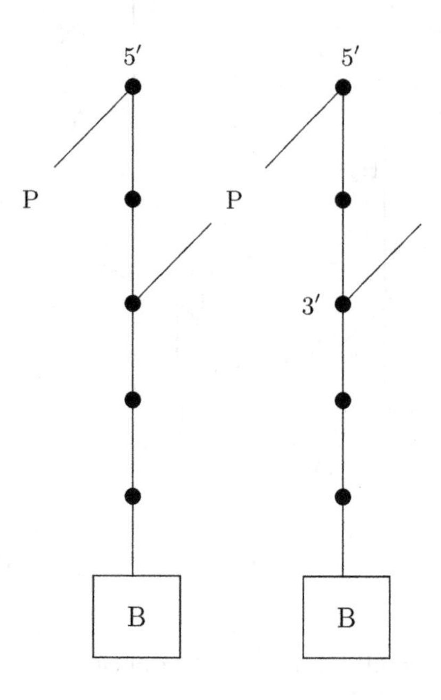

Figure 1.3: Phosphodiester bond

This pairing principle is called the *Watson–Crick complementarity* (named after James D. Watson and Francis H. C. Crick who deduced the famous double helix structure of DNA in 1953, and won the Nobel Prize for this discovery). It is the cornerstone of understanding the structure and functioning of DNA. The principle is illustrated in Fig. 1.4, where a thin wiggly line between the bases represents the fact that the hydrogen bond is (much) weaker than the phosphodiester bond.

As a matter of fact, the A – T pairing involves the formation of two hydrogen bonds between the two nucleotides, while the C – G pairing involves the formation of three hydrogen bonds between the two nucleotides. Consequently, the C – G pairing is stronger than the A – T pairing; one needs more energy (e.g., higher temperature) to separate the C – G pairing. To reflect this difference, we could use two wiggly lines for the A – T pairing, and three wiggly lines for the C – G pairing, but this is not necessary for the considerations of this book.

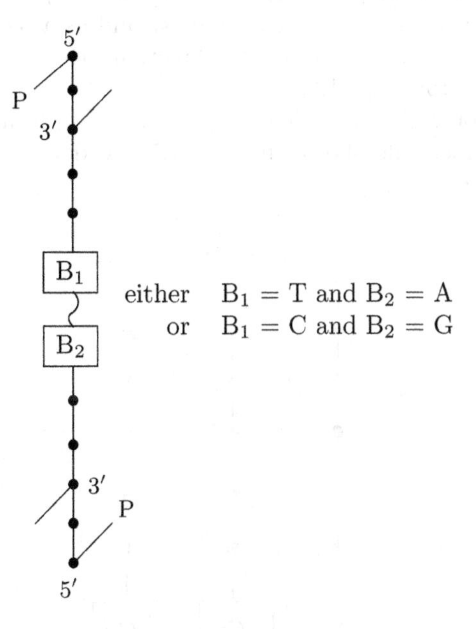

Figure 1.4: Hydrogen bond

Using phosphodiester bonds we can form single stranded DNA (Fig. 1.5). It is a standard convention that when we draw a single stranded

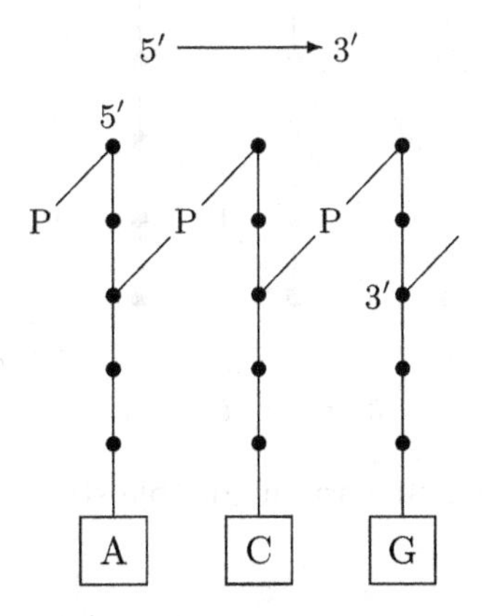

Figure 1.5: Single stranded DNA

molecule as in Fig. 1.5, the nucleotide with the free 5′-phosphate is the leftmost and the nucleotide with the free 3′-hydroxyl end is the rightmost.

Since in naming nucleotides (A, G, C, or T nucleotide) we identify them with their bases, we can also represent a single strand as a sequence of letters (a word), providing that we indicate the direction. Hence, 5′-ACG represents the single strand from Fig. 1.5.

Using Watson–Crick complementarity, we can form from the single stranded DNA molecule shown in Fig. 1.5 the double stranded molecule shown in Fig. 1.6.

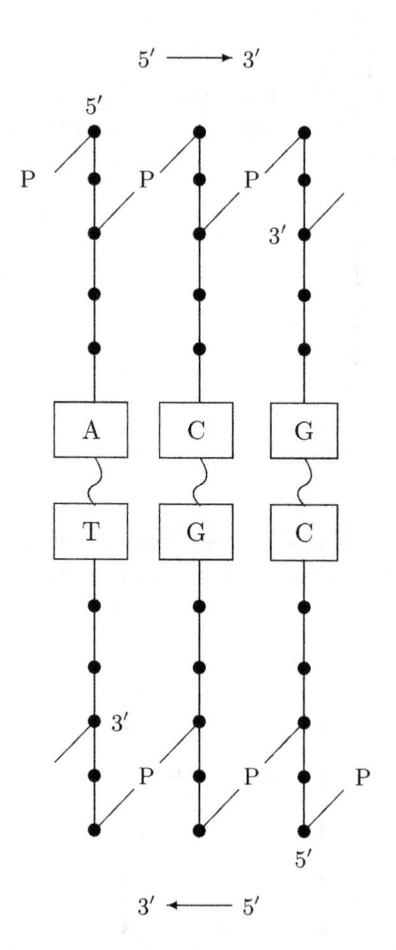

Figure 1.6: Forming double strands

As a matter of fact, in practice the hydrogen bond between single nucleotides as illustrated in Fig. 1.4 is too weak to keep the two nucleotides together – one really needs longer stretches to keep them bonded together. It is the cumulative effect (the sum) of hydrogen bonds between complementary bases in a DNA molecule that makes it a stable bond.

Figure 1.6 illustrates the general rule for joining two single stranded molecules using hydrogen bonds (the Watson–Crick complementarity). In the double stranded molecule the two single strands have opposite directions: the nucleotide at the $5'$ end of one strand is bonded to the nucleotide at the $3'$ end of the other strand. It is a standard convention that when a double stranded molecule (also referred to as a *duplex*) is drawn, then the upper strand runs from left to right in the $5' - 3'$ direction, and (consequently) the lower strand runs from left to right in the $3' - 5'$ direction. Thus the upper strand from Fig. 1.6 is $5'$-ACG, and the lower strand is $3'$-TGC.

Representing a (double stranded) DNA molecule as two linear strands bound together by Watson–Crick complementarity is already a major simplification of reality, because in a DNA molecule the two strands are wound around each other to form the famous double helix – see Fig. 1.7.

In vivo the situation is much more complicated, because a very large DNA molecule has to fit in a very small cell (in a typical bacterium the DNA molecule is 10^4 times longer than the host cell!). Such a packing is quite intricate, and in more complex cells (eukaryotes) this packing is done "hierarchically" in several stages. The actual shape of a DNA molecule is of crucial importance in considering processes in living cells. However, for the purpose of this book, we may assume that a DNA molecule has a double string-like structure.

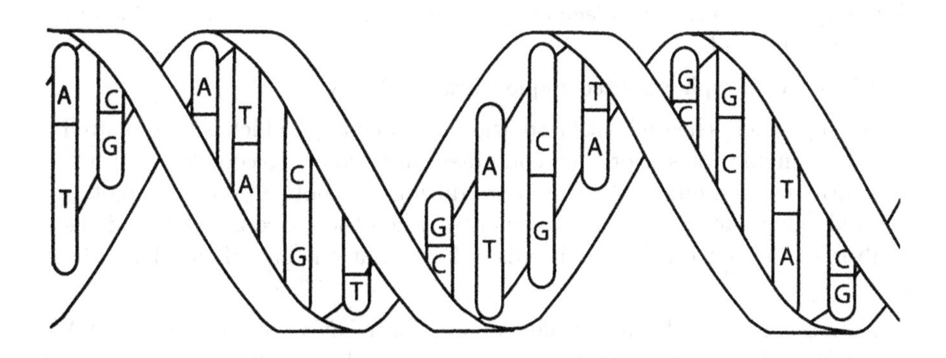

Figure 1.7: The double helix

Also, we have described above the structure of linear DNA molecules, while, e.g., bacterial DNA is very often circular. One can construct circular molecules simply by establishing a phospodiester bond between the "first" and the "last" nucleotide.

The ability to process (to manipulate) DNA is central to genetic engineering and in particular to DNA computing. We move now to describe methods for all kinds of manipulation of DNA – a basic tool box for the processing of DNA. We begin by discussing how to measure DNA.

Measuring the length of DNA molecules

The *length* of a single stranded molecule is the number of nucleotides comprising the molecule. Thus if a molecule consists of 12 nucleotides, then we say that it is a 12 mer (it is a polymer consisting of 12 monomers). The length of a double stranded molecule (where each nucleotide is base paired with a "partner") is counted in the number of base pairs. Thus if we make a double stranded DNA from a single stranded 12 mer, then the length of the double stranded molecule is 12 base pairs, also written 12 bp. If the length is, e.g., 12 000 base pairs, then we write that the length is 12 kbp (where "k" stands for "kilo").

To measure the length of a DNA molecule one can use *gel electrophoresis.* The *electrophoresis* technique is based on the fact that DNA molecules are negatively charged. Thus if they are placed in an electric field, they will move (migrate) towards the positive electrode. While the negative charge of a DNA molecule is proportional to its length, the force needed to move the molecule is also proportional to its length. Thus these two forces cancel each other, and in an ideal solution all molecules travel with the same speed. Hence, in order to cause molecules of different length to move with different speed, we need gel.

The gel electrophoresis technique works as follows.

A gel powder is heated with a solution, forming a gel which is then poured into a rectangular plastic or glass container, and allowed to cool down. It will then form a slab filling in the container; during the cooling process a comb is inserted along one side of the container, so when the gel cools down and the comb is removed a row of small wells is formed at one end of the slab (Fig. 1.8).

Now a small (really minuscule) amount of a DNA solution, with DNA molecules to be measured, is brought into the wells, and the electric field is activated. DNA molecules will move through the gel toward positive electrodes. Since the mesh of the gel acts as a molecular sieve, small molecules move easier (faster) through the gel than big ones, and obviously groups of the same length move with the same speed. When the first molecules reach the (positive) end of the gel, the electric field is deactivated. Clearly, in a given time span, the small molecules will travel a longer distance than the long ones.

Since DNA molecules are colorless, and hence invisible in the resulting gel, they must be marked in some way before they are put into the gel. There are two main methods for marking DNA molecules:

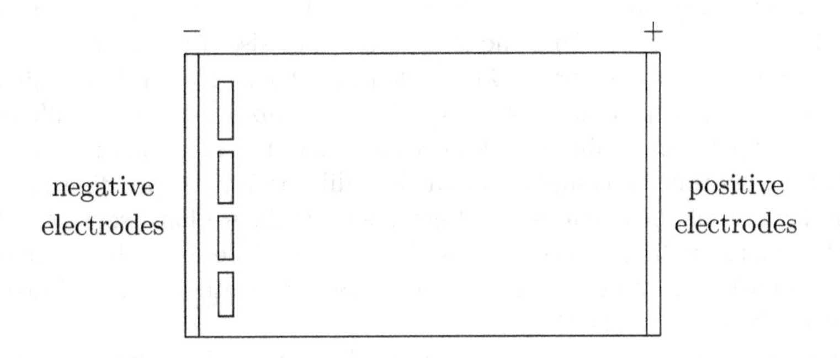

Figure 1.8: Gel prepared for electrophoresis

(1) Staining with ethidium bromide which fluoresces under an ultraviolet light when bound to DNA. When the gel is viewed under ultraviolet light, one sees bright fluorescent bands of groups of DNA fragments of the same length (Fig. 1.9). This method works best for double stranded DNA, because ethidium bromide really uses the double strand structure to stick to a molecule.

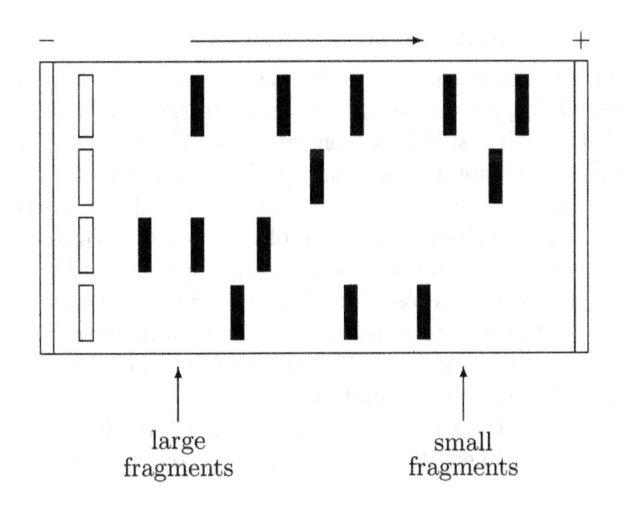

Figure 1.9: Gel electrophoresis

(2) Attaching radioactive markers to the ends of DNA molecules, so that when a film is exposed to the resulting gel the bands corresponding to various groups of DNA molecules will show on the film.

Now, knowing the distance travelled by a molecule, one can compute its length. Rather than computing the length, one can also use one of the wells for depositing there fragments of known length. Then the bands visualized on the path from this well may be used as a calibration path: different bands on this path mark different (known) lengths. The location of bands on other paths may then be compared with the calibration path, yielding in this way the lengths of those bands. In Fig. 1.8 and 1.9, we have several wells (with the corresponding spectra of visualized lengths) because often, for the sake of comparison, several sample solutions are run together (the calibration solution is often one of them).

Two kinds of gels are mostly used in gel electrophoresis: *agarose gel* and *polyacrylamide gel*. Agarose gel electrophoresis is the standard technique for resolving large fragments (longer than 500 bp). The resolution power of the gel clearly depends on its porosity and in this respect polyacrylamide gel is much better: it can resolve DNA fragments differing in length by only one base! This gel is the preferred method for determining the length of small fragments of DNA.

The DNA molecules present in the gel after electrophoresis can be recovered if needed. For example, a slice containing the DNA to be recovered is cut from the gel and frozen (in liquid nitrogen). This freezing breaks up the structure of the gel, and so if the solution (after it thaws out) is centrifuged through a special filter, only the DNA will get through.

Fishing for known molecules

Annealing of complementary single strands can be used for fishing out known molecules (called *target* molecules). Unless the target molecules are already single stranded, the first step is to denature double stranded molecules.

Suppose that we want to take out single stranded molecules α from a solution S containing them as well as many other single stranded molecules. We then attach $\overline{\alpha}$ molecules ($\overline{\alpha}$ is the molecule complementary to α and is called a *probe*) to a filter and pour the solution S through the filter. Then α molecules will bind to $\overline{\alpha}$ molecules while the other molecules will just flow through the filter. In this way we get a collection of double stranded molecules (resulting from annealing of α and $\overline{\alpha}$) fixed to the filter, and the solution S' resulting from S by removing α molecules.

Then the filter is transferred to a container where the double stranded DNA is denatured. When the filter is removed, only the target molecules remain.

The *filter method* as described above is conceptually very simple, but not used much any more (since better methods exist).

One can also attach probes to tiny glass beads and have them placed "tightly" in a glass column C. When a solution S containing target molecules is poured through C, target molecules will stay in C annealed to probes.

Yet another way of catching target molecules is to attach probes to tiny magnetic beads and throw them into a solution S containing target molecules.

When this mixture is well shaken, target molecules attach to probes, and hence to magnetic beads. By placing a magnet to a side of the glass container where this takes place, one gets all the target molecules grouped in one place where they are easy to extract.

1.2 Operations on DNA Molecules

Two extremes are possible in describing the tool box of techniques for manipulating DNA: a dictionary-like listing of techniques with a short definitional description of each of them, or a (very) detailed description of each technique. We have chosen for a middle ground, where we give (mostly oversimplified) descriptions which should provide the reader with a clear intuition about the nature of the techniques involved. We feel that this style is best suited for readers of this book.

Separating and fusing DNA strands

As we have mentioned already, the hydrogen bonding between complementary bases is (much) weaker than the phosphodiester bond between consecutive nucleotides within one strand. This allows us to separate the two strands of a DNA molecule without breaking the single strands. One way of doing this is to heat a DNA solution until the DNA melts, meaning that the two strands come apart – this is called *denaturation.* Melting temperatures are from 85° C up to 95° C (just below boiling); the melting temperature of a DNA molecule is the temperature at which half of the molecule separates.

Now if this heated solution is cooled down again, the separated strands fuse again by the hydrogen bonds (this cooling down must be done slowly so that the corresponding complementary bases have enough time to find each other). This process is called *renaturation.* Fusing two single stranded molecules by complementary base pairing is also called *annealing*, so renaturation is also called *reannealing.*

Another term used for fusing is *hybridization*, although originally it was used for describing the complementary base pairing of single strands of different origin (e.g., DNA with RNA, or DNA with radioactively tagged DNA, or strands coming from different organisms). Imprecise use of terminology is more common in biology than in mathematics.

Finally, we would like to mention that the denaturation of a double stranded molecule can be also facilitated by exposing it to certain chemicals. A commonly used chemical for this purpose is formamide – the melting temperature in the presence of formamide is much lower.

We move now to consider various manipulations of DNA that are mediated by enzymes.

Enzymes are proteins that catalyze chemical reactions taking place in living cells. They are very specific – most of them catalyze just a single chemical reaction, and they do this extremely efficiently (speeding up chemical reactions by as much as a trillion times). Without enzymes, chemical reactions going on in living cells would be much too slow to support life.

Since enzymes are so crucial for the life of a cell, nature has created a multitude of enzymes that are very useful in processing DNA. They are used very extensively in genetic engineering.

Lengthening DNA

A class of enzymes called (DNA) *polymerases* is able to add nucleotides to an existing DNA molecule. To do so, they require (1) an existing single stranded *template* which prescribes (by Watson–Crick complementarity) the chain of nucleotides to be added, and (2) an already existing sequence (*primer*) which is bonded (by Watson–Crick complementarity) to a part of the template, with the $3'$ end (the $3'$-hydroxyl) available for extension.

As a matter of fact, polymerase can extend only in the $5'-3'$ direction – see Fig. 1.10.

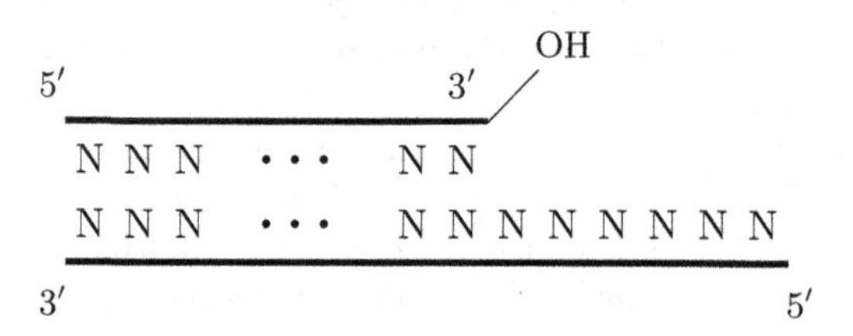

Figure 1.10: A DNA molecule with an incomplete upper strand

We use here and in the following figures various graphic representations different from the "stick representation" that we have used until now. By now they should be self-explanatory. There is no need to fix one notation for representing DNA, and we use the graphic representations that we feel best fit the discussed situation. The letter N in Fig. 1.10 means that any of the four possible nucleotides can be at the position labeled N (of course, providing that base pair complementarity is preserved).

Polymerase will then extend repeatedly the $3'$ end of the "shorter strand" complementing the sequence on the template strand, providing that required nucleotides are available in the solution where the reaction takes place – see Fig. 1.11.

As usual in biology, the rules have exceptions. Thus, whereas indeed all polymerases require the $3'$ end for extension (a primer), there are some polymerases that will extend a DNA molecule without a prescribed template.

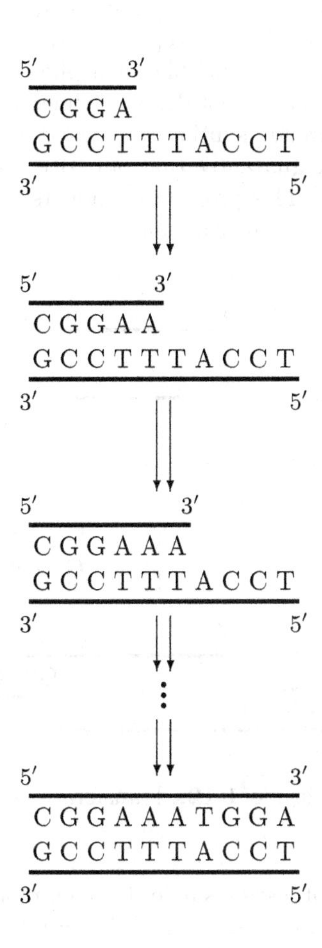

Figure 1.11: A polymerase in action

Terminal transferase is such a polymerase. It is useful when we want to add single stranded "tails" to both ends of a double stranded molecule – see Fig. 1.12.

If we want to make a specific double stranded molecule for which we have one strand (a template) already, then we can do it by priming the given strand and then using polymerase to extend the primer according to the template. The direction of this synthesis is $5' - 3'$: this is the direction favored by nature, since also *in vivo* enzymatic synthesis of DNA follows this direction.

One can chemically synthesize single stranded molecules following a prescribed sequence of nucleotides. For a number of technical reasons, the chemical synthesis that adds nucleotide by nucleotide to the already synthesized chain, proceeds in the $3'-5'$ direction: the $3'$ end of the first molecule is fixed to a solid support, so that at each step of the synthesis only the $5'$ end of the already synthesized chain is available for a phospodiester bond with the $3'$ end of the "incoming" new nucleotide. Well timed blocking and unblocking of the free $5'$ end of the already synthesized strand and of the $3'$ and $5'$ ends of the incoming nucleotide guarantee that only one specific nucleotide is added at one step of synthesis. This procedure lends itself to automatation – many "synthesizing robots" are now available.

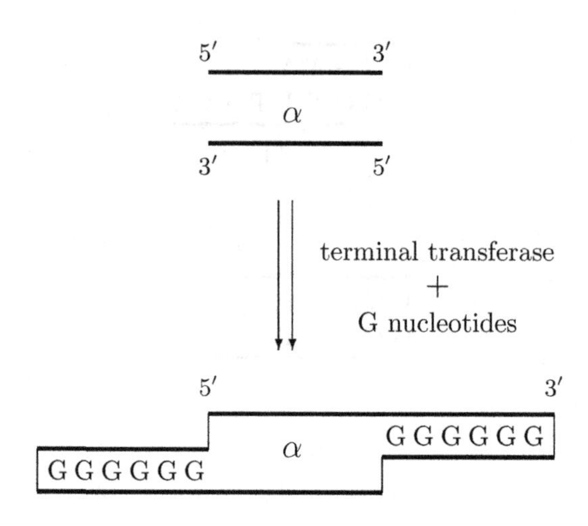

Figure 1.12: Transferase activity

Short chemically synthesized single stranded molecules are called *oligonucleotides* or simply *oligos*. Oligonucleotides are very useful in genetic engineering, e.g., they are used as primers.

Shortening DNA

DNA nucleases are enzymes that degrade DNA. They are divided into (DNA) exonucleases and (DNA) endonucleases.

Exonucleases shorten DNA by cleaving (removing) nucleotides one at a time from the ends of the DNA molecule. They are more flexible (less uniform) than polymerases, because some exonucleases will remove nucleotides from the $5'$ end while other will do this from the $3'$ end. Also some exonucleases may be specific for single stranded molecules while other will be specific for double stranded ones (and some can degrade both).

For example, *Exonuclease* III is a $3'$-nuclease (degrading strands in the $3' - 5'$ direction) – see Fig. 1.13. In this way a molecule is obtained with overhanging $5'$ ends.

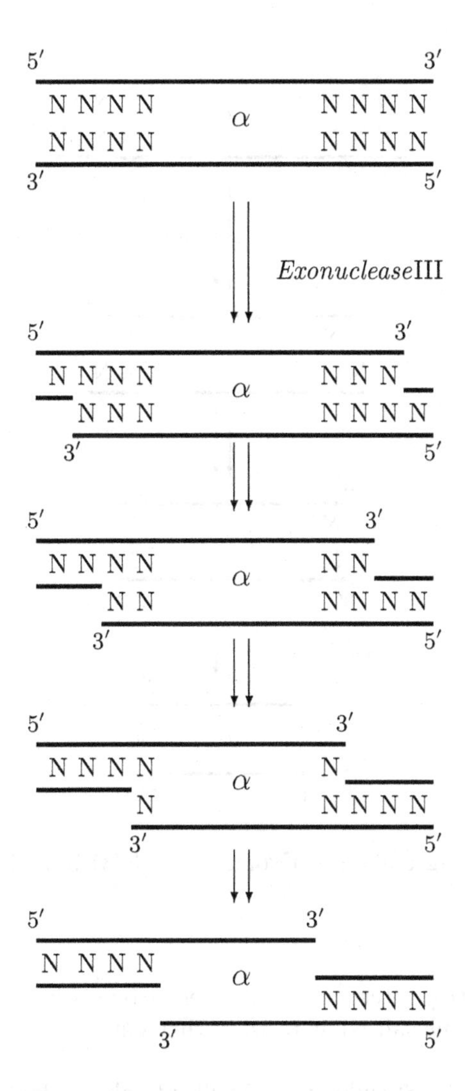

Figure 1.13: Exonuclease III in action

Another exonuclease, *Bal*31 removes nucleotides from both strands of a double stranded molecule – see Fig. 1.14.

As a matter of fact, many polymerases have also exonuclease activities. This is quite crucial in the DNA replication process ("performed" by polymerases) as a mistake correcting activity. While polymerase extending activity is always in $5' - 3'$ direction, the associated exonuclease can be both $5' - 3'$ and $3' - 5'$.

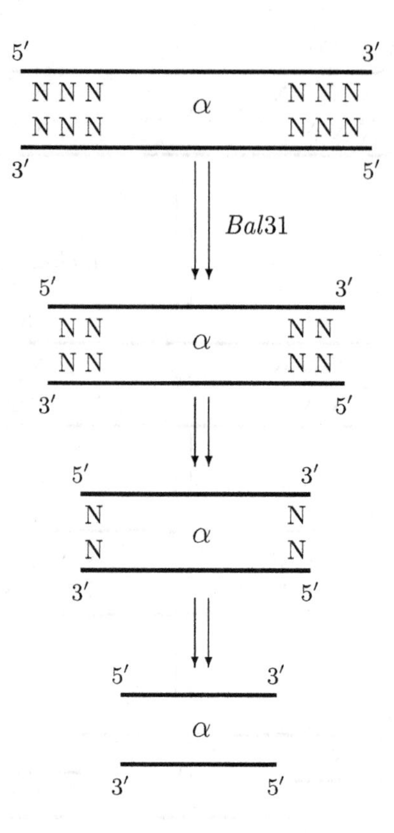

Figure 1.14: Exonuclease *Bal*31 in action

Cutting DNA

Endonucleases destroy internal phosphodiester bonds in the DNA molecule. They can be quite specialized as to what they cut, where they cut, and how they cut.

Thus, for example, *S*1 endonuclease will cut only single strands (Fig. 1.15) or within single strand pieces of a mixed DNA molecule containing single stranded and double stranded pieces (Fig. 1.16). Such cuts may happen at any place (any phosphodiester bond); we say that *S*1 endonuclease is not site specific.

On the other hand, endonuclease *DNase*I cuts both single stranded and double stranded molecules; it is also not site specific.

Restriction endonucleases are much more specific: they cut only double stranded molecules, and moreover only at a specific (for a given restriction endonuclease) set of sites.

A restriction enzyme will bind to DNA at a specific *recognition site* and then cleave DNA mostly within, but sometimes outside of this recognition site. It will cut the phosphodiester bond between adjacent nucleotides in such a way that it generates the OH group on the 3′ end of one nucleotide and the phosphate group on the 5′ end of the other nucleotide.

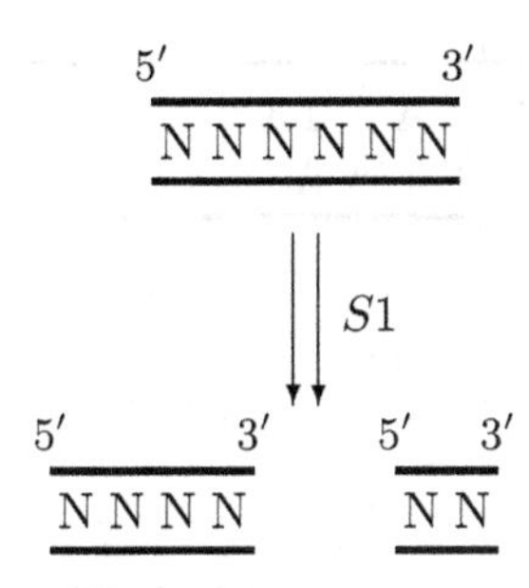

Figure 1.15: $S1$ endonuclease in action (i)

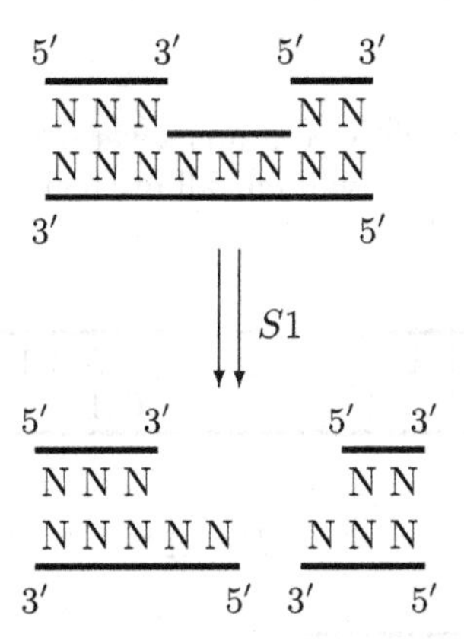

Figure 1.16: $S1$ endonuclease in action (ii)

The cut itself can be *blunt* (straight through both strands) or *staggered.* Here are some examples.

Restriction enzyme *Eco*RI – see Fig. 1.17.
The recognition site is 5′-GAATTC, so *Eco*RI will bind to it. The directionality is very important here: *Eco*RI will not bind to 3′-GAATTC. The cut is staggered, leaving two overhanging 5′ ends.

Note that the recognition site is a *palindrome* in the sense that reading one of the strands in the $5'-3'$ direction one gets the same result (GAATTC) as reading the other strand in the $5'-3'$ direction. This is often the case for restriction enzymes.

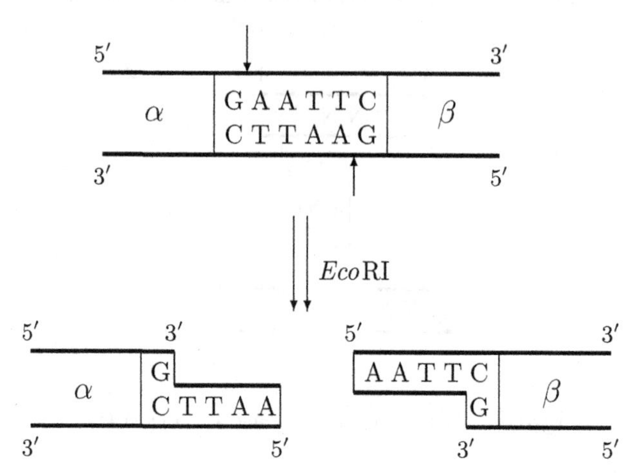

Figure 1.17: *Eco*RI in action

Clearly, if a stretch of DNA contains several recognition sites, then the restriction enzyme in principle will cut all of them – see Fig. 1.18.

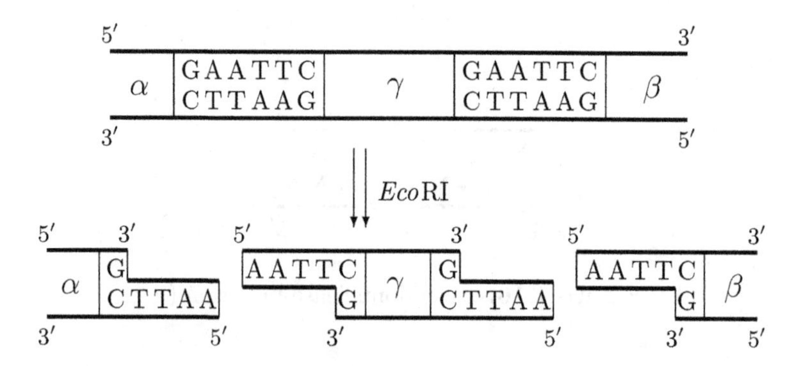

Figure 1.18: Multiple cut by *Eco*RI

For the reader who finds the names of restriction enzymes (like *Eco*RI above) strange, we would like to add that they follow precise rules of nomenclature. The name is always based on the organism from which the restriction enzyme was originally isolated. The first three letters are: the first letter of the genus name followed by the first two letters of the species name. Then, if necessary, the letter indicating the strain is given. The last character is the number indicating the order in which this enzyme was discovered in the given organism. Thus, *Eco*RI denotes the first (I) restriction enzyme isolated from the bacterium *Escherichia coli*, strain serotype R.

Restriction endonuclease *Xma*I – see Fig. 1.19.

The recognition site is 5′-CCCGGG. The cut is staggered, leaving two overhanging 5′ ends.

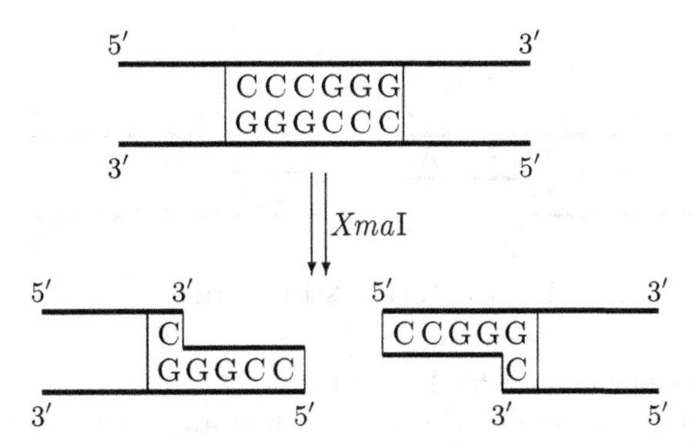

Figure 1.19: *Xma*I in action

Restriction endonuclease *Sma*I – see Fig. 1.20.

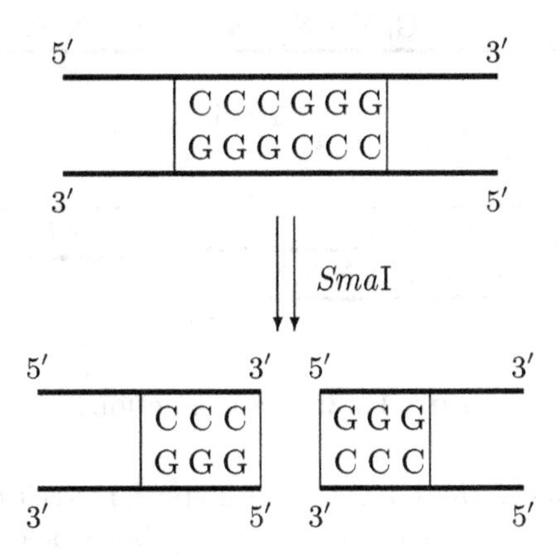

Figure 1.20: *Sma*I in action

The recognition site is the same as for *Xma*I: 5′-CCCGGG, but the cut is different – this is a blunt cut.

There exist also different restriction endonucleases that have the same recognition sites and the same cut (e.g., *Hpa*II and *Msp*I).

Restriction endonuclease *Pst*I – see Fig. 1.21.

The recognition site is 5′-CTGCAG. The cut is staggered, leaving two overhanging 3′ ends.

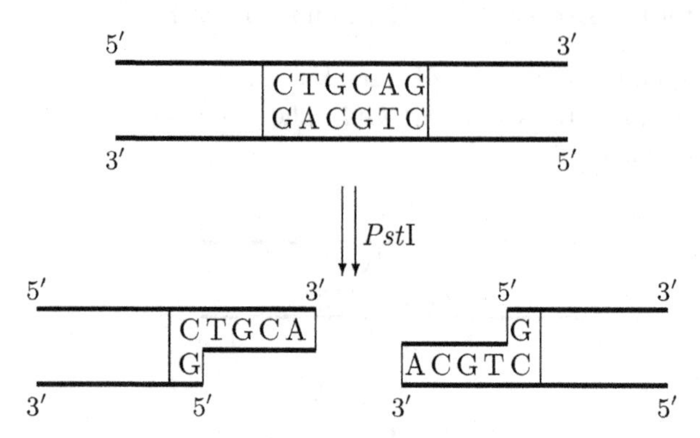

Figure 1.21: *Pst*I in action

Restriction endonuclease *Hga*I – see Fig. 1.22.

The recognition site is 5′-GACGC. The cut is staggered, leaving two overhanging 5′ ends.

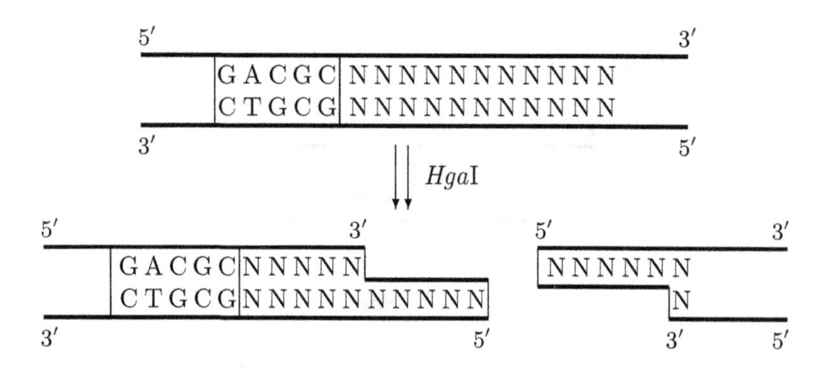

Figure 1.22: *Hga*I in action

As a matter of fact, *Hga*I belongs to Type I restriction endonucleases, while the restriction endonucleases discussed above belong to Type II restriction endonucleases (which cut within the recognition site). A discussion of differences between Type I and Type II (and also Type III) restriction endonucleases is beyond the scope of this chapter, but it is worthwhile to point out the following. Type I restriction endonucleases are rather imprecise, e.g., their cutting sites may be quite unpredictable. For this reason the use of Type I restriction endonucleases in genetic engineering is quite limited.

Linking (pasting) DNA

DNA molecules may be linked together through a process called *ligation* which is mediated by enzymes called *ligases.* This can be done in several ways.

Consider, e.g., the restriction enzyme *Xma*I and the situation shown in Fig. 1.19. The two molecules produced by the *Xma*I cut have overhanging ends. As a matter of fact, if they stay close enough they may reanneal (stick together) by hydrogen bonding of complementary bases – this is why such overhanging ends are also called *sticky ends.* In the situation of Fig. 1.19 the complementary sticky ends are 5′-CCGG and 3′-GGCC.

While the hydrogen bond keeps complementary sticky ends together, there is a gap in each of the strands, called a *nick.* A nick is a lack of a phosphodiester bond between consecutive nucleotides. Such a bond can be established by a ligase providing that the 3′ end to be connected has the hydroxyl (3′-hydroxyl) and the 5′ end to be connected has the phosphate group (5′-phosphate); see Fig. 1.23. Fortunately, when a restriction enzyme cuts the phosphodiester bond between adjacent nucleotides it generates the hydroxyl on the 3′ end and the phosphate on the 5′ end (as indicated in Fig. 1.23).

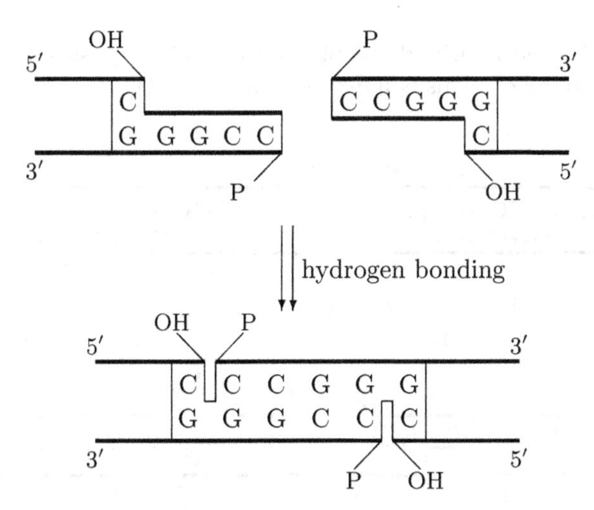

Figure 1.23: Complementary base pairing

For the resulting molecule of Fig. 1.23, the effect of ligation is illustrated in Fig. 1.24.

Note that here the work of a DNA ligase is made "easy" because the hydrogen bond is keeping the nucleotides to be ligated close to each other.

In the situation illustrated in Figs. 1.19, 1.23, and 1.24, a molecule cut by a restriction enzyme has restored itself using the sticky ends produced by the restriction enzyme cut.

But we could also have two different molecules M_1 and M_2 cut by the same restriction enzyme (or by different restriction enzymes that produce the

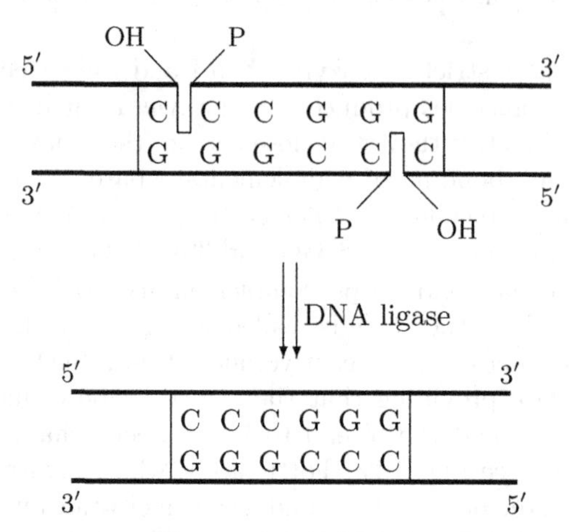

Figure 1.24: Ligation

same sticky ends) and then find the pieces recombining in such a way that we get hybrid molecules. This is illustrated in Fig. 1.25.

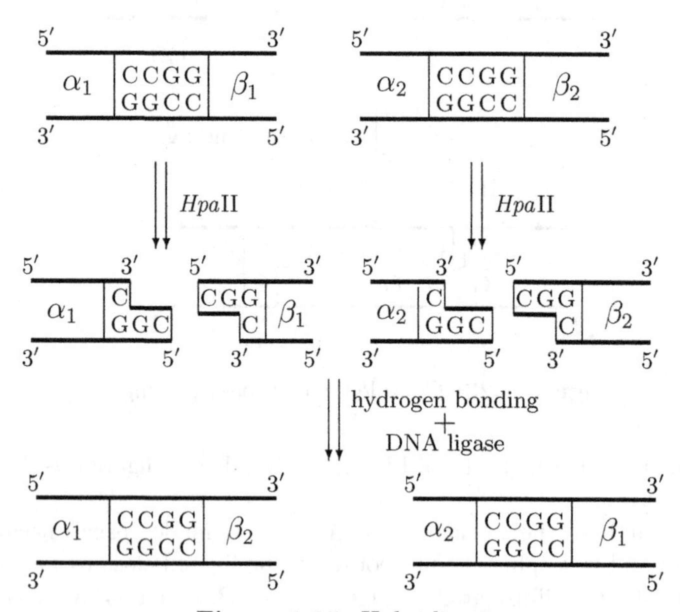

Figure 1.25: Hybridization

Note also that one of the molecules resulting from a multiple cut, illustrated in Fig. 1.18, has two sticky ends, which can anneal and then be ligated, thus forming a circular double stranded molecule.

In the blunt end ligation a DNA ligase will join together the $3'$ end and the $5'$ end of one molecule with the $5'$ end and the $3'$ end, respectively, of another molecule. For example, pieces cut by *Sma*I (see Fig. 1.20) can be ligated together – as shown in Fig. 1.26.

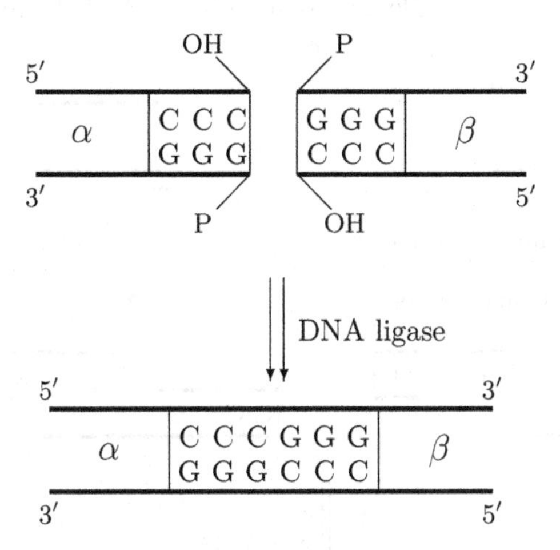

Figure 1.26: Blunt ligation

Blunt end ligation is much less efficient than sticky end ligation. The reason is that DNA ligase cannot bring the ends needed for ligation close together (in sticky end ligation the required ends were kept in proximity by hydrogen bonds between sticky ends). The advantage of blunt end ligation is that it joins DNA molecules independently of the specific nucleotide sequences at their ends.

Here is another way of performing blunt end ligation. Consider the terminal transferase enzyme which is $3'$-end extending, see Fig. 1.12. The situation illustrated in Fig. 1.12 is called, for obvious reasons, a "homopolymer tailing." This can be used for blunt end ligation in the way illustrated in Fig. 1.27.

Although the term "ligation" means technically just the sealing of a nick, it is often used also to describe the combined process of the annealing of sticky ends and then ligating the nicks.

Modifying nucleotides of DNA

Enzymes that modify DNA molecules by adding or deleting certain chemical components are very useful in controlling various operations on DNA (these enzymes are thus called *modifying enzymes*).

Methylases are enzymes that are used *in vivo* as partners of restriction enzymes. The main role of restriction enzymes *in vivo* is the defence of the host organism (e.g., bacteria) against the invading organism (e.g., virus). Restriction enzymes will digest (cut in pieces) the DNA of the invader – a big variety of restriction sites allows the destruction of a big variety of invaders. However, the DNA of the host itself may contain recognition sites of some of the restriction enzymes – if these sites are not protected, the host organism will destroy its own DNA while destroying the DNA of the invader.

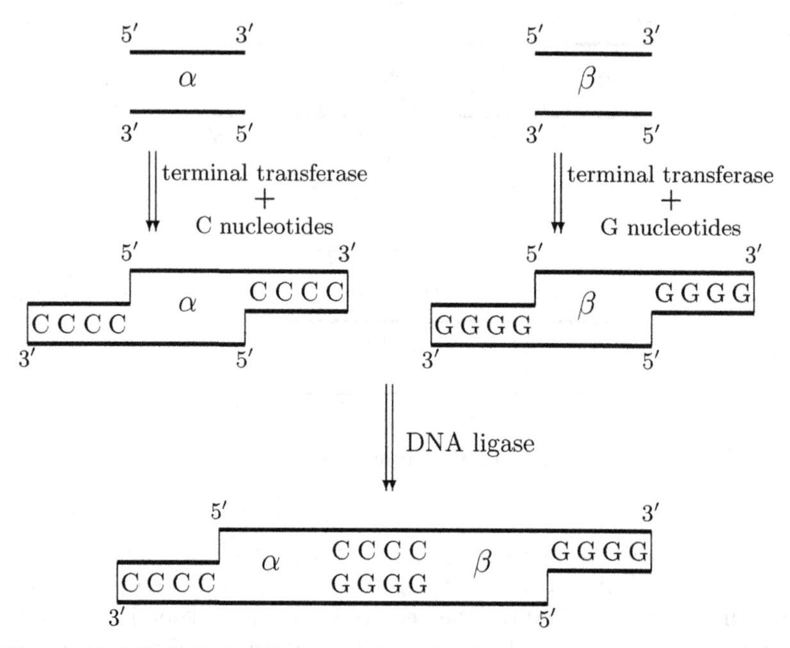

Figure 1.27: Joining blunt ended molecules using homopolymer tailing

The methylase, that is the partner of a restriction enzyme, has the same recognition site as the restriction enzyme; when it binds to this recognition site, it will modify one of the nucleotides within the restriction site (by adding a methyl group to it). In this way this recognition site becomes inaccessible for the corresponding restriction enzyme, and so the DNA molecule is protected against destruction (digestion) by it.

Alkaline phosphatase removes phosphate group from $5'$ ends of DNA, leaving there the $5'$-OH groups – see Fig. 1.28. Clearly, the molecule so obtained cannot ligate with itself (forming a circular molecule) – a phosphodiester bond cannot be formed. This is very important if you want to make sure that, given molecules α and molecules β, you get ligations of α and β, but not of α with α or β with β.

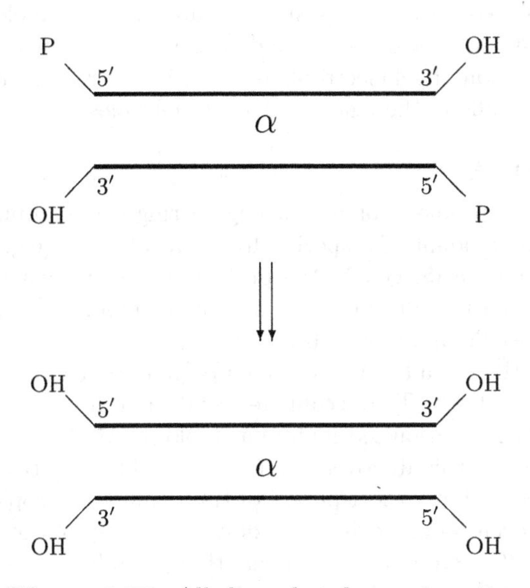

Figure 1.28: Alkaline phosphatase in action

Polynucleotide kinase has the opposite effect: it transfers phosphate groups (from available ATP molecules) onto the 5′-OH ends (of a molecule treated by alkaline phosphatase) – this is illustrated in Fig. 1.29.

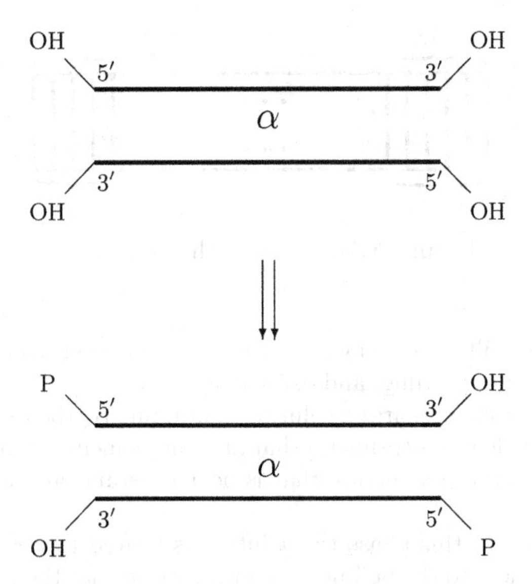

Figure 1.29: Polynucleotide kinase in action

If the transferred phosphate groups are radioactively labeled, then the so obtained molecule may be detected by detection methods using radioactivity (recall our discussion of gel electrophoresis). Moreover, restoring a (missing) 5′-phosphate end allows the ligation of such molecules.

Multiplying DNA

One of the central problems of genetic engineering is the amplification of the available "small" amount of a specific fragment of DNA (e.g., the fragment that encodes for a specific gene). The problem is especially acute if the small amount of the known fragment is lost in a huge amount of other pieces (like the proverbial needle in a haystack).

Fortunately there is a technique, called *polymerase chain reaction* (*PCR*), that solves this problem. This technique was devised in 1985 by Kary Mullis, and it has really revolutionized molecular biology (Mullis has been awarded the Nobel Prize for this discovery). It is incredibly sensitive and efficient: one can produce within a short period of time millions of copies of a desired DNA molecule even if one begins with only one strand of the molecule. The applications of PCR are really enormous; they include areas such as genetic engineering, forensic analysis, genome analysis, archeology, paleontology, and clinical diagnosis.

The beauty of PCR is that it is very simple and really elegant. Here is how it works.

Assume that we want to amplify a DNA molecule α with known borders (flanking sequences) β and γ – see Fig. 1.30.

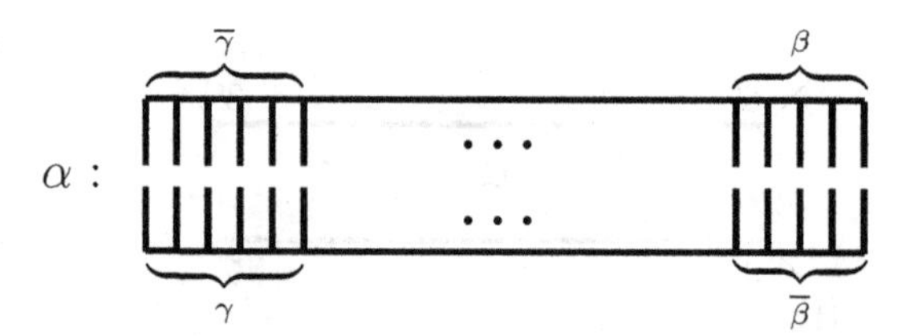

Figure 1.30: DNA with borders

Amplifying α will be done by repeating the basic cycle consisting of three steps: denaturation, priming, and extension.

To start with, one prepares a solution containing α (the target molecule), synthetic oligonucleotides (primers) that are complementary to β (β-primers) and to γ (γ-primers), polymerase that is heat resistant, and nucleotides.

Denaturation. In this phase the solution is heated to a really high temperature (often close to the boiling temperature), so that the hydrogen bonds between the two strands are destroyed, and α separates (denatures) into two strands α_1 and α_2 – see Fig. 1.31.

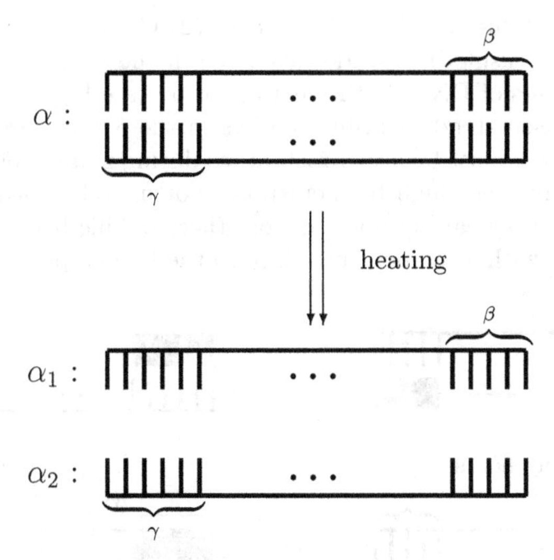

Figure 1.31: Denaturation

Priming. Now the solution is cooled down (usually to about 55° C) so that the primers will anneal to their complementary borders: β-primers to β, and γ-primers to γ – see Fig. 1.32.

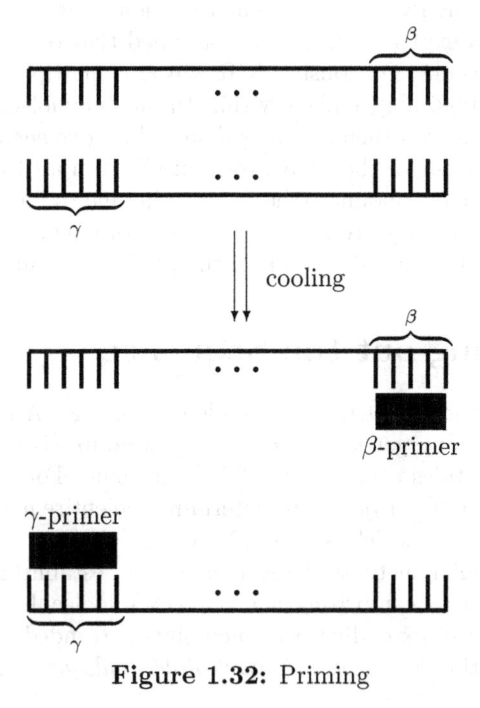

Figure 1.32: Priming

Extension. The solution is heated again (to 72° C) and a polymerase will extend the primers (using nucleotides available in the solution) to produce two complete strands of DNA, both identical to α – see Fig. 1.33. Remember that polymerase extends a primer always in the $5' - 3'$ direction. The polymerase used here must be heat resistant, as during many repeated cycles they have to survive very high temperatures. Fortunately, nature provides such polymerases: they can be isolated from thermophilic bacteria that live in thermal springs with a temperature close to the boiling point.

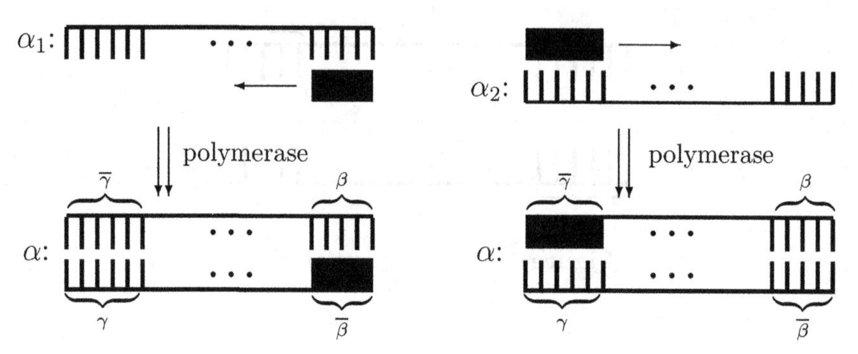

Figure 1.33: Extension

Obviously, repeating the basic cycle n times will yield 2^n copies of α, at least in theory. Thus PCR is a very efficient molecular Xerox machine!

For the ease of explanation we have assumed that our target sequence α is a separate molecule. Obviously, PCR will also multiply α, even if α is a part of a larger molecule (flanked within the larger molecule by borders β and γ). The explanation then becomes more subtle (we encourage the reader to analyze the working of the PCR procedure in such a situation).

Before the reader pronounces the PCR technique to be a real miracle, we need to stress that in order to amplify a DNA molecule (α) we need to know the borders (β and γ) in order to make the primers (β- and γ-primers).

1.3 Reading out the Sequence

We know already how to determine the length of a DNA molecule, but the ultimate goal in many genetic engineering procedures is to learn the exact sequence of nucleotides comprising a DNA molecule. For example, the goal of the Human Genome Project is to determine the entire nucleotide sequence of the human genome, which is about $3 \cdot 10^9$ bp long!

The most popular method of *sequencing* (i.e., establishing the exact sequence of nucleotides comprising a given DNA molecule) is based on the polymerase action of extending a primed single stranded template, and on the use of nucleotide analogues. A *nucleotide analogue* is a nucleotide that has been chemically modified *in vitro*. One can chemically modify the sugar and/or the phosphate group and/or the base of a nucleotide. The modification that is mostly used in sequencing is the modification of the sugar that changes the 3′-hydroxyl group (3′-OH) into the 3′-hydrogen atom (3′-H); such nucleotides are called *dideoxynucleotides* and they are denoted by ddA, ddT, ddC, and ddG. The sequencing method based on such nucleotides is called accordingly the *dideoxy enzymatic method*, or the *Sanger method* (named after its inventor).

It works as follows. Assume that we want to sequence a single stranded molecule α. We extend it at the 3′ end by a short sequence γ (say of length 20) so that we get the molecule 3′-$\gamma\alpha$. For example, if α = 3′-AGTACGTGACGC, then the resulting molecule is β = 3′-γAGTACGTGACGC.

The reason for adding γ "in front of" α is that in this way we can add the primer $\overline{\gamma}$ (complemented γ) so that a polymerase enzyme can start to extend such a molecule following (complementing) the template α, see Fig. 1.34. Let β' be the so primed β molecule. Usually, the primer $\overline{\gamma}$ is labeled (e.g., radioactively, or fluorescently marked) so that later in the procedure we can easily identify single strands beginning with $\overline{\gamma}$.

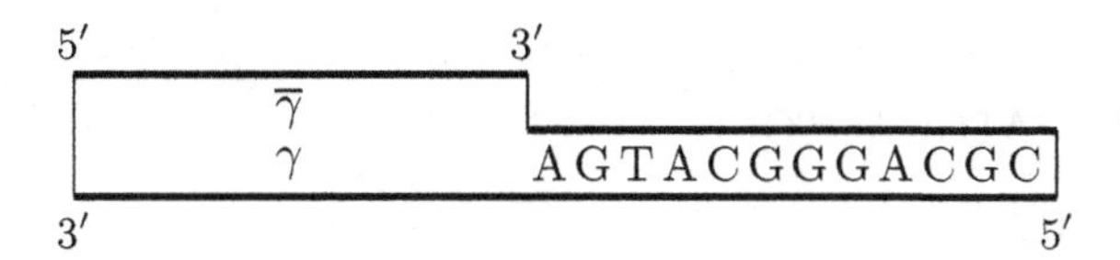

Figure 1.34: β' molecule

We now prepare four tubes (called Tube A, Tube T, Tube C, and Tube G) so that each of them will contain β molecules and primers (so that β' molecules will form), polymerase, and nucleotides A, T, C, and G. Moreover, Tube A contains a limited amount of ddA, Tube T a limited amount of ddT, Tube C a limited amount of ddC, and Tube G a limited amount of ddG.

Let us analyze the reaction going on in Tube A. The polymerase enzyme will extend the primer $\overline{\gamma}$ of β' using the nucleotides present in Tube A. Using only A, T, C, G nucleotides, β' is extended to the full duplex (Fig. 1.35).

5′ 3′
$\overline{\gamma}$ TCATGCACTGCG
γ AGTACGTGACGC
3′ 5′

Figure 1.35: Full duplex

Sometimes the polymerase enzyme will use a ddA rather than an A nucleotide. When this happens, then the complementing of the template will end at this position because ddA does not have the 3′-OH end needed for the phosphodiester bond. Hence, beside the full duplexes, we will also get the molecules shown in Fig. 1.36.

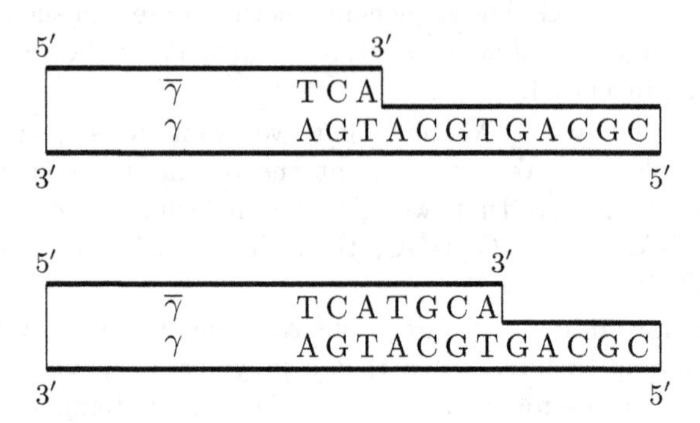

Figure 1.36: Incomplete molecules in Tube A

Hence the $5' - 3'$ fragments made by polymerase in Tube A (according to the template β) are:

$5' - \overline{\gamma}$TCATGCACTGCG,

$5' - \overline{\gamma}$TCA,

$5' - \overline{\gamma}$TCATGCA.

We can easily get these single stranded sequences by denaturing the (partially) double stranded sequences and selecting only those single strands that begin with the primer $\overline{\gamma}$ (remember that we have labeled $\overline{\gamma}$ for this purpose).

Reasoning in the same way, the $5' - 3'$ fragments made by polymerase in Tube T are:

$5' - \overline{\gamma}$TCATGCACTGCG,

$5' - \overline{\gamma}$T,

$5' - \overline{\gamma}$TCAT,

$5' - \overline{\gamma}$TCATGCACT.

In Tube C we get:

$5' - \overline{\gamma}$TCATGCACTGCG,

$5' - \overline{\gamma}$TC,

$5' - \overline{\gamma}$TCATGC,

$5' - \overline{\gamma}$TCATGCAC,

$5' - \overline{\gamma}$TCATGCACTGC.

In Tube G we get:

$5' - \overline{\gamma}$TCATGCACTGCG,

$5' - \overline{\gamma}$TCATG,

$5' - \overline{\gamma}$TCATGCACTG.

Now we perform the polyacrylamide gel electrophoresis using four wells (one for each tube), and we get the size separation (we ignore here the prefix $\overline{\gamma}$ which is the same for all fragments) shown in Fig. 1.37. Note that the distribution of the (overlapping) fragments forms a sequencing ladder where the rungs are the fragments, and a rung r directly precedes a rung r' if r' is longer than r by one nucleotide.

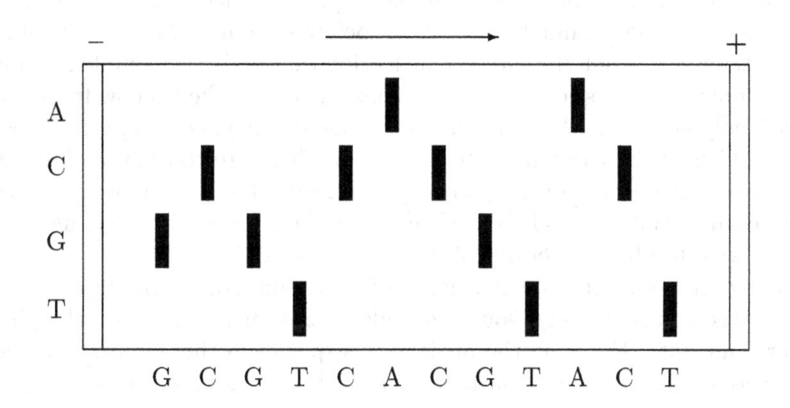

Figure 1.37: Sequencing ladder

Clearly, we read the bands (the ladder) from right to left, climbing the ladder, because the length of (molecules in the) bands increases from right to left. In Fig. 1.37 we have indicated under each band the nucleotide that is at the $3'$ end of molecules in this band. The molecules themselves that are in the bands (ordered by length, hence from right to left in Fig. 1.37) are:

T,

TC,

TCA,

TCAT,

TCATG,

TCATGC,

TCATGCA,

TCATGCAC,

TCATGCACT,

TCATGCACTG,

TCATGCACTGC,

TCATGCACTGCG.

Hence the original molecule α is: $3'$-AGTACGTGACGC.

We stress once again that our descriptions are very simplified. Thus, e.g., there are many subtle points in the sequencing described above. The polymerase used here cannot have the associated exonuclease activity since it could then cleave out the dideoxynucleotide that ends the complementing of the template. The "standard" enzyme used here was the Klenow fragment of DNA polymerase I. Also, the amount of the dideoxynucleotides in Tubes A, T, C, G must be carefully determined – if there are too many of them, then the polymerase action may always end within the proper prefix of the sequenced molecule – the chance of getting to the end of the template will decrease with too high a density of dideoxynucleotides.

Finally, one can run only molecules of quite limited length through the gel. But one can represent a long molecule by a sequence of its overlapping shorter fragments. Knowing the nucleotide sequence of these fragments and their overlappings, one can construct the nucleotide sequence of the whole molecule.

1.4 Bibliographical Notes

There are very many good books on molecular biology and genetic engineering, written for readers with different backgrounds.

The book by K. Drlica [48] is a beautiful introduction to molecular biology and genetic engineering that does not assume any background in either biochemistry or biology (our "stick" representation for nucleotides is from this book). Also [208] is a very nice and easy introduction to genetic engineering, although somewhat outdated now. [43] is written for the uninitiated reader – it is more sketchy than the other two books, but quite good as a quick reference.

The book by B. Alberts et al. [5] is a standard textbook on molecular biology and is very good also as a comprehensive reference book. The book by M. R. Walker and R. Rapley [214] represents a new concept in book writing. The reader may determine himself/herself a "route" through the topics he/she likes. It is a wonderful reference book. It is especially recommended for the reader who after reading some more popular books on molecular biology and genetic engineering would like to bring more order to the acquired information.

Chapter 2

Beginnings of Molecular Computing

2.1 Adleman's Experiment

"We can see only a short distance ahead, but we can see plenty there that needs to be done." These words of Turing [213] can be taken as an underlying principle of any program for scientific development. Such an underlying principle is very characteristic for research programs in computer science. Advances in computer science are often shown by and remembered from some unexpected demonstration, rather than from a dramatic experiment as in physical sciences. As pointed out by Hartmanis [83], it is the role of such a *demo* to show the possibility or feasibility of doing what was previously thought to be impossible or not feasible. Often, the ideas and concepts brought about and tested in such demos determine or at least influence the research agenda in computer science. Adleman's experiment [1] constituted such a demo. This book is about the short distance we can see ahead, and about the theoretical work already done concerning various aspects of molecular computing. The ultimate impact of DNA computing cannot yet be seen; this matter will be further discussed in Sect. 2.4.

Already when computers were generally referred to as "giant brains" and when nothing short of room-size could be visualized as a powerful computing device, some visionary remarks were made about possible miniaturizations. Often quoted is the view of Feynman from 1959 [55], describing the possibility of building "sub-microscopic" computers. Since then, remarkable progress in computer miniaturization has been made but the goal of sub-microscopic computers has not yet been achieved. Two major approaches, *quantum computing* and *DNA computing*, have been proposed and already widely discussed. Adleman's experiment, which we now start to describe, was a powerful demo in DNA computing. To keep our presentation on a realistic level, we will discuss here the example given in [1].

Adleman's experiment solves the *Hamiltonian Path Problem, HPP*, for a given directed graph. We consider the problem in the following formulation. Let G be a directed graph with designated input and output vertices, v_{in} and v_{out}. A path from v_{in} to v_{out} is termed *Hamiltonian* if it involves every vertex exactly once. (This implies that $v_{in} \neq v_{out}$ because $v_{in} = v_{out}$ would be in the path twice.)

For example, the graph depicted in Fig. 2.1 has the designated input vertex 0 and output vertex 6. The path consisting of the directed edges $0 \longrightarrow 1$, $1 \longrightarrow 2$, $2 \longrightarrow 3$, $3 \longrightarrow 4$, $4 \longrightarrow 5$, $5 \longrightarrow 6$ is Hamiltonian.

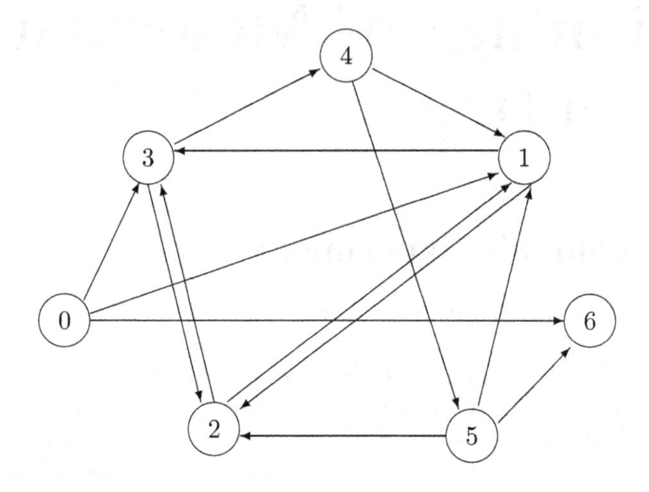

Figure 2.1: The graph in Adleman's experiment

We have chosen the numbering of the vertices in such a way that the Hamiltonian path comes out in the numerical order. Of course, the numbering can always be chosen in this fashion once a Hamiltonian path has been found. In this particular example the path mentioned turns out to be the only Hamiltonian path. Indeed, it is easy to exhaust all the possibilities. The beginning $0 \longrightarrow 3$ gives only the maximal paths $0 \longrightarrow 3$, $3 \longrightarrow 2$, $2 \longrightarrow 1$; $0 \longrightarrow 3$, $3 \longrightarrow 4$, $4 \longrightarrow 1$, $1 \longrightarrow 2$, and $0 \longrightarrow 3$, $3 \longrightarrow 4$, $4 \longrightarrow 5$, $5 \longrightarrow 2$, $2 \longrightarrow 1$, before the only possibility for continuation is a repetition of a vertex, and $0 \longrightarrow 3$, $3 \longrightarrow 4$, $4 \longrightarrow 5$, $5 \longrightarrow 6$. The beginnings $0 \longrightarrow 1$, $1 \longrightarrow 3$ and $0 \longrightarrow 6$ are also immediately seen to be unsuccessful. This argument also shows that if any edge from the path $0 \longrightarrow 1$, $1 \longrightarrow 2$, $2 \longrightarrow 3$, $3 \longrightarrow 4$, $4 \longrightarrow 5$, $5 \longrightarrow 6$ is removed, then the resulting graph has no Hamiltonian path. Clearly, if some vertex other than 0 is the input vertex, or some vertex other than 6 is the output vertex, then the resulting graph (with the same edges) has no Hamiltonian path. This follows because there are no edges entering 0 and no edges emanating from 6.

In general, the Hamiltonian Path Problem, HPP, consists of deciding whether or not an arbitrarily given graph has a Hamiltonian path. It is obvious that HPP can be solved by an exhaustive search. Moreover, various algorithms have been developed for solving HPP. Although the algorithms are successful for some special classes of graphs, they all have an exponential worst-case complexity for general directed graphs. This means that, in the general case, all known algorithms essentially amount to exhaustive search. Indeed, HPP has been shown to be an NP-complete problem, which means that it is unlikely to possess an efficient (that is, working in polynomial time) algorithm. HPP is *intractable* in the sense that the decision about graphs of modest size may require an altogether impractical amount of computer time. In his experiment Adleman solved the HPP of the example given above, a small graph by all standards. However, the solution is, at least in principle, applicable to bigger graphs as well. *Massive parallelism* and *complementarity* are the key issues in the solution.

Adleman's solution is based on the following nondeterministic algorithm for solving HPP.

Input: A directed graph G with n vertices, among which are designated vertices v_{in} and v_{out}.

Step 1: Generate paths in G randomly in large quantities.

Step 2: Reject all paths that do not begin with v_{in} and end in v_{out}.

Step 3: Reject all paths that do not involve exactly n vertices.

Step 4: For each of the n vertices v, reject all paths that do not involve v.

Output: "Yes" if any path remains, "No" otherwise.

Essentially, this algorithm carries out an exhaustive search. In Adleman's solution, the massive *parallelism* of the DNA strands takes care of the undesirable nondeterminism. Watson–Crick *complementarity* is applied to assure that the constructed sequences of edges are indeed paths in the graph G. We will now look at the details of Adleman's experiment.

Each vertex i of the graph is associated with a random 20-mer strand of DNA denoted $s_i, 0 \leq i \leq 6$. For instance, for $i = 2, 3, 4$, Adleman used the following oligonucleotides of length 20:

$$s_2 = \text{TATCGGATCGGTATATCCGA},$$
$$s_3 = \text{GCTATTCGAGCTTAAAGCTA},$$
$$s_4 = \text{GGCTAGGTACCAGCATGCTT}.$$

As regards orientation, all of these oligonucleotides are written $5'$ to $3'$.

It will be convenient for us to use a function h mapping each of the DNA bases to its Watson–Crick complement:

$$h(\mathrm{A}) = \mathrm{T},\ h(\mathrm{T}) = \mathrm{A},\ h(\mathrm{C}) = \mathrm{G},\ h(\mathrm{G}) = \mathrm{C}.$$

For DNA strands, h is applied letter by letter:

$$h(\mathrm{CATTAG}) = \mathrm{GTAATC}.$$

Thus, h produces the Watson–Crick complement of a strand. (The orientation is changed by h in this way: if the original strand is written $5'$ to $3'$, then the Watson–Crick complement will be written $3'$ to $5'$.) The mapping h is a *morphism* according to the terminology of language theory (described in detail in Chap. 3). It will be referred to as the *Watson–Crick morphism*. For instance,

$$\begin{aligned} h(s_2) &= \mathrm{ATAGCCTAGCCATATAGGCT}, \\ h(s_3) &= \mathrm{CGATAAGCTCGAATTTCGAT}. \end{aligned}$$

Decompose now each $s_i, 0 \leq i \leq 6$, into two strands, each of length 10: $s_i = s_i' s_i''$. Thus, s_i' (resp. s_i'') can be viewed as the first (resp. second) half of s_i. An edge from the vertex i to the vertex j, provided one exists in the graph G, is encoded as $h(s_i'' s_j')$. Thus, also an edge will be encoded as a 20-mer, obtainable as the Watson–Crick complement of the second and the first halves of the oligonucleotides encoding the vertices touching the edge. The encodings of three particular edges are given below:

$$\begin{aligned} e_{2\to 3} &= \mathrm{CATATAGGCTCGATAAGCTC}, \\ e_{3\to 2} &= \mathrm{GAATTTCGATATAGCCTAGC}, \\ e_{3\to 4} &= \mathrm{GAATTTCGATCCGATCCATG}. \end{aligned}$$

An important observation is that this construction preserves edge orientation; $e_{2\to 3}$ and $e_{3\to 2}$ are entirely different.

We are now ready to describe the main phase of Adleman's experiment. For each vertex i in the graph and for each edge $i \longrightarrow j$ in the graph, large quantities of oligonucleotides s_i and $e_{i\to j}$ were mixed together in a single ligation reaction. Here the oligonucleotides s_i served as splints to bring oligonucleotides associated with compatible edges together for ligation. Consequently, the ligation reaction caused the formation of DNA molecules that could be viewed as encodings of random paths through the graph. (Adleman used in his experiment also some ligase buffers, and the whole mixture was incubated for 4 hours at room temperature. For readers familiar with Adleman's paper [1], we want to mention that our notation above is slightly different. We put the oligos s_i in the "soup," whereas Adleman puts there the oligos $h(s_i)$. The corresponding complementarity change concerns the oligos $e_{i\to j}$.)

In Adleman's experiment, the scale of the ligation reaction far exceeded what was necessary for the graph of this size. Indeed, for each edge, a number of magnitude 10^{13} copies of the encoding oligonucleotide were present in the soup. This means that many DNA molecules encoding the Hamiltonian path were probably created, although the existence of a single such molecule would prove the existence of a Hamiltonian path.

In other words, quantities of oligonucleotides considerably smaller than those used by Adleman would probably have been sufficient, or a much larger graph could have been processed with the quantities he used.

As an illustration, we depict in Fig. 2.2 some of the DNA double strands that might have been produced in the experiment. We use the notations s_i and $e_{i \to j}$ introduced above. Observe that the double strands are open-ended.

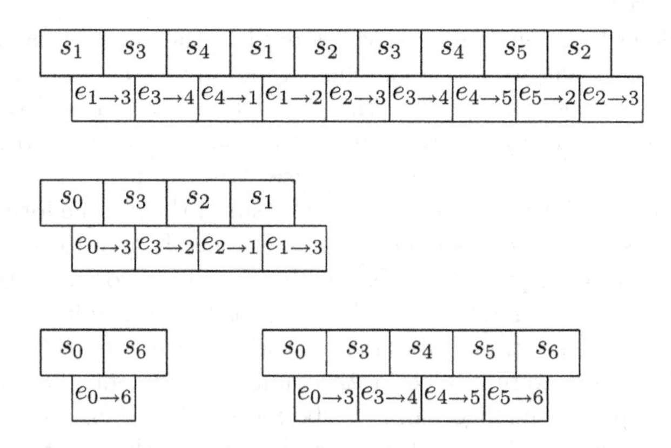

Figure 2.2: Examples of paths in Adleman's graph

Let us go back to the nondeterministic algorithm consisting of 4 steps, described above. We have already explained how Step 1 can be carried out. The remaining steps, as well as the conclusion in the output, are filtering or screening procedures that require biochemical techniques lying outside the scope of this book. (The interested reader is referred to [1] for details. For instance, Step 2 can be implemented by amplifying the product of Step 1 using the polymerase chain reaction (PCR) with primers $h(s_0)$ and s_6. This means that only molecules encoding paths that begin with the vertex 0 and end with 6 are amplified.)

From our point of view, the following considerations are more important than the filtering procedures based on biochemical techniques. Adleman's experiment took approximately 7 days of lab work. The screening procedure required in Step 4 was the most time-consuming. However, one should not draw negative conclusions too quickly from the seemingly slow handling of this small example. The molecular algorithm used in the experiment was rather primitive and inefficient. As in connection with traditional computers, improved algorithms will extend the applicability of the method. In particular, from a graph-theoretical point of view, the use of equal quantities of each oligonucleotide is by far not optimal. For instance, it leads to the formation of large amounts of molecules encoding paths that either do not start at vertex 0 or do not end at vertex 6. One should first calculate a flow in the graph and use the results to determine the quantities of each oligonucleotide required.

In general, the optimal quantity of each nucleotide needed is quite hard to determine or even approximate. Rather tricky graph-theoretic issues are involved. Intuitively, the quantity should be sufficient to insure that a molecule encoding a Hamiltonian path, provided one exists, will be formed with a high probability. This implies that the quantity should grow exponentially with the number of vertices.

Also, the length of the oligonucleotides used in the encoding is a matter of choice and optimization. Adleman chose randomly some 20-mer oligonucleotides, of which there exist altogether 4^{20}. The random choice made it unlikely that oligonucleotides associated with different vertices would share long common subsequences that might cause "unintended" binding in the course of the ligation. The choice of 20-mers assured that in the formation of double strands 10 nucleotide pairs between oligos encoding vertices and edges were involved and, consequently, the binding was stable at room temperature. Longer oligonucleotides might have to be used for larger graphs.

As already pointed out, probably many DNA molecules encoding the correct Hamiltonian path were formed in Adleman's experiment. The screening procedure in Step 4 of the algorithm can be modified in such a way [1] that actually an explicit description of the Hamiltonian path (or of one of the Hamiltonian paths in case there are several of them) is produced. The experiment has enormous potential for further development and modifications. For instance, consider the well-known "traveling salesman" problem. It amounts to finding the shortest Hamiltonian path in a directed graph, where the edges are provided with lengths. This could perhaps be accomplished by encoding path length information using oligonucleotides of different lengths. The shortest product (representing the shortest Hamiltonian cycle) has to be screened out.

We conclude this section with an abstract formulation of the key issues in Adleman's experiment. The abstract formulation will be also needed in the next section. It allows the use of a "programming language."

By definition, a (*test*) *tube* is a multiset of words (finite strings) over the alphabet {A, C, G, T}. (Intuitively, a tube is a collection of DNA single strands. Strands occur in the tube with a multiplicity, that is, several copies of the same strand may be contained in the tube). The following basic operations are initially defined for tubes, that is, multisets of DNA single strands [2]. However, appropriate modifications of them will be applied for DNA double strands as well.

Merge. Given tubes N_1 and N_2, form their union $N_1 \cup N_2$ (understood as a multiset).

Amplify. Given a tube N, produce two copies of it. (Observe that this operation makes sense for multisets only.)

Detect. Given a tube N, return *true* if N contains at least one DNA strand, otherwise return *false.*

Separate (or *Extract*). Given a tube N and a word w over the alphabet $\{\mathrm{A, C, G, T}\}$, produce two tubes $+(N, w)$ and $-(N, w)$, where $+(N, w)$ consists of all strands in N which contain w as a (consecutive) substring and, similarly, $-(N, w)$ consists of all strands in N which do not contain w as a substring.

The four operations of *merge, amplify, detect,* and *separate* allow us to program simple questions concerning the occurrence and non-occurrence of subwords. For instance, the following program

(1) $\mathit{input}(N)$
(2) $N \leftarrow +(N, \mathrm{A})$
(3) $N \leftarrow +(N, \mathrm{G})$
(4) $\mathit{detect}(N)$

finds out whether or not a given tube contains strands, where both of the purines A and G occur. The following program extracts from a given test tube all strands containing at least one of the purines A and G, preserving at the same time the multiplicity of such strands:

(1) $\mathit{input}(N)$
(2) $\mathit{amplify}(N)$ to produce N_1 and N_2
(3) $N_{\mathrm{A}} \leftarrow +(N_1, \mathrm{A})$
(4) $N_{\mathrm{G}} \leftarrow +(N_2, \mathrm{G})$
(5) $N'_{\mathrm{G}} \leftarrow -(N_{\mathrm{G}}, \mathrm{A})$
(6) $\mathit{merge}(N_{\mathrm{A}}, N'_{\mathrm{G}})$

Iterations of the operation *amplify* produce an exponential (with respect to the number of iterations) replication of the number of strands in the given tube.

Besides the four operations listed above and in [2], Adleman's experiment makes use of Watson–Crick complementarity and modifications of the operation *separate* that can be formulated as follows.

Length-separate. Given a tube N and and integer n, produce the tube $(N, \leq n)$ consisting of all strands in N with length less than or equal to n.

Position-separate. Given a tube N and a word w, produce the tube $B(N, w)$ (resp. $E(N, w)$) consisting of all strands in N which begin (resp. end) with the word w.

We will not introduce at this stage any formalism for Watson–Crick complementarity, since many such formalisms will be considered later on in this book. Coming back to Adleman's experiment, we now describe the filtering procedure using the operations introduced above. Thus, we start with the input tube N, consisting of the result of the basic step, the ligation reaction. Since double strands can again be dissolved into single strands by heating (melting) the solution, we may assume that N consists of single strands, that is, strings of oligonucleotides $s_i, 0 \leq i \leq 6$. (We ignore here the fragility of single strands, an issue of definite concern when dealing with larger graphs.) The filtering or screening part of Adleman's experiment can now be described in terms of the following program. Recall that each of the oligonucleotides $s_i, 0 \leq i \leq 6$, is of length 20.

(1) *input*(N)
(2) $N \leftarrow B(N, s_0)$
(3) $N \leftarrow E(N, s_6)$
(4) $N \leftarrow (N, \leq 140)$
(5) **for** $i = 1$ **to** 5 **do begin** $N \leftarrow +(N, s_i)$ **end**
(6) *detect*(N).

We will go on in this chapter to discuss the feasibility of the operations.

2.2 Can We Solve the Satisfiability Problem and Break the DES Code?

We now take a major step forward by presenting a solution, due originally to Lipton [115], of a very general problem by means of DNA computing. The problem we refer to is the *satisfiability problem for propositional formulas*. We give here only a brief description of the problem. For its great importance and versatility, we refer to [191] and [198].

We consider (well-formed) formulas α built from propositional *variables* $x_1, x_2, \ldots$, by the use of *connectives* $\sim, \vee, \wedge$ (negation, disjunction, conjuction). Thus,

$$\alpha = (x_1 \vee \sim x_2 \vee x_3) \wedge (x_2 \vee x_3) \wedge (\sim x_1 \vee x_3) \wedge \sim x_3$$

is such a formula.

A *truth-value assignment* for such a formula α is a mapping f of the set of variables occurring in α into the set $\{0,1\}$. Here 0 and 1 denote the truth-values "false" and "true", respectively. Thus, a truth-value assignment means the association of a truth-value to each of the variables. For any given truth-value assignment f, the truth-value assumed by the formula α can be computed using the *truth-tables* of the connectives:

$\vee$	0	1
0	0	1
1	1	1

$\wedge$	0	1
0	0	0
1	0	1

x	0	1
$\sim x$	1	0

The formula α is *satisfiable* if it assumes the truth-value 1 for at least one truth-value assignment. Clearly, α is not satisfiable exactly in case its negation $\sim \alpha$ is a *tautology*, that is, assumes the truth-value 1 for all assignments.

The following simple argument shows that the formula α mentioned above is not satisfiable. Assume the contrary: an assignment f gives α the value 1. Then f gives the value 1 to each of the four components (referred to as *clauses* in the sequel) of the conjunction. In particular, $f(\sim x_3) = 1$, implying $f(x_3) = 0$. From the third clause we see that $f(x_1) = 0$ and, from the second, that $f(x_2) = 1$. But for this assignment the first clause assumes the value 0, a contradiction. In special cases such as the one mentioned above, various *ad hoc* methods can be used to settle the satisfiability problem of a given propositional formula. However, in the general case no method essentially better than the *exhaustive search* is known: one has to search through all possible 2^k truth-value assignments, given a formula with k variables. This makes the task computationally intractable. It is already computationally infeasible, say, in the case of 200 variables. The satisfiability problem is known to be NP-complete. Indeed, it is intuitively very basic among NP-complete problems in the sense that it constitutes perhaps the most suitable reference point for NP-complete problems. The reduction of a given problem to the satisfiability problem is in many cases very natural.

Lipton's DNA-based solution of the satisfiability problem [115] uses some of the basic operations described in Sect. 2.1. Indeed, it consists of the exhaustive search made computationally feasible by the massive parallelism of DNA strands. We begin with a graphical description of truth-value assignments. Assume that we are dealing with a propositional formula containing k variables. Consider the directed graph in Fig. 2.3.

There are 2^k paths from v_{in} to v_{out} (none of the paths is Hamiltonian). Indeed, there are two choices in each of the vertices $v_{in}, v_1, \ldots, v_{k-1}$, the choices being independent of each other. Moreover, the paths and the truth-value assignments for the variables $x_1, x_2, \ldots, x_k$ have a natural one-to-one correspondence. For instance, the path $v_{in}a_1^0v_1a_2^0v_2 \ldots v_{k-1}a_k^0v_{out}$ corresponds to the truth-value assignment, where each of the variables gets the value 0. In general, the path $v_{in}a_1^{i_1}v_1a_2^{i_2}v_2 \ldots v_{k-1}a_k^{i_k}v_{out}$ corresponds to the truth-value assignment, where the variable x_j gets the value i_j, for $j = 1, \ldots, k$.

We now proceed with the graph exactly as in Adleman's experiment. Each vertex is encoded by a random oligonucleotide, say, of length 20. Consider the encodings s_i and s_j of two vertices such that there is an edge $e_{i,j}$ from the former to the latter. Write s_i in the form $s_i = s_i' s_i''$, where s_i' and s_i'' are of equal length, and similarly, $s_j = s_j' s_j''$. Then the edge $e_{i,j}$ is encoded by the oligonucleotide $h(s_i'' s_j')$, where h is the Watson–Crick morphism.

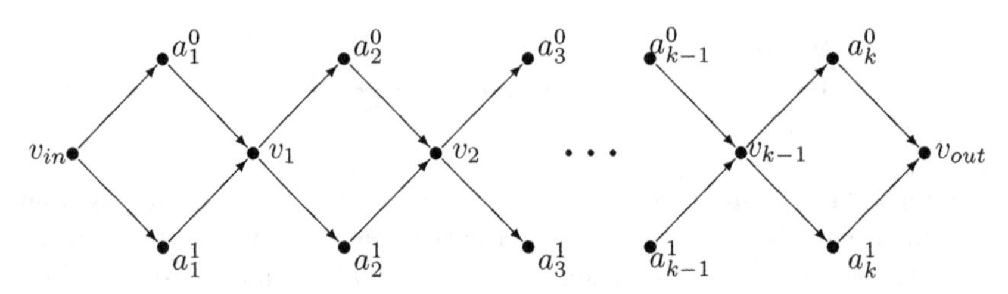

Figure 2.3: A graph associated with a truth-value assignment

The continuation of the procedure also happens in the same way as in Adleman's experiment. For each vertex and edge in the graph, large quantities of oligonucleotides encoding them are mixed together in a single ligation reaction. Again the oligonucleotides s_i serve as splints to bring oligonucleotides associated with compatible edges together for ligation. The end of the (oligonucleotide encoding the) vertex and the beginning of an edge can anneal because they are Watson–Crick complements. Similarly, the end of the edge and the beginning of the next vertex can also anneal. Since the encoding oligonucleotides are chosen randomly and are of sufficient length (with the number k of variables increasing, the length 20 might not be sufficient), no inadvertent paths are likely to form. This means that, after the annealing has been completed, the "soup" will contain a DNA double strand encoding an arbitrary path through the graph. As previously explained, we will also have encodings of arbitrary truth-value assignments for k variables. As the graph is very symmetric, there is no reason to believe that some paths will be more likely to appear than others.

We now come to an interesting and significant difference between Adleman's experiment and Lipton's solution of the satisfiability problem, concerning the basic ligation reaction. In the latter case, the graph is always the same and independent of the given propositional formula, provided the number of variables is fixed. Thus, one may always start with the *same test tube* that encodes all possible truth-value assignments. Taking several copies of this test tube, one is able to handle several propositional formulas simultaneously. The setup is different in Adleman's experiment. Since the graph is the actual input of the problem, the initial test tube cannot stay the same but varies with the input.

Thus, in the sequel when describing the solution of the satisfiability problem, we will speak of the *initial test tube*. The test tube is constructed in the way described above, and contains encodings of all possible truth-value assignments. Operations described in the preceding section will be performed. Following Lipton (see [115], page 544), we assume that the strands of DNA are actually *single strands*. (It is a matter of molecular biology whether it is better to actually separate the double strands obtained in the process above, or just understand the operations as being performed on one half of the double strands.)

Operations *separate, merge*, and *detect* will be used. We will first consider the example given in [115]. Consider the propositional formula

$$\beta = (x_1 \vee x_2) \wedge (\sim x_1 \vee \sim x_2).$$

Thus, we have two variables, in which case the corresponding graph is as shown in Fig. 2.4.

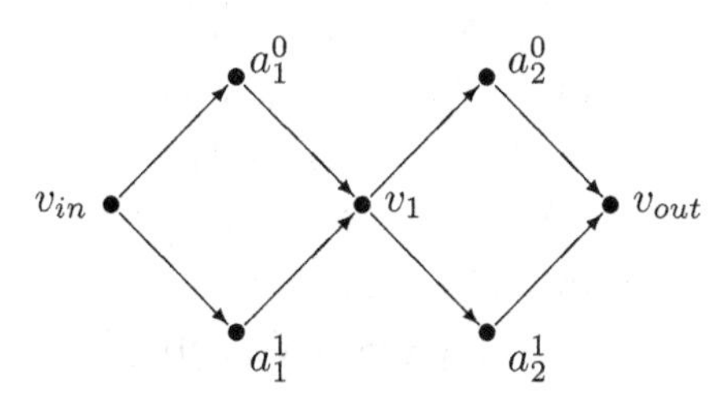

Figure 2.4: The graph associated to formula β

Each of the 4 paths through this graph corresponds to one of the 4 truth-value assignments for the variables x_1 and x_2. The initial test tube, say N_0, constructed as indicated above, contains strands for each of the paths and, consequently, for each of the truth-value assignments. Given the length of the oligonucleotides encoding the vertices a_j^i, these oligonucleotides will be easily distinguishable from each other, even in the case of a much larger number of variables. This means that, for instance, the oligonucleotide encoding a_1^1 does not appear in the paths elsewhere than in the intended position. If we apply the operation *separate*, forming the test tube $+(N_0, a_1^1)$, we get those truth-value assignments where x_1 assumes the value 1 (true). (Recall that $+(N_0, a_1^1)$ consists of those strands in N_0, where the oligonucleotide a_1^1 appears as a consecutive substring.) This simple observation is the basis of the whole procedure.

We denote the truth-value assignments by two-bit sequences in the natural way. Thus, 01 stands for the assignment $x_1 = 0, x_2 = 1$. Similar notation is also used if there are more than two variables. This simple notation of bit sequences is extended to the DNA strands resulting from our basic graphs. Thus, the strand $v_{in}a_1^0v_1a_2^1v_{out}$ is denoted simply by 01. Indeed, among the strands resulting from paths through our basic graph, this notation gives an exhaustive characterization. Finally, given a test tube N consisting of strands of the kind mentioned and denoted by bit sequences, we denote by $S(N,i,j)$ the test tube of such strands in N, where the ith bit equals $j, j = 0,1$. According to the simple observation made above, $S(N,i,j)$ results from N by the operation *separate*:

$$S(N,i,j) = +(N, a_i^j).$$

We consider also the tube of such strands in N, where the ith bit equals the complement of j:

$$S^-(N,i,j) = -(N, a_i^j).$$

The following program solves the satisfiability problem for the propositional formula β:

$$\begin{array}{rl}
(1) & input(N_0) \\
(2) & N_1 = S(N_0,1,1) \\
(3) & N_1' = S^-(N_0,1,1) \\
(4) & N_2 = S(N_1',2,1) \\
(5) & merge(N_1,N_2) = N_3 \\
(6) & N_4 = S(N_3,1,0) \\
(7) & N_4' = S^-(N_3,1,0) \\
(8) & N_5 = S(N_4',2,0) \\
(9) & merge(N_4,N_5) = N_6 \\
(10) & detect(N_6)
\end{array}$$

Observe that the steps (2), (4), (6), (8) apply the operation *separate* in the sense of $+(N,w)$, whereas the steps (3), (7) apply it in the sense of $-(N,w)$. The following table summarizes the contents of the tubes at the different steps of the program.

Step	1	2	3	4	5	6	7	8	9
Tube	00,01,10,11	10,11	00,01	01	10,11,01	01	10,11	10	01,10

Thus, the return is *true* at the step (10).

The program is based on exhaustive search. The initial tube at step (1) contains all possible truth-value assignments. The tube at step (5) contains the assignments satisfying the *first clause* of the propositional formula β. (Either x_1 or x_2 must assume the value 1. At step (2) we have those assignments for which x_1 is 1. Of the remaining ones we still take, at step (4), those for which x_2 is 1.) The assignments in this tube, N_3, are filtered further to yield at step (9) those assignments that also satisfy the *second clause* of the propositional formula β.

The method of this simple example can be directly applied in the general case. We consider propositional formulas in *conjunctive normal form.* This means that the formulas look like α and β above; they are conjunctions

$$(\dots) \wedge (\dots) \wedge \dots \wedge (\dots),$$

where each of the *clauses* in the parentheses is a disjunction of terms, each of which is a variable or its negation. Fast algorithms are available to transform an arbitrary propositional formula into conjunctive normal form.

Thus, consider a propositional formula

$$\gamma = C_1 \wedge C_2 \wedge \dots \wedge C_m,$$

where each of the m clauses C_i is a disjunction consisting of variables and their negations. Assume that altogether k variables $x_1, x_2, \dots, x_k$ appear in γ. This leads to the directed graph already depicted above in Fig. 2.3, as well as to the initial test tube N_0 containing all of the k-bit sequences, provided the strands in N_0 are denoted in the way described above. Starting with N_0, we go through the clauses of γ, extracting all the time strands from N_0, until after getting through C_m, only those strands remain that encode assignments satisfying γ.

Explicitly, we show inductively how this is done. Assume that each of the assignments encoded by the strands in $N_i, 0 \leq i < m$, satisfies the subformula

$$\gamma_i = C_1 \wedge \dots \wedge C_i$$

and that

$$C_{i+1} = y_1 \vee \dots \vee y_l,$$

where each y_j is one of the variables x or its negation. Initially, we have the test tube N_0 of all possible truth-value assignments and the empty formulaγ_0.

Using the operations *separate* and *merge*, we now transform N_i into N_{i+1} by the same procedure as in the example. Consider y_1. We form $S(N_i, n, 1)$ or $S(N_i, n, 0)$, depending on whether $y_1 = x_n$ or $y_1 = \sim x_n, 1 \leq n \leq k$. Thus, we extract from N_i the subtube $S(N_i, n, j)$ satisfying also y_1. The remainder from N_i, that is, $S^-(N_i, n, j)$ is now investigated with respect to the satisfaction of y_2 and the positive part (that is, the strands satisfying y_2) is merged with $S(N_i, n, j)$. The negative part is still useful and is investigated with respect to the satisfaction of y_3, and so on, until we have exhausted the clause by taking y_l into account.

When we have constructed the tube N_m in this fashion, a single application of the operation *detect* suffices to settle the problem. As in connection with Adleman's experiment, this final step can be modified to actually read the solution, provided one exists.

The complexity of the process described is feasible: m steps are needed, each consisting of some applications of *separate* and *merge.* The number of such applications does not exceed the number of variables in a clause.

Quite a different matter is that the process assumes that the operations are perfect, that is, they are performed without error. This is of course far from being obvious, and the microbiological grounds need to be studied. One can also take a different point of view, where perfection is not called for. If the initial test tube contains many copies of each truth-value assignment, then something may be lost in the extractions, and the correct answer is still reached with a high probability.

We still consider another example, a propositional formula δ. The formula β considered above is unsatisfactory because of two reasons. First, all variables appear in all clauses. Secondly, β has too few clauses even with respect to the very small number of variables. (A formula having only few clauses is satisfiable independently of the clauses themselves; numerical lower bounds can be easily computed.) Although the formula δ is still small and its satisfiability can be detected without any difficulty, it does not have these two defects.

The propositional formula δ, defined as follows, has 5 variables and 11 conjunctive clauses:

$$\begin{aligned}\delta = &(\sim x_1 \vee \sim x_2 \vee \sim x_3) \wedge (x_1 \vee x_2 \vee \sim x_4) \wedge (\sim x_1 \vee x_2 \vee x_4)\\ &\wedge(x_1 \vee x_2 \vee \sim x_5) \wedge (x_1 \vee \sim x_2 \vee \sim x_5) \wedge (x_1 \vee x_3 \vee x_4)\\ &\wedge(\sim x_1 \vee x_3 \vee \sim x_5) \wedge (x_1 \vee \sim x_4 \vee x_5) \wedge (x_2 \vee \sim x_3 \vee x_4)\\ &\wedge(x_3 \vee x_4 \vee x_5) \wedge (x_3 \vee \sim x_4 \vee x_5).\end{aligned}$$

The initial test tube N_0 contains all the 32 possible truth-value assignments: 00000, 00001, 00010,..., 11111. The following table gives the contents of the tubes $N_0, N_1, \ldots, N_{11}$, defined in the process described above. Each $N_i, i = 1, 2, \ldots, 11$, is characterized by listing the strands extracted from N_{i-1}.

Test tube	Strands extracted
N_0	none
N_1	11100, 11101, 11110, 11111
N_2	00010, 00011, 00110, 00111
N_3	10000, 10001, 10100, 10101
N_4	00001, 00101
N_5	01001, 01011, 01101, 01111
N_6	00000, 01000
N_7	10011, 11001, 11011
N_8	01010, 01110
N_9	00100
N_{10}	11000
N_{11}	10010, 11010

This means that the strands encoding the truth-value assignments 01100, 10110, and 10111 still remain in the final test tube N_{11}. Thus, the final *detect* operation returns *true*.

The method described above can be easily modified to work for any propositional formula, not necessarily in the conjunctive normal form, [115]. Thus, the formula results by applying to the variables the unary operation of negation and the binary operations of conjunction and disjunction. (Other propositional connectives could also be taken into account here but the matter is rather irrelevant for our purposes.) Only the number m of *binary* operations is significant for the complexity of the process. (The number of negations and the number of variables are irrelevant.) After $m+1$ separations and m merges two test tubes have been produced, the first of which contains molecules representing (that is, encoding) satisfying truth-value assignments, whereas the second contains molecules representing unsatisfying assignments. Thus, the satisfiability problem is solved by a single application of the operation *detect* to the first tube.

When facing the physical obstructions in creating a practical molecular computer, attention has to be focused on the possible realizations of the various operations. Programs are easy to write in terms of the operations we have considered but the feasibility of the implementation is a matter of microbiological technique. Any detailed discussion about such techniques lies outside the scope of this book. However, some overall remarks can be made.

A natural way to realize the operation *merge* is to pour the contents of one tube into another. At least intuitively, this is faster and less error prone than the operation *separate* which certainly requires much more sophisticated techniques. The same holds true with respect to the operation *detect*. However, it appears that in standard programs *detect* is rarely done. Consequently, the realization of *detect* does not affect much the complexity of the process, yet its error rate is important. Realizations of the operation *separate* can have errors of both inclusion and exclusion. By the former we mean that an item that should go to $+(N, w)$ actually ends up somewhere else, maybe in $-(N, w)$. Similarly, an error of exclusion means that an item which should go to $-(N, w)$ does not go there. It might be useful to consider different probabilities for these two types of errors. We will still discuss in Sect. 2.4 some matters concerning error rates and the feasibility of operations.

In the remainder of this section we consider a model of molecular computation that was introduced and called the *sticker model* in [189]. A method based on this model, [3], for breaking the most widely used cryptosystem DES (Data Encryption Standard) will also be discussed. The sticker model is based on the paradigm of Watson–Crick complementarity. It makes use of DNA strands as the physical substrate in which information is represented. Basically, the sticker model has a random access memory, where no strand extension is required. The materials are reusable, at least in theory.

We first describe a way, based on complementarity, of representing information in DNA. It will use two basic kinds of single-stranded DNA molecules, referred to as *memory strands* and *sticker strands* or shortly *stickers*. A memory strand is n bases in length and contains k non-overlapping substrands, each of which is m bases long. Thus, we must have $n \geq mk$. Although this is not necessarily the case, we assume in the following illustrations that each substrand follows another consecutively, without any bases lying between them. During the course of a computation, each substrand is identified with exactly one boolean variable (or equivalently one bit position). The substrands should be significantly different from one another: any two of them should differ with respect to several base positions. (This is intended to ensure a sufficient identification for each bit position.) Each sticker is m bases long and complementary to exactly one of the k substrands in the memory strand.

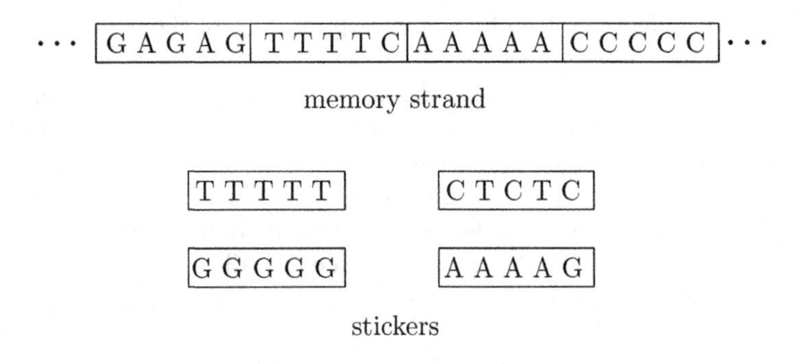

Figure 2.5: Example of a sticker memory

A specific substrand of a memory strand is either *on* or *off*. If a sticker is annealed to its matching substrand on a memory strand, then the particular substrand is said to be *on*. Otherwise, if no sticker is annealed to a substrand, then the substrand is said to be *off*. A *memory complex* is the general term used for memory strands, where the substrands are on or off. Memory complexes represent binary numbers, where a substring being on (resp. off) represents the bit 1 (resp. 0). Thus, memory complexes are DNA strands that are partially double.

In the illustration in Fig. 2.5, we consider a memory strand of length $n = 20$, divided into $k = 4$ substrands, each of length $m = 5$.

Thus, in this case the necessary complexes are interpreted as containing four bits of information. In particular, consider the memory complexes in Fig. 2.6.

In the first memory complex, all substrands are off, whereas in the last complex the last two substrands are on. The binary numbers represented by these four memory complexes are 0000, 0100, 1001, and 0011, respectively.

In the memory strand used in Fig. 2.5, the substrands corresponding to odd-numbered (resp. even-numbered) bit positions consist entirely of purines (resp. pyrimidines). An advantage of such a choice is the natural creation of borders between the substrands intended as encoding substrands. In other words, it is not possible that a sticker is bonded with a substrand overlapping two of the intended substrands. (Such an annealing could cause confusion in the outcome of the operations described below.) Indeed, for any extensive applications of sticker systems, a careful study of ideal encodings is important. Such a study would have to combine the microbiological feasibility with theoretical advantages, trying to achieve an optimal trade-off between the two.

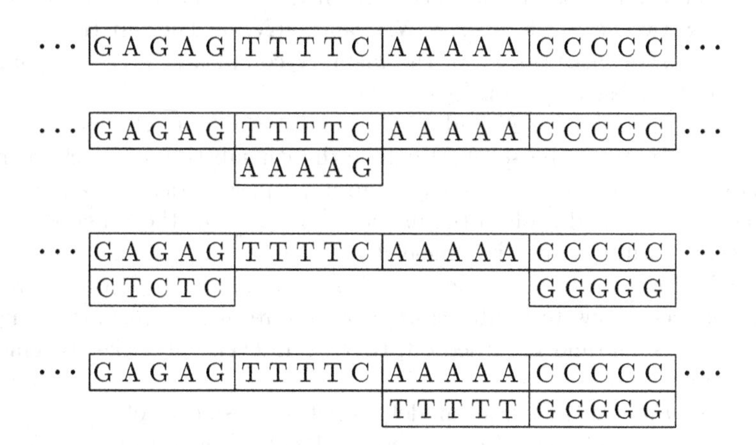

Figure 2.6: Examples of memory complexes

It is also instructive to compare the basic representation mechanisms of Adleman's experiment and sticker systems, and in particular, the idea of constructing double strands. The paradigm of complementarity is present in both cases. In sticker systems, one starts with a long single memory strand to which short stickers are annealed, to yield a memory complex, a partially double strand. In Adleman's experiment (as well as in Lipton's solution of the satisfiability problem) there is no long strand to start with, but short single strands are annealed in a step by step fashion, leaving a sticky end after each step. The double strand created in this fashion is supposed to have no single-stranded gaps.

The information density in both cases (sticker systems and Adleman's experiment) can be considered to be the same, $\frac{1}{m}$ bits per base. While the theoretical maximum in DNA representations is two bits per base, such a high value would render any separation-based molecular computer dangerously error prone.

We are now ready to introduce the *operations* used in sticker systems. While they resemble the operations considered above, they are simple yet flexible for implementing general algorithms. As before, a *test tube* or *tube* is a multiset, its elements now being memory complexes. (The actual representation of memory complexes as two-dimensional data structures is irrelevant for our purposes.) The operations we consider are *merge, separate, set,* and *clear.*

The operation *merge* is exactly as before: two test tubes are combined into one. Thus the memory complexes from the two input tubes, with their annealed stickers undisturbed, are combined to form the multiset union of the two inputs.

The operation *separate* produces, given a test tube N and an integer $i, 1 \leq i \leq k$, two new test tubes $+(N,i)$ and $-(N,i)$. The test tube $+(N,i)$ (resp. $-(N,i)$) consists of all of the memory complexes in the original N, where the ith substrand is on (resp. off).

For a given test tube N and an integer $i, 1 \leq i \leq k$, the operation *set* produces a new test tube *set*(N,i), where the ith substrand of each memory complex in N is turned on. (That is, an appropriate sticker is annealed to it if the ith substrand is off in the memory complex but the ith substrand is left unchanged if it is already annealed.)

Finally, for a given test tube N and integer $i, 1 \leq i \leq k$, the operation *clear* produces a new test tube *clear*(N,i), where in each memory complex of N the ith substrand is turned off, that is, an eventual sticker is removed from it.

Computations in the sticker model consist of a sequence of the operations *merge, separate, set,* and *clear. Inputs* and *outputs* will be test tubes. To *read* the *output,* one memory complex must be isolated from the output test tube and its annealed stickers determined, or else it must be reported that the output test tube contains no memory complexes.

The input or *initial test tube* will be a *library* of memory complexes. In particular, a (k,l) *library,* $1 \leq l \leq k$, consists of memory complexes with k substrands, the last $k-l$ of which are off, whereas the first l substrands are on or off in all possible ways. Thus, viewed as a multiset, a (k,l) library contains 2^l different kinds of memory complexes. The represented binary sequences are of the form $w0^{k-l}$, where w is an arbitrary binary sequence of length l. In the initial test tube, the first l substrands of the memory complexes represent the actual input, whereas the remaining $k-l$ substrands are used for intermediate storage and output.

The computational paradigm associated with the sticker model is to settle hard problems by exhaustive combinatorial searches over inputs of length l. All possible 2^l inputs are processed in parallel. One can also say that this paradigm is the essence of DNA computing in general.

Following [189], we now illustrate the sticker model by presenting a solution to the *Minimal Set Cover* problem. The problem can be formulated as follows. Given a finite set $S = \{1,2,\ldots,p\}$ and a finite collection $\{C_1,\ldots,C_q\}$ of subsets of S, find the smallest subset I of $\{1,2,\ldots,q\}$ such that

$$\bigcup_{i \in I} C_i = S.$$

Of course, an exhaustive search through all the 2^q subsets of I will solve the problem.

We will now describe a solution using the sticker model. Memory strands will have $k = p + q$ substrands. The initial test tube N_0 will be a $(p + q, q)$ library. (It should be emphasized that any widespread applications of the sticker model would assume that libraries of specific sizes are readily available. With an increasing k, the construction of memory strands with k substrands becomes more challenging.)

We denote by $card(X)$ the cardinality of a set X (that is, the number of elements in X). The elements of the set $C_i, 1 \leq i \leq q$, are denoted by $c_i^j, 1 \leq j \leq card(C_i)$. Thus, each c_i^j is an integer between 1 and p.

The memory complexes in the initial test tube N_0 represent all possible subsets I of the set $\{1, 2, \ldots, q\}$. In particular, the first q substrands in each memory complex tell, by being on or off, which of the numbers $1, 2, \ldots, q$ belong to the particular subset I represented by the memory complex. The last p substrands are initially off in each memory complex M. Those of the substrands $q + j, 1 \leq j \leq p$, are eventually turned on, for which the number j belongs to some set C_i, where i is in the index set I represented by M. Thus, given M, we proceed as follows: we look through the first q substrands of M; whenever we encounter a substrand that is turned on (let it be the ith substrand, the operation *separate* is used to differentiate between on and off), then we use the operation *set* to turn on those among the last p substrands that come from the elements of C_i. After having gone through all of the first q substrands of M in this fashion, we look to see whether or not each of the last p substrands has been turned on. This can again be done by the operation *separate.* The last p substrands being on means that the index set I represented by M indeed leads to a cover of the set $S = \{1, 2, \ldots, p\}$. Thus, we may discard the memory complexes not satisfying this condition, and must find the smallest index set among those satisfying it.

It is fairly obvious that the described procedure works. In standard sequential computation, however, the amount of work is enormous: for $q = 100$, we have to apply the procedure for each of the 2^{100} memory complexes. Things are different in DNA computing and the sticker model. All memory complexes in N_0, where the first substrand is on (that is, C_1 is one of the sets in the proposed cover of S), are now processed *simultaneously.* The result is brought over to the next step, where the memory complexes having the second substrand on are processed. In this fashion, the overall procedure will have only q steps, instead of 2^q.

Recall the notations $+(N, i), -(N, i)$, and $set(N, i)$ used in connection with the operations *separate* and *set.* The following simple program, where the initial test tube N_0 is a $(p + q, q)$ library, formalizes the ideas described above:

(1) **for** $i = 1$ **to** q

separate $+(N_0, i)$ **and** $-(N_0, i)$

for $j = 1$ **to** $card(C_i)$

$set(+(N_0, i), q + c_i^j)$

$N_0 \leftarrow merge(+(N_0, i), -(N_0, i))$

(2) **for** $i = q + 1$ **to** $q + p$

$N_0 \leftarrow +(N_0, i)$

The test tube N_0 resulting as the output of this program contains only memory complexes, where each of the p last substrands is on. To get a solution to the problem of the Minimal Set Cover, we still have to detect from N_0 a memory complex which has the smallest number of the first q substrands turned on. At the end of the outer loop in (3) in the following program, the test tube $N_i, i \geq 1$, contains all the memory complexes, where exactly i among the first q substrands are turned on. Thus, the output will give a solution to the Minimal Set Cover problem.

(3) **for** $i = 0$ **to** $q - 1$

for $j = i$ **down to** 0

separate $+(N_j, i + 1)$ **and** $-(N_j, i + 1)$

$N_{j+1} \leftarrow merge(+(N_j, i + 1), N_{j+1})$

$N_j \leftarrow -(N_j, i + 1)$

(4) *read* N_1;

else if it was empty *read* N_2;

else if it was empty *read* N_3;

....

The reader might want to consider the double loop in step (3) in terms of a simple example. For instance, assume that $q = 4$ and that C_3 covers the set S together with any of the other sets C_i, whereas C_1, C_2, C_4 do not, even together, cover S. This means that the initial test tube N_0 contains the memory complexes representing the covers (1, 3), (2, 3), (3, 4), as well as all covers containing any of them. The table on the next page describes the situation after each step in the outer loop.

Observe that the above procedure does not use the operation *clear* at all. Another more far-reaching observation is that the entire algorithm can eventually be executed by a *robotic system*. A robot would perform the experiments needed in the operations automatically. In this case it should be assumed that the operations are performed "blindly", that is, without getting any feedback from the DNA during the experiments. Such a feedback does not seem possible using present techniques. The operations discussed above (*merge, separate, set, clear*) can feasibly be executed by a robotic system [189].

The most dramatic potential application of the sticker model was presented in [3], for attacking the *Data Encryption Standard, DES.* (See [201] for a detailed description of it.) The analysis presented in [3] suggests that such an attack might be mounted on a table-top machine, based on DNA computing but using also robotic parts. Approximately one gram of DNA would be needed. Quite importantly, the attack is likely to succeed even in the presence of a large number of errors. This is an aspect that might render DNA computing especially suitable for cryptanalytic tasks. One can never expect a 100% success rate. But even if some DNA operations are error prone, the cryptanalytic attack might succeed with a reasonable probability.

	N_0	N_1	N_2	N_3	N_4
initial	(1,3),(2,3), (3,4),(1,2,3), (1,3,4),(2,3,4), (1,2,3,4)	empty	empty	empty	empty
$i = 0$, *separate* on 1	(2,3),(3,4), (2,3,4)	(1,3),(1,2,3), (1,3,4), (1,2,3,4)	empty	empty	empty
$i = 1$, *separate* on 2	(3,4)	(1,3),(1,3,4), (2,3),(2,3,4)	(1,2,3), (1,2,3,4)	empty	empty
$i = 2$, *separate* on 3	empty	(3,4)	(1,3),(2,3), (1,3,4), (2,3,4)	(1,2,3), (1,2,3,4)	empty
$i = 3$, *separate* on 4	empty	empty	(1,3),(2,3), (3,4)	(1,2,3), (1,3,4), (2,3,4)	(1,2,3,4)

The cryptosystem DES translates plaintext blocks 64 bits in length into 64-bit cryptotext blocks. The encryption happens under the control of a 56-bit key. The same key is used for both encryption and decryption. (DES is a classical two-way cryptosystem, in contrast to one-way or public-key cryptosystems.) We consider the "known plaintext" attack of cryptanalysis, [201]. It means that the cryptanalyst knows some of the pairs consisting of plaintext and the corresponding cryptotext and, on the basis of this information, is supposed to find the key. Of course, this must happen within a reasonable amount of time. It has been suggested that special-purpose electronic hardware or massively parallel supercomputers might do the job in case of DES. However, there has been no breakthrough. While DNA computing based on the sticker model might be the right approach, the real feasibility of such an attack should ultimately be decided in the laboratory.

An immediate approach to the problem is an exhaustive search through all the 2^{56} different keys. It is an indication of the strength of DES that no significantly better approach is known. This brute force approach is the one taken in DNA computing.

We now describe the cryptanalytic attack presented in [3]. Thus, the sticker model is used. The initial test tube will be a (579, 56) library. The substrands in the memory complexes will be oligonucleotides of length 20. Thus, the memory strands will be 11580 nucleotides long. This is still a safe size but oligonucleotides longer that 15000 bases might be fragmented by shear forces of pouring and mixing the test tubes.

In the memory complexes, a region of 56 substrands will store the 56-bit key. Another region of 64 substrands will, after the computation, encode the corresponding cryptotext. The remaining 459 substrands are needed to store intermediate results during the computation. The known pair (plaintext, cryptotext) is not represented in the memory complexes. It always remains the same; each of the keys works on this particular plaintext, and the resulting cryptotext is compared with the known fixed cryptotext. Thus, the whole procedure consists of the following three steps.

1. Construct the initial (579, 56) library, representing all possible 2^{56} keys.

2. On each memory complex, compute the cryptotext obtained by encrypting the known plaintext by the key represented by the memory complex.

3. Select the memory complex whose cryptotext matches the known cryptotext, and read its key.

The main part of the work is step (2). The "sticker machine" implementing the algorithm can be envisioned as a *parallel robotic workstation.* It consists of a rack of tubes (data tubes, sticker tubes, operator tubes), some robotics (arms, pumps, heaters or coolers), as well as a microprocessor that controls the robotics. The robotics are arranged to perform any of the four operations discussed above in connection with the sticker model: *merge, separate, set,* and *clear.* Moreover, the robotics are capable of performing the operations in the following extended *parallel* sense.

Robotics can *merge* the DNA from 64 data tubes into one data tube. They can *separate* the DNA from each of 32 data tubes into two more data tubes by using 32 specific "separation operator" tubes. The robotics can *set* the ith substrand on, in all memory complexes of 64 data tubes. For this it uses a sticker tube containing stickers for the ith substrand, as well as a sticker operator tube. Similarly, the robotics can *clear* specific substrands. The reader is referred to [3] concerning further details about the implementation of the operations, as well as the algorithm for step (2). We conclude this section by explaining the construction in step (1), the creation of the initial library. How can one obtain all the possible 2^{56} keys? The technique is also of general interest in DNA computing.

We begin with approximately 2^{56} identical memory strands (single strands) of the correct length, and divide them equally into two tubes N_1 and N_2. Large amounts of each of the 56 stickers are added to N_1, so that in the ligation reaction all of the 56 appropriate substrands in N_1 are turned on. The unused stickers are washed away from N_1, after which N_1 and N_2

are merged into one tube N. Finally, N is heated and cooled, to randomly reanneal the stickers. Roughly, 63% of the keys will be represented after this process. If we begin with three times the necessary amount of DNA, the percentage is increased to 95%.

2.3 Paradigm of Computing – Some Rethinking

Will an eventual large-scale realization of DNA computers change or significantly alter the general paradigm of computing? This will be the topic of our subsequent discussion. We apply here the word "paradigm," widely in use nowadays, to mean the "set of beliefs and opinions common to a scientific community." When speaking about the "paradigm of computing," the scientific community we mean apparently consists of computer scientists, understood in a very wide sense.

There can be no doubt about the fact that the Turing machine has already been an incarnation of the paradigm of computing for roughly half a century. So let us go back to Turing's original paper, [212], and see how he argued that his definition of "computable" numbers actually included all numbers which would naturally be regarded as computable. In present terminology, such an argument would defend the Church–Turing Thesis, that is, explain why a Turing machine actually computes everything.

In [212], Turing used arguments of three kinds:

(i) A direct appeal to intuition.

(ii) A proof of the equivalence of two models.

(iii) Giving examples of large classes of computable numbers, as well as showing the closure of computable numbers under various operations.

For our purposes the arguments (ii) and (iii) are irrelevant, whereas it is important to recall what Turing says about (i). His idea of a computer as a diligent clerk has to be contrasted with the idea of a computer as a multitude of DNA strands. The latter are in no way smarter than the diligent clerk – in fact it might be the other way round. But their massive, theoretically unbounded, parallelism changes the paradigm at least on some level.

Turing opens the argument (i) as follows. "Computing is normally done by writing certain symbols on paper. We may suppose that this paper is divided into squares like a child's arithmetic book. In elementary arithmetic the two-dimensional character of the paper is sometimes used. But such a use is always avoidable, and I think it will be agreed that the two-dimensional character of paper is no essential of computation. I assume then that computation is carried out on one-dimensional paper, i.e. on a tape divided into squares. I shall also suppose that the number of symbols which may be printed is finite. If we were to allow an infinity of symbols, then there would be symbols differing to an arbitrarily small extent. The effect of this restriction of the number of symbols is not very serious. It is always possible to use sequences of symbols in place of single symbols. ... The behaviour of the computer at any moment is determined by the symbols which he is observing, and the 'state of mind' at that moment. We may suppose that there is a bound to the number of symbols or squares which the computer can observe at one moment. If he wishes to observe more, he must use successive observations. We will also suppose that the number of states of mind which need to be taken into account is finite. The reason for this are of the same character as those which restrict the number of symbols. If we admitted an infinity of states of mind, some of them will be 'arbitrarily close' and will be confused. Again, the restriction is not one which seriously affects computation, since the use of more complicated states of mind can be avoided by writing more symbols on the tape. Let us imagine the operations performed by the computer to be split up into 'simple operations' which are so elementary that it is not easy to imagine them further divided. Every such operation consists of some change of the physical system consisting of the computer and his tape. We know the state of the system if we know the sequence of symbols on the tape, which of these are observed by the computer (possibly with a special order), and the state of mind of the computer."

The analysis continues along the same lines. It makes no difference if only observed squares are changed and in a simple operation not more than one symbol is altered. Because the new observed squares must be immediately recognizable by the computer, their distance from the squares observed at the preceding step should not exceed a certain fixed amount. By invoking the simplicity of the individual operations and the resulting necessity to replace more complicated operations by a sequence of simple ones, it can be concluded that the most general single operation is either a possible change of the observed symbol together with a possible change of the state of mind, or else a possible change of the observed square also together with a possible change of the state of mind. Thus, this intuitive analysis has brought forward the standard notion of a Turing machine which will be discussed formally in Chap. 3.

The operation of Turing's computer, the diligent clerk, is fundamentally *sequential.* He works step by step, always inspecting some part of his eventually huge piles of data. (We observe in passing that, in Turing's days, no fuss was made in mathematical writing about the "he-she" distinction. Turing refered to the clerk as "he".) Nothing happened in *parallel.* Turing could have of course thought of several clerks working simultaneously but then apparently the idea would have been to simulate their work by one clerk doing all the individual workloads in succession. This would have increased the time needed, maybe enormously, but all complexity considerations are irrelevant for Turing's clerk. Notions such "tractable" or "feasible", let alone "practical", do not enter the discussion. What is said about the multitude of clerks applies also to the multitude of DNA strands. We can always simulate, in one way or another, the massive parallelism of DNA molecules by doing all the parallel work in successive sequential steps. It seems clear that DNA computers cannot violate the Church–Turing Thesis. If something has been computed by a DNA computer, then we can call Turing's clerk and ask him to carry out the same computation. Computability, or the set of computable numbers as in Turing's terminology, is not affected by DNA computing. The paradigm of computing, when associated with the *a posteriori* notion of computability, seems to be highly invariant.

Things look different if the paradigm of computing is viewed in the *a priori* sense and, in particular, if complexity is taken into consideration. We are facing a problem and speculate *a priori* our possibilities for solving it. Then we might come to a different conclusion, depending on whether we have Turing's clerk or a test tube of DNA strands at our disposal. In this sense DNA computers, if successful, will surely change the paradigm of computing. This is surely reflected also in theoretical studies concerning complexity classes. Among the early examples are the *genetic Turing machines* introduced in [181]. In this model, the class of problems which can be solved in polynomial time (and which because of this consists of computationally tractable problems) coincides with the class *PSPACE* (which includes the class *NP* and, thus, very many intractable problems). Further examples will be still quoted below.

In the second part of this book we will investigate various mathematical models, asking the question whether it is possible to construct a universal computing machine out of biological macromolecular components and perform arbitrary computations by biological techniques. An overall, quite striking observation is that, at least theoretically, there seem to be many diverse ways of constructing DNA based universal computers. We will now try to explain the compelling mathematical reason behind this overall observation.

We claim that *Watson–Crick complementarity guarantees universal computations* in any model of DNA computers having sufficient capabilities for handling inputs and outputs. This view was first presented and discussed in [192]. Watson–Crick complementarity is closely related to the *twin-shuffle language*, [54], [200]. The basic variant of this language uses two letters 0 and 1, as well as their complementary letters $\bar{0}$ and $\bar{1}$. (The definition of the twin-shuffle language and formal mathematical details guaranteeing the universality will be presented in Sect. 3.2. In particular, see Corollary 3.4 and Theorem 3.18.) This is in complete analogy of DNA being made up of four nucleotides that can be divided into two complementary pairs: (A, T) and (C, G). The two letters 0 and 1 are used for the necessary encodings, whereas their complementary versions $\bar{0}$ and $\bar{1}$ provide the structure needed to describe arbitrary computations in terms of words in the twin-shuffle language. This state of affairs is the essence of computational universality and can be viewed also as a *mathematical explanation to the number of nucleotides in DNA being four.* Three nucleotides would not be enough for the twin-shuffle language, whereas five would perhaps make too many matters superfluous, especially in view of the fact that the twin-shuffle language involves a considerable amount of redundancy in itself.

As already mentioned, we will return to the formal details in Sect. 3.2 below. However, because of the central role of this matter in the theory of DNA computing, we want to give at this stage some idea about the interrelation between the Watson–Crick complementarity and the twin-shuffle language. The latter will be given in its basic variant over the four letter alphabet $\{0, 1, \bar{0}, \bar{1}\}$ and denoted by TS.

Consider a word w over the alphabet $\{0, 1\}$, that is, w is a string built from 0's and 1's. Let $\overline{w}$ be the complementary string, built from $\bar{0}$'s and $\bar{1}$'s. For instance, $w = 00101$ and $\overline{w} = \bar{0}\bar{0}\bar{1}\bar{0}\bar{1}$. We denote by *shuffle*$(w, \overline{w})$ the set of words obtained by shuffling w and $\overline{w}$, quite arbitrarily but without changing the order of letters in w or $\overline{w}$. For instance, each of the words

$$0\bar{0}0\bar{0}1\bar{1}0\bar{0}\bar{1}1, \quad \bar{0}\bar{0}\bar{1}\bar{0}\bar{1}00101, \quad 00\bar{0}10\bar{0}\bar{1}\bar{0}\bar{1}1$$

is in the set *shuffle*$(w, \overline{w})$, whereas $0\bar{0}0\bar{0}1\bar{1}0\bar{1}\bar{0}1$ is not. By definition, the language TS consists of all words in *shuffle*$(w, \overline{w})$, where w runs over all words over $\{0, 1\}$. The following is a simple way of finding out whether or not a given word x, built from four letters $0, 1, \bar{0}, \bar{1}$, actually is in TS. Erase first from x all letters $\bar{0}$ and $\bar{1}$, leaving a word x'. Erase next from x all letters 0 and 1, as well as the bars from the remaining letters, leaving a word x''. Then the original x is in TS exactly in case $x' = x''$. The reader might want to try out this method on the examples given above.

Consider now the following association between the "DNA alphabet" and the four-letter alphabet discussed above:

$$\mathrm{A} = 0, \ \mathrm{G} = 1, \ \mathrm{T} = \bar{0}, \ \mathrm{C} = \bar{1}.$$

If we view the letters in the pairs $(0, \bar{0})$ and $(1, \bar{1})$ as being complementary, then this complementarity is the same as the Watson–Crick complementarity, via the association indicated.

The interconnection of the twin-shuffle language TS with the double strands of DNA can now be seen as follows. Consider a double strand, say

TAGCATCAT
ATCGTAGTA

We first rewrite the letters according to the association indicated:

$$\bar{0}01\bar{1}0\bar{0}\bar{1}0\bar{0}$$
$$0\bar{0}\bar{1}1\bar{0}01\bar{0}0$$

Taking letters from both strands by turns, we obtain the string $\bar{0}00\bar{0}1\bar{1}\bar{1}10\bar{0}\bar{0}0\bar{1}10\bar{0}\bar{0}0$ which belongs to TS. The result holds in general: this method always produces from an arbitrary DNA double strand, a string in TS. That we do not get all strings of TS in this fashion is irrelevant because it depends only on our convention of taking letters from the two strands by turns.

The following is an interesting observation from the point of view of mathematics of computation. The universality of DNA computations would not be affected if one of the DNA strands would consist, say, entirely of purines and the other strand of pyrimidines. In our extended binary notation this would mean that the barred and non-barred letters always occur on different strands. That this does not actually happen in nature, surely provides more computational leeway and efficiency.

Another way to "read" strings in TS by scanning the nucleotides of DNA molecules is based on the encoding suggested below:

	upper strand	lower strand
A, T	0	$\bar{0}$
C, G	1	$\bar{1}$

In other words, both nucleotides A and T are identified with 0, without a bar when appearing in the upper strand and barred when appearing in the lower strand; the nucleotides C, G are identified with 1 in the upper strand and with $\bar{1}$ in the lower strand. Given a DNA (double-stranded) molecule, by reading the two strands from left to right, with non-deterministic non-correlated speeds in the two strands, we get a string in TS. The reader might try with the molecule considered above. Conversely, we can obtain *all* strings in TS if we consider *all* molecules (complete double stranded sequences) and *all* possibilities to read them as specified above. The same result is obtained if we use molecules containing in the upper strand only nucleotides in any of the pairs

$$(\text{A, C}), (\text{A, G}), (\text{T, C}), (\text{T, G}).$$

The universality of the language TS can be briefly described as follows. According to the commonly accepted Church–Turing Thesis, every computation can be performed by a Turing machine and, thus, all computations are characterized by such Turing-machine-acceptable languages L_0. On the other hand, every such L_0 can be represented in the form $L_0 = f(TS)$, where f is a so-called *gsm mapping.* The mapping f depends on the language L_0. (The abbreviation "gsm" comes from "generalized sequential machine"; the acronym was used in computer science long before the era of mobile phones.) Thus, TS remains always the same, whereas f must be specified according to the needs of each particular L_0. The mapping f can be viewed to represent the input-output facilities. The situation is analogous in DNA based computing. The Watson–Crick complementarity always remains the same and guarantees universality in the same sense as TS. The key problem in developing DNA based computers is to determine which types of computational techniques or, theoretically, which aspects of gsm-mappings are adaptable to DNA computing.

We will return to the mathematical details in Chap. 3, especially in Sect. 3.2. The representation result $L_0 = f(TS)$ is very fundamental, yet the basic idea behind its proof is very conspicuous. Perhaps this also reflects the simplicity of the basic DNA structure.

It should have already become apparent to the reader that theoretical studies about DNA computing must make use of the following two advantages stemming from DNA molecules: (i) Watson–Crick complementarity which renders the power of the twin-shuffle language available, and (ii) the multitude of DNA molecules which brings massive parallelism to the computing scene. We already discussed the impact of (ii) to the paradigm of computing. As regards (i), the general *paradigm of complementarity* can be formulated in one of the following two ways.

(a) A string induces the complementary string, either randomly or guided by a control device.

(b) The complementarity of two strings leads to some phenomenon such as bonding. Conversely, the occurrence of this phenomenon guarantees that the strings involved indeed are complementary.

We have given here a conspicuously abstract formulation of the paradigm of complementarity. The alphabet of the strings can be bigger than the four-letter DNA alphabet, we only assume that complementarity is present among the letters. More general alphabets will be considered in the mathematical theory presented in the second part of this book.

The version (b) is an abstraction of the idea present already in Adleman's experiment. The "free availability" of the twin-shuffle language, as well as the resulting universality of many models of DNA computing, can also be explained using (b): the bonding guarantees that the opposing nucleotides are complementary, which again leads to a word in the twin-shuffle language, as pointed out before.

On the other hand, the version (a) of the paradigm of complementarity presents complementarity as an operation: from a string (strand), go to the complementary one. This might become an important operation at least in certain phases of DNA computing. The operation is certainly very interesting from the theoretical point of view. Sometimes something surprisingly new can be created when a classical structure is supplemented with the operation of complementarity. For instance, this happens when a Lindenmayer system is furnished with complementarity [137].

2.4 DNA Computing: Hopes and Warnings

"For the long term, one can speculate about the prospect for molecular computation. It seems likely that a single molecule of DNA can be used to encode the instantaneous description of a Turing machine and that currently available protocols and enzymes could (at least under idealized conditions) be used to induce successive sequence modifications, which would correspond to the execution of the machine. In the future, research in molecular biology may provide improved techniques for manipulating macromolecules. Research in chemistry may allow for the development of synthetic designer enzymes. One can imagine the eventual emergence of a general purpose computer consisting of nothing more than a single macromolecule conjugated to a ribosomelike collection of enzymes that act on it."

These words of Adleman [1] are a compact formulation of the great expectations concerning DNA computing. Even in these optimistic words the need for further research in molecular biology, as well as in chemistry, is clearly expressed. Indeed, it has not yet been finalized in any way whether DNA computing will become an important reality or remain a footnote in history books. In this section we will discuss the positive and negative prospects of DNA computing. For instance, we will return to the possibility of some of the simple operations essential in DNA computing, such as the ones discussed in Sects. 2.1 and 2.2. It is important to mention already at this stage that some areas of the mathematical theory presented in the second part of this book go far beyond the reach of these simple observations. Some of the stronger results in this theory presuppose new kinds of techniques in molecular biology, before they can be properly applied. However, a lot can be accomplished using only the simple operations discussed in Sects. 2.1 and 2.2.

Let us consider first a very specific task, namely, the breaking of the Data Encryption Standard, DES. A technique due to [3] was outlined already in Sect. 2.2; [3] gives also a detailed analysis of the feasibility of the procedure. The analysis is of importance to the general evaluation of DNA computing. It shows that "real problems" (certainly the breaking of the most widely used cryptosystem is a real problem!) can be solved with small machines which do not require huge amounts of DNA (and use few enzymes if any). At present cryptographic tasks seem to be the most suitable ones for DNA computing, since error rates much greater than those normally required of electronic computers will suffice.

The computation to break DES is estimated to run five days. This is under the assumption that each individual operation can be completed in one minute, perhaps using an auxiliary robotic machine. If a graduate student has to perform each operation, it might take a day, and then the whole computation will take 18 years. An operation per hour gives a total duration of 9 months.

How big will the machinery be and what are the expectations for success? Some operations are more prone to errors than others. For each specific operation, its *error rate* is the fraction of molecules that commit an error during that operation. To say that the overall error rate is E means that E is the error rate of the worst operation, that is, all operations occurring in a computation have an error rate less than or equal to E. If E is the error rate, $1-E$ is customarily called the *yield*. Thus, an error rate of 10^{-4} corresponds to a yield of 99.99%.

In the cryptanalytic setup for breaking DES that was considered in Sect. 2.2, the cryptanalyst knows a pair consisting of a plaintext and the corresponding cryptotext and has to determine the key. All keys mapping the given plaintext to the given cryptotext are referred to as *winning keys*. It is conceivable that there are several winning keys, although their number is not likely to be large in connection with DES. Under ideal conditions, the algorithm produces a "final tube" containing, for each winning key, at least one molecule encoding it. Moreover, the final tube should contain no *distractors*, that is molecules which do not encode a winning key but have ended up in the final tube because of errors. That a winning key is missing from the final tube is either because it was not created during the initialization, or else because it was created but met an error during the computation.

Very interesting numerical results have been presented in [3] about this setup. The results concern the amount of DNA required and the number of distractors in the final tube. Specifically, the amount of DNA required is computed to ensure a "reasonable chance," 63% or more, of getting at least one winning key in the final tube. (The figure 63% comes from the Poisson distribution associated to the method of creating the keys during the initialization.)

If an error rate of 10^{-4} is attainable, only a little more than 1 gram of DNA is needed; the calculation gives the value 1.4g. Moreover, for the error rate of 10^{-4}, the probability of a distractor ending up in the final tube is only 8%.

Things are dramatically different if only an error rate of 10^{-2} is attainable. The figures tell us quite clearly where the borderline of the feasibility of DNA computing lies, at least in the case where the task is particularly suitable for DNA computing, as we already observed cryptanalytic tasks to be. An error rate of 10^{-2}, let alone an even bigger error rate, would make DNA computing definitely unfeasible. Then large amounts of DNA would be needed, approximately 23 Earth masses, to have a 63% chance that a winning key ends up in the final tube, and it would still have to be distinguished from a colossal number of distractors. An error rate of 10^{-3} would call for less than a kilogram of DNA, something that could still be considered feasible.

The size of the machinery is dictated by the amount of DNA used. The robotics must operate on a rack of test tubes, in fact, on 96 tubes in parallel. The estimates given in [3] make it reasonable to conclude that, under an achievable error rate of 10^{-4}, the entire machinery would fit on a desktop.

This very specific task of breaking the DES cryptosystem shows quite explicitly the feasibility borderline in DNA computing. It is essential how successfully, with low error rate, the operations can be performed. Assuming that low error rates are achievable, one may study the complexity of various tasks. Then the massive parallelism of DNA computing renders many of the *exponential* time complexity estimates in sequential computing, such as those dealing with some well-known NP-complete problems, to *linear* time. This is true of two of the problems discussed in Sects. 2.1 and 2.2, HPP and the satisfiability problem. At the same time, the number of DNA strands that may appear in a test tube during the course of the algorithm tends to be exponential; in fact it is of the order of $n!$ in connection with HPP. The reader is referred to [17] for complexity estimates of various problems in terms of the two parameters: the number of (biological) steps taken by the algorithm and the number of DNA strands used.

Complexity estimates of this kind have already been considered earlier in the area of parallel algorithms. In these studies the trade-off between the number of steps and the number of parallel processors is important. As we have observed, in DNA computing the number of steps can be drastically reduced, at the cost of the number of processors becoming exponential. Since the processors are DNA strands, this state of affairs can still be acceptable.

The *length* of the DNA strands should also be taken into consideration. In most cases the length will cause no problem since it is linear in the size of the problem.

Let us now summarize the operations of DNA computing, discussed earlier in this chapter. We have observed that the operations are basic in many of the algorithms in DNA computing and, consequently, further laboratory studies about their reliability, efficiency, and error rates are quite essential. (The operation of *splicing* is not included here; it will be explained in connection with the mathematical theory in the second part of the book.) We also remind the reader of the basic currently used techniques, described in Chap. 1 of this book, for carrying out each operation.

Melting. Double-stranded DNA is dissolved into single strands by heating the solution to a specific temperature. In this way the hydrogen bonds between complementary strands are broken.

Annealing. This is an operation reverse to melting. A solution of single strands is cooled, whereby strands complementary according to Watson–Crick can bind together.

Merge. This means pouring the contents of two test tubes into one tube.

Separate (or *Extract*). Recall that this operation produces from a tube N and a strand w a new tube $+(N, w)$, consisting of all strands in N which contain w as a substrand. Separation by hybridization uses a multitude of strands complementary to w, anchored to a matrix in a certain way. To these the strands in N containing w will anneal.

Amplify. This is an application of Polymerase Chain Reaction, PCR. At each step, the number of strands is doubled, resulting in an exponential growth.

Detect and *Length-separate.* Both operations apply the technique of gel electrophoresis.

Finally, DNA polymerases perform various functions, including the repair of DNA and forming complementary strands.

Specified oligonucleotides can be synthesized. However, it is still a largely open research problem to determine optimal oligonucleotides for DNA computation, both as regards their length and overall composition.

In conclusion, we feel it justified already to claim that at this stage biomolecular techniques are advanced enough and sufficiently adaptable to basic programming tasks occurring in DNA computing. This is the positive side of the matter. Many *caveats* still remain; it is no wonder that strong warnings have been expressed. (See, for instance, the correspondence section in *Science.*) Since this book is mainly about the mathematical theory of DNA computing, we do not discuss here all possible sources of troubles arising in laboratory realizations. For instance, sticking of strands to approximate matches, as opposed to exact matches, might lead the overall computation astray. Such problems should definitely be addressed before DNA computing can become a reality.

Perhaps the most constructive attitude at the moment is to think that DNA computers will supplement current computers in important aspects, not replace them. Certain classes of tasks and problems seem particularly apt for DNA computing. Features characteristic of such problems are that an exhaustive search is the best known method of solving the problem and that a high probability of success is almost as good as certainty. Such is the setup in typical cryptanalytic tasks. Advances in robotics might also open new vistas for building computers with both molecular and electronic components.

Part II

Mathematical Theory

Chapter 3

Introduction to Formal Language Theory

The mathematical theory of DNA computing presented in Part II of this book is developed in the framework of formal language theory. As we have seen in Chap. 1, DNA molecules have a natural representation through "double" strings satisfying certain assumptions (Watson–Crick complementarity and opposite directionality). Also, various enzymatic operations on DNA molecules can be naturally represented as operations on (double) strings. Consequently, using DNA molecules and their manipulation for the purpose of DNA computing can be conveniently and naturally expressed in the framework of (double) strings and operations on them. This leads to formal language theory as a natural framework for formalizing and investigating DNA computing.

In this chapter we introduce formal language theory to the extent needed for this book.

For additional information, the reader is referred to the many monographs in this area, such as: [4], [29], [40], [42], [93], [195], [197], [198]. A comprehensive source of information is [193]. We suggest that a reader already familiar with language theory consult Chap. 3 only when need arises.

3.1 Basic Notions, Grammars, Automata, Grammar Systems

Basic notations. The family of subsets of a set X is denoted by $\mathcal{P}(X)$; if X is an infinite set, then we denote by $\mathcal{P}_f(X)$ the family of finite subsets of X. The cardinality of X is denoted by $card(X)$. The set of natural numbers, $\{0, 1, 2, \ldots\}$ is denoted by $\mathbf{N}$. The empty set is denoted by $\emptyset$.

An *alphabet* is a finite nonempty set of abstract symbols. For an alphabet

V we denote by V^* the set of all strings of symbols in V. The empty string is denoted by λ. Mathematically speaking, V^* is the free monoid generated by V under the operation of *concatenation.* (The unit element of this monoid is λ.) The set of nonempty strings over V, that is $V^* - \{\lambda\}$, is denoted by V^+. Each subset of V^* is called a *language* over V. A language which does not contain the empty string (hence being a subset of V^+) is said to be λ-*free.*

If $x = x_1x_2$, for some $x_1, x_2 \in V^*$, then x_1 is called a *prefix* of x and x_2 is called a *suffix* of x; if $x = x_1x_2x_3$ for some $x_1, x_2, x_3 \in V^*$, then x_2 is called a *substring* of x. The sets of all prefixes, suffixes, substrings of a string x are denoted by $Pref(x), Suf(x), Sub(x)$, respectively.

The *length* of a string $x \in V^*$ (the number of symbol occurrences in x) is denoted by $|x|$. The number of occurrences of a given symbol $a \in V$ in $x \in V^*$ is denoted by $|x|_a$. If $x \in V^*$, $U \subseteq V$, then by $|x|_U$ we denote the length of the string obtained by erasing from x all symbols not in U, that is,

$$|x|_U = \sum_{a \in U} |x|_a.$$

For a language $L \subseteq V^*$, the set $length(L) = \{|x| \mid x \in L\}$ is called the *length set* of L.

The set of symbols occurring in a string x is denoted by $alph(x)$. For a language $L \subseteq V^*$, we denote $alph(L) = \bigcup_{x \in L} alph(x)$. Observe that $alph(L)$ may be a proper subset of V.

The *Parikh vector* associated to a string $x \in V^*$ with respect to the alphabet $V = \{a_1, \ldots, a_n\}$ is $\Psi_V(x) = (|x|_{a_1}, |x|_{a_2}, \ldots, |x|_{a_n})$. For $L \subseteq V^*$ we define $\Psi_V(L) = \{\Psi_V(x) \mid x \in L\}$.

A set M of vectors in $\mathbf{N}^n$ is said to be *linear* if there are $v_i \in \mathbf{N}^n$, $0 \leq i \leq m$, such that

$$M = \{v_0 + \sum_{i=1}^{m} \alpha_i v_i \mid \alpha_1, \ldots, \alpha_m \in \mathbf{N}\}.$$

A finite union of linear sets is said to be *semilinear.*

A language $L \subseteq V^*$ is semilinear if $\Psi_V(L)$ is a semilinear set.

Operations with strings and languages. The boolean operations (with languages) are denoted as usual: $\cup$ – union, $\cap$ – intersection, C – complementation.

The *concatenation* of L_1, L_2 is $L_1L_2 = \{xy \mid x \in L_1, y \in L_2\}$.

We define further:

$$L^0 = \{\lambda\},$$
$$L^{i+1} = LL^i, \ i \geq 0,$$
$$L^* = \bigcup_{i=0}^{\infty} L^i \text{ (the } * \text{-Kleene closure)},$$

$$L^+ = \bigcup_{i=1}^{\infty} L^i \text{ (the } + \text{-Kleene closure).}$$

A mapping $s : V \longrightarrow \mathcal{P}(U^*)$, extended to $s : V^* \longrightarrow \mathcal{P}(U^*)$ by $s(\lambda) = \{\lambda\}$ and $s(x_1x_2) = s(x_1)s(x_2)$, for $x_1, x_2 \in V^*$, is called a *substitution.* For a language $L \subseteq V^*$ we define $s(L) = \bigcup_{x \in L} s(x)$.

If $card(s(a))$ is finite for each $a \in V$, then s is called a *finite substitution*; if $card(s(a)) = 1$ for each $a \in V$, then s is called a *morphism.* If $\lambda \notin s(a)$, for each $a \in V$, then s is a λ*-free* substitution (λ-free morphism, respectively).

A morphism $h : V^* \longrightarrow U^*$ is called a *coding* if $h(a) \in U$ for each $a \in V$ and a *weak coding* if $h(a) \in U \cup \{\lambda\}$ for each $a \in V$. If $h : (V_1 \cup V_2)^* \longrightarrow V_1^*$ is the morphism defined by $h(a) = a$ for $a \in V_1$, and $h(a) = \lambda$ otherwise, then we say that h is a *projection* (associated to V_1) and we denote it by pr_{V_1}. For a morphism $h : V^* \longrightarrow U^*$, we define a mapping $h^{-1} : U^* \longrightarrow \mathcal{P}(V^*)$ (and we call it an *inverse morphism*) by $h^{-1}(w) = \{x \in V^* \mid h(x) = w\}$.

If $L \subseteq V^*, k \geq 1$, and $h : V^* \longrightarrow U^*$ is a morphism such that $h(x) \neq \lambda$ for each $x \in Sub(L), |x| = k$, then we say that h is k*-restricted* on L.

In general, when we have an alphabet V and we consider some given variants $g(a)$ of symbols $a \in V$ (primed, barred, etc.), then we denote $V^g = \{g(a) \mid a \in V\}$ and for $w \in V^*$ we write $w^g = g(w)$. (Thus, when considering primed symbols, $V' = \{a' \mid a \in V\}$ and for $w \in V^*$, $w = a_1 \ldots a_k$, with $a_i \in V$, $1 \leq i \leq k$, we have $w' = a'_1 \ldots a'_k$.)

For $x, y \in V^*$ we define their *shuffle* by

$$\begin{aligned} x \sqcup\!\sqcup\, y = \{&x_1y_1 \ldots x_ny_n \mid x = x_1 \ldots x_n, y = y_1 \ldots y_n, \\ &x_i, y_i \in V^*, 1 \leq i \leq n, n \geq 1\}. \end{aligned}$$

The *mirror image* of a string $x = a_1a_2 \ldots a_n$, for $a_i \in V, 1 \leq i \leq n$, is the string $mi(x) = a_n \ldots a_2a_1$.

In general, if we have an n-ary operation for strings, $g : V^* \times \ldots \times V^* \longrightarrow \mathcal{P}(U^*)$, we extend it to languages over V by

$$g(L_1, \ldots, L_n) = \bigcup_{\substack{x_i \in L_i \\ 1 \leq i \leq n}} g(x_1, \ldots, x_n).$$

For instance, $mi(L) = \{mi(x) \mid x \in L\}$.

The *left quotient* of a language $L_1 \subseteq V^*$ with respect to $L_2 \subseteq V^*$ is

$$L_2 \backslash L_1 = \{w \in V^* \mid \text{there is } x \in L_2 \text{ such that } xw \in L_1\}.$$

The *left derivative* of a language $L \subseteq V^*$ with respect to a string $x \in V^*$ is

$$\partial^l_x(L) = \{w \in V^* \mid xw \in L\}.$$

The *right quotient* and the *right derivative* are defined in a symmetric manner:

$$L_1/L_2 = \{w \in V^* \mid \text{there is } x \in L_2 \text{ such that } wx \in L_1\},$$
$$\partial_x^r(L) = \{w \in V^* \mid wx \in L\}.$$

A family FL of languages is *closed* under an n-ary operation g if, for all languages $L_1, \dots, L_n$ in FL, the language $g(L_1, \dots, L_n)$ is also in FL.

A language that can be obtained from the letters of an alphabet V and λ by using finitely many times the operations of union, concatenation, and Kleene $*$ is called *regular*; also the empty language is said to be regular.

A family of languages is *nontrivial* if it contains at least one language different from $\emptyset$ and $\{\lambda\}$. (We use here the word "family" synonymously with "set" or "collection".) A nontrivial family of languages is called a *trio* if it is closed under λ-free morphisms, inverse morphisms, and intersection with regular languages. A trio closed under union is called a *semi-AFL* (AFL = abstract family of languages). A semi-AFL closed under concatenation and Kleene $+$ is called an *AFL*. A trio/semi-AFL/AFL is said to be *full* if it is closed under arbitrary morphisms (and Kleene $*$ in the case of AFL's). A family of languages closed under none of the six AFL operations is called an *anti-AFL*.

Several facts about the operations defined above are useful when investigating the closure properties of a given family of languages (for instance, in order to prove that a family of languages is an AFL it is not necessary to check the closure under all the six AFL operations):

1. The family of regular languages is the smallest full trio.

2. Each (full) semi-AFL closed under Kleene $+$ is a (full) AFL.

3. If FL is a family of λ-free languages which is closed under concatenation, λ-free morphisms, and inverse morphisms, then FL is closed under intersection with regular languages and union, hence FL is a semi-AFL. (If FL is also closed under Kleene $+$, then it is an AFL.)

4. If FL is a family of languages closed under intersection with regular languages, union with regular languages, and substitution with regular languages, then FL is closed under inverse morphisms.

5. Every semi-AFL is closed under substitution with λ-free regular languages. Every full semi-AFL is closed under substitution with arbitrary regular languages and under left and right quotients with regular languages.

6. A family of λ-free languages is an AFL if it is closed under concatenation, λ-free morphisms, inverse morphisms, and Kleene $+$.

7. A family of languages that is closed under intersection with regular languages, union with regular languages, substitution by λ-free regular languages, and restricted morphisms is closed also under inverse morphisms.

Chomsky grammars. Generally speaking, a *grammar* is a (finite) device *generating* in a well specified sense the strings of a language (hence defining a set of syntactically correct strings). Many types of grammars are particular cases of rewriting systems.

A *rewriting system* is a pair $\gamma = (V, P)$, where V is an alphabet and P is a finite subset of $V^* \times V^*$; the elements (u, v) of P are written in the form $u \rightarrow v$ and are called *rewriting rules/productions* (or simply *rules* or *productions*). For $x, y \in V^*$ we write $x \Longrightarrow_\gamma y$ if $x = x_1ux_2, y = x_1vx_2$, for some $u \rightarrow v \in P$ and $x_1, x_2 \in V^*$. If the rewriting system γ is understood, then we write $\Longrightarrow$ instead of $\Longrightarrow_\gamma$. The reflexive and transitive closure of $\Longrightarrow$ is denoted by $\Longrightarrow^*$.

If an *axiom* is added to a rewriting system and all rules $u \rightarrow v$ have $u \neq \lambda$, then we obtain the notion of a *pure grammar*. For a pure grammar $G = (V, w, P)$, where $w \in V^*$ is the axiom, we define the *language generated* by G by

$$L(G) = \{x \in V^* \mid w \Longrightarrow^* x\}.$$

A *Chomsky grammar* is a quadruple $G = (N, T, S, P)$, where N, T are disjoint alphabets, $S \in N$, and P is a finite subset of $(N \cup T)^*N(N \cup T)^* \times (N \cup T)^*$.

The alphabet N is called the *nonterminal alphabet*, T is the *terminal alphabet*, S is the *axiom*, and P is the set of *production rules* of G. The rules (we also say *productions*) (u, v) of P are written in the form $u \rightarrow v$. Note that $|u|_N \geq 1$.

For $x, y \in (N \cup T)^*$ we write

$$x \Longrightarrow_G y \quad \text{iff} \quad x = x_1ux_2, y = x_1vx_2, \\ \text{for some } x_1, x_2 \in (N \cup T)^* \text{ and } u \rightarrow v \in P.$$

One says that x *directly derives* y (with respect to G). Each string $w \in (N \cup T)^*$ such that $S \Longrightarrow_G^* w$ is called a *sentential form*.

The language generated by G, denoted by $L(G)$, is defined by

$$L(G) = \{x \in T^* \mid S \Longrightarrow^* x\}.$$

Two grammars G_1, G_2 are called *equivalent* if $L(G_1) - \{\lambda\} = L(G_2) - \{\lambda\}$ (the two languages coincide modulo the empty string).

In general, in this book we consider two generative mechanisms equivalent if they generate the same language when we ignore the empty string.

If in $x \Longrightarrow y$ above we have $x_1 \in T^*$, then the derivation step is *leftmost* and we write $x \Longrightarrow_{left} y$. The leftmost language generated by the grammar

G is obtained by derivations where every step is leftmost and is denoted by $L_{left}(G)$.

According to the form of their rules, the Chomsky grammars are classified as follows. A grammar $G = (N, T, S, P)$ is called:

- *monotonous/length-increasing*, if for all $u \to v \in P$ we have $|u| \leq |v|$.
- *context-sensitive*, if each $u \to v \in P$ has $u = u_1 A u_2, v = u_1 x u_2$, for $u_1, u_2 \in (N \cup T)^*, A \in N$, and $x \in (N \cup T)^+$. (In monotonous and context-sensitive grammars the production $S \to \lambda$ is allowed, providing that S does not appear in the right-hand members of rules in P.)
- *context-free*, if each production $u \to v \in P$ has $u \in N$.
- *linear*, if each rule $u \to v \in P$ has $u \in N$ and $v \in T^* \cup T^*NT^*$.
- *right-linear*, if each rule $u \to v \in P$ has $u \in N$ and $v \in T^* \cup T^*N$.
- *left-linear*, if each rule $u \to v \in P$ has $u \in N$ and $v \in T^* \cup NT^*$.
- *regular*, if each rule $u \to v \in P$ has $u \in N$ and $v \in T \cup TN \cup \{\lambda\}$.

The arbitrary, monotonous, context-free, and regular grammars are also said to be of *type* 0, *type* 1, *type* 2, and *type* 3, respectively.

The family of languages generated by monotonous grammars is equal to the family of languages generated by context-sensitive grammars; the families of languages generated by right- or by left-linear grammars coincide and they are equal to the family of languages generated by regular grammars, as well as with the family of regular languages.

We denote by *RE, CS, CF, LIN,* and *REG* the families of languages generated by arbitrary, context-sensitive, context-free, linear, and regular grammars, respectively (RE stands for *recursively enumerable*). By FIN we denote the family of finite languages.

The following strict inclusions hold:

$$FIN \subset REG \subset LIN \subset CF \subset CS \subset RE.$$

This is *the Chomsky hierarchy*, the constant reference in the investigations in the following chapters.

The closure properties of the families listed above are indicated in Table 3.1 (Y stands for **yes** and N for **no**).

Therefore, *RE, CF, REG* are full AFL's, *CS* is an AFL (not full), and *LIN* is a full semi-AFL.

A context-free grammar $G = (N, T, S, P)$ is called *reduced* if for each $A \in N$ there is a derivation $S \Longrightarrow^* xAy \Longrightarrow^* xwy$, where $x, w, y \in T^*$ (each nonterminal is reachable from the axiom and it can be rewritten into a terminal string). Given a context-free grammar, an equivalent reduced context-free grammar can be found.

A linear grammar $G = (N, T, S, P)$ is said to be *minimal* if $N = \{S\}$ (it has only one nonterminal symbol).

Table 3.1. Closure properties of the families in the Chomsky hierarchy

	RE	*CS*	*CF*	*LIN*	*REG*
Union	Y	Y	Y	Y	Y
Intersection	Y	Y	N	N	Y
Complement	N	Y	N	N	Y
Concatenation	Y	Y	Y	N	Y
Kleene $*$	Y	Y	Y	N	Y
Intersection with regular languages	Y	Y	Y	Y	Y
Substitution	Y	N	Y	N	Y
λ-free substitution	Y	Y	Y	N	Y
Morphisms	Y	N	Y	Y	Y
λ-free morphisms	Y	Y	Y	Y	Y
Inverse morphisms	Y	Y	Y	Y	Y
Left/right quotient	Y	N	N	N	Y
Left/right quotient with regular languages	Y	N	Y	Y	Y
Left/right derivative	Y	Y	Y	Y	Y
Shuffle	Y	Y	N	N	Y
Mirror image	Y	Y	Y	Y	Y

Normal forms. Reducing grammars to a specified form, without losing generative power, is in general useful. There are several results which guarantee the existence of such normal forms. We mention here only four of them, which will be useful below.

Theorem 3.1. (Chomsky normal form) *For every context-free grammar G, an equivalent grammar $G' = (N, T, S, P)$ can be effectively constructed, with the rules in P of the forms $A \rightarrow a$ and $A \rightarrow BC$, for $A, B, C \in N$ and $a \in T$.*

Theorem 3.2. (Strong Chomsky normal form) *For every context-free grammar G, an equivalent grammar $G' = (N, T, S, P)$ can be effectively constructed, with the rules in P of the forms $A \rightarrow a$ and $A \rightarrow BC$, for $A, B, C \in N$ and $a \in T$, subject to the further restrictions:*

1. *if $A \rightarrow BC$ is in P, then $B \neq C$,*

2. *if $A \rightarrow BC$ is in P, then for each rule $A \rightarrow DE$ in P we have $E \neq B$ and $D \neq C$.*

If we also want to generate the empty string, then in the theorems above we also allow a completion rule $S \rightarrow \lambda$.

Theorem 3.3. (Kuroda normal form) *For every type-0 grammar G, an equivalent grammar $G' = (N, T, S, P)$ can be effectively constructed, with the rules in P of the forms $A \rightarrow BC, A \rightarrow a, A \rightarrow \lambda, AB \rightarrow CD$, for $A, B, C, D \in N$ and $a \in T$.*

Theorem 3.4. (Penttonen normal form) *For every type-0 grammar G, an equivalent grammar $G' = (N, T, S, P)$ can be effectively constructed, with the rules in P of the forms $A \rightarrow x$, $x \in (N \cup T)^*, |x| \leq 2$, and $AB \rightarrow AC$ with $A, B, C \in N$.*

Similar results hold true for length-increasing grammars; then rules of the form $A \rightarrow \lambda$ are no longer allowed, but only a completion rule $S \rightarrow \lambda$ if the generated language should contain the empty string.

Theorem 3.5. (Geffert normal forms) (1) *Each recursively enumerable language can be generated by a grammar $G = (N, T, S, P)$ with $N = \{S, A, B, C\}$ and the rules in P of the forms $S \rightarrow uSv, S \rightarrow x$, with $u, v, x \in (T \cup \{A, B, C\})^*$, and only one non-context-free rule, $ABC \rightarrow \lambda$.*

(2) *Each recursively enumerable language can be generated by a grammar $G = (N, T, S, P)$ with $N = \{S, A, B, C, D\}$ and the rules in P of the forms $S \rightarrow uSv, S \rightarrow x$, with $u, v, x \in (T \cup \{A, B, C, D\})^*$, and only two non-context-free rules, $AB \rightarrow \lambda, CD \rightarrow \lambda$.*

Otherwise stated, each recursively enumerable language can be obtained from a minimal linear language by applying the reduction rule $ABC \rightarrow \lambda$, or the reduction rules $AB \rightarrow \lambda, CD \rightarrow \lambda$.

Necessary conditions. For a language $L \subseteq V^*$, we define the equivalence relation $\sim_L$ over V^* by $x \sim_L y$ iff $(uxv \in L \Leftrightarrow uyv \in L)$ for all $u, v \in V^*$. Then $V^*/\sim_L$ is called the *syntactic monoid* of L.

Theorem 3.6. (Myhill–Nerode theorem) *A language $L \subseteq V^*$ is regular iff $V^*/\sim_L$ is finite.*

Theorem 3.7. (Bar-Hillel/$uvwxy$/pumping lemma for context-free languages) *If $L \in CF, L \subseteq V^*$, then there are $p, q \in \mathbf{N}$ such that every $z \in L$ with $|z| > p$ can be written in the form $z = uvwxy$, with $u, v, w, x, y \in V^*$, $|vwx| \leq q, vx \neq \lambda$, and $uv^iwx^iy \in L$ for all $i \geq 0$.*

Theorem 3.8. (Pumping lemma for linear languages) *If $L \in LIN, L \subseteq V^*$, then there are $p, q \in \mathbf{N}$ such that every $z \in L$ with $|z| > p$ can be written in the form $z = uvwxy$, with $u, v, w, x, y \in V^*$, $|uvxy| \leq q, vx \neq \lambda$, and $uv^iwx^iy \in L$ for all $i \geq 0$.*

Theorem 3.9. (Pumping lemma for regular languages) *If $L \in REG, L \subseteq V^*$, then there are $p, q \in \mathbf{N}$ such that every $z \in L$ with $|z| > p$ can be written* *in the form* $z = uvw$, *with* $u, v, w \in V^*$, $|uv| \leq q, v \neq \lambda$, *and* $uv^iw \in L$ *for all* $i \geq 0$.

Theorem 3.10. (Parikh theorem) *Every context-free language is semilinear.*

Corollary 3.1. (i) *Every context-free language over a one-letter alphabet is regular.*

(ii) *The length set of a context-free language is a finite union of arithmetical progressions.*

The conditions of Theorems 3.7 – 3.10 are only necessary, not sufficient for a language to be in the corresponding family.

Using these necessary conditions the following relations can be proved:

$$\begin{aligned}
L_1 &= \{a^n b^n \mid n \geq 1\} \in LIN - REG, \\
L_2 &= L_1 L_1 \in CF - LIN, \\
L_3 &= \{a^n b^n c^n \mid n \geq 1\} \in CS - CF, \\
L_4 &= \{xcx \mid x \in \{a, b\}^*\} \in CS - CF, \\
L_5 &= \{a^{2^n} \mid n \geq 1\} \in CS - CF, \\
L_6 &= \{a^n b^m c^n d^m \mid n, m \geq 1\} \in CS - CF, \\
L_7 &= \{a^n b^m \mid n \geq 1, 1 \leq m \leq 2^n\} \in CS - CF, \\
L_8 &= \{a^n b^m c^p \mid 1 \leq n \leq m \leq p\} \in CS - CF, \\
L_9 &= \{x \in \{a, b\}^* \mid |x|_a = |x|_b\} \in CF - LIN, \\
L_{10} &= \{x \in \{a, b, c\}^* \mid |x|_a = |x|_b = |x|_c\} \in CS - CF.
\end{aligned}$$

The *Dyck language*, D_n, over $T_n = \{a_1, a_1', \ldots, a_n, a_n'\}$, $n \geq 1$, is the context-free language generated by the grammar

$$G = (\{S\}, T_n, S, \{S \to \lambda, S \to SS\} \cup \{S \to a_i S a_i' \mid 1 \leq i \leq n\}, S).$$

Intuitively, the pairs (a_i, a_i'), $1 \leq i \leq n$, can be viewed as parentheses, left and right, of different kinds. Then D_n consists of all strings of correctly nested parentheses.

Theorem 3.11. (Chomsky–Schützenberger) *Every context-free language* L *can be written in the form* $L = h(D_n \cap R)$, *where* h *is a morphism,* D_n, $n \geq 1$, *is a Dyck language, and* R *is a regular language.*

Lindenmayer systems. Because (like the generative mechanisms introduced in the subsequent sections) *Lindenmayer systems* or *L systems* are introduced with biological motivation and because we shall mention them occasionally, we provide here the basic definitions.

Basically, a 0L (0-interactions Lindenmayer) system is a context-free pure grammar with parallel derivations: $G = (V, w, P)$, where V is an alphabet, $w \in V^*$ (axiom), and P is a finite set of rules of the form $a \to v$ with

$a \in V, v \in V^*$, such that for each $a \in V$ there is at least one rule $a \rightarrow v$ in P (we say that P is *complete*). For $w_1, w_2 \in V^*$ we write $w_1 \Longrightarrow w_2$ if $w_1 = a_1 \dots a_n, w_2 = v_1 \dots v_n$, for $a_i \rightarrow v_i \in P, 1 \leq i \leq n$. The generated language is $L(G) = \{x \in V^* \mid w \Longrightarrow^* x\}$.

If for each rule $a \rightarrow v \in P$ we have $v \neq \lambda$, then we say that G is *propagating* (non-erasing); if for each $a \in V$ there is only one rule $a \rightarrow v$ in P, then G is said to be *deterministic*. If we distinguish a subset T of V and we define $L(G)$ as $L(G) = \{x \in T^* \mid w \Longrightarrow^* x\}$, then we say that G is *extended*. The family of languages generated by 0L systems is denoted by $0L$; we add the letters P, D, E in front of 0L if propagating, deterministic, or extended 0L systems are used, respectively.

A *tabled* 0L system, abbreviated T0L, is a system $G = (V, w, P_1, \dots, P_n)$, such that each triple $(V, w, P_i), 1 \leq i \leq n$, is a 0L system; each P_i is called a *table*, $1 \leq i \leq n$. The generated language is defined by

$$\begin{aligned} L(G) = \{x \in V^* \mid w \Longrightarrow_{P_{j_1}} w_1 \Longrightarrow_{P_{j_2}} \dots \Longrightarrow_{P_{j_m}} w_m = x, \\ m \geq 0, 1 \leq j_i \leq n, 1 \leq i \leq m\}. \end{aligned}$$

(Each derivation step is performed by the rules of the same table.)

A T0L system is deterministic when each of its tables is deterministic. The propagating and the extended features are defined in the usual way.

The family of languages generated by T0L systems is denoted by $T0L$; the $ET0L, EDT0L$, etc. families are obtained in the same way as $E0L, ED0L$, etc.

The $D0L$ family is incomparable with *FIN, REG, LIN, CF*, whereas $E0L$ strictly includes the CF family; $ET0L$ is the largest family of Lindenmayer languages with 0-interactions, it is strictly included in CS, and it is a full AFL. The idea of 0-interactions corresponds to context-freeness: the letters develop independently of their neighbours.

An interesting feature of a D0L system, $G = (V, w, P)$, is that it generates its language in a *sequence*, $L(G) = \{w = w_0, w_1, w_2, \dots\}$, such that $w_0 \Longrightarrow w_1 \Longrightarrow w_2 \Longrightarrow \dots$. Thus, we can define the *growth function* of G, denoted by $growth_G : \mathbf{N} \longrightarrow \mathbf{N}$, by

$$growth_G(n) = |w_n|, \ n \geq 0.$$

Descriptional complexity. A given language can be generated by many, often infinitely many, different grammars. It is natural to look for grammars which are as simple as possible and to this end we need measures of grammar complexity.

Having a class X of grammars, a descriptional complexity measure (we also say *measure of syntactical complexity*) is a mapping $K : X \longrightarrow \mathbf{N}$ which is extended to languages generated by elements of X by $K(L) = \min\{K(G) \mid L = L(G), G \in X\}$. If necessary, then we also write $K_X(L)$, to specify the class of grammars used.

The following are three basic measures for context-free languages. For a context-free grammar $G = (N, T, S, P)$ we define

$$Var(G) = \text{card}(N),$$
$$Prod(G) = \text{card}(P),$$
$$Symb(G) = \sum_{r \in P} Symb(r), \text{ where } Symb(r : A \to x) = |x| + 2.$$

A complexity measure K is called *non-trivial* if for each n there is a grammar G_n such that $K(L(G_n)) > n$; K is said to be *connected* if there is n_0 such that for each $n \geq n_0$ there is G_n with $K(L(G_n)) = n$.

All the measures *Var, Prod, Symb* are connected (even with respect to the family of regular languages). Two measures of syntactical complexity cannot generally be simultaneously improved: if we find a grammar which is simpler from the point of view of one measure, then most of the time this grammar is more complex from the point of view of the other measure.

An important complexity measure is the *index*. Let $G = (N, T, S, P)$ be a grammar of any type. For a derivation

$$D : S = w_0 \Longrightarrow w_1 \Longrightarrow \ldots \Longrightarrow w_n = x \in T^*,$$

we denote

$$Ind(D) = max\{|w_i|_N \mid 0 \leq i \leq n\}.$$

For $x \in L(G)$, we define

$$Ind(x, G) = min\{Ind(D) \mid D : S \Longrightarrow^* x \in G\}.$$

Further,

$$Ind(G) = sup\{Ind(x, G) \mid x \in L(G)\}.$$

For a language L we denote

$$Ind(L) = min\{Ind(G) \mid L = L(G)\}.$$

Clearly, $Ind(L) = 1$ for each linear language L. It is known that $Ind(D_n) = \infty$, $n \geq 1$, and that Ind is a connected measure with respect to the family of context-free languages. Moreover, the family

$$CF_{fin} = \{L \in CF \mid Ind_{CF}(L) < \infty\}$$

is a full AFL.

Automata and transducers. Automata are language defining devices which work in the direction opposite to grammars. They start from the strings over a given alphabet and *analyze* them (we also say *recognize*), telling us whether or not the input string belongs to a specified language.

The five basic families of languages in the Chomsky hierarchy, *REG, LIN, CF, CS, RE*, are also characterized by recognizing automata. These automata are: the finite automaton, the one-turn pushdown automaton, the pushdown automaton, the linear-bounded automaton, and the Turing machine, respectively. We present here only the basic variants of these devices; we refer to [93], [138], [195], [198] for the many existing variants.

A (nondeterministic) *finite automaton* is a construct

$$M = (K, V, s_0, F, \delta),$$

where K and V are disjoint alphabets, $s_0 \in K, F \subseteq K$, and $\delta : K \times V \longrightarrow \mathcal{P}(K)$; K is the set of states, V is the alphabet of the automaton, s_0 is the initial state, F is the set of final states, and δ is the transition mapping. If $card(\delta(s,a)) \leq 1$ for all $s \in K, a \in V$, then we say that the automaton is *deterministic*. A relation $\vdash$ is defined in the following way on the set $K \times V^*$: for $s, s' \in K, a \in V, x \in V^*$, we write $(s, ax) \vdash (s', x)$ if $s' \in \delta(s, a)$; by definition, $(s, \lambda) \vdash (s, \lambda)$. If $\vdash^*$ is the reflexive and transitive closure of the relation $\vdash$, then the language of the strings recognized by automaton M is defined by

$$L(M) = \{x \in V^* \mid (s_0, x) \vdash^* (s, \lambda), s \in F\}.$$

It is known that both deterministic and nondeterministic finite automata characterize the same family of languages, namely *REG*. The power of finite automata is not increased if we also allow λ-*transitions*, that is δ is defined on $K \times (V \cup \{\lambda\})$ (the automaton can also change state when reading no symbol on its tape) or when the input string is scanned in a two-way manner, going along it to right or to left, without changing its symbols.

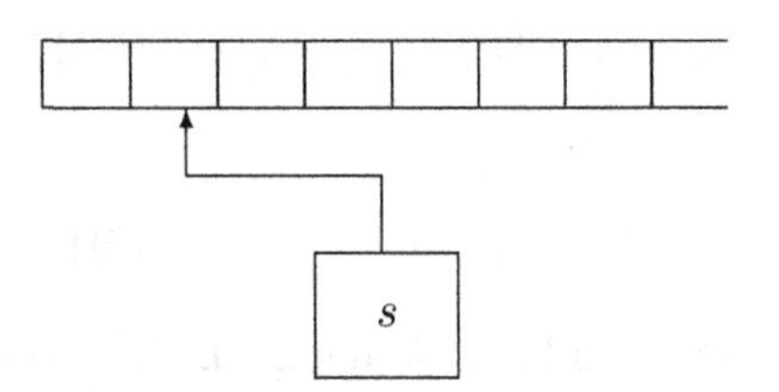

Figure 3.1: A finite automaton

An important related notion is that of a *sequential transducer*; we shall use the abbreviation *gsm*, from "generalized sequential machine". Such a device is a system $g = (K, V_1, V_2, s_0, F, \delta)$, where K, s_0, F are the same as in a finite automaton, V_1, V_2 are alphabets (the input and the output alphabet, respectively), and $\delta : K \times V_1 \longrightarrow \mathcal{P}_f(V_2^* \times K)$. If $\delta(s, a) \subseteq V_2^+ \times K$ for all $s \in K, a \in V_1$, then g is said to be λ-*free*. If $card(\delta(s,a)) \leq 1$ for each

$s \in K, a \in V_1$, then g is said to be *deterministic*. For $s, s' \in K, a \in V_1, y \in V_1^*, x, z \in V_2^*$, we write $(x, s, ay) \vdash (xz, s', y)$ if $(z, s') \in \delta(s, a)$. Then, for $w \in V_1^*$, we define

$$g(w) = \{z \in V_2^* \mid (\lambda, s_0, w) \vdash^* (z, s, \lambda), s \in F\}.$$

The mapping g is extended in the natural way to languages over V_1.

A gsm can be seen as a finite automaton with outputs. It is also easy to see that if a family of languages is closed under gsm mappings, then it is also closed under finite substitutions (and therefore under morphisms, too), as well as under the operations *Sub, Pref, Suf.*

We can imagine a finite automaton as in Fig. 3.1, where we distinguish the input tape, on whose cells we write the symbols of the input alphabet, the read head, which scans the tape from the left to the right, and the memory, able to hold a state from a finite set of states. In the same way, a gsm is a device as in Fig. 3.2, where we also have an output tape, where the write head can write the string obtained by translating the input string.

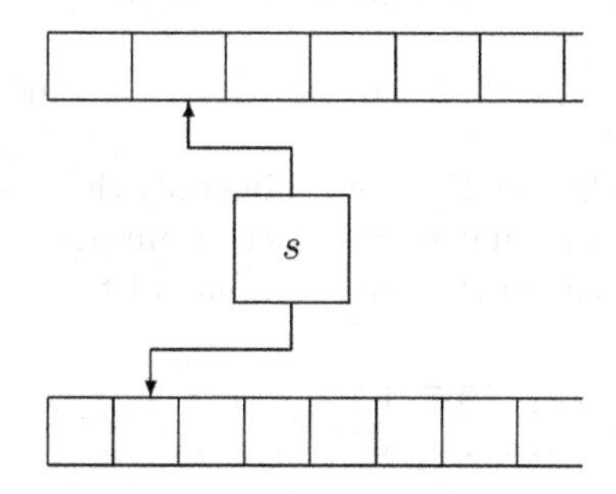

Figure 3.2: A sequential transducer

Sometimes it is useful to present the transition mapping of finite automata and of gsm's as a set of rewriting rules: we write $sa \to as'$ instead of $s' \in \delta(s, a)$ in the case of finite automata and $sa \to zs'$ instead of $(z, s') \in \delta(s, a)$ in the case of gsm's. Then the relations $\vdash, \vdash^*$ are exactly the same as $\Longrightarrow, \Longrightarrow^*$ in the rewriting system obtained in this way and, for a gsm g and a language $L \in V_1^*$, we get

$$g(L) = \{z \in V_2^* \mid s_0 w \Longrightarrow^* zs,\ w \in L, s \in F\}.$$

For finite automata we have a special case: $L(M) = \{x \in V^* \mid s_0 x \Longrightarrow^* xs, s \in F\}$.

A *pushdown* automaton is a construct

$$M = (K, V, U, s_0, Z_0, F, \delta),$$

where K, V, U are alphabets (of states, input symbols, and pushdown list symbols), K being disjoint from V and U, $s_0 \in K$ (initial state), $Z_0 \in U$ (initial pushdown list symbol), $F \subseteq K$ (final states), and $\delta : K \times (V \cup \{\lambda\}) \times U \longrightarrow \mathcal{P}_f(K \times U^*)$ (the transition mapping).

A *configuration* of M is a triple (s, w, z), where $s \in K$ is the current state, $w \in V^*$ is the input string not yet scanned, and $z \in U^*$ is the contents of the pushdown list. For two configurations $(s, w, z), (s', w', z')$ we define

$$\begin{aligned}(s, w, z) \vdash (s', w', z') \;\text{ iff }\; & w = aw', a \in V \cup \{\lambda\},\\ & z = \alpha z_1, z' = z_2 z_1, \text{ for } \alpha \in U, z_1, z_2 \in U^*,\\ & \text{and } (s', z_2) \in \delta(s, a, \alpha).\end{aligned}$$

We say that the leftmost symbol of the input, a, is scanned in state s, and we pass to state s' depending on the symbol in the top of the pushown list, α, which is replaced by z_2. Note that a can be λ. When $z_2 = \lambda$, we say that α is popped from the list.

We define the reflexive and transitive closure $\vdash^*$ of $\vdash$ in the natural way. Then the language recognized by M is defined by

$$L(M) = \{x \in V^* \mid (s_0, x, Z_0) \vdash^* (s, \lambda, z), \text{ for some } s \in F, z \in U^*\}.$$

(We start with the pushdown list containing only the symbol Z_0, in the initial state, and we finish in a final state, after scanning the whole input string. There are no restrictions on the final contents of the pushdown list.)

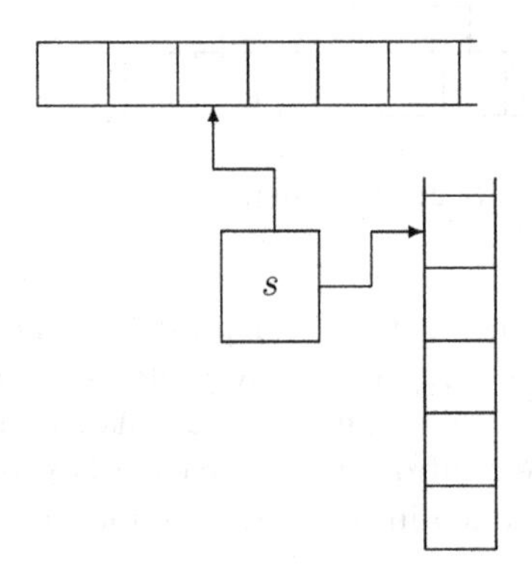

Figure 3.3: A pushdown automaton

A pushdown automaton can be represented as in Fig. 3.3, where we distinguish the input tape, the pushdown store and the memory with its two heads, a read-only head scanning the input tape and a read-write head always pointing to the first cell of the pushdown memory.

The (nondeterministic) pushdown automata recognize exactly the context-free languages. A *deterministic* pushdown automaton has only one possible behavior in each configuration. It is known that the family of languages recognized by deterministic pushdown automata is strictly included in CF.

A *Turing machine* is a construct

$$M = (K, V, T, B, s_0, F, \delta),$$

where K, V are disjoint alphabets (the set of states and the tape alphabet), $T \subseteq V$ (the input alphabet), $B \in V - T$ (the blank symbol), $s_0 \in K$ (the initial state), $F \subseteq K$ (the set of final states), and δ is a partial mapping from $K \times V$ to $\mathcal{P}(K \times V \times \{L, R\})$ (the move mapping; if $(s', b, d) \in \delta(s, a)$, for $s, s' \in K, a, b \in V$, and $d \in \{L, R\}$, then the machine reads the symbol a in state s and passes to state s', replaces a with b, and moves the read-write head to the left when $d = L$ and to the right when $d = R$). If $card(\delta(s, a)) \leq 1$ for all $s \in K, a \in V$, then M is said to be *deterministic.*

An *instantaneous description* of a Turing machine as above is a string xsy, where $x \in V^*, y \in V^*(V - \{B\}) \cup \{\lambda\}$, and $s \in K$. In this way we identify the contents of the tape, the state, and the position of the read-write head: it scans the first symbol of y. Observe that the blank symbol may appear in x, y, but not in the last position of y; both x and y may be empty. We denote by ID_M the set of all instantaneous descriptions of M.

On the set ID_M one defines the *direct transition* relation $\vdash_M$ as follows:

$$\begin{array}{l} xsay \vdash_M xbs'y \text{ iff } (s', b, R) \in \delta(s, a), \\ xs \vdash_M xbs' \text{ iff } (s', b, R) \in \delta(s, B), \\ xcsay \vdash_M xs'cby \text{ iff } (s', b, L) \in \delta(s, a), \\ xcs \vdash_M xs'cb \text{ iff } (s', b, L) \in \delta(s, B), \end{array}$$

where $x, y \in V^*, a, b, c \in V, s, s' \in K$.

The language recognized by a Turing machine M is defined by

$$L(M) = \{w \in T^* \mid s_0 w \vdash_M^* xsy \text{ for some } s \in F, x, y \in V^*\}.$$

(This is the set of all strings such that the machine reaches a final state when starting to work in the initial state, scanning the first symbol of the input string.)

It is also customary to define the language accepted by a Turing machine as consisting of the input strings $w \in T^*$ such that the machine, starting from the configuration $s_0 w$, reaches a configuration where no further move is possible (we say that the machine *halts*). The two modes of defining the language $L(M)$ are equivalent, the identified families of languages are the same, namely RE.

Graphically, a Turing machine can be represented as a finite automaton (Fig. 3.1). The difference between a finite automaton and a Turing machine is visible only in their functioning: the Turing machine can move its head in both directions and it can rewrite the scanned symbol, possibly erasing it (replacing it with the blank symbol).

Both the deterministic and the nondeterministic Turing machines characterize the family of recursively enumerable languages.

A Turing machine can be also viewed as a mapping-defining device, not only as a mechanism defining a language. Specifically, consider a Turing machine $M = (K, V, T, B, s_0, F, \delta)$. If $\beta \in ID_M$ such that $\beta = x_1 s a x_2$ and $\delta(s, a) = \emptyset$, then we write $\beta\downarrow$ (β represents a halting configuration). We define the mapping $F_M : ID_M \longrightarrow \mathcal{P}(ID_M)$ by $F_M(\alpha) = \{\beta \in ID_M \mid \alpha \vdash_M^* \beta$ and $\beta\downarrow\}$. If M is deterministic, then F_M is a mapping from ID_M to ID_M.

Given a mapping $f : U_1^* \longrightarrow U_2^*$, where U_1, U_2 are arbitrary alphabets, we say that f is *computed* by a Turing machine M if there are two (recursive) mappings C and D (of *coding* and *decoding*),

$$C : U_1^* \longrightarrow ID_M, \ D : ID_M \longrightarrow U_2^*,$$

such that

$$D(F_M(C(x))) = f(x).$$

In the next section, when discussing and presenting universal Turing machines, we shall use this interpretation of Turing machines (as well as the termination of a computation by halting configurations, not by using final states).

When working on an input string a Turing machine is allowed to use as much tape as it needs. Note that finite automata and pushdown automata use (in the read only manner) only the cells where the input string is written. In addition, a pushdown automaton has an unlimited pushdown tape. A Turing machine allowed to use only a working space linearly bounded with respect to the length of the input string is called a *linearly bounded automaton.* These machines characterize the family CS.

Regulated rewriting. As the context-free grammars are not powerful enough for covering most of the important syntactic constructions in natural and artificial languages, while the context-sensitive grammars are too powerful (for instance, the family CS has many negative decidability properties and the derivations in a non-context-free grammar cannot be described by a derivation tree), it is of interest to increase the power of context-free grammars by controlling the use of their rules. This leads to considering regulated context-free grammars. We present here some variants, which will be useful for our investigations.

A context-free *matrix* grammar is a construct $G = (N, T, S, M)$, where N, T are disjoint alphabets (of nonterminals and terminals, respectively), $S \in N$ (axiom), and M is a finite set of matrices, that is, sequences of the form $(A_1 \rightarrow z_1, \ldots, A_n \rightarrow z_n)$, $n \geq 1$, of context-free rules over $N \cup T$. For a string x, an element $m = (r_1, \ldots, r_n)$ is executed by applying productions

$r_1, \ldots, r_n$ one after the other, following the strict order they are listed in. The resulting string y is said to be directly derived from the original x and we write $x \Longrightarrow y$. Then, the generated language is defined in the usual way. The family of languages generated by context-free matrix grammars is denoted by MAT^λ (the superscript indicates that λ-rules are allowed); when using only λ-free rules, we denote the corresponding family by MAT.

A context-free *programmed* grammar is a construct $G = (N, T, S, P)$, where N, T, S are as above, the set of nonterminals, the set of terminals and the start symbol, and P is a finite set of productions of the form $(b : A \rightarrow z, E, F)$, where b is a label, $A \rightarrow z$ is a context-free production over $N \cup T$, and E, F are two sets of labels of productions of G. (E is said to be the *success field*, and F is the *failure field* of the production.) A production of G is applied as follows: if the context-free part can be successfully executed, then it is applied and the next production to be executed is chosen from those with the label in E, otherwise, we choose a production labeled by some element of F, and try to apply it. This type of programmed grammars is said to be with *appearance checking*; if no failure field is given for any of the productions, then a programmed grammar without appearance checking is obtained.

Sometimes it is useful to write a programmed grammar in the form $G = (N, T, S, P, \sigma, \varphi)$, where N, T, S are as above, P is a set of usual context-free rules and σ, φ are mappings from P to the power set of P; $\sigma(p), p \in P$, is the success field of the rule p (a rule in $\sigma(p)$ must be used after successfully applying the rule p), $\varphi(p), p \in P$, is the failure field (a rule from $\varphi(p)$ must be considered when p cannot be applied).

A context-free *ordered* grammar is a system $G = (N, T, S, P, >)$, where N, T, S are as above, P is a finite set of context-free productions, and $>$ is a partial ordering over P. A production p can be applied to a sentential form x only if it can be applied as a context-free rule and there is no production $r \in P$ such that r is applicable and $r > p$ holds.

Regulated applications of productions can also be based on checking context conditions.

A *generalized semi-conditional* grammar is a construct $G = (N, T, S, P)$, where N, T, S are as above and P is a finite set of triples of the form $p = (A \rightarrow w; E, F)$, where $A \rightarrow w$ is a context-free production over $N \cup T$ and E, F are finite subsets of $(N \cup T)^+$. Then, p can be applied to a string $x \in (N \cup T)^*$ only if A appears in x, each element of E and no element of F is a subword of x. If E or F is the empty set, then no condition is imposed by E, or respectively, F. E is said to be the set of *permitting* and F is said to be the set of *forbidding* context conditions of p.

If both $card(E) \leq 1$ and $card(F) \leq 1$ hold, then we speak of a *semi-conditional* grammar. If $E, F \subseteq N$, then we speak of a *random context* grammar.

Two other well-known versions of grammars with context conditions are the *conditional* and the *weakly conditional* grammars. A conditional grammar is a construct $G = (N, T, S, P)$, where P is a finite set of productions of the form $p = (A \to w; R_p)$, where R_p is a regular language over $N \cup T$. For a string x we can apply p to x only if A appears in x and $x \in R_p$.

If for every $p \in P$ we have $R = R_p$, for a fixed regular language R, then we speak of a weakly conditional grammar.

Conditional (weakly conditional) and semi-conditional grammars are of the same generative power; they generate all recursively enumerable or all context-sensitive languages, depending on whether λ-rules are used or not, respectively.

Finally, let us consider the notion of a *simple matrix* grammar. Such a grammar (of degree $n \geq 1$) is a system $G = (N_1, \ldots, N_n, T, S, M)$, with $N_1, \ldots, N_n, T$ mutually disjoint alphabets, $S \notin V_G$, for $V_G = T \cup \bigcup_{i=1}^{n} N_i$, and M is a set of matrices of the following forms:

(*i*) $(S \to w_1 w_2 \ldots w_n), w_i \in (T \cup N_i)^*, 1 \leq i \leq n,$
such that $|w_i|_{N_i} = |w_j|_{N_j}, 1 \leq i, j \leq n,$

(*ii*) $(A_1 \to w_1, \ldots, A_n \to w_n), A_i \in N_i, w_i \in (T \cup N_i)^*, 1 \leq i \leq n,$
such that $|w_i|_{N_i} = |w_j|_{N_j}, 1 \leq i, j \leq n.$

For each matrix in M, the derivation is leftmost for each rule ($A_i \to w_i$ rewrites the leftmost occurrence of a symbol in N_i in the current string).

It is known that the simple matrix grammars generate a family of semilinear languages which is strictly intermediate between CF and CS.

Useful for our purposes in the sequel is the idea of controlling the application of context-free rules to increase the power of context-free grammars.

Grammar systems. Another very fruitful idea for increasing the power of context-free grammars (in certain cases, also of regular grammars), is to consider distributed generative devices: constructs composed of several grammars working together according to a well-specified cooperation protocol. This leads to the idea of a *grammar system.* Two main classes of grammar systems have been investigated, the sequential ones (introduced in [28] under the name of *cooperating distributed grammar systems*) and the *parallel communicating grammar systems* introduced in [172]. For our investigations the second class is more useful, hence we present its definition only.

A *parallel communicating* (PC, for short) *grammar system* of degree n, $n \geq 1$, is a construct

$$\Gamma = (N, T, K, (S_1, P_1), \ldots, (S_n, P_n)),$$

where N, T, K are pairwise disjoint alphabets, with $K = \{Q_1, \ldots, Q_n\}$, $S_i \in N$, and P_i are finite sets of rewriting rules over $N \cup T \cup K, 1 \leq i \leq n$;

the elements of N are *nonterminal* symbols, those of T are *terminals*; the elements of K are called *query symbols*; the pairs (S_i, P_i) are the *components* of the system (often, the sets P_i are called components). Note that the query symbols are associated in a one-to-one manner with the components. When discussing the type of the components in the Chomsky hierarchy, the query symbols are interpreted as nonterminals.

For $(x_1, \ldots, x_n), (y_1, \ldots, y_n)$, with $x_i, y_i \in (N \cup T \cup K)^*, 1 \leq i \leq n$ (we call such an n-tuple a *configuration*), and $x_1 \notin T^*$, we write $(x_1, \ldots, x_n) \Longrightarrow_r (y_1, \ldots, y_n)$ if one of the following two cases holds:

(i) $|x_i|_K = 0$ for all $1 \leq i \leq n$; then $x_i \Longrightarrow_{P_i} y_i$ or $x_i = y_i \in T^*, 1 \leq i \leq n$;

(ii) there is $i, 1 \leq i \leq n$, such that $|x_i|_K > 0$; we write such a string x_i as

$$x_i = z_1 Q_{i_1} z_2 Q_{i_2} \ldots z_t Q_{i_t} z_{t+1},$$

for $t \geq 1, z_j \in (N \cup T)^*, 1 \leq j \leq t+1$; if $|x_{i_j}|_K = 0$ for all $1 \leq j \leq t$, then

$$y_i = z_1 x_{i_1} z_2 x_{i_2} \ldots z_t x_{i_t} z_{t+1},$$

[and $y_{i_j} = S_{i_j}, 1 \leq j \leq t$]; otherwise $y_i = x_i$. For all unspecified i we have $y_i = x_i$.

Point (i) defines a *rewriting* step (componentwise, synchronously, using one rule in all components whose current strings are not terminal); (ii) defines a *communication* step: the query symbols Q_{i_j} introduced in some x_i are replaced by the associated strings x_{i_j}, providing that these strings do not contain further query symbols. The communication has priority over rewriting (a rewriting step is allowed only when no query symbol appears in the current configuration). The work of the system is blocked when circular queries appear, as well as when no query symbol is present but point (i) is not fulfilled because a component cannot rewrite its sentential form, although it is a nonterminal string.

The relation $\Longrightarrow_r$ considered above is said to be performed in the *returning* mode: after communicating, a component resumes working from its axiom. If the brackets, [and $y_{i_j} = S_{i_j}, 1 \leq i \leq t$], are removed, then we obtain the *non-returning* mode of derivation: after communicating, a component continues the processing of the current string. We denote by $\Longrightarrow_{nr}$ the obtained relation.

The language generated by Γ is the language generated by its first component, when starting from $(S_1, \ldots, S_n)$, that is

$$L_f(\Gamma) = \{w \in T^* \mid (S_1, \ldots, S_n) \Longrightarrow_f^* (w, \alpha_2, \ldots, \alpha_n), \text{ for } \alpha_i \in (N \cup T \cup K)^*, 2 \leq i \leq n\}, \ f \in \{r, nr\}.$$

(No attention is paid to strings in the components $2, \ldots, n$ in the last configuration of a derivation; moreover, it is supposed that the work of Γ stops when a terminal string is obtained by the first component.)

Such systems communicate *on request.* A class of parallel communicating grammar systems with communication *by command* has been considered in [33]. In such a system, each component has an associated regular language. In any moment, each component sends its current sentential form to all other components, but the transmitted string is accepted only if it is an element of the regular language associated with the receiving component. Thus, these regular languages act as filters, controlling the communication in a way similar to the control of derivations in conditional grammars.

We present formally here only a particular class of such systems. We consider systems working with *maximal* derivations as rewriting steps, communicating *without splitting* the strings, *replacing* the string of the target component by a *concatenation* of the received messages, in the order of the system components, and *returning* to axioms after communicating; the generated language will be the language of the first component (which is the *master* of the system). The *filters* will be regular languages.

Formally, such a system is a construct

$$\Gamma = (N, T, (S_1, P_1, R_1), \ldots, (S_n, P_n, R_n)),$$

where N, T are disjoint alphabets (the nonterminal and the terminal one), $S_i \in N$, P_i are finite sets of context-free rules over $N \cup T$, and R_i are regular languages over $N \cup T$, $1 \leq i \leq n$.

With respect to Γ above, we define a *rewriting step* by

$$\begin{aligned}(x_1, \ldots, x_n) \Longrightarrow (y_1, \ldots, y_n) \text{ iff}& \\ & x_i \Longrightarrow^* y_i \text{ in } P_i \text{ and there is no } z_i \in (N \cup T)^* \\ & \text{such that } y_i \Longrightarrow z_i \text{ in } P_i\end{aligned}$$

(thus, if $x_i \in T^*$, then $y_i = x_i$, otherwise $x_i \Longrightarrow^+ y_i$), whereas a *communication step*, denoted by,

$$(x_1, \ldots, x_n) \vdash (y_1, \ldots, y_n)$$

is defined as follows:

Let

$$\delta_i(x_i, j) = \begin{cases} \lambda, & \text{if } x_i \notin R_j \text{ or } i = j, \\ x_i, & \text{if } x_i \in R_j \text{ and } i \neq j, \end{cases}$$

for $1 \leq i, j \leq n$,

$$\Delta(j) = \delta(x_1, j)\delta(x_2, j) \ldots \delta(x_n, j),$$

for $1 \leq j \leq n$ (this is the "total message" to be received by the j-th component), and

$$\delta(i) = \delta(x_i, 1)\delta(x_i, 2) \ldots \delta(x_i, n),$$

for $1 \leq i \leq n$ (this is the "total message" sent by the i-th component, a power of x_i indicating to how many targets the i-th component sends a message).

Then, for $1 \leq i \leq n$, we define

$$y_i = \begin{cases} \Delta(i), & \text{if } \Delta(i) \neq \lambda, \\ x_i, & \text{if } \Delta(i) = \lambda \text{ and } \delta(i) = \lambda, \\ S_i, & \text{if } \Delta(i) = \lambda \text{ and } \delta(i) \neq \lambda. \end{cases}$$

In words, y_i is either the concatenation of the received messages, if any exist, or it is the previous string, when this component is not involved in communications, or it is equal to S_i, if this component sends messages but it does not receive messages. Observe that a component cannot send messages to itself.

The generated language is defined as follows:

$$\begin{aligned} L(\Gamma) = \{w \in T^* \mid (S_1, \dots, S_n) & \Longrightarrow (x_1^{(1)}, \dots, x_n^{(1)}) \vdash (y_1^{(1)}, \dots, y_n^{(1)}) \\ & \Longrightarrow (x_1^{(2)}, \dots, x_n^{(2)}) \vdash (y_1^{(2)}, \dots, y_n^{(2)}) \Longrightarrow \dots \\ & \dots \Longrightarrow (x_1^{(s)}, \dots, x_n^{(s)}), \\ & \text{for some } s \geq 1 \text{ such that } w = x_1^{(s)}\}. \end{aligned}$$

Here is an **example**.

Let

$$\begin{aligned} \Gamma &= (N, T, (S_1, P_1, R_1), (S_2, P_2, R_2), (S_3, P_3, R_3)), \\ N &= \{S_1, S_2, S_2', S_3, S_3', X\}, \\ T &= \{a, b, c\}, \\ P_1 &= \{S_1 \to aS_1, S_1 \to bS_1, S_1 \to X\}, \\ R_1 &= \{a, b\}^* c, \\ P_2 &= \{S_2 \to S_2', X \to c\}, \\ R_2 &= \{a, b\}^* X, \\ P_3 &= \{S_3 \to S_3', X \to c\}, \\ R_3 &= \{a, b\}^* X. \end{aligned}$$

We start from (S_1, S_2, S_3). A componentwise maximal derivation is of the form

$$(S_1, S_2, S_3) \Longrightarrow (xX, S_2', S_3'),$$

for some $x \in \{a, b\}^*$. The string xX will be communicated to both the second and the third component, hence we have

$$\begin{aligned} &(xX, S_2', S_3') \vdash (S_1, xX, xX) \Longrightarrow (yX, xc, xc) \vdash \\ &\vdash (xcxc, yX, yX) \Longrightarrow (xcxc, yc, yc), \end{aligned}$$

for some $y \in \{a, b\}^*$. The string $xcxc$ is terminal, hence we have

$$L(\Gamma) = \{xcxc \mid x \in \{a, b\}^*\}.$$

Therefore, the very simple system Γ, with only three right-linear components, is able to generate the non-context-free (replication) language above. Observe that each derivation in Γ contains exactly two communication steps (and three rewriting steps, the last one being considered only for the sake of consistency with the definition of $L(\Gamma)$ as written above, where the last step is supposed to be a rewriting one).

We do not discuss here the power of these grammar systems. As in the case of regulated rewriting, the ideas these systems are based on (distribution, cooperation, communication, parallelism) are more important for us.

3.2 Characterizations of Recursively Enumerable Languages

The unchanging landmarks in the investigations in the following chapters will be the following two borders of computability. The power of finite automata constitutes the lowest interesting level of computability. The power of Turing machines is the highest level of computability – according to the Church–Turing Thesis. Finite automata stand for regular languages, Turing machines stand for recursively enumerable languages. In order to prove that a given generative device is able to generate each regular language, it is in general an easy task to start from a finite automaton or from a regular grammar and to simulate it by a device of the desired type. Sometimes, we can do the same when we have to characterize the recursively enumerable languages. Very useful in this sense are the normal forms discussed in the previous section. However, in many cases such a direct simulation is not so straightforward. Then the representation results available for recursively enumerable languages can be of great help. Some of them are quite non-intuitive, which makes their consequences rather interesting. As several results in the subsequent chapters essentially rely on such representations, we present here some results of this type, also giving proofs of them.

The first result is rather simple.

Theorem 3.12. *For every language $L \subseteq T^*, L \in RE$, there are $L' \in CS$ and $c_1, c_2 \notin T$, such that $L' \subseteq L\{c_1\}\{c_2\}^*$, and for each $w \in L$ there is $i \geq 0$ such that $wc_1c_2^i \in L'$.* (Thus, L is equal to L' modulo a tail of the form $c_1c_2^i, i \geq 0$.)

Proof. For a type-0 grammar $G = (N, T, S, P)$ we construct the grammar

$$G' = (N \cup \{S', X\}, T \cup \{c_1, c_2\}, S', P'),$$

with

$$\begin{aligned} P' &= \{S' \to Sc_1\} \\ &\cup \{u \to v \mid u \to v \in P, |u| \leq |v|\} \end{aligned}$$

$$\cup \{u \to vX^n \mid u \to v \in P, |u| > |v|, n = |u| - |v|\}$$
$$\cup \{X\alpha \to \alpha X \mid \alpha \in N \cup T\}$$
$$\cup \{Xc_1 \to c_1c_2\}.$$

It is easy to see that G' simulates the derivations in G, the only difference being that instead of the length-decreasing rules of P one uses rules which introduce the symbol X; this symbol is moved to the right and transformed into the terminal c_2 at the right hand of c_1. Thus, taking $L = L(G)$, the properties of $L' = L(G')$ as specified in the theorem are satisfied. □

Corollary 3.2. (i) *Each recursively enumerable language is the projection of a context-sensitive language.*

(ii) *For each $L \in RE$ there is a language $L_1 \in CS$ and a regular language L_2 such that $L = L_1/L_2$.*

Proof. The first assertion is obtained by taking the projection which erases the symbols c_1, c_2 above, whereas the second assertion follows by using the regular language $L_2 = c_1c_2^*$. (In both cases the context-sensitive language is L' as in Theorem 3.12.) □

Of course, the assertions above are valid also in a "mirrored" version: with $L' \subseteq \{c_2\}^*\{c_1\}L$ in Theorem 3.12, and with a left quotient by a regular language in point (ii) of Corollary 3.2.

These results prove that the two families RE and CS are "almost equal," the difference lies in a tail of arbitrary length to be added to the strings of a language; being of the form $c_1c_2^i, i \geq 1$, this tail carries no information other than its length, hence from a syntactical point of view the two languages L and L' in Theorem 3.12 can be considered indistinguishable.

The results below are of a rather different nature: we represent the recursively enumerable languages starting from "small" subfamilies of RE, but, instead, we use powerful operations (such as intersection, quotients, etc).

Theorem 3.13. *Each recursively enumerable language is the quotient of two linear languages.*

Proof. Take a language $L \in RE, L \subseteq T^*$. Consider a type-0 grammar $G = (N, T, S, P)$ for the language $mi(L)$ and add to P the rule $S \to S$. (In this way we may assume that each derivation in G has at least two steps.) Take a symbol c not in $N \cup T$ and construct the languages

$$\begin{aligned} L_1 = \{&x_nu_ny_ncx_{n-1}u_{n-1}y_{n-1}c\dots cx_1u_1y_1cc\ mi(y_1)mi(v_1)mi(x_1)c \\ &mi(y_2)mi(v_2)mi(x_2)c\dots c\ mi(y_{n-1})\ mi(v_{n-1})\ mi(x_{n-1})ccc \\ &mi(y_n)mi(v_n)mi(x_n) \mid n \geq 2, x_i, y_i \in (N \cup T)^*, \\ &u_i \to v_i \in P, 1 \leq i \leq n \text{ and } x_nv_ny_n \in T^*\}, \\ L_2 = \{&w_ncw_{n-1}c\dots cw_1cScc\ mi(w_1)c\ mi(w_2)c\dots c\ mi(w_n)ccc \mid \\ &n \geq 1, w_i \in (N \cup T)^*, 1 \leq i \leq n\}. \end{aligned}$$

Both these languages are linear. Here is a grammar for L_1:

$$\begin{aligned} G &= (\{X_0, X_1, X_2, X_3\}, N \cup T \cup \{c\}, X_0, P_1), \\ P_1 &= \{X_0 \to aX_0a \mid a \in T\} \\ &\cup \{X_0 \to uX_1\ mi(v) \mid u \to v \in P, v \in T^*\} \\ &\cup \{X_1 \to aX_1a \mid a \in T\} \\ &\cup \{X_1 \to cX_2ccc\} \\ &\cup \{X_2 \to aX_2a \mid a \in N \cup T\} \\ &\cup \{X_2 \to uX_3\ mi(v) \mid u \to v \in P\} \\ &\cup \{X_3 \to aX_3a \mid a \in N \cup T\} \\ &\cup \{X_3 \to cX_2c,\ X_3 \to cc\}. \end{aligned}$$

We have the equality $L = L_2 \backslash L_1$.
Indeed, each string in L_1 is of the form

$$w = w_n c w_{n-1} c \ldots c w_1 c c w'_1 c w'_2 c \ldots c w'_{n-1} c c c w'_n,$$

with $n \geq 2, w_i \Longrightarrow mi(w'_i), 1 \leq i \leq n$, in grammar G, and $w'_n \in T^*$. Each string in L_2 is of the form

$$z = z_m c z_{m-1} c \ldots c z_1 c S c c\ mi(z_1) c \ldots c\ mi(z_m) c c c,$$

with $m \geq 1$, $z_i \in (N \cup T)^*, 1 \leq i \leq m$. Therefore, $w = zz'$ if and only if $n = m + 1, S = w_1, z_i = w_{i+1}$ for $1 \leq i \leq m$, $w'_i = mi(z_i), 1 \leq i \leq m$, and $z' = w'_n$. This implies $w'_i = mi(z_i) = mi(w_{i+1}), 1 \leq i \leq n$, that is

$$S = w_1 \Longrightarrow w_2 \Longrightarrow \ldots \Longrightarrow w_n \Longrightarrow mi(z')$$

in the grammar G. Thus, $mi(z') \in L(G)$, that is $L_2 \backslash L_1 = mi(L(G)) = L$, which completes the proof. □

Corollary 3.3. *Each recursively enumerable language is the weak coding of the intersection of two linear languages.*

Proof. We repeat the construction above, taking the block $x_n v_n y_n$ in the writing of language L_1 as composed of primed versions of symbols in T. Let T' be the set of such symbols. Instead of L_2 we take the language $L_2 T'^*$. Denote the obtained languages by L'_1, L'_2, respectively. Obviously, they are linear, and $x \in L'_1 \cap L'_2$ if and only if $x = x_1 ccc x'_2, x_1 \in (N \cup T \cup \{c\})^*, x'_2 \in T'^*$, such that $x_2 \in L_2 \backslash L_1$. For the weak coding h defined by $h(a) = \lambda, a \in N \cup T \cup \{c\}$, and $h(a') = a, a \in T$, we obviously get $h(L'_1 \cap L'_2) = mi(L(G)) = L$. □

For a gsm $g = (K, V_1, V_2, s_0, F, P)$ with $V_1 = V_2$ we can define a mapping $g^* : V_1^* \longrightarrow \mathcal{P}(V_1^*)$ by

$$\begin{aligned} g^*(w) = \{z \in V_1^* \mid &\text{ there are } w_1, \ldots, w_n \text{ in } V_1^*, n \geq 2, \\ &\text{such that } w_i \in g(w_{i-1}), 2 \leq i \leq n, \text{ and} \\ &w = w_1, z = w_n\} \cup \{w\}. \end{aligned}$$

(We iterate the gsm g, repeatedly translating the current string.)

Theorem 3.14. *Each language $L \in RE, L \subseteq T^*$, can be written in the form $L = g^*(a_0) \cap T^*$, where $g = (K, V, V, s_0, F, P)$ is a gsm and $a_0 \in V$.*

Proof. Take a type-0 grammar $G = (N, T, S, P)$. Without loss of the generality we may suppose that S does not appear in the right hand member of rules in P. We construct the gsm

$$\begin{aligned}
g &= (K, N \cup T, N \cup T, s_0, F, P), \\
K &= \{s_0, s_1\} \cup \{[x] \mid x \in Pref(u) - \{\lambda\}, u \rightarrow v \in P\}, \\
F &= \{s_1\}, \\
P &= \{s_0 a \rightarrow a s_0 \mid a \in N \cup T\} \\
&\cup \{s_0 a_1 \rightarrow [a_1] \mid u \rightarrow v \in P, u = a_1 u', a_1 \in N \cup T, u' \in (N \cup T)^*\} \\
&\cup \{[x]a \rightarrow [xa] \mid xa \in Pref(u) - \{u\}, u \rightarrow v \in P, a \in N \cup T, \\
&\quad x \in (N \cup T)^*\} \\
&\cup \{[x]a \rightarrow v s_1 \mid xa = u, u \rightarrow v \in P, a \in N \cup T, x \in (N \cup T)^*\} \\
&\cup \{s_0 a \rightarrow v s_1 \mid a \rightarrow v \in P, a \in N\} \\
&\cup \{s_1 a \rightarrow a s_1 \mid a \in N \cup T\}.
\end{aligned}$$

It is clear that at each translation step the gsm g simulates the application of a rule in P. Therefore, $g^*(S) \cap T^* = L(G)$. □

We now move on to consider some quite powerful (and useful for some of the next chapters) representations of recursively enumerable languages, starting from *equality sets* of morphisms.

For two morphisms $h_1, h_2 : V^* \longrightarrow U^*$, the set

$$EQ(h_1, h_2) = \{w \in V^* \mid h_1(w) = h_2(w)\}$$

is called the *equality set* of h_1, h_2.

Theorem 3.15. *Every recursively enumerable language $L \subseteq T^*$ can be written in the form $L = pr_T(EQ(h_1, h_2) \cap R)$, where h_1, h_2 are two morphisms, R is a regular language, and pr_T is the projection associated to the alphabet T.*

Proof. Consider a type-0 grammar $G = (N, T, S, P)$ and assume the productions in P labeled in a one-to-one manner with elements of a set *Lab*.

Consider the alphabets

$$\begin{aligned}
V_1 &= N \cup T \cup \{c\}, \\
V_2 &= N \cup T \cup Lab \cup T' \cup \{B, F, c\},
\end{aligned}$$

where $T' = \{a' \mid a \in T\}$.

We define the morphisms $h_1, h_2 : V_2^* \longrightarrow V_1^*$ by

$$\begin{array}{lll} h_1(B) = Sc, & h_2(B) = \lambda, & \\ h_1(c) = c, & h_2(c) = c, & \\ h_1(p) = v, & h_2(p) = u, & \text{for } p : u \to v \in P, \\ h_1(A) = A, & h_2(A) = A, & \text{for } A \in N, \\ h_1(a') = a, & h_2(a') = a, & \text{for } a \in T, \\ h_1(a) = \lambda, & h_2(a) = a, & \text{for } a \in T, \\ h_1(F) = \lambda, & h_2(F) = c. & \end{array}$$

Moreover, we consider the regular language

$$R = \{B\}((N \cup T')^* Lab(N \cup T')^* \{c\})^+ T^* \{F\}.$$

The idea behind this construction is as follows. Every string in $L(G)$ appears as the last string in a derivation D according to G. For a string $w(D)$ codifying the derivation D, the morphisms h_1, h_2 satisfy $h_1(w(D)) = h_2(w(D))$. However, h_1 "runs faster" than h_2 on prefixes of $w(D)$, and h_2 "catches up" only at the end. The projection pr_T (defined by $h(a) = a$ for $a \in T$ and $h(\alpha) = \lambda$ for $\alpha \in V_1 - T$) erases everything else except the end result. The language R is used to check that only strings of the proper form are taken into account.

Using these explanations, the reader can verify that we obtain the equality $L(G) = pr_T(EQ(h_1, h_2) \cap R)$. (Complete details can be found in [200].) □

A variant of this result, useful in Chap. 4, is the following one.

Theorem 3.16. *For each recursively enumerable language $L \subseteq T^*$, there exist two λ-free morphisms h_1, h_2, a regular language R, and a projection pr_T such that $L = pr_T(h_1(EQ(h_1, h_2)) \cap R)$.*

Proof. Consider a type-0 grammar $G = (N, T, S, P)$ with the rules in P labeled in a one-to-one manner with elements in a set Lab. Without loss of generality, we assume that for each production $p : u \to v$ in P we have $v \neq \lambda$, except for the production $S \to \lambda$ if $\lambda \in L(G)$.

Define $T' = \{a' \mid a \in T\}$, $T'' = \{a'' \mid a \in T\}$, and $Lab' = \{p' \mid p \in Lab\}$. For notational purposes, we also define a morphism $d : (N \cup T)^* \longrightarrow (N \cup T')^*$ by $d(A) = A$ for $A \in N$ and $d(a) = a'$ for $a \in T$. Note that d is a bijection; thus, the inverse of d, d^{-1}, is well defined.

Let

$$\begin{aligned} V_1 &= N \cup T \cup T' \cup \{B, F, c\}, \\ V_2 &= N \cup T \cup T'' \cup Lab \cup Lab' \cup \{B, F, c, c'\}, \end{aligned}$$

where B, F, c, and c' are new symbols. The morphisms $h_1, h_2 : V_2^* \longrightarrow V_1^*$, depending on G, are defined as follows:

$$\begin{array}{lll}
h_1(B) = BSc, & h_2(B) = B, & \\
h_1(c) = c, & h_2(c) = c, & \\
h_1(p) = d(v), & h_2(p) = d(u), & \text{for } p : u \to v \in P, \\
h_1(p') = v, & h_2(p') = d(u), & \text{for } p : u \to v \in P, \\
h_1(A) = A, & h_2(A) = A, & \text{for } A \in N, \\
h_1(a') = a', & h_2(a') = a', & \text{for } a' \in T', \\
h_1(a'') = a, & h_2(a'') = a', & \text{for } a'' \in T'', \\
h_1(a) = F, & h_2(a) = a, & \text{for } a \in T, \\
h_1(c') = F, & h_2(c') = c, & \\
h_1(F) = F, & h_2(F) = FF. &
\end{array}$$

Consider also the regular language

$$R = \{BS\}(\{c\}(N \cup T')^*)^*\{c\}T^*\{F\}^+.$$

Note that $u, v \neq \lambda$ for all $u \to v \in P$. So, both h_1 and h_2 are λ-free morphisms. If $\lambda \in L$, then we introduce an additional symbol d to V_2 and define

$$h_1(d) = h_2(d) = BScF.$$

It is easy to see that in this fashion we will not introduce any new words to $h_1(EQ(h_1, h_2)) \cap R$. Therefore, we assume that $\lambda \notin L$ in the following arguments.

The proof of the fact that $x \in L(G)$ implies $x \in pr_T(h_1(EQ(h_1, h_2)) \cap R)$ is similar to the proof of the corresponding inclusion in Theorem 3.15.

Conversely, let $w \in pr_T(h_1(EQ(h_1, h_2)) \cap R)$, i.e., $w = pr_T(y)$ for some $y \in h_1(EQ(h_1, h_2)) \cap R$. Then by the definition of R, y is of the form

$$BScy_1cy_2c \dots cy_tF^l,$$

where $y_1, \dots, y_{t-1} \in (N \cup T')^*$, $y_t \in T^*$, and $l > 0$. Let $y = h_1(x)$ for some $x \in EQ(h_1, h_2)$. Then

$$x = Bx_1cx_2c \dots cx_tc'x_{t+1}F^m$$

such that $h_2(x_1) = S$, $h_1(x_i) = h_2(x_{i+1}) = y_i$, for $1 \leq i \leq t$, and $l = 2m$ and $h_1(x_{t+1}) = F^{m-1}$. Note that if $x_j = x_{j+1}$ for some j, $1 \leq j < t$, then we can construct a new word x' by deleting x_jc from x so that $pr_T(h_1(x') \cap R) = pr_T(h_1(x) \cap R) = w$. So, without loss of generality, we assume that $x_j \neq x_{j+1}$ for all j, $1 \leq j < t$. (It is clear that $x_t \neq x_{t+1}$.)

The following assertions are clear:

(1) $x_1 = p$ (or $x_1 = p'$ if $t = 1$) for some $p : S \to z$ in P,

(2) $x_i \in (N \cup T' \cup Lab)^* Lab(N \cup T' \cup Lab)^*$, for $2 \leq i < t$,

(3) $x_t \in (N \cup T' \cup Lab')^*$,

(4) $x_{t+1} \in T^*$,

(5) $h_1(x_i) = y_i$ and $h_2(x_i) = y_{i-1}$, for $1 \leq i \leq t$ (letting $y_0 = S$).

By (2) and (5) above and the definition of h_1 and h_2, it follows that

$$d^{-1}(y_{i-1}) \Longrightarrow_G^+ d^{-1}(y_i),$$

$2 \leq i \leq t-1$. Note also that $S \Longrightarrow_G d^{-1}(y_1)$ and $d^{-1}(y_{t-1}) \Longrightarrow_G^+ y_t$. Therefore, we have $S \Longrightarrow_G^+ y_t$, i.e., $y_t \in L(G)$. Since $w = pr_T(y) = y_t$, we have proved that $w \in L$. □

Note the difference between the representations in Theorems 3.15, 3.16: in the first theorem the language L is obtained as a projection of the intersection of the equality set with a regular language, whereas in the second case the language L is the projection of the intersection of a regular language of *the image of the equality set under one of the morphisms defining the equality set.*

A very useful consequence of Theorem 3.15 is the following one.

Consider an alphabet V and its barred variant, $\overline{V} = \{\bar{a} \mid a \in V\}$. The language

$$TS_V = \bigcup_{x \in V^*} (x \sqcup\!\sqcup \bar{x})$$

is called the *twin-shuffle* language over V. (For a string $x \in V^*$, $\bar{x}$ denotes the string obtained by replacing each symbol in x with its barred variant.)

For the morphism $h : (V \cup \overline{V})^* \longrightarrow V^*$ defined by

$$h(a) = \lambda, \text{ for } a \in V,$$
$$h(\bar{a}) = a, \text{ for } a \in V,$$

we clearly have the equality $TS_V = EQ(h, pr_V)$. This makes the following result plausible.

Theorem 3.17. *Each recursively enumerable language $L \subseteq T^*$ can be written in the form $L = pr_T(TS_V \cap R')$, where V is an alphabet and R' is a regular language.*

Proof. Consider again the proof of Theorem 3.15. We may assume that V_1 and V_2, the range and the target alphabets of the morphisms h_1, h_2 are disjoint. (We simply rename the letters of V_1; this does not affect the proof above.)

Consider now the alphabet $V = V_1 \cup V_2$ and let g be the morphism satisfying

$$g(a) = ah_1(a)\overline{h_2(a)}, \text{ for every } a \in V_2.$$

Let also R' be the regular language

$$R' = g(R) \sqcup\!\sqcup \overline{V}_2^*.$$

The following equality follows directly from the definitions above

$$pr_T(TS_V \cap R') = pr_T(EQ(h_1, h_2) \cap R),$$

therefore the representation in Theorem 3.17 is a consequence of the representation in Theorem 3.15. □

Note the similarity of the representation above of recursively enumerable languages with the Chomsky–Schützenberger representation of context-free languages (Theorem 3.11); the role of the Dyck languages is now played by the twin-shuffle languages.

In this representation, the language TS_V depends on the language L. This can be avoided in the following way. Take a coding, $f : V \longrightarrow \{0,1\}^*$, for instance, with $f(a_i) = 01^i0$, where a_i is the ith symbol of V in a specified ordering. The language $f(R')$ is regular. A gsm can simulate the intersection with a regular language, the projection pr_T, as well as the decoding of elements in $f(TS_V)$. Thus we obtain

Corollary 3.4. *For each recursively enumerable language L there is a gsm g_L such that $L = g_L(TS_{\{0,1\}})$.*

Therefore, each recursively enumerable language can be obtained by a sequential transducer starting from the unique language $TS_{\{0,1\}}$. One can also see that this transducer can be a deterministic one.

Somewhat surprisingly, the result above is true also for a "mirror" variant of the twin-shuffle language.

For an alphabet V, consider the language

$$RTS_V = \bigcup_{x \in V^*} (x \sqcup\!\sqcup\, mi(\bar{x})).$$

This is the *reverse twin-shuffle language* associated to V.

Theorem 3.18. *For each recursively enumerable language L there is a deterministic gsm g_L such that $L = g_L(RTS_{\{0,1\}})$.*

Proof. Since the deterministic gsm's are closed under composition, it suffices to find a deterministic gsm g such that $g(RTS_{\{0,1\}}) = TS_{\{0,1\}}$. The idea of constructing such a gsm g is to let g to select twins x that are palindromes of the form $x = y00\, mi(y)$, with $y \in \{0,1\}^*$, with 0 and 1 coded as 01, 10, respectively.

Then g is the gsm which maps $w = h(u)00\bar{0}\bar{0}v$ into $g(w) = u$, for all strings $u, v \in \{0, 1, \bar{0}, \bar{1}\}^*$, where h is the morphism defined by $h(0) = 01, h(1) = 10, h(\bar{0}) = \bar{0}\bar{1}, h(\bar{1}) = \bar{1}\bar{0}$. Furthermore, the gsm g is defined for strings in

$$D = \{01, 10, \bar{0}\bar{1}, \bar{1}\bar{0}\}^* 00\bar{0}\bar{0}\{010\bar{1}, 101\bar{0}\}^*.$$

It is easy to see that g is a deterministic gsm mapping. It remains to show that $g(RTS_{\{0,1\}}) = TS_{\{0,1\}}$.

Consider a string $w = h(u)00\bar{0}\bar{0}v$ in $RTS_{\{0,1\}} \cap D$. Since $pr_{\{\bar{0},\bar{1}\}}(w) = mi(\overline{pr_{\{0,1\}}(w)})$, the specific form of strings in D (with the "marker" $00\bar{0}\bar{0}$ in the middle) implies that $pr_{\{\bar{0},\bar{1}\}}(h(u)) = mi(\overline{pr_{\{0,1\}}(v)})$ and $pr_{\{\bar{0},\bar{1}\}}(v) = mi(\overline{pr_{\{0,1\}}(h(u))})$. Since $pr_{\{\bar{0},\bar{1}\}}(v) = \overline{pr_{\{0,1\}}(v)}$ (because $v \in \{01\bar{0}\bar{1}, 10\bar{1}\bar{0}\}^*$), this implies that $pr_{\{\bar{0},\bar{1}\}}(h(u)) = \overline{pr_{\{0,1\}}(h(u))}$. Hence $h(pr_{\{\bar{0},\bar{1}\}}(u)) = h(\overline{pr_{\{0,1\}}(u)})$ and so, since h is injective, $pr_{\{\bar{0},\bar{1}\}}(u) = \overline{pr_{\{0,1\}}(u)}$, i.e., $g(w) = u \in TS_{\{0,1\}}$. This shows that $g(RTS_{\{0,1\}}) \subseteq TS_{\{0,1\}}$.

Conversely, if $u \in \{0, 1, \bar{0}, \bar{1}\}^*$ such that $pr_{\{\bar{0},\bar{1}\}}(u) = \overline{pr_{\{0,1\}}(u)}$, then consider the string $w = h(u)00\bar{0}\bar{0}v$, where v is the unique element of $\{01\bar{0}\bar{1}, 10\bar{1}\bar{0}\}^*$ such that $pr_{\{0,1\}}(v) = mi(pr_{\{0,1\}}(h(u)))$. Then $w \in D$, $g(w) = u$, and $w \in RTS_{\{0,1\}}$. This ends the proof. □

Results similar to Corollary 3.4 and Theorem 3.18 hold also for a weaker variant of the twin-shuffle language.

Consider the coding $c : \{0,1\}^* \longrightarrow \{\bar{0},1\}^*$ defined by $c(0) = \bar{0}$ and $c(1) = 1$. The *semi-twin-shuffle* language over $\{0,1\}$ is denoted by $STS_{\{0,1\}}$ and defined by

$$STS_{\{0,1\}} = \bigcup_{x \in \{0,1\}^*} (x ⧢ c(x)).$$

Theorem 3.19. *For each recursively enumerable language L there is a gsm g_L such that $L = g_L(STS_{\{0,1\}})$.*

Proof. In view of Corollary 3.4, it is enough to prove that $TS_{\{0,1\}} = g(STS_{\{0,1\}})$ for a gsm g.

Consider the morphism $h : \{0, 1, \bar{0}, \bar{1}\}^* \longrightarrow \{0, 1, \bar{0}\}^*$ defined by

$$h(0) = 00, \quad h(1) = 01,$$
$$h(\bar{0}) = \bar{0}\bar{0}, \quad h(\bar{1}) = \bar{0}1,$$

as well as the regular language

$$R = \{00, 01, \bar{0}\bar{0}, \bar{0}1\}^*.$$

The following equality holds:

$$TS_{\{0,1\}} = h^{-1}(STS_{\{0,1\}} \cap R).$$

Consider a string $y \in TS_{\{0,1\}}$. There is $x \in \{0,1\}^*$ such that $y \in x ⧢ \bar{x}$. We clearly have $h(y) \in h(x) ⧢ h(\bar{x}) = h(x) ⧢ c(h(x))$. Consequently, $h(y) \in STS_{\{0,1\}}$. Obviously, $h(y) \in R$, hence the inclusion $TS_{\{0,1\}} \subseteq h^{-1}(STS_{\{0,1\}} \cap R)$ follows.

Conversely, take a string $z \in STS_{\{0,1\}} \cap R$ and look for $h^{-1}(z)$. Because h is injective, $h^{-1}(z)$ is a singleton (it is non-empty for all $z \in R$). By the definition of $STS_{\{0,1\}}$ we have $z \in x ⧢ c(x)$ for some $x \in \{0,1\}^*$. Because $z \in R$, we must have $x \in \{00, 01\}^*$, that is $c(x) \in \{\bar{0}\bar{0}, \bar{0}1\}^*$. Consider the strings

$y = h^{-1}(x)$ and $\bar{y}$. We have $\bar{y} = \overline{h^{-1}(x)} = h^{-1}(c(x))$. Consequently, $h^{-1}(z) \in h^{-1}(x \sqcup\!\sqcup\, c(x)) = h^{-1}(x) \sqcup\!\sqcup\, h^{-1}(c(x)) = h^{-1}(x) \sqcup\!\sqcup\, \overline{h^{-1}(x)} \subseteq TS_{\{0,1\}}$. This proves that also the inclusion $h^{-1}(STS_{\{0,1\}} \cap R) \subseteq TS_{\{0,1\}}$ is true.

Now, the intersection with a regular language and the inverse morphism can be computed at the same time by a gsm. □

Also a counterpart of Theorem 3.18 can be obtained. The *reverse semi-twin-shuffle* language over $\{0,1\}$ is denoted by $RSTS_{\{0,1\}}$ and is defined by

$$RSTS_{\{0,1\}} = \bigcup_{x \in \{0,1\}^*} (x \sqcup\!\sqcup\, c(mi(x))).$$

By a proof similar to that of Theorem 3.19 we obtain the following result.

Corollary 3.5. *For each recursively enumerable language L there is a gsm g_L such that $L = g_L(RSTS_{\{0,1\}})$.*

3.3 Universal Turing Machines and Type-0 Grammars

A computer is a *programmable* machine, able to execute any program it receives. From a theoretical point of view, this corresponds to the notion of a *universal Turing machine*, and in general, to the notion of a machine which is universal for a given class, in the following sense.

Consider an alphabet T and a Turing machine $M = (K, V, T, B, s_0, F, \delta)$. As we have seen above, M starts working with a string w written on its tape and reaches or not a final state (and then halts), depending on whether or not $w \in L(M)$. A Turing machine can be also codified as a string of symbols over a suitable alphabet. Denote such a string by *code*(M). Imagine a Turing machine M_u which starts working from a string which contains both $w \in T^*$ and *code*(M) for a given Turing machine M, and stops in a final state if and only if $w \in L(M)$.

In principle, the construction of M_u is simple. M_u only has to simulate the way of working for Turing machines, and this is clearly possible: look for a transition, as defined by the mapping δ, depending on the current state and the current position of the read-write head (this information is contained in the instantaneous descriptions of the particular machine); whenever several choices are possible, make copies of the current instantaneous description and branch the machine evolution; if two copies of the same instantaneous description appear, delete one of them; if at least one of the evolution variants leads to an accepting configuration, stop and accept the input string, otherwise continue.

Such a machine M_u is called *universal*. It can simulate any given Turing machine, providing that a code of a particular one is written on the tape of the universal one, together with a string to be dealt with by the particular machine.

The parallelism with a computer, as we know the computers in their general form, is clear: the code of a Turing machine is its *program*, the strings to be recognized are the input data, the universal Turing machine is the computer itself.

Let us stress here an important distinction, that between computational *completeness* and *universality*. Given a class $\mathcal{C}$ of computability models, we say that $\mathcal{C}$ is *computationally complete* if the devices in $\mathcal{C}$ can characterize the power of Turing machines (or of any other type of equivalent devices). This means that given a Turing machine M we can find an element C in $\mathcal{C}$ such that C is equivalent with M. Thus, completeness refers to the capacity of covering the level of computability (in grammatical terms, this means to generate all recursively enumerable languages). Universality is an internal property of $\mathcal{C}$ and it means the existence of a fixed element of $\mathcal{C}$ which is able to simulate any given element of $\mathcal{C}$, in the way described above for Turing machines.

Of course, we can define the completeness in a relative way, not referring to the whole class of Turing machines but to a subclass of them. For instance, we can look for context-free completeness (the possibility of generating all context-free languages). Accordingly, we can look for universal elements in classes of computing devices which are computationally complete for smaller families of languages than the recursively enumerable languages. However, important for any theory which attempts to provide general models of computing are the completeness and universality with respect to Turing machines, and this will be the level we shall consider in this book.

The idea of a universal Turing machine was introduced by Turing himself, who has also produced such a machine [212]. Many universal Turing machines are now available in the literature, in general looking for simple (if not minimal) examples from different points of view. We present below some of them, for the case when Turing machines are considered as devices which compute mappings (see again Sect. 3.1). In such a framework, we say that a Turing machine is *universal* if it computes a universal partial recursive function (modulo the coding-decoding "interface" mentioned in Sect. 3.1). Similarly, a Turing machine M_1 *simulates* a Turing machine M_2 if there are two coding-decoding mappings

$$C : ID_{M_2} \longrightarrow ID_{M_1},\ D : ID_{M_1} \longrightarrow ID_{M_2},$$

such that for each $\alpha \in ID_{M_2}$ we have

$$D(F_{M_1}(C(\alpha))) = F_{M_2}(\alpha).$$

The complexity of a Turing machine can be evaluated from various points of view: the number of *states*, the number of *tape symbols* (the blank symbol included), or the number of *moves* (quintuples (s, a, b, d, s') such that $(s', b, d) \in \delta(s, a)$).

We denote by $UTM(m,n)$ the class of universal deterministic Turing machines with m states and n symbols (because we must have halting configurations, there can exist at most $m \cdot n - 1$ moves).

Small universal Turing machines were produced already in [205] (with two states) and [138] (with seven states and four symbols). The up-to-date results in this area are summarized in [185]:

Theorem 3.20. (i) *The classes* $UTM(2,3), UTM(3,2)$ *are empty.* (ii) *The following classes are non-empty:* $UTM(24,2)$, $UTM(10,3)$, $UTM(7,4)$, $UTM(5,5)$, $UTM(4,6)$, $UTM(3,10)$, $UTM(2,18)$.

Therefore, the problem is open for 51 classes $UTM(m,n)$.

We recall from [185] three examples of universal Turing machines, from the classes $UTM(7,4), UTM(5,5), UTM(4,6)$; the last one has the smallest number of moves. Because the machines are deterministic, we present them in a tabular way: for a quintuple (s,a,b,d,s'), an entry bds' will appear at the intersection of the row marked with s and the column marked with a. The states will be always denoted with $s_0, s_1, \ldots, s_m$ and the blank symbol with B. (We do not present here the coding-decoding mappings C, D; the reader is referred to [185] for details.)

One sees that these machines contain 26, 23, and 22 moves, respectively. These are the best results known ([185]).

Table 3.2. A Turing machine in $UTM(7,4)$

	B	1	a	b
s_0	BLs_0	BLs_0	bRs_1	aLs_0
s_1	$1Rs_1$	BLs_0	bRs_1	$1Rs_4$
s_2	$1Ls_3$	$1Rs_2$	bRs_2	aRs_2
s_3	$1Ls_6$	$1Ls_3$	bLs_3	aLs_3
s_4	bLs_3	$1Rs_4$	bRs_4	aRs_4
s_5	BRs_4	BRs_5	aRs_5	BRs_0
s_6	BRs_2	–	aLs_5	–

Table 3.3. A Turing machine in $UTM(5,5)$

	B	0	1	a	b
s_0	bRs_0	$1Rs_0$	$0Ls_0$	$0Rs_1$	BLs_0
s_1	$0Ls_3$	$0Rs_1$	$0Rs_1$	aRs_1	bRs_1
s_2	BRs_4	aLs_3	$0Rs_2$	aRs_2	bRs_2
s_3	bLs_2	$1Ls_3$	$0Rs_1$	aLs_3	bLs_3
s_4	–	–	$1Rs_4$	$1Rs_0$	BRs_4

Table 3.4. A Turing machine in $UTM(4,6)$

	B	1	a	b	b'	b''
s_0	$b''Ls_0$	$b''Ls_0$	$0Rs_3$	$b'Rs_0$	bLs_0	$0Rs_0$
s_1	$1Ls_1$	BRs_1	bRs_1	$b'Ls_2$	$b''Rs_1$	$b'Ls_1$
s_2	aRs_0	$1Rs_2$	$1Rs_0$	$b''Rs_3$	bRs_2	–
s_3	aLs_1	$0Rs_3$	bRs_3	aLs_1	$b''Rs_3$	–

In most of the constructions on which the proofs in the subsequent chapters are based, we shall start from a Chomsky type-0 grammar.

Given a Turing machine M we can effectively construct a type-0 grammar G such that $L(M) = L(G)$. (Similarly, we can produce a type-0 grammar G such that G computes, in a natural way and using appropriate coding-decoding mappings, the same mapping F_M as M. So, a grammar can be considered a function computing device, not only a language generating mechanism.)

The idea is very simple. Take a Turing machine $M = (K, V, T, B, s_0, F, \delta)$ and construct a non-restricted Chomsky grammar G working as follows: starting from its axiom, G nondeterministically generates a string w over V, then it makes a copy of w (of course, the two copies of w are separated by a suitable marker; further markers, scanners and other auxiliary symbols are allowed, because they can be erased when they are no longer necessary). On one of the copies of w, G can simulate the work of M, choosing nondeterministically a computation as defined by δ; if a final state is reached, then the witness copy of w is preserved, everything else is erased.

For the sake of the completeness, we present the details of such a construction.

Consider a deterministic Turing machine $M = (K, V, T, B, s_0, F, \delta)$ and construct the grammar

$$G = (N, T, S, P),$$

where

$$N = \{[a, b] \mid a \in T \cup \{\lambda\}, b \in V\} \cup \{S, X, Y\} \cup K,$$

and P contains the following rules:

1) $S \to s_0 X$,
2) $X \to [a, a]X$, for $a \in T$,
3) $X \to Y$,
4) $Y \to [\lambda, B]Y$,
5) $Y \to \lambda$,
6) $s[a, \alpha] \to [a, \beta]s'$, for $a \in T \cup \{\lambda\}, s, s' \in K, \alpha, \beta \in V$, such that $\delta(s, \alpha) = (s', \beta, R)$,
7) $[b, \gamma]s[a, \alpha] \to s'[b, \gamma][a, \beta]$, for $\alpha, \beta, \gamma \in V, a, b \in T \cup \{\lambda\}, s, s' \in K$,

such that $\delta(s, \alpha) = (s', \beta, L)$,

8) $[a, \alpha]s \rightarrow sas$,

$s[a, \alpha] \rightarrow sas$,

$s \rightarrow \lambda$, for $a \in T \cup \{\lambda\}, \alpha \in V, s \in F$.

The reader can easily check that $L(G) = L(M)$.

Applying this construction to a universal Turing machine M_u, we obtain a *universal type-0 Chomsky grammar* G_u, a grammar which is universal in the following sense: the language generated by G_u consists of strings of the form, say, $w\#code(M)$, such that $w \in L(M)$. (We can call the language $\{w\#code(M) \mid w \in L(M)\}$ itself universal, and thus any grammar generating this language is universal.) However, we are interested in a "more grammatical" notion of universality, and this leads to the following definition.

A triple $G = (N, T, P)$, where the components N, T, P are as in a usual Chomsky grammar is called a *grammar scheme*. For a string $w \in (N \cup T)^*$ we define the language $L(G, w) = \{x \in T^* \mid w \Longrightarrow^* x\}$, the derivation being performed according to the productions in P.

A *universal type-0 grammar* is a grammar scheme $G_u = (N_u, T_u, P_u)$, where N_u, T_u are disjoint alphabets, and P_u is a finite set of rewriting rules over $N_u \cup T_u$, with the property that for any type-0 grammar $G = (N, T_u, S, P)$ there is a string $w(G)$ such that $L(G_u, w(G)) = L(G)$.

Therefore, the universal grammar simulates any given grammar, provided a code $w(G)$ of the given grammar is taken as a starting string of the universal one.

There are universal type-0 grammars in the sense specified above. Because this assertion is fundamental for the investigations in the following chapters, we prove it with full details.

Let $G = (N, T, S, P)$ be a type-0 grammar. Without loss of the generality, we may assume that N contains only three nonterminals, $N = \{S, A, B\}$. (If we have more nonterminals, say $S, X_1, X_2, \ldots, X_n$, for $n \geq 3$, then we systematically replace each X_i appearing in a rule of P with AB^iA, $1 \leq i \leq n$, for some new symbols A, B. The obtained grammar is obviously equivalent with the original grammar.)

We construct the grammar scheme

$$G_u = (N_u, T, P_u),$$

with

$$\begin{aligned} N_u &= \{A, B, C, D, E, F, H, R, Q, S, X, Y\} \\ &\cup \{[a, i] \mid a \in T, 1 \leq i \leq 9\}, \end{aligned}$$

and the set P_u contains the following rules:

$$
\begin{array}{lll}
(I) & (1)\ C \to BQ, & \\
 & (2)\ Q\alpha \to \alpha Q, & \text{for } \alpha \in N \cup T \cup \{D,E\}, \\
(II) & (3)\ QD\alpha \to [\alpha,2]D[\alpha,1], & \text{for } \alpha \in N \cup T, \\
 & (4)\ \alpha[\beta,2] \to [\beta,2]\alpha, & \text{for } \alpha \in N \cup T \cup \{D,E\}, \\
 & & \beta \in N \cup T, \\
 & (5)\ B[\alpha,2] \to [\alpha,3]B, & \text{for } \alpha \in N \cup T, \\
 & (6)\ \alpha[\beta,3] \to [\beta,3]\alpha, & \text{for } \alpha,\beta \in N \cup T, \\
 & (7)\ \alpha[\alpha,3] \to [\alpha,4], & \text{for } \alpha \in N \cup T, \\
(III) & (8)\ [\alpha,1]\beta \to [\beta,5][\alpha,1][\beta,1], & \text{for } \alpha,\beta \in N \cup T, \\
 & (9)\ \alpha[\beta,5] \to [\beta,5]\alpha, & \text{for } \alpha \in N \cup T \cup \{D,E\}, \\
 & & \beta \in N \cup T, \\
 & (10)\ B[\alpha,5] \to [\alpha,6]B, & \text{for } \alpha \in N \cup T, \\
 & (11)\ \alpha[\beta,6] \to [\beta,6]\alpha, & \text{for } \alpha,\beta \in N \cup T, \\
 & (12)\ [\alpha,4]\beta[\beta,6] \to [\alpha,4], & \text{for } \alpha,\beta \in N \cup T, \\
(IV) & (13)\ [\alpha,1]E\beta \to [\alpha,7]E[\beta,9], & \text{for } \alpha,\beta \in N \cup T, \\
 & (14)\ [\alpha,1][\beta,7] \to [\alpha,7]\beta, & \text{for } \alpha,\beta \in N \cup T, \\
 & (15)\ D[\alpha,7] \to D\alpha, & \text{for } \alpha \in N \cup T, \\
 & (16)\ [\alpha,9]\beta \to [\beta,8][\alpha,9][\beta,9], & \text{for } \alpha,\beta \in N \cup T, \\
 & (17)\ \alpha[\beta,8] \to [\beta]\alpha, & \text{for } \alpha \in N \cup T \cup \{B,D,E\}, \\
 & & \beta \in N \cup T, \\
 & (18)\ [\alpha,9][\beta,8] \to [\beta,8][\alpha,9], & \text{for } \alpha,\beta \in N \cup T, \\
 & (19)\ [\alpha,4][\beta,8] \to \beta[\alpha,4], & \text{for } \alpha,\beta \in N \cup T, \\
(V) & (20)\ [\alpha,1]ED \to [\alpha,7]RED, & \text{for } \alpha \in N \cup T, \\
(VI) & (21)\ [\alpha,9]D \to R\alpha D, & \text{for } \alpha \in N \cup T, \\
 & (22)\ [\alpha,9]R \to R\alpha, & \text{for } \alpha \in N \cup T, \\
 & (23)\ \alpha R \to R\alpha, & \text{for } \alpha \in N \cup T \cup \{D,E\}, \\
 & (24)\ BR \to RC, & \\
 & (25)\ [\alpha,4]R \to \lambda, & \text{for } \alpha \in N \cup T \\
(VII) & (26)\ A\alpha \to \alpha A, & \text{for } \alpha \in T, \\
 & (27)\ AC \to H, & \\
 & (28)\ H\alpha \to H, & \text{for } \alpha \in N \cup T \cup \{D,E\}, \\
 & (29)\ HF \to \lambda. &
\end{array}
$$

Assume that $P = \{u_i \to v_i \mid 1 \le i \le k\}$ and consider the string

$$code(G) = ASCDu_1Ev_1Du_2Ev_2D\dots Du_kEv_kDF.$$

Let us first examine how the grammar scheme G_u works on a string of the form $AwCDu_1Ev_1D\dots Du_kEv_kDF$.

Group (I) of rules introduces the nonterminal Q which selects a rule $u_i \to v_i$ occurring in the right hand of a nonterminal D (by rule (3)). By the second group of rules, the first symbol α in u_i is transformed in $[\alpha,1]$ and the copy nonterminal $[\alpha,2]$ is moved to the left hand of B, where it becomes

$[\alpha, 3]$. If in w there exists an occurrence of α, then by rule (7) we introduce the nonterminal $[\alpha, 4]$ in order to encode this information.

The rules of group (III) transform all symbols β from u_i in $[\beta, 1]$ and then any such symbol β is erased from the right hand of $[\alpha, 4]$, if this can be performed in the correct order. In this way, the occurrence of u_i is identified in w.

By rules of group (IV) each symbol β of $v_i \neq \lambda$ is transformed into $[\beta, 9]$ and a copy $[\beta, 8]$ is introduced which is moved to the left of the symbol B. When $[\beta, 8]$ reaches the symbol $[\alpha, 4]$ one introduces the symbol β. In this way, the string u_i, erased by the rules in group (III), is replaced by v_i. If $v_i = \lambda$, then rule (20) is used instead of rules in group (IV). In this way we obtain a derivation step $w \Longrightarrow w'$ using the rule $u_i \rightarrow v_i$.

This procedure can be repeated, due to the rules in group (VI). If w' contains no nonterminal occurrence, then by rules in group (VII) we erase all auxiliary symbols, leaving only the terminal string w'.

Consequently, $L(G) \subseteq L(G_u, code(G))$.

In order to prove the reverse inclusion, let us observe that the nonterminal A can be eliminated only when in between A and C there is a terminal string. Every derivation must begin by the introduction of the nonterminal Q. The erasing of this symbol implies the introduction of a nonterminal $[\alpha, 1]$, which can be eliminated only by replacing it with the nonterminal $[\alpha, 9]$. These operations are possible if and only if a string u_i was erased from w. The removing of $[\alpha, 9]$ is possible after writing v_i instead of the erased u_i. The symbol R introduced in this way can be eliminated only when the string reaches the form $Aw'CDu_1Ev_1D \dots Du_kEv_kDF$. In this way we simulate a derivation using the rule $u_i \rightarrow v_i$. All derivations in G_u which are not of this form are blocked. Thus, the inclusion $L(G_u, code(G)) \subseteq L(G)$ is obtained, completing the proof of the equality $L(G_u, code(G)) = L(G)$. □

Note that the universal grammar G_u constructed above codifies the way of using a grammar in a derivation process: choose a rule, remove an occurrence of its left hand member, introduce instead an occurrence of its right hand member, check whether or not a terminal string is obtained.

A natural question here, also important for molecular computing, is whether or not universality results hold also for other classes of automata and grammars than Turing machines and type-0 grammars, in particular for finite automata.

If the question is understood in the strict sense, then the answer is negative for finite automata: no finite automaton can be universal for the class of all finite automata. There are two reasons for this. Firstly, one cannot encode the way of using a finite automaton in terms of a finite automaton (we have to remember symbols in the input string without marking them, and this cannot be done with a finite set of states). Secondly, we cannot codify the states of any given finite automaton in such a way that the finite set of states of the universal automaton can handle them (an arbitrarily large set of states will lead to arbitrarily long codes, hence again the information carried by them cannot be handled by a finite set of states).

However, for our purposes it is sufficient to construct a universal automaton in the following relaxed sense. Consider the class of finite automata whose state set and input alphabet are subsets of fixed finite sets K and V, respectively. A universal finite automaton for this class can be constructed.

We consider the following finite automaton

$$M_u = (K_u, V \cup K \cup \{c_1, c_2\}, q_{0,u}, F_u, P_u),$$

where

$$\begin{aligned} K_u &= \{q_{0,u}, q'_{0,u}, q''_{0,u}, q_f\} \\ &\cup \{[s], (s), (s)', (s)'', \overline{(s)}, \overline{(s)}', \overline{(s)}'', \overline{(s)}''' \mid s \in K\} \\ &\cup \{[sa], [sas'], [sas']', [sas']'', [sas']''', [sas']^{iv} \mid s, s' \in K, a \in V\}, \\ F_u &= \{q_f\}, \end{aligned}$$

and P_u contains the following transition rules:

1. $q_{0,u}s \to sq'_{0,u}, \ s \in K,$
   $q'_{0,u}a \to aq''_{0,u}, \ a \in V,$
   $q''_{0,u}s \to sq_{0,u}, \ s \in K,$
2. $q_{0,u}s_0 \to s_0[s_0],$
   $[s_0]a \to a[s_0a], \ a \in V,$
   $[s_0a]s \to s[s_0as], \ s \in K, a \in V,$
3. $[sas']s'' \to s''[sas']', \ s, s', s'' \in K, a \in V,$
   $[sas']'b \to b[sas']'', \ s, s' \in K, a, b \in V,$
   $[sas']''s'' \to s''[sas'], \ s, s', s'' \in K, a \in V,$
   $[sas']c_1 \to c_1[sas']''', \ s, s' \in K, a \in V,$
   $[sas']'''s'' \to s''[sas']''', \ s, s', s'' \in K, a \in V,$
   $[sas']'''c_2 \to c_2[sas']^{iv}, \ s, s' \in K, a \in V,$
4. $[sas']^{iv}a \to a(s'), \ s, s' \in K, a \in V,$
5. $(s)s' \to s'(s)', \ s, s' \in K,$
   $(s)'a \to a(s)'', \ s \in K, a \in V,$
   $(s)''s' \to s'(s), \ s, s' \in K,$
6. $(s)s \to s[s], \ s \in K,$
   $[s]a \to a[sa], \ s \in K, a \in V,$
   $[sa]s' \to s'[sas'], \ s, s' \in K, a \in V,$
7. $(s)s' \to s'\overline{(s)}', \ s, s' \in K,$
   $\overline{(s)}s' \to s'\overline{(s)}', \ s, s' \in K,$

$$\overline{(s)}'a \to a\overline{(s)}'', \ s \in K, a \in V,$$
$$\overline{(s)}''s' \to s'\overline{(s)}, \ s, s' \in K,$$
$$\overline{(s)}c_1 \to c_1\overline{(s)}''', \ s \in K,$$
$$\overline{(s)}'''s' \to s'\overline{(s)}''', \ s, s' \in K, s \neq s',$$
$$\overline{(s)}'''s \to sq_f, \ s \in F,$$
$$q_f s \to sq_f, \ s \in K,$$
$$q_f c_2 \to c_2 q_f.$$

For a finite automaton $M = (K, V, s_0, F, P)$ let us consider the string

$$code(M) = s_1 a_1 s_1' s_2 a_2 s_2' \dots s_n a_n s_n' c_1 s_{f1} s_{f2} \dots s_{fm} c_2,$$

where $s_i a_i \to a_i s_i' \in P, 1 \leq i \leq n$, each string $s_i a_i s_i'$ appears only once, $s_{f1}, s_{f2}, \dots, s_{fm}$ are the elements of F, and c_1, c_2 are new symbols.

For two strings $z, x \in V^*$ with $x = a_1 a_2 \dots a_p, a_i \in V, 1 \leq i \leq p$, we define the *block shuffle operation* of z, x by

$$bls(z, x) = za_1 za_2 \dots za_p z.$$

The automaton M_u is universal for the class of finite automata of the form $M = (K', V', s_0, F, P)$ with $K' \subseteq K, V' \subseteq V$, in the following sense:

$$bls(code(M), x) \in L(M_u) \text{ iff } x \in L(M).$$

Indeed, M_u works as follows: in the initial state $q_{0,u}$ and in each state (s), one parses $code(M)$ in such a way that some blocks $s_i a_i s_i'$ are skipped, then one block of this type is memorized (when starting, we must have $s_i = s_0$) in the state $[s_i a_i s_i']$, then further blocks $s_j a_j s_j'$ are skipped after passing also over $c_1 s_{f_1} \dots s_{f_m} c_2$ (thus reaching the state $[s_i a_i s_i']^{iv}$); the rule $s_i a_i \to a_i s_i'$ of M is simulated by a rule of type 4, returning to a state of the form (s); the process can be iterated; using rules in group 7, we reach q_f only when we have simulated in M_u a parsing in M ending in a state of F.

We conclude with the following two observations concerning the universal automaton M_u. Firstly, the above construction remains unaltered if K and V are infinite sets. Thus, M_u can be considered universal for all finite automata. However, this modified M_u is not a finite automaton, although its schematic definition is very simple. (Such modifications of finite automata were often discussed in the early days of automata theory.)

Secondly, the description of the individual automata to be simulated, $code(M)$, appears numerous times in the input for M_u. This drawback can be remedied by making M_u a two-way automaton with two tapes. The details of such a construction are omitted.

3.4 Bibliographical Notes

Several monographs in formal language theory were mentioned in Sect. 3.1. Others can be found in the bibliography.

We do not specify here the origin of the results mentioned in Sect. 3.1.

Theorem 3.12 is classic, it appears in most formal language theory monographs. Theorem 3.13 appears in [110]; Corollary 3.3 is first established in [11], where a previous result is strenghtened (saying that each recursively enumerable language is the morphic image of the intersection of two context-free languages). Results like Theorem 3.14 were given in [145], [188], [218].

Characterizations of recursively enumerable languages starting from equality sets of morphisms were given in [199], [34]; the construction in the proof of Theorem 3.15 is from [200], where complete details can be found. The variant of Theorem 3.15 given in Theorem 3.16 is proved in [101]. Theorem 3.17 and its Corollary 3.4 were proved in [54]. Theorem 3.18 is from [53]. Similar (sometimes slightly weaker) results appear also in [18]. Semi-twin-shuffle languages were considered in [131], where Theorem 3.19 and Corollary 3.5 are given.

Universal Turing machines can be found in [138], [205], and, mainly, in [185]. The construction of the universal type-0 grammar from Sect. 3.3 is from [20]; it also appears in [19].

Chapter 4

Sticker Systems

Data structures basic in language theory are *words*, that is, strings of elements, letters. Here the idea of a "string" entails a *linear order* among the elements. The double helix of DNA, when presented in two dimensions as we have already done, constitutes a data structure of a new kind: a double strand. While both strands still are linear strings of elements, the double strand possesses an important additional property. The *paired* elements in the strands are *complementary* with respect to a given symmetric relation. We have already discussed the interconnection between this Watson–Crick complementarity and the *twin-shuffle language*. The computational capacity of the latter has also been pointed out. In the next two chapters intensive use will be made of these two facts, the interconnection and computational capacity, for DNA computing. Our previous characterizations of recursively enumerable languages, based on equality sets and twin-shuffle languages, find here very natural applications.

4.1 The Operation of Sticking

We start by a formalization of the basic operation we shall use, in which we can build double stranded sequences starting from "DNA dominoes," sequences with sticky ends at one or at both their ends, or single stranded sequences, which, by ligation and annealing, stick to each other.

Consider an alphabet V and a symmetric relation $\rho \subseteq V \times V$ over V (of *complementarity*).

The property of symmetry is not used below, but we consider it because Watson–Crick complementarity is symmetric (in general, the intuitive idea of complementarity assumes the symmetry).

Besides the monoid V^*, of strings over V, we associate with V also the monoid $V^* \times V^*$, of pairs of strings. In accordance with the way of representing DNA molecules, where one considers the two strands placed one over the other, we also write the elements $(x_1, x_2) \in V^* \times V^*$ in the form $\binom{x_1}{x_2}$. Therefore, the concatenation of two pairs $\binom{x_1}{x_2}, \binom{y_1}{y_2}$ is $\binom{x_1 y_1}{x_2 y_2}$. We also write $\binom{V^*}{V^*}$ instead of $V^* \times V^*$.

The identity element $\binom{\lambda}{\lambda}$ of $\binom{V^*}{V^*}$ is often identified with λ, and omitted when it is not significant in a given context.

We also denote

$$\begin{bmatrix} V \\ V \end{bmatrix}_\rho = \{ \begin{bmatrix} a \\ b \end{bmatrix} \mid a, b \in V, (a, b) \in \rho \},$$

$$WK_\rho(V) = \begin{bmatrix} V \\ V \end{bmatrix}_\rho^* .$$

The set $WK_\rho(V)$ is called the *Watson–Crick domain* associated to the alphabet V and the complementarity relation ρ. The elements $\begin{bmatrix} a_1 \\ b_1 \end{bmatrix} \begin{bmatrix} a_2 \\ b_2 \end{bmatrix} \cdots \begin{bmatrix} a_n \\ b_n \end{bmatrix} \in WK_\rho(V)$ are also written in the form $\begin{bmatrix} w_1 \\ w_2 \end{bmatrix}$, for $w_1 = a_1 a_2 \ldots a_n, w_2 = b_1 b_2 \ldots b_n$. We call such elements $\begin{bmatrix} w_1 \\ w_2 \end{bmatrix} \in WK_\rho(V)$ well-formed double stranded sequences, or simply *double stranded sequences*, or *molecules*, in order to remind us of the reality they are modeling. The two component strings, w_1, w_2, are also called *strands*; w_1 is the *upper* strand and w_2 is the *lower* strand.

By the definition of $WK_\rho(V)$, $\begin{bmatrix} \lambda \\ \lambda \end{bmatrix}$ is also a molecule (although it has no biochemical representation). In this way, $WK_\rho(V)$ is a monoid. For any two elements $\begin{bmatrix} x_1 \\ x_2 \end{bmatrix}, \begin{bmatrix} y_1 \\ y_2 \end{bmatrix}$ in $WK_\rho(V)$, the sequence $\begin{bmatrix} x_1 y_1 \\ x_2 y_2 \end{bmatrix}$ is well formed, hence it is in $WK_\rho(V)$.

Note the essential difference between $\binom{x}{y}$ and $\begin{bmatrix} x \\ y \end{bmatrix}$: $\binom{x}{y}$ is just another notation for the pair (x, y), that is, no relation is assumed between the symbols appearing in x and y, whereas $\begin{bmatrix} x \\ y \end{bmatrix}$ represents a molecule, with a precise bonding between the corresponding symbols in the two strands. This bonding is defined by the complementarity relation ρ on the alphabet V, hence in order to specify all these details we shall usually write $\begin{bmatrix} x \\ y \end{bmatrix} \in WK_\rho(V)$, although the expression $\begin{bmatrix} x \\ y \end{bmatrix}$ tells us that we have a molecule.

We emphasize the two properties characterizing the elements $\begin{bmatrix} w_1 \\ w_2 \end{bmatrix}$ of

$WK_\rho(V)$, because they are essential for the models considered in this and the next chapter:

- the two strands w_1, w_2 are of the same length,
- the corresponding symbols in the two strands are complementary in the sense of the relation ρ.

These properties are rather strong. We shall see that using elements of $WK_\rho(V)$ one can easily obtain characterizations of RE. As a matter of fact, an intersection is "incorporated" in the definition of a Watson–Crick domain. However, this strength is provided to us "for free" by the DNA molecules: they *are* well-formed double stranded sequences, with the correctness checked in a natural way, where "natural" refers directly to the nature.

We shall also use below "incomplete molecules," that is elements in the set

$$W_\rho(V) = L_\rho(V) \cup R_\rho(V) \cup LR_\rho(V),$$

where

$$L_\rho(V) = (\binom{\lambda}{V^*} \cup \binom{V^*}{\lambda}) \begin{bmatrix} V \\ V \end{bmatrix}_\rho^*,$$

$$R_\rho(V) = \begin{bmatrix} V \\ V \end{bmatrix}_\rho^* (\binom{\lambda}{V^*} \cup \binom{V^*}{\lambda}),$$

$$LR_\rho(V) = (\binom{\lambda}{V^*} \cup \binom{V^*}{\lambda}) \begin{bmatrix} V \\ V \end{bmatrix}_\rho^+ (\binom{\lambda}{V^*} \cup \binom{V^*}{\lambda}).$$

Here, when we write, for instance, $\binom{u}{\lambda} \begin{bmatrix} x \\ y \end{bmatrix}$, this is just an expression obtained by concatenating the two symbols $\binom{u}{\lambda}$ and $\begin{bmatrix} x \\ y \end{bmatrix}$. This cannot be replaced, say, by $\binom{ux}{y}$, because we lose the complementarity relations between the symbols in x, y; $\binom{ux}{y}$ is only a pair of strings, whereas $\begin{bmatrix} x \\ y \end{bmatrix}$ is a molecule. If $u \neq \lambda$, then $\begin{bmatrix} ux \\ y \end{bmatrix}$ is simply undefined.

The possible shapes of elements in $W_\rho(V)$ are illustrated in Fig. 4.1. In all cases, we have a well-formed double stranded sequence x and overhangs y, z in one or two sides of x. These overhangs (sticky ends) can be placed in the upper strand or in the lower one. Note that in the case of $L_\rho(V)$ and $R_\rho(V)$, the block x may be empty, but in the elements of $LR_\rho(V)$ we have $x \in \begin{bmatrix} V \\ V \end{bmatrix}_\rho^+$, hence it contains at least one element $\begin{bmatrix} a \\ b \end{bmatrix}$ with $(a, b) \in \rho$. In turn, the overhangs can also be empty; what remains is then an element of $WK_\rho(V)$, therefore $WK_\rho(V)$ is included in each set $L_\rho(V), R_\rho(V), LR_\rho(V)$.

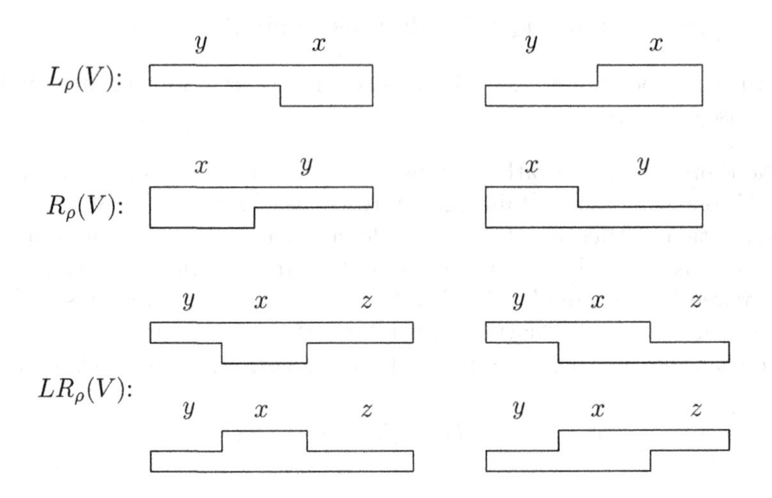

Figure 4.1: Possible shapes of "dominoes"

Any element of $W_\rho(V)$ which contains at least a position $\begin{bmatrix} a \\ b \end{bmatrix}$, $a \neq \lambda, b \neq \lambda$, is called a *well-started double stranded sequence*; of course, when several "columns" $\begin{bmatrix} a \\ b \end{bmatrix}$, with $(a, b) \in \rho$, appear, they appear consecutively. In general, the elements of $W_\rho(V)$ are also called *dominoes* (*polyominoes* could be more rigorous).

Among the elements of $W_\rho(V)$ we can define a partial operation, modeling the ligation or annealing operation: a well-started molecule (hence a sequence having at least a position filled in both of the two strands) can be prolonged to the right or to the left with a domino, providing that the sticky ends match, that is they are complementary in the corresponding positions. The result should always be a well-started molecule, hence a sequence which does not have empty places surrounded by symbols from V.

Specifically, consider a well-started molecule $x \in W_\rho(V)$, and a domino $y \in W_\rho(V)$. Being well-started, x can be written (obviously, in a unique way) in the form

$$x = x_1 x_2 x_3,$$

where $x_2 \in WK_\rho(V) - \{\begin{bmatrix} \lambda \\ \lambda \end{bmatrix}\}$ and $x_1, x_3 \in \begin{pmatrix} V^* \\ \lambda \end{pmatrix} \cup \begin{pmatrix} \lambda \\ V^* \end{pmatrix}$.

The *sticking* of x, y (in this order, non-commutatively) is defined and denoted by $\mu(x, y)$ in the following cases:

1) $x_3 = \begin{pmatrix} u \\ \lambda \end{pmatrix}, y = \begin{pmatrix} \lambda \\ v \end{pmatrix} y'$, for $u, v \in V^*$ such that

$\begin{bmatrix} u \\ v \end{bmatrix} \in WK_\rho(V)$ and $y' \in R_\rho(V)$; then $\mu(x,y) = x_1 x_2 \begin{bmatrix} u \\ v \end{bmatrix} y'$;

2) $x_3 = \begin{pmatrix} \lambda \\ v \end{pmatrix}, y = \begin{pmatrix} u \\ \lambda \end{pmatrix} y'$, for $u, v \in V^*$ such that

$\begin{bmatrix} u \\ v \end{bmatrix} \in WK_\rho(V)$ and $y' \in R_\rho(V)$; then $\mu(x,y) = x_1 x_2 \begin{bmatrix} u \\ v \end{bmatrix} y'$;

3) $x_3 = \begin{pmatrix} u_1 \\ \lambda \end{pmatrix}, y = \begin{pmatrix} u_2 \\ \lambda \end{pmatrix}$, for $u_1, u_2 \in V^*$;

then $\mu(x,y) = x_1 x_2 \begin{pmatrix} u_1 u_2 \\ \lambda \end{pmatrix}$;

4) $x_3 = \begin{pmatrix} u_1 u_2 \\ \lambda \end{pmatrix}, y = \begin{pmatrix} \lambda \\ v \end{pmatrix}$, for $u_1, u_2, v \in V^*$ such that

$\begin{bmatrix} u_1 \\ v \end{bmatrix} \in WK_\rho(V)$; then $\mu(x,y) = x_1 x_2 \begin{bmatrix} u_1 \\ v \end{bmatrix} \begin{pmatrix} u_2 \\ \lambda \end{pmatrix}$;

5) $x_3 = \begin{pmatrix} u \\ \lambda \end{pmatrix}, y = \begin{pmatrix} \lambda \\ v_1 v_2 \end{pmatrix}$, for $u, v_1, v_2 \in V^*$such that

$\begin{bmatrix} u \\ v_1 \end{bmatrix} \in WK_\rho(V)$; then $\mu(x,y) = x_1 x_2 \begin{bmatrix} u \\ v_1 \end{bmatrix} \begin{pmatrix} \lambda \\ v_2 \end{pmatrix}$;

6) $x_3 = \begin{pmatrix} \lambda \\ v_1 \end{pmatrix}, y = \begin{pmatrix} \lambda \\ v_2 \end{pmatrix}$, for $v_1, v_2 \in V^*$;

then $\mu(x,y) = x_1 x_2 \begin{pmatrix} \lambda \\ v_1 v_2 \end{pmatrix}$,

7) $x_3 = \begin{pmatrix} \lambda \\ v_1 v_2 \end{pmatrix}, y = \begin{pmatrix} u \\ \lambda \end{pmatrix}$, for $u, v_1, v_2 \in V^*$such that

$\begin{bmatrix} u \\ v_1 \end{bmatrix} \in WK_\rho(V)$; then $\mu(x,y) = x_1 x_2 \begin{bmatrix} u \\ v_1 \end{bmatrix} \begin{pmatrix} \lambda \\ v_2 \end{pmatrix}$;

8) $x_3 = \begin{pmatrix} \lambda \\ v \end{pmatrix}, y = \begin{pmatrix} u_1 u_2 \\ \lambda \end{pmatrix}$, for $u_1, u_2, v \in V^*$such that

$\begin{bmatrix} u_1 \\ v \end{bmatrix} \in WK_\rho(V)$; then $\mu(x,y) = x_1 x_2 \begin{bmatrix} u_1 \\ v \end{bmatrix} \begin{pmatrix} u_2 \\ \lambda \end{pmatrix}$.

These eight cases are illustrated in Fig. 4.2.

In the symmetric way we can define $\mu(y,x)$, the prolongation of a well-started molecule x, by a sequence y, to the left. Note that we do not need to distinguish the "left prolongation" from the "right prolongation" by denoting them in different ways: in any case at least one of the terms of the operation must be a well-started molecule and the result – it is a well-started molecule, too – entirely depends on the order of the two sequences and on their sticky ends.

Note that in all cases we also allow the prolongation of "blunt" ends, with empty overhangs. We always obtain a well-started double stranded molecule (with the subsequence in $WK_\rho(V)$ not necessarily strictly longer than the subsequence $x_2 \in WK_\rho(V)$ in x).

In cases 3 and 6 we do not use annealing (hence the complementarity relation); when $\mu(x, y)$ is defined without allowing such cases, to the right or to the left, we denote it by $\mu'(x, y)$ and we call this operation *restricted sticking*.

The maximal length of an overhang in a sequence $z \in W_\rho(V)$ is also called the *delay* of z and it is denoted by $d(z)$; it represents the delay in completing the two strands with symbols in V. (Hence, in cases 1, 2 in Fig. 4.2, the "right delay" of x and the "left delay" of y should coincide when y is also a well-started double stranded sequence.)

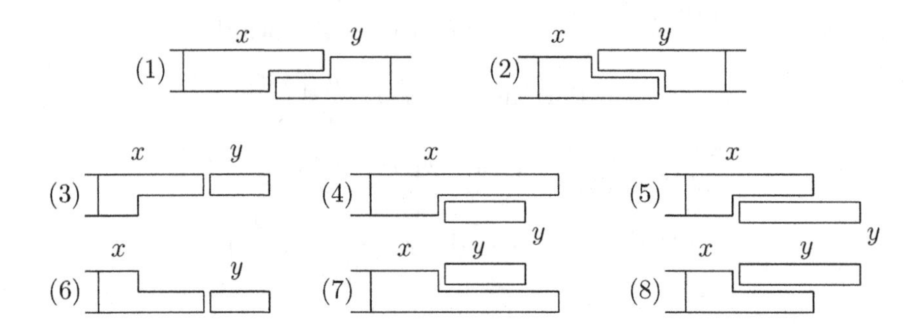

Figure 4.2: The sticking operation

In the same way that rewriting is the underlying operation for Chomsky grammars, the sticking operation is the underlying one for *sticker systems*, investigated in subsequent sections.

4.2 Sticker Systems; Classifications

We define here the sticker systems in their most general form: when building molecules, we start from well-started sequences and prolong them in both directions, using dominoes of arbitrary forms; the prolongation is done by means of the operation μ. We shall see below that systems of particular forms are equally powerful to general ones – modulo squeezing mechanisms such as weak codings or deterministic gsm mappings.

A *sticker system* is a construct

$$\gamma = (V, \rho, A, D),$$

where V is an alphabet, $\rho \subseteq V \times V$ is a symmetric relation, A is a finite subset of $LR_\rho(V)$, and D is a finite subset of $W_\rho(V) \times W_\rho(V)$.

The relation ρ is the complementarity relation on V, the elements of A are called *axioms*. Starting from these axioms and using the pairs (u, v) of dominoes in D, we can obtain a set of double stranded sequences in $WK_\rho(V)$, hence complete molecules, by using the operation μ of sticking.

Formally, for a given sticker system $\gamma = (V, \rho, A, D)$ and two sequences $x, y \in LR_\rho(V)$, we write

$$x \Longrightarrow y \text{ iff } y = \mu(u, \mu(x, v)), \text{ for some } (u, v) \in D.$$

Obviously,

$$\mu(u, \mu(x, v)) = \mu(\mu(u, x), v),$$

because the prolongation to the left is independent of the prolongation to the right.

A sequence $x_1 \Longrightarrow x_2 \Longrightarrow \ldots \Longrightarrow x_k$, with $x_1 \in A$, is called a *computation* in γ. A computation $\sigma : x_1 \Longrightarrow^* x_k$ is *complete* when $x_k \in WK_\rho(V)$ (no sticky end – hence blank symbol – is present in the last sequence).

The set of all molecules over V produced at the end of complete computations in γ is denoted by $LM_n(\gamma)$ (LM stands for "language of molecules" and the subscript n stands for "non-restricted": there is no restriction on the computations except that they are complete):

$$LM_n(\gamma) = \{w \in WK_\rho(V) \mid x \Longrightarrow^* w, x \in A\}.$$

In what follows we consider the sticker systems as generating languages of strings. To this aim, we associate with $LM_n(\gamma)$ the language

$$L_n(\gamma) = \{w \in V^* \mid \begin{bmatrix} w \\ w' \end{bmatrix} \in LM_n(\gamma) \text{ for some } w' \in V^*\}.$$

We say that $L_n(\gamma)$ is the language *generated* by γ (at the end of non-restricted complete computations).

Several types of restricted computations in γ are of interest.

A complete computation $x_1 \Longrightarrow x_2 \Longrightarrow \ldots \Longrightarrow x_k$ (hence with $x_1 \in A$ and $x_k \in WK_\rho(V)$) is said to be:

- *primitive*, if for no $i, 1 \leq i < k$, we have $x_i \in WK_\rho(V)$ (x_k is the first molecule in this computation);
- *of delay d*, if $d(x_i) \leq d$, for each $1 \leq i \leq k$.

We denote by $L_p(\gamma)$ and $L_d(\gamma), d \geq 1$, the language of strings generated by γ at the end of primitive computations and at the end of computations of delay at most d, respectively.

As above in the case of $L_n(\gamma)$, the languages $L_p(\gamma), L_d(\gamma)$ consist of strings in the upper strands of molecules generated by γ, but we do not elaborate on the difference between languages $L_\alpha(\gamma)$ and $LM_\alpha(\gamma)$, of molecules, $\alpha \in \{n, p, d\}$, because we do not investigate the languages of molecules here. On the other hand, the relation between $L_\alpha(\gamma)$ and $LM_\alpha(\gamma)$ depends on ρ: if we work with an injective mapping ρ, then $LM_\alpha(\gamma)$ is precisely identified by

$L_\alpha(\gamma)$; if ρ is arbitrary, but symmetric, then $L_\alpha(\gamma)$ is a coding of $LM_\alpha(\gamma)$ and $LM_\alpha(\gamma)$ is the image of $L_\alpha(\gamma)$ through an inverse coding.

Clearly, we have the following relations:

i) $L_p(\gamma) \subseteq L_n(\gamma)$,

ii) $L_{d_1}(\gamma) \subseteq L_{d_2}(\gamma)$, if $1 \leq d_1 \leq d_2$,

iii) $L_d(\gamma) \subseteq L_n(\gamma)$, for all $d \geq 0$.

A sticker system γ is said to have a *bounded delay* if there is $d \geq 1$ such that $L_d(\gamma) = L_n(\gamma)$.

Several restricted variants of sticker systems are also of interest. A system $\gamma = (V, \rho, A, D)$ is said to be:

- *one-sided*, if for each pair $(u, v) \in D$ we have either $u = \lambda$ or $v = \lambda$,
- *regular*, if for each pair $(u, v) \in D$ we have $u = \lambda$,
- *simple*, if all pairs $(u, v) \in D$ have either $u, v \in \begin{pmatrix} V^* \\ \lambda \end{pmatrix}$, or $u, v \in \begin{pmatrix} \lambda \\ V^* \end{pmatrix}$.

In one-sided systems, the prolongation to the left is independent of the prolongation to the right; in regular systems we only prolong the sequences to the right (hence the axioms must be of the form x_1x_2, with $x_1 \in WK_\rho(V)$ and $x_2 \in \begin{pmatrix} V^* \\ \lambda \end{pmatrix} \cup \begin{pmatrix} \lambda \\ V^* \end{pmatrix}$). In a computation in a simple sticker system we add symbols only to one of the two strands.

We denote by $ASL(\alpha)$ the family of languages of the form $L_\alpha(\gamma), \alpha \in \{n, p\}$, for γ a sticker system of an arbitrary form (SL stands for "sticker language" and A indicates the use of sticker systems of "arbitrary forms"); the family of languages generated by sticker systems of bounded delay is denoted by $ASL(b)$. When only sticker systems which are one-sided, regular, simple, simple and one-sided, or simple and regular are used, we replace A in front of $SL(\alpha)$ by O, R, S, SO, SR, respectively. We stress the fact that these families contain *string languages*, not languages of molecules, hence we can discuss their relationships with families in the Chomsky hierarchy without further precautions. This might not be the case with families of languages of the form $LM_\alpha(\gamma)$, because we must take care of the complementarity relation. As we have mentioned above, for an injective ρ, the language $LM_\alpha(\gamma)$ is isomorphic with $L_\alpha(\gamma)$, but if ρ is not injective then we have to pay attention to the coding relating $LM_\alpha(\gamma)$ and $L_\alpha(\gamma)$.

From the definitions, we obtain:

Lemma 4.1. *For each $\alpha \in \{n, p, b\}$ we have the relations in the diagram in Fig. 4.3, where the arrows indicate inclusions which are not necessarily proper.*

Lemma 4.2. *For each* $X \in \{A, O, R, S, SO, SR\}$ *we have* $XSL(b) \subseteq XSL(n)$.

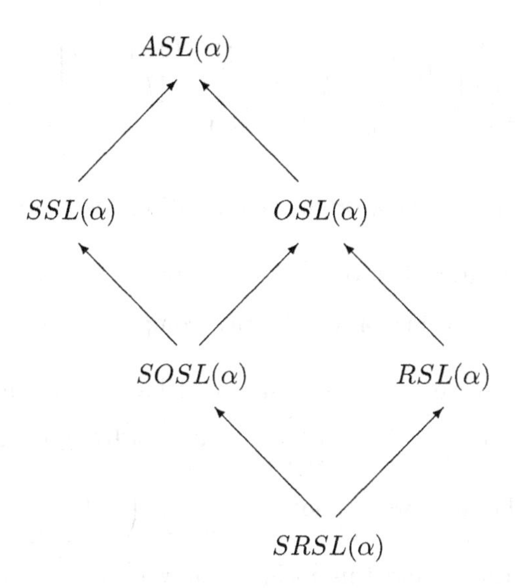

Figure 4.3: Relationships between families of languages generated by sticker systems (preliminary results)

Because we do not have erasing operations, we obtain in a straightforward way the following result.

Lemma 4.3. $XSL(\alpha) \subseteq CS$, *for all* X *and* α *as above.*

Before starting to investigate the size of the families $XSL(\alpha)$, let us examine two examples; consider first the simple sticker system

$$\begin{aligned}
&\gamma_1 = (V, \rho, A, D),\\
&V = \{a, b, c\},\\
&\rho = \{(a, a), (b, b), (c, c)\},\\
&A = \{\begin{bmatrix} a \\ a \end{bmatrix}\},\\
&D = \{(\begin{pmatrix} b \\ \lambda \end{pmatrix}, \begin{pmatrix} b \\ \lambda \end{pmatrix}), (\begin{pmatrix} c \\ \lambda \end{pmatrix}, \begin{pmatrix} \lambda \\ \lambda \end{pmatrix}), (\begin{pmatrix} \lambda \\ b \end{pmatrix}, \begin{pmatrix} \lambda \\ \lambda \end{pmatrix}), (\begin{pmatrix} \lambda \\ c \end{pmatrix}, \begin{pmatrix} \lambda \\ b \end{pmatrix})\}.
\end{aligned}$$

We have

$$WK_\rho(V) = \{\begin{bmatrix} a \\ a \end{bmatrix}, \begin{bmatrix} b \\ b \end{bmatrix}, \begin{bmatrix} c \\ c \end{bmatrix}\}^*,$$

but, because of the form of the pairs in D, the generated sequences can be only of the form $x \begin{bmatrix} a \\ a \end{bmatrix} \begin{bmatrix} b \\ b \end{bmatrix}^m$, with $x \in \{\begin{bmatrix} b \\ b \end{bmatrix}, \begin{bmatrix} c \\ c \end{bmatrix}\}^*$; moreover, x should contain exactly m occurrences of $\begin{bmatrix} b \\ b \end{bmatrix}$ and exactly m occurrences of $\begin{bmatrix} c \\ c \end{bmatrix}$. Consequently, $L_n(\gamma_1)$ is not context-free:

$$LM_n(\gamma_1) \cap \begin{bmatrix} c \\ c \end{bmatrix}^+ \begin{bmatrix} b \\ b \end{bmatrix}^+ \begin{bmatrix} a \\ a \end{bmatrix} \begin{bmatrix} b \\ b \end{bmatrix}^+ = \{\begin{bmatrix} c \\ c \end{bmatrix}^m \begin{bmatrix} b \\ b \end{bmatrix}^m \begin{bmatrix} a \\ a \end{bmatrix} \begin{bmatrix} b \\ b \end{bmatrix}^m \mid m \geq 1\},$$
$$L_n(\gamma_1) \cap c^+b^+ab^+ = \{c^m b^m a b^m \mid m \geq 1\}.$$

Working first in the upper strand, that is using the pairs $(\begin{pmatrix} b \\ \lambda \end{pmatrix}, \begin{pmatrix} b \\ \lambda \end{pmatrix})$, $(\begin{pmatrix} c \\ \lambda \end{pmatrix}, \begin{pmatrix} \lambda \\ \lambda \end{pmatrix})$, and after that on the lower strand, we can produce every sequence in $LM_n(\gamma_1)$ using only primitive computations, hence $L_p(\gamma_1) = L_n(\gamma_1)$.

This is not the case with the bounded delay restriction. Specifically, sequences of the form $\begin{bmatrix} c \\ c \end{bmatrix}^m \begin{bmatrix} b \\ b \end{bmatrix}^m \begin{bmatrix} a \\ a \end{bmatrix} \begin{bmatrix} b \\ b \end{bmatrix}^m$ cannot be produced with a delay smaller than m, because we cannot use $(\begin{pmatrix} \lambda \\ c \end{pmatrix}, \begin{pmatrix} \lambda \\ b \end{pmatrix})$ before having used $(\begin{pmatrix} \lambda \\ b \end{pmatrix}, \begin{pmatrix} \lambda \\ \lambda \end{pmatrix})$ m times, and this means that we have already produced a sequence of the form

$$\begin{pmatrix} \lambda \\ b^p \end{pmatrix} \begin{bmatrix} b \\ b \end{bmatrix}^q \begin{bmatrix} a \\ a \end{bmatrix} \begin{pmatrix} b^q \\ \lambda \end{pmatrix},$$

for $p + q = m$; that is, the delay is at least $\frac{m}{2}$.

We obtain $L_d(\gamma_1) \subset L_n(\gamma_1)$, for all $d \geq 1$. All these languages $L_d(\gamma_1)$ are linear. We do not prove this assertion for γ_1 above, but give a general result of this type in the next section.

Consider one more sticker system (this time not a simple one):

$$\gamma_2 = (V, \rho, A, D),$$
$$V = U \cup \overline{U} \cup U', \text{ for some alphabet } U,$$
$$\rho = \{(a, a), (\bar{a}, \bar{a}), (a', a') \mid a \in U\},$$
$$A = \{\begin{bmatrix} a_0' \\ a_0' \end{bmatrix}\}, \text{ for some fixed } a_0 \in U,$$
$$D = \{(\begin{pmatrix} a' \\ \lambda \end{pmatrix}, \begin{bmatrix} a \\ a \end{bmatrix}), (\begin{pmatrix} \lambda \\ a' \end{pmatrix}, \begin{bmatrix} \bar{a} \\ \bar{a} \end{bmatrix}) \mid a \in U\}.$$

We start from $\begin{bmatrix} a_0' \\ a_0' \end{bmatrix}$ for the fixed $a_0 \in U$ and we build a molecule by adding columns $\begin{bmatrix} a \\ a \end{bmatrix}, \begin{bmatrix} \bar{a} \\ \bar{a} \end{bmatrix}$ to the right hand side of it and, simultaneously with $\begin{bmatrix} a \\ a \end{bmatrix}$ and $\begin{bmatrix} \bar{a} \\ \bar{a} \end{bmatrix}$, respectively, we add symbols a' in the upper strand and symbols a' in the lower strand to the left of $\begin{bmatrix} a'_0 \\ a'_0 \end{bmatrix}$. This means that, modulo the primes and the bars, the sequence of columns $\begin{bmatrix} a' \\ a' \end{bmatrix}$ to the left of $\begin{bmatrix} a'_0 \\ a'_0 \end{bmatrix}$ is identical to the reversed sequence of columns $\begin{bmatrix} a \\ a \end{bmatrix}$ to the right of $\begin{bmatrix} a'_0 \\ a'_0 \end{bmatrix}$, and the same sequence $\begin{bmatrix} a' \\ a' \end{bmatrix}$ to the left of $\begin{bmatrix} a'_0 \\ a'_0 \end{bmatrix}$ is identical to the reversed sequence of columns $\begin{bmatrix} \bar{a} \\ \bar{a} \end{bmatrix}$ to the right of $\begin{bmatrix} a'_0 \\ a'_0 \end{bmatrix}$. Consequently,

$$LM_n(\gamma_2) = \{\begin{bmatrix} x' \\ x' \end{bmatrix} \begin{bmatrix} a'_0 \\ a'_0 \end{bmatrix} \begin{bmatrix} w \\ w \end{bmatrix} \mid x \in U^*, mi(w) \in x \sqcup\!\sqcup \bar{x}\},$$
$$L_n(\gamma_2) = \{x'a'_0w \mid x \in U^*, mi(w) \in x \sqcup\!\sqcup \bar{x}\},$$

where $x', \bar{x}$ are the primed and the barred versions of $x \in U^*$, respectively.

Therefore, the twin-shuffle language over U is obtained to the right of $\begin{bmatrix} a'_0 \\ a'_0 \end{bmatrix}$, on both strands, together with the copy of the shuffled strings which is present in a primed version in both strands to the left of $\begin{bmatrix} a'_0 \\ a'_0 \end{bmatrix}$. (Because $mi(U^*) = U^*$, the mirror image operation can be ignored.)

Using Theorem 3.17, we get in this way a representation of recursively enumerable languages as gsm images of languages in the family $ASL(n)$. In Sect. 4.4 we shall give a stronger variant of this result.

4.3 The Generative Capacity of Sticker Systems

In this section we are looking for the relationships between the families $XSL(\alpha)$, $X \in \{A, O, R, S, SO, SR\}$, $\alpha \in \{n, p, b\}$, and for their relationships with families in the Chomsky hierarchy.

We start with some estimations from above.

Theorem 4.1. $OSL(n) \subseteq REG$.

Proof. Consider a one-sided sticker system $\gamma = (V, \rho, A, D)$. Let us denote by d the length of the longest sticky end or of the longest single stranded sequence appearing in A or in the pairs of D.

We construct the context-free grammar

$$G = (N, T, S, P),$$

where

$$N = \{\langle \binom{u}{\lambda} \rangle_l, \langle \binom{u}{\lambda} \rangle_r, \langle \binom{\lambda}{u} \rangle_l, \langle \binom{\lambda}{u} \rangle_r \mid u \in V^*, 0 \leq |u| \leq d\}$$

$$\cup \{S\},$$

$$T = \begin{bmatrix} V \\ V \end{bmatrix}_\rho,$$

and P contains the following rules:

1. $S \to \langle \binom{u_1}{u_2} \rangle_l \begin{bmatrix} x_1 \\ x_2 \end{bmatrix} \langle \binom{v_1}{v_2} \rangle_r$, for $\binom{u_1}{u_2} \begin{bmatrix} x_1 \\ x_2 \end{bmatrix} \binom{v_1}{v_2} \in A$, with $\binom{u_1}{u_2}, \binom{v_1}{v_2} \in \binom{\lambda}{V^*} \cup \binom{V^*}{\lambda}$ and $\begin{bmatrix} x_1 \\ x_2 \end{bmatrix} \in WK_\rho(V)$.

2. $\langle \binom{u_1}{u_2} \rangle_l \to \langle \binom{u_1'}{u_2'} \rangle_l \begin{bmatrix} w_1 \\ w_2 \end{bmatrix}$, where $\binom{u_1}{u_2}, \binom{u_1'}{u_2'} \in \binom{\lambda}{V^*} \cup \binom{V^*}{\lambda}$, $\begin{bmatrix} w_1 \\ w_2 \end{bmatrix} \in WK_\rho(V)$, and there is a pair in D of the form $(\binom{u_1'}{u_2'} \begin{bmatrix} x_1 \\ x_2 \end{bmatrix} \binom{y_1}{y_2}, \binom{\lambda}{\lambda})$ with $\binom{y_1}{y_2} \in \binom{V^*}{\lambda} \cup \binom{\lambda}{V^*}$, and $\begin{bmatrix} x_1 \\ x_2 \end{bmatrix} \in WK_\rho(V)$, such that $\begin{bmatrix} x_1 y_1 u_1 \\ x_2 y_2 u_2 \end{bmatrix} = \begin{bmatrix} w_1 \\ w_2 \end{bmatrix}$.
(We prolong the sequence to the left, using the pairs with an empty right hand member, in accordance with the sticky end; we remember which is the sticky end by means of the nonterminal $\langle \binom{u_1}{u_2} \rangle_l$; the subscript l stands for "left". Note that $\begin{bmatrix} w_1 \\ w_2 \end{bmatrix}$ above can be equal to $\begin{bmatrix} \lambda \\ \lambda \end{bmatrix}$.)

3. $\langle \binom{u_1}{u_2} \rangle_r \to \begin{bmatrix} w_1 \\ w_2 \end{bmatrix} \langle \binom{u_1'}{u_2'} \rangle_r$, where $\binom{u_1}{u_2}, \binom{u_1'}{u_2'} \in \binom{\lambda}{V^*} \cup \binom{V^*}{\lambda}$, $\begin{bmatrix} w_1 \\ w_2 \end{bmatrix} \in WK_\rho(V)$, and there is a pair in D of the form $(\binom{\lambda}{\lambda}, \binom{x_1}{x_2} \begin{bmatrix} y_1 \\ y_2 \end{bmatrix} \binom{u_1'}{u_2'})$, with $\binom{x_1}{x_2} \in \binom{V^*}{\lambda} \cup \binom{\lambda}{V^*}$, and $\begin{bmatrix} y_1 \\ y_2 \end{bmatrix} \in WK_\rho(V)$, such that $\begin{bmatrix} u_1 x_1 y_1 \\ u_2 x_2 y_2 \end{bmatrix} = \begin{bmatrix} w_1 \\ w_2 \end{bmatrix}$.
(The same idea as above, but prolonging the sequence to the right.)

4. $\langle \binom{\lambda}{\lambda} \rangle_l \to \lambda$,
$\langle \binom{\lambda}{\lambda} \rangle_r \to \lambda$.
(When no sticky end is present, we can finish the derivation.)

It is easy to see that $L(G) = LM_n(\gamma) = LM_d(\gamma)$: because we only use one-sided pairs in order to build sequences, the operation of prolonging sequences to the left is independent of the operation of prolonging sequences to the right and conversely; consequently, we can always use that pair $(\binom{z_1}{z_2}, \binom{\lambda}{\lambda})$ or $(\binom{\lambda}{\lambda}, \binom{z_1}{z_2})$ from D which sticks to the existing overhanging ends, which means that the overhanging ends are not longer than those already existing in A or in D. Thus, the nonterminals in N can control the process in the same way as the sticky ends do this.

In the grammar G there is no derivation of the form $X \Longrightarrow^* uXv$ with both u and v being non-empty strings. Consequently ([77], Exercise 9, page 55), the language $L(G)$ is regular. Because $L(G) = LM_n(\gamma)$ and $L_n(\gamma)$ is a coding of $LM_n(\gamma)$, we also have $L_n(\gamma) \in REG$. □

Corollary 4.1. $OSL(p) \subseteq REG$.

Proof. If in the proof above we replace d in the definition of N by $2d$ (hence the nonterminals remember sticky ends of length at most twice the longest sticky end in A or in D) and in rules of types 2 and 3 we take the left hand nonterminal $\langle\binom{u_1}{u_2}\rangle_l, \langle\binom{u_1}{u_2}\rangle_r$ with $u_1u_2 \neq \lambda$, then we get a grammar G' such that $L(G') = LM_p(\gamma)$. Indeed, when a blunt end is obtained, the grammar G' cannot continue to prolong the sequence in that direction, hence $L(G') \subseteq LM_p(\gamma)$. Conversely, each primitive computation in γ can be simulated by G', including those derivations where the overhanging strand is prolonged first in order to prevent a blunt end which could lead to a complete computation. □

From the first example at the end of the previous section we obtain the following result.

Theorem 4.2. *The families $SSL(n), SSL(p)$ contain non-context-free languages.*

Corollary 4.2. *The inclusions $SOSL(\alpha) \subset SSL(\alpha), \alpha \in \{n, p\}$, are proper.*

In the case of two-sided sticker systems, even simple, the bounded delay property cannot be forced, as in the proof of Theorem 4.1. More precisely, the following result holds.

Theorem 4.3. $ASL(b) \subseteq LIN$.

Proof. Consider a sticker system $\gamma = (V, \rho, A, D)$ of unrestricted form and let d be an integer such that $L_d(\gamma) = L_n(\gamma)$. We construct the linear grammar

$$G = (N, T, S, P),$$

where

$$N = \{\langle\binom{u_1}{u_2}, \binom{v_1}{v_2}\rangle \mid \binom{u_1}{u_2}, \binom{v_1}{v_2} \in \binom{\lambda}{V^*} \cup \binom{V^*}{\lambda},$$

$$|u_1|, |u_2|, |v_1|, |v_2| \leq d\} \cup \{S\},$$

$$T = \begin{bmatrix} V \\ V \end{bmatrix}_\rho,$$

and P contains the following rules:

1. $S \to \begin{bmatrix} w_1 \\ w_2 \end{bmatrix} \langle \begin{pmatrix} u_1 \\ u_2 \end{pmatrix}, \begin{pmatrix} v_1 \\ v_2 \end{pmatrix} \rangle \begin{bmatrix} z_1 \\ z_2 \end{bmatrix}$, for
$\begin{bmatrix} w_1 \\ w_2 \end{bmatrix}, \begin{bmatrix} z_1 \\ z_2 \end{bmatrix} \in WK_\rho(V), \langle \begin{pmatrix} u_1 \\ u_2 \end{pmatrix}, \begin{pmatrix} v_1 \\ v_2 \end{pmatrix} \rangle \in N,$
and $(\begin{bmatrix} w_1 \\ w_2 \end{bmatrix} \begin{pmatrix} u_1 \\ u_2 \end{pmatrix}, \begin{pmatrix} v_1 \\ v_2 \end{pmatrix} \begin{bmatrix} z_1 \\ z_2 \end{bmatrix}) \in D.$
(The computations in γ are simulated in G in a reversed order, starting from the last used pair in D and progressing towards the "center" of the sequence, where an axiom in A will be used.)

2. $\langle \begin{pmatrix} u_1 \\ u_2 \end{pmatrix}, \begin{pmatrix} v_1 \\ v_2 \end{pmatrix} \rangle \to \begin{bmatrix} w_1 \\ w_2 \end{bmatrix} \langle \begin{pmatrix} u_1' \\ u_2' \end{pmatrix}, \begin{pmatrix} v_1' \\ v_2' \end{pmatrix} \rangle \begin{bmatrix} z_1 \\ z_2 \end{bmatrix}$, for
$\begin{bmatrix} w_1 \\ w_2 \end{bmatrix}, \begin{bmatrix} z_1 \\ z_2 \end{bmatrix} \in WK_\rho(V), \langle \begin{pmatrix} u_1 \\ u_2 \end{pmatrix}, \begin{pmatrix} v_1 \\ v_2 \end{pmatrix} \rangle, \langle \begin{pmatrix} u_1' \\ u_2' \end{pmatrix}, \begin{pmatrix} v_1' \\ v_2' \end{pmatrix} \rangle \in N,$
and $(\begin{pmatrix} x_1 \\ x_2 \end{pmatrix} \begin{bmatrix} x_1' \\ x_2' \end{bmatrix} \begin{pmatrix} u_1' \\ u_2' \end{pmatrix}, \begin{pmatrix} v_1' \\ v_2' \end{pmatrix} \begin{bmatrix} y_1 \\ y_2 \end{bmatrix} \begin{pmatrix} y_1' \\ y_2' \end{pmatrix}) \in D$ such that
$\begin{bmatrix} u_1 x_1 x_1' \\ u_2 x_2 x_2' \end{bmatrix} = \begin{bmatrix} w_1 \\ w_2 \end{bmatrix}, \begin{bmatrix} y_1 y_1' v_1 \\ y_2 y_2' v_2 \end{bmatrix} = \begin{bmatrix} z_1 \\ z_2 \end{bmatrix}.$
(We proceed towards to the "center" of the sequence, adjoining blocks to the left and to the right, as provided by the pairs in D and controlled by the nonterminals in N. The control provided by the nonterminals suffices for correct simulations of the computations in γ by derivations in G, because of the bounded delay property.)

3. $\langle \begin{pmatrix} u_1 \\ u_2 \end{pmatrix}, \begin{pmatrix} v_1 \\ v_2 \end{pmatrix} \rangle \to \begin{bmatrix} w_1 \\ w_2 \end{bmatrix} \begin{bmatrix} x_1 \\ x_2 \end{bmatrix} \begin{bmatrix} z_1 \\ z_2 \end{bmatrix}$, for
$\begin{bmatrix} w_1 \\ w_2 \end{bmatrix}, \begin{bmatrix} x_1 \\ x_2 \end{bmatrix}, \begin{bmatrix} z_1 \\ z_2 \end{bmatrix} \in WK_\rho(V), \begin{bmatrix} x_1 \\ x_2 \end{bmatrix} \neq \begin{bmatrix} \lambda \\ \lambda \end{bmatrix},$
$\langle \begin{pmatrix} u_1 \\ u_2 \end{pmatrix}, \begin{pmatrix} v_1 \\ v_2 \end{pmatrix} \rangle \in N$, and there is $\begin{pmatrix} w_1' \\ w_2' \end{pmatrix} \begin{bmatrix} x_1 \\ x_2 \end{bmatrix} \begin{pmatrix} z_1' \\ z_2' \end{pmatrix} \in A$
such that $\begin{pmatrix} w_1' \\ w_2' \end{pmatrix}, \begin{pmatrix} z_1' \\ z_2' \end{pmatrix} \in \begin{pmatrix} \lambda \\ V^* \end{pmatrix} \cup \begin{pmatrix} V^* \\ \lambda \end{pmatrix}$
and $\begin{bmatrix} w_1 \\ w_2 \end{bmatrix} = \begin{bmatrix} u_1 w_1' \\ u_2 w_2' \end{bmatrix}, \begin{bmatrix} z_1 \\ z_2 \end{bmatrix} = \begin{bmatrix} z_1' v_1 \\ z_2' v_2 \end{bmatrix}.$
(When an axiom in A has sticky ends which fit both the left and the right sticky ends memorized by the element of N currently present in the sentential form, the derivation can be terminated.)

4. $S \to \begin{bmatrix} w_1 \\ w_2 \end{bmatrix}$, for $\begin{bmatrix} w_1 \\ w_2 \end{bmatrix} \in A.$

From the previous explanations, it is easy to see that $L(G) = LM_d(\gamma)$. Since $L_d(\gamma)$ is a coding of $LM_d(\gamma)$, we obtain $L_d(\gamma) \in LIN$. □

Corollary 4.3. *For every sticker system γ and integer d we have $L_d(\gamma) \in LIN$.*

Proof. This is a direct consequence of the previous proof. □

Surprisingly enough, because the sticker systems use no auxiliary symbols, the inclusions reverse to those in Theorems 4.1, 4.3 also hold true, even in stronger forms. We give first the proof for the regular case, because it is easier and it provides a good background for proving the result for the linear case.

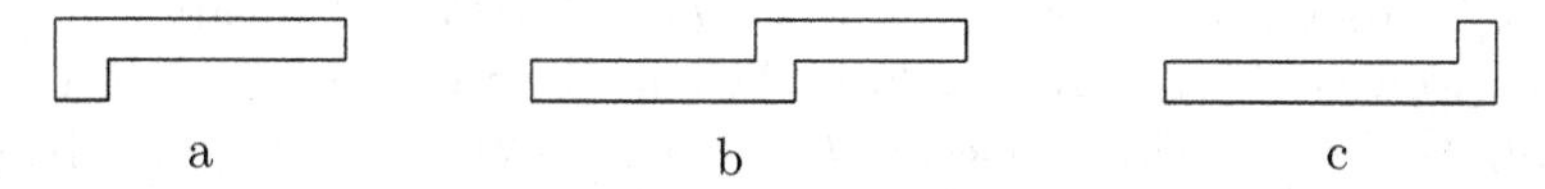

Figure 4.4: Dominoes used in the proof of Theorem 4.4

Theorem 4.4. *$REG \subseteq RSL(b) \cap RSL(p)$.*

Proof. Consider a finite automaton $M = (K, V, s_0, F, \delta)$ with $K = \{s_0, s_1, \ldots, s_k\}$, $k \geq 0$. We construct the regular sticker system

$$\gamma = (V, \rho, A, D),$$

with

$$\begin{aligned}
\rho &= \{(a, a) \mid a \in V\},\\
A &= \{\begin{bmatrix} x \\ x \end{bmatrix} \mid x \in L(M), |x| \leq k+2\}\\
&\cup \{\begin{bmatrix} x \\ x \end{bmatrix}\begin{pmatrix} u \\ \lambda \end{pmatrix} \mid |xu| = k+2, |x| \geq 1, |u| = i, \text{ for}\\
&\quad 1 \leq i \leq k+1 \text{ such that } s_0 xu \Longrightarrow^* xus_{i-1}\},\\
D &= \{(\begin{pmatrix} \lambda \\ \lambda \end{pmatrix}, \begin{pmatrix} \lambda \\ v \end{pmatrix}\begin{bmatrix} x \\ x \end{bmatrix}\begin{pmatrix} u \\ \lambda \end{pmatrix}) \mid 1 \leq |v| \leq k+1, |xu| = k+2, |x| \geq 1,\\
&\quad |u| = i, \text{ for } 1 \leq i \leq k+1, \text{ such that } s_j xu \Longrightarrow^* xus_{i-1},\\
&\quad \text{and } j = |v| - 1\}\\
&\cup \{(\begin{pmatrix} \lambda \\ \lambda \end{pmatrix}, \begin{pmatrix} \lambda \\ v \end{pmatrix}\begin{bmatrix} x \\ x \end{bmatrix}) \mid 1 \leq |v| \leq k+1, 1 \leq |x| \leq k, \text{ and}\\
&\quad s_j x \Longrightarrow^* xs_f, s_f \in F, \text{ where } j = |v| - 1\}.
\end{aligned}$$

The idea is to start with a domino of the form shown in Fig. 4.4a, to iteratively use dominoes of the form shown in Fig. 4.4b, and to end the computation with a domino of the form shown in Fig. 4.4c.

The overhangs codify the states of M by their lengths. The axioms in A which are not already in $WK_\rho(V)$ and the dominoes of the form in Fig. 4.4b appearing in the right hand member of pairs in D have overhangs of lengths $i, 1 \leq i \leq k+1$, which identify the state s_{i-1} by the length i. This state is reached by M when receiving the string in the upper strand of the well-started molecule which is obtained using the domino. All dominoes of the forms in Fig. 4.4b and Fig. 4.4c have a non-empty left overhang, hence a molecule in $WK_\rho(V)$ cannot be prolonged. Thus, after using a domino of type c), the computation must stop. Since the system γ has a delay at most $k+1$, we have $L_n(\gamma) = L_p(\gamma) = L_{k+1}(\gamma) = L(M)$, which completes the proof. □

Corollary 4.4. $RSL(\alpha) = OSL(\alpha) = REG,\ \alpha \in \{n, p, b\}$.

Proof. From Theorem 4.1 we have $OSL(n) \subseteq REG$. Corollary 4.1 gives the inclusion $OSL(p) \subseteq REG$. From Lemma 4.2 we also have $OSL(b) \subseteq OSL(n) \subseteq REG$. The inclusions $RSL(\alpha) \subseteq OSL(\alpha), \alpha \in \{n, p, b\}$, are pointed out in Lemma 4.1. The previous theorem proves the inclusions $REG \subseteq RSL(b)$, $REG \subseteq RSL(p)$. With $RSL(b) \subseteq RSL(n)$ (Lemma 4.2), we also get $REG \subseteq RSL(n)$. □

Theorem 4.5. $LIN \subseteq ASL(b) \cap ASL(p)$.

Proof. Consider a linear grammar $G = (N, T, S, P)$. There is an equivalent grammar $G' = (N', T, S, P')$ with P' containing only rules of the forms $X \to aY, X \to Ya, X \to a$, for $X, Y \in N', a \in T$.

Assume that $N' = \{X_1, X_2, \ldots, X_k\}, k \geq 1$. We construct the sticker system

$$\gamma = (T, \rho, A, D),$$

where

$$\begin{aligned}
\rho = {} & \{(a, a) \mid a \in T\}, \\
A = {} & \{\begin{bmatrix} x \\ x \end{bmatrix} \mid x \in L(G), |x| \leq 3k+1\} \\
& \cup \{\binom{u}{\lambda}\begin{bmatrix} x \\ x \end{bmatrix} \mid |ux| \leq 3k+1, |x| \geq 1, |u| = i, \text{ for } 1 \leq i \leq k \\
& \quad \text{such that } X_i \Longrightarrow^* ux\} \\
& \cup \{\begin{bmatrix} x \\ x \end{bmatrix}\binom{u}{\lambda} \mid |xu| \leq 3k+1, |x| \geq 1, |u| = i, \text{ for } 1 \leq i \leq k \\
& \quad \text{such that } X_i \Longrightarrow^* xu\},
\end{aligned}$$

and D contains the following groups of domino pairs:

1) $(\binom{u}{\lambda}\begin{bmatrix} x \\ x \end{bmatrix}\binom{\lambda}{v}, \begin{bmatrix} z \\ z \end{bmatrix})$, for $1 \leq |u| \leq k, 1 \leq |v| \leq k, |ux| = k+1$, $0 \leq |z| \leq k$, and $X_{|u|} \Longrightarrow^* uxX_{|v|}z$,

2) $(\begin{bmatrix} x \\ x \end{bmatrix}\binom{\lambda}{v}, \begin{bmatrix} z \\ z \end{bmatrix}\binom{u}{\lambda})$, for $1 \leq |v| \leq k, 1 \leq |u| \leq k, 1 \leq |x| \leq k$, $|zu| = k+1$, and $X_{|u|} \Longrightarrow^* xX_{|v|}zu$,

3) $(\begin{bmatrix} x \\ x \end{bmatrix}\binom{\lambda}{v}, \begin{bmatrix} z \\ z \end{bmatrix})$, for $1 \leq |v| \leq k, |x| \geq 1, |xz| \leq 2k+1$, $|z| \geq 0$, and $S \Longrightarrow^* xX_{|v|}z$,

4) $(\begin{bmatrix} z \\ z \end{bmatrix}, \binom{\lambda}{v}\begin{bmatrix} x \\ x \end{bmatrix}\binom{u}{\lambda})$, for $1 \leq |v| \leq k, 1 \leq |u| \leq k, |xu| = k+1$, $0 \leq |z| \leq k$, and $X_{|u|} \Longrightarrow^* zX_{|v|}xu$,

5) $(\binom{u}{\lambda}\begin{bmatrix} z \\ z \end{bmatrix}, \binom{\lambda}{v}\begin{bmatrix} x \\ x \end{bmatrix})$, for $1 \leq |u| \leq k, 1 \leq |v| \leq k, 1 \leq |x| \leq k$, $|uz| = k+1$, and $X_{|u|} \Longrightarrow^* uzX_{|v|}x$,

6) $(\begin{bmatrix} z \\ z \end{bmatrix}, \binom{\lambda}{v}\begin{bmatrix} x \\ x \end{bmatrix})$, for $1 \leq |v| \leq k, |x| \geq 1, |xz| \leq 2k+1$, $|z| \geq 0$, and $S \Longrightarrow^* zX_{|v|}x$.

The thought behind this construction is as follows. We intend to simulate the derivations in G', backwards, by computations in γ which introduce first a block in the center of the string and continue by adding blocks at the two ends of the string. The nonterminals in N' are again identified by the length of overhanging ends, at the left hand or at the right hand of the currently produced sequence; the other end of the sequence is blunt. Using domino pairs from group 1 we continue to update the information about the current nonterminal in the left hand of the sequence; group 2 changes the sticky end in the right hand end of the sequence, completing a blunt end in the left hand. With pairs of dominoes of type 3 we complete a molecule. Symmetrically, groups 4, 5, 6 of domino pairs continue to encode the current nonterminal in the length of the right hand sticky end, move this information to the left hand end, and finish the computation, respectively. A sequence with both ends being blunt (a molecule) cannot be continued, because all pairs in D have a non-empty sticky end towards the "inside" of its domino pair.

Thus, it is clear that all complete computations in γ correspond to derivations in G'. Conversely, every derivation in G can be simulated by a complete computation in γ.

Indeed, consider a derivation $\delta : S \Longrightarrow^* w$ in G. If $|w| \leq 3k+1$, then $\begin{bmatrix} w \\ w \end{bmatrix} \in A$. Assume that $|w| > 3k+1$. Because all rules in P' introduce exactly one terminal symbol each, we can decompose the derivation δ as follows:

$$S \Longrightarrow^* u_1X_{i_1}v_1 \Longrightarrow^* u_1u_2X_{i_2}v_2v_1 \Longrightarrow^* \ldots \Longrightarrow^* u_1 \ldots u_rX_{i_r}v_r \ldots v_1$$
$$\Longrightarrow^* u_1 \ldots u_ryv_r \ldots v_1,$$

with

(1) $|u_j| = k + 1$ and $0 \leq |v_j| \leq k$, or
$0 \leq |u_j| \leq k$ and $|v_j| = k + 1$, for each $j = 1, 2, \ldots, r$,

(2) $k + 1 \leq |y| \leq 3k + 1$,

(3) $r \geq 1$.

Then, for each (u_j, v_j) with $|u_j| = k + 1$, we can find a pair of dominoes of type 1 or 5, encoding $X_{i_{j-1}}$ in a left sticky end of length i_{j-1}, and for each (u_j, v_j) with $|v_j| = k + 1$ we can find a pair of dominoes of type 2 or 4, encoding $X_{i_{j-1}}$ in a right sticky end of length i_{j-1}. Clearly, for y we can find an axiom encoding X_{i_r} in one of its ends and similarly for (u_1, v_1) we can find a pair of type 3 or 6, producing blunt ends in both directions. Consequently, we also have $L(G) \subseteq L_n(\gamma)$.

Obviously, $L_n(\gamma) = L_p(\gamma)$ and the delay of γ is at most k, hence $L_n(\gamma) = L_k(\gamma)$, completing the proof. □

Corollary 4.5. $LIN = ASL(b)$.

Proof. Combine Theorems 4.3 and 4.5. □

In the proof of Theorem 4.1 we have pointed out that if $\gamma = (V, \rho, A, D)$ is a one-sided sticker system, then $L_n(\gamma) = L_d(\gamma)$ for some integer d depending on A and D (the length of the longest sticky end in A or in the dominoes of D). This is obviously true also for simple and for simple regular systems, and so we obtain the following result.

Theorem 4.6. $SOSL(n) = SOSL(b)$, $SRSL(n) = SRSL(b)$.

Proof. The inclusions $\subseteq$ were discussed above, the reverse inclusions are mentioned in Lemma 4.2. □

Summarizing the previous results for families $XSL(b), XSL(n)$, we get the diagram in Fig. 4.5; as usual, the arrows indicate inclusions, not necessarily proper.

Theorem 4.7. $REG - SOSL(\alpha) \neq \emptyset$, $\alpha \in \{n, b\}$.

Proof. Consider the regular language

$$L = ba^+b,$$

and assume that $L = L_n(\gamma)$ for some simple one-sided sticker system $\gamma = (V, \rho, A, D)$. Because A is a finite set (of well-started molecules) and L is an infinite language, there are two pairs in D of one of the following forms

(1) $(\binom{\lambda}{\lambda}, \binom{\lambda}{y_2})$ and $(\binom{\lambda}{\lambda}, \binom{y_1'}{\lambda})$, with $y_2 \in V^+, y_1' \in a^+$,

(2) $(\binom{\lambda}{\lambda}, \binom{y_1}{\lambda})$ and $(\binom{\lambda}{\lambda}, \binom{\lambda}{y_2'})$, with $y_1 \in a^+, y_2' \in V^+$,

(3) $(\binom{\lambda}{x_2}, \binom{\lambda}{\lambda})$ and $(\binom{x_1'}{\lambda}, \binom{\lambda}{\lambda})$, with $x_2 \in V^+, x_1' \in a^+$,

(4) $(\binom{x_1}{\lambda}, \binom{\lambda}{\lambda})$ and $(\binom{\lambda}{x_2'}), \binom{\lambda}{\lambda})$, with $x_1 \in a^+, x_2' \in V^+$,

that are used arbitrarily many times in the generation of strings in L of arbitrarily large length.

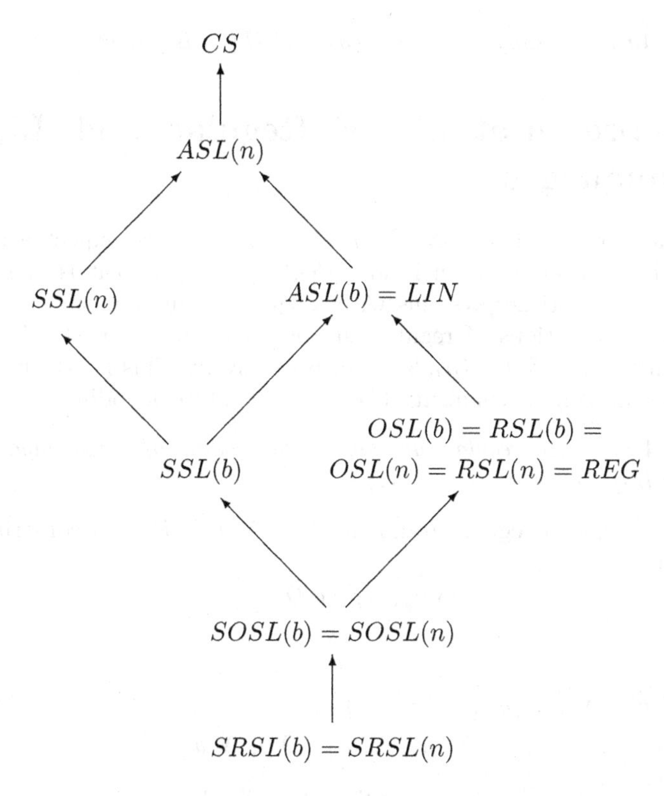

Figure 4.5: Relationships between families of languages generated by sticker systems

All the four cases can be treated in the same way. Assume that we have the first case, hence $(\binom{\lambda}{\lambda}, \binom{\lambda}{y_2}) \in D$, $(\binom{\lambda}{\lambda}, \binom{y_1'}{\lambda}) \in D$. Clearly, $y_1' = a^i$ for some $i \geq 1$ and y_2 is composed of symbols c such that $(a, c) \in \rho$.

Assume that $|y_2| = j, j \geq 1$. A complete computation

$$\binom{x_1}{x_2}\begin{bmatrix} y_1 \\ y_2 \end{bmatrix}\binom{z_1}{z_2} \Longrightarrow^* \begin{bmatrix} w_1 \\ w_2 \end{bmatrix},$$

for $\binom{x_1}{x_2}\begin{bmatrix} y_1 \\ y_2 \end{bmatrix}\binom{z_1}{z_2} \in A$, $\binom{x_1}{x_2}, \binom{z_1}{z_2} \in \binom{V^*}{\lambda} \cup \binom{\lambda}{V^*}$ and $\begin{bmatrix} y_1 \\ y_2 \end{bmatrix}, \begin{bmatrix} w_1 \\ w_2 \end{bmatrix}$

$\in WK_\rho(V), w_1 = ba^n b$, can be continued as follows:

$$\binom{x_1}{x_2}\begin{bmatrix}y_1\\y_2\end{bmatrix}\binom{z_1}{z_2} \Longrightarrow^* \begin{bmatrix}w_1\\w_2\end{bmatrix} \Longrightarrow^* \begin{bmatrix}w_1 y_1'^j\\w_2 y_2^i\end{bmatrix}.$$

This is a complete computation, producing the string $w_1 y_1'^j = ba^n ba^{ij}$, which is not in L, a contradiction. □

Corollary 4.6. *The inclusion $SOSL(n) \subseteq OSL(n)$ is proper.*

4.4 Representations of Regular and Linear Languages

We have the strict inclusion $SOSL(n) \subset REG$ and we expect a similar result in what concerns the inclusion $SSL(b) \subseteq LIN$, so it is of interest to supplement the sticker systems with a squeezing mechanism and to try to obtain representations of regular and of linear languages starting from languages in families $SOSL(n), SSL(b)$, respectively. This is possible, even using weak squeezing mechanisms, like codings and weak codings.

Theorem 4.8. *Each regular language is the coding of a language in the family $SRSL(\alpha)$, for each $\alpha \in \{n, b, p\}$.*

Proof. Consider a regular grammar $G = (N, T, S, P)$ and construct the sticker system

$$\gamma = (V, \rho, A, D),$$

with

$$\begin{aligned}
V &= \{[X,a]_i \mid X \in N, a \in T, i = 1,2\},\\
\rho &= \{([X,a]_1,[X,a]_1),\ ([X,a]_2,[X,a]_2) \mid X \in N, a \in T\},\\
A &= \{\begin{bmatrix}[S,a]_1\\ [S,a]_1\end{bmatrix}\binom{\lambda}{[X,b]_2} \mid S \to aX \in P, \text{ and either}\\
&\quad X \to bY \in P, \text{ or } X \to b \in P, a,b \in T, X,Y \in N\}\\
&\cup \{\begin{bmatrix}[S,a]_1\\ [S,a]_1\end{bmatrix} \mid S \to a \in P, a \in T\},\\
D &= \{(\binom{\lambda}{\lambda}, \binom{\lambda}{[X,a]_1[Y,b]_2}) \mid X \to aY \in P \text{ and either } Y \to bY' \in P,\\
&\quad \text{or } Y \to b \in P, \text{ for } a,b \in T \text{ and } X,Y,Y' \in N\}\\
&\cup \{(\binom{\lambda}{\lambda}, \binom{\lambda}{[X,a]_1}) \mid X \to a \in P, X \in N, a \in T\}\\
&\cup \{(\binom{\lambda}{\lambda}, \binom{[X,a]_2[Y,b]_1}{\lambda}) \mid X \to aY \in P \text{ and either } Y \to bY',
\end{aligned}$$

or $Y \to b \in P$, for $a, b \in T$ and $X, Y, Y' \in N\}$

$$\cup \{(\binom{\lambda}{\lambda}, \binom{[X,a]_2}{\lambda}) \mid X \to a \in P, X \in N, a \in T\}.$$

Each derivation in G of the form

$$S \Longrightarrow a_1X_1 \Longrightarrow a_1a_2X_2 \Longrightarrow \ldots \Longrightarrow a_1a_2\ldots a_kX_k \Longrightarrow a_1a_2\ldots a_ka_{k+1},$$

$k \geq 1$, corresponds to a computation in γ of the form

$$\begin{bmatrix}[S,a_1]_1\\ [S,a_1]_1\end{bmatrix}\begin{pmatrix}\lambda\\ [X_1,a_2]_2\end{pmatrix} \Longrightarrow \begin{bmatrix}[S,a_1]_1[X_1,a_2]_2\\ [S,a_1]_1[X_1,a_2]_2\end{bmatrix}\begin{pmatrix}[X_2,a_3]_1\\ \lambda\end{pmatrix} \Longrightarrow \ldots$$

with alternating sticky ends $\begin{pmatrix}[X_{2i},a_{2i+1}]_1\\ \lambda\end{pmatrix}$ and $\begin{pmatrix}\lambda\\ [X_{2i-1},a_{2i}]_2\end{pmatrix}$. When k is even, we can finish by using the block $\begin{pmatrix}\lambda\\ [X_k,a_{k+1}]_1\end{pmatrix}$ and when k is odd we can finish by using the block $\begin{pmatrix}[X_k,a_{k+1}]_2\\ \lambda\end{pmatrix}$. All the lower blocks in dominoes of D start with symbols of the form $[X,a]_1$, all upper blocks start with symbols of the form $[X,a]_2$. Therefore, a complete computation cannot be continued (because of the relation ρ). It is now clear that $L_n(\gamma) = L_p(\gamma) = L_1(\gamma)$ and that by the coding h defined by

$$h([X,a]_i) = a, \text{ for } X \in N, a \in T, i = 1, 2,$$

we obtain $L(G) = h(L_n(\gamma))$. □

For a family of languages FL, let us denote by $Cod(FL)$ the family of languages of the form $h(L)$, for $L \in FL$ and h a coding.

Corollary 4.7. $Cod(SOSL(\alpha)) = Cod(SRSL(\alpha)) = REG$, $\alpha \in \{n, b, p\}$.

Proof. All families $SOSL(\alpha), SRSL(\alpha), \alpha \in \{n, b, p\}$, are included in REG (Theorem 4.1, Corollary 4.1) and REG is closed under (arbitrary) morphisms. Therefore, $Cod(SOSL(\alpha))$ and $Cod(SRSL(\alpha))$ are also included in REG. The previous theorem proves the reverse inclusions. □

Theorem 4.9. *Each linear language is the weak coding of a language in the family $SSL(b)$.*

Proof. Consider a linear grammar $G = (N, T, S, P)$. Without loss of generality, we may assume that all rules in P are of the forms $X \to aY$, $X \to Ya, X \to a$, $a \in T, X, Y \in N$. Take a new symbol, $c \notin T$, and modify the rules above to $X \to aYc, X \to cYa, X \to cac$. Thus, we may assume that for each linear language $L \subseteq T^*$ there is a symbol $c \notin T$ and a linear grammar G' with rules of the forms mentioned above such that $L = g(L(G'))$, for g the weak coding erasing the symbol c and leaving the symbols in T unchanged.

Suppose that already $G = (N, T \cup \{c\}, S, P)$ is such a grammar, for a given language $L \in LIN$.

We construct the simple sticker system

$$\gamma = (V, \rho, A, D),$$

with

$$\begin{aligned}
V &= \{[X,a]_i \mid X \in N, a \in T \cup \{c\}, i = 1,2,3\},\\
\rho &= \{([X,a]_i, [X,a]_i) \mid X \in N, a \in T \cup \{c\}, i = 1,2,3\},\\
A &= \{\binom{[X,a_1]_1}{\lambda}\begin{bmatrix}[X,a_2]_3\\ [X,a_2]_3\end{bmatrix}\binom{[X,a_3]_1}{\lambda} \mid X \to a_1a_2a_3 \in P, X \in N,\\
&\quad a_1, a_2, a_3 \in T \cup \{c\}\}
\end{aligned}$$

and D contains the following pairs of dominoes:

1) $(\binom{\lambda}{[Y,a_2]_2[X,a_1]_1}, \binom{\lambda}{[X,a_3]_1[Y,a_4]_2})$, for $Y \to a_2Xa_4 \in P$,
$a_1, a_2, a_3, a_4 \in T \cup \{c\}, X, Y \in N,$ and there is a rule $X \to a_1X'a_3$ or $X \to a_1a_5a_3$ in $P, X' \in N, a_5 \in T \cup \{c\}$,

2) $(\binom{[Y,a_2]_1[X,a_1]_2}{\lambda}, \binom{[X,a_3]_2[Y,a_4]_1}{\lambda})$, for $Y \to a_2Xa_4 \in P$,
$a_1, a_2, a_3, a_4 \in T \cup \{c\}, X, Y \in N,$ and there is a rule $X \to a_1X'a_3$ or $X \to a_1a_5a_3$ in $P, X' \in N, a_5 \in T \cup \{c\}$,

3) $(\binom{\lambda}{[S,a_1]_1}, \binom{\lambda}{[S,a_2]_1})$, for $S \to a_1Xa_2 \in P, X \in N, a_1, a_2 \in T \cup \{c\}$,

4) $(\binom{[S,a_1]_2}{\lambda}, \binom{[S,a_2]_2}{\lambda})$, for $S \to a_1Xa_2 \in P, X \in N, a_1, a_2 \in T \cup \{c\}$.

We simulate the derivations in G from their end to the beginning, starting to grow the string from the center. The pairs of dominoes in group 1 add lower level blocks, all of them having symbols $[X,a]_1$ toward the sequence to which these pairs are adjoined, the pairs of type 2 add upper level blocks, all of them having symbols $[X,a]_2$ toward the sequence to which these pairs are adjoined. Thus, the obtained molecule will consist of a column of the form $\begin{bmatrix}[X,a]_3\\ [X,a]_3\end{bmatrix}$ in the center, and then, alternating both to the left and to the right, columns of the form $\begin{bmatrix}[X,a]_1\\ [X,a]_1\end{bmatrix}$ and $\begin{bmatrix}[X,a]_2\\ [X,a]_2\end{bmatrix}$. This alternation ensures the fact that all computations in γ correspond to correct derivations in G. The computations can lead to complete molecules by using pairs of types 3 or 4, depending on the parity of the step at which we want to stop. This corresponds to using an S-rule in P, hence to a correctly started derivation in G. Conversely, due to the form of the rules in P, all derivations in G can be simulated by complete computations in γ.

Clearly, the delay of γ is 1 and no complete computation can be continued, hence $L_n(\gamma) = L_1(\gamma) = L_p(\gamma)$.

By the weak coding h defined by

$$h([X,a]_i) = a, \text{ for } X \in N, a \in T, i = 1,2,3,$$
$$h([X,c]_i) = \lambda, \text{ for } X \in N, i = 1,2,3,$$

we clearly obtain $h(L_n(\gamma)) = L$, which completes the proof. □

For a family of languages FL, let us denote by $CodW(FL)$ the family of languages of the form $h(L)$, for $L \in FL$ and h a weak coding.

Corollary 4.8. $CodW(SSL(b)) = LIN$.

Proof. The inclusion $SSL(b) \subseteq LIN$ follows from Theorem 4.3; the family LIN is closed under arbitrary morphisms, hence $CodW(SSL(b)) \subseteq LIN$. The reverse inclusion is proved in the previous theorem. □

4.5 Characterizations of Recursively Enumerable Languages

From the point of view of DNA computing, of more interest is the possibility of representing (hence characterizing) the recursively enumerable languages by means of sticker languages. We have already presented such a possibility at the end of Sect. 4.2, when we have discussed an example of a sticker system (denoted there by γ_2) such that

$$h(L_n(\gamma_2)) = TS_V,$$

for a weak coding h. Combining this with Theorem 3.17 (the weak coding can be simulated by a gsm), we obtain the following representation result.

Theorem 4.10. *Every language $L \in RE$ can be written in the form $L = g(L')$, for $L' \in ASL(n)$ and g a deterministic gsm mapping.*

In view of the results in Sect. 4.3, such a representation cannot be obtained for L' in any other family than $ASL(n), ASL(p), SSL(n), SSL(p)$, because all other families contain only linear languages (see again the diagram in Figure 4.5), and the family LIN is closed under arbitrary gsm mappings. However, $SSL(n)$ – and even $SSL(p)$ – can be used to obtain a representation of recursively enumerable languages, thus strengthening the result in Theorem 4.10.

Theorem 4.11. *Every language $L \in RE$ can be written in the form $L = h(L')$, where h is a weak coding and $L' \in SSL(n) \cap SSL(p)$.*

Proof. Consider a language $L \subseteq T^*$, $L \in RE$. According to Theorem 3.16, there exist two λ-free morphisms $h_1, h_2 : V_1^* \longrightarrow V_2^*$, a regular language $R \subseteq V_2^*$ and a projection $pr_T : V_2^* \longrightarrow T^*$ for $T \subseteq V_2$, such that $L = pr_T(h_1(EQ(h_1, h_2)) \cap R)$.

Consider a deterministic finite automaton $M = (K, V_2, s_0, F, \delta)$ recognizing the language R.

We construct the simple sticker system

$$\gamma = (V, \rho, A, D),$$

with

$$V = V_2 \cup \overline{V}_2 \cup K \cup \{\$, E, E', C, Z\},$$
$$\rho = \{(X, X) \mid X \in V\},$$
$$A = \{\binom{s_0}{\lambda} \begin{bmatrix} \$ \\ \$ \end{bmatrix} \binom{Z}{\lambda}\},$$

and D contains the following pairs of dominoes:

1. For every $a \in V_1$ such that $h_1(a) = b_1 \dots b_k, k \geq 1$, and $h_2(a) = c_1 \dots c_m, m \geq 1$, with $b_1, \dots, b_k, c_1, \dots, c_m \in V_2$, and for $s_{i_j} \in K, 0 \leq j \leq m$, such that $\delta(s_{i_j}, c_i) = s_{i_{j+1}}, 0 \leq j \leq m$, we introduce in D the pair
$$(\binom{s_{i_m}\bar{c}_m s_{i_{m-1}} \dots s_{i_2}\bar{c}_2 s_{i_1} s_{i_1}\bar{c}_1 s_{i_0}}{\lambda}, \binom{b_1CZb_2CZ \dots CZb_kCZ}{\lambda}).$$
(To the left of $\begin{bmatrix} \$ \\ \$ \end{bmatrix}$ we produce the reversed image of some $h_2(a)$, for $a \in V_1$, and at the same time we guess a valid path through M over $h_2(a)$: $s_{i_0}c_1c_2 \dots c_m \Longrightarrow^* s_{i_m}$. To the right we produce the image of a through h_1, with the symbols of $h_1(a)$ separated by the auxiliary symbols CZ.)

2. $(\binom{E's_f}{\lambda}, \binom{E}{\lambda})$, for $s_f \in F$.
(The recognition of the string in the upper strand of the left part of the sequence by means of M is finished correctly.)

3. $(\binom{\lambda}{ss}, \binom{\lambda}{Z})$, for all $s \in Q$.
(These rules check the correct continuation of the recognition path through M: if $s_1x \Longrightarrow^* xs_2$ is followed by $s_3y \Longrightarrow^* ys_4$, then we must have $s_2 = s_3$, otherwise the complementarity is not observed when using the block $\binom{\lambda}{ss}$.)

4. $(\binom{\lambda}{\bar{b}}, \binom{\lambda}{bC})$, for $b \in V_2$.

(The string of symbols $\bar{b}$ generated to the left of $\begin{bmatrix} \$ \\ \$ \end{bmatrix}$ in the upper strand is compared with the string of symbols b generated to the right of $\begin{bmatrix} \$ \\ \$ \end{bmatrix}$ in the upper strand. Note that the symbols Z are "consumed" together with the pairs of states, by rules of type 3; now we also "consume" the symbols C introduced in the upper strand, to the right of $\begin{bmatrix} \$ \\ \$ \end{bmatrix}$.)

5. $(\binom{\lambda}{E'}, \binom{\lambda}{E})$.

(Only in this way we can get a complete molecule.)

From the explanations above, one can see that the complete molecules produced by γ are of the form

$$\begin{bmatrix} E's_f s_f \bar{c}_t s_t s_t \dots \bar{c}_2 s_1 s_1 \bar{c}_1 s_0 s_0 \\ E's_f s_f \bar{c}_t s_t s_t \dots \bar{c}_2 s_1 s_1 \bar{c}_1 s_0 s_0 \end{bmatrix} \begin{bmatrix} \$ \\ \$ \end{bmatrix} \begin{bmatrix} Zb_1CZb_2CZ \dots CZb_tCZE \\ Zb_1CZb_2CZ \dots CZb_tCZE \end{bmatrix},$$

for

$$c_1c_2 \dots c_t = h_1(w) = h_2(w) = b_1b_2 \dots b_t,$$

for some $w \in V_1^*$, and $s_0c_1 \dots c_t \Longrightarrow^* c_1 \dots c_t s_f$ in M for $s_f \in F$, hence $h_1(w) \in R$.

No complete computation can be continued, because the upper strands of dominoes (in groups 1 and 2) have one state only in the left end of the left domino, whereas the lower strands of dominoes (in groups 3, 4, 5) have either two states or a symbol $\bar{b}$, $b \in T$, or the symbol E' in that position. Therefore, $L_n(\gamma) = L_p(\gamma)$.

Consider now the weak coding (in fact, a projection) h defined by

$$\begin{aligned} h(a) &= a, \text{ for } a \in T, \\ h(\bar{a}) &= \lambda, \text{ for } a \in T, \\ h(s) &= \lambda, \text{ for } s \in Q, \\ h(E) &= h(E') = h(\$) = h(C) = h(Z) = \lambda. \end{aligned}$$

Clearly, we get $L = h(L_n(\gamma))$, which completes the proof. □

The construction above has a rather interesting consequence for classic formal language theory: a strenghtening of the representation of recursively enumerable languages in Corollary 3.3:

Corollary 4.9. *Every recursively enumerable language is the projection of the intersection of two minimal linear languages.*

Proof. With the notations in the previous proof, we construct two minimal linear grammars

$$G_i = (\{S\}, V, S, P_i),\ i = 1, 2,$$

with

$$\begin{aligned}
P_1 = \{ & S \to s_{i_m}\bar{c}_m s_{i_m} s_{i_m} \dots s_{i_2}\bar{c}_2 s_{i_1} s_{i_1}\bar{c}_1 s_{i_1} S b_1 C Z b_2 C \dots Z b_k C Z \mid \\
& \text{for } b_1 \dots b_k = h_1(a), c_1 \dots c_m = h_2(a) \text{ for some } m \geq 1, \\
& a \in V_1, \text{ and } \delta(s_{i_j}, c_j) = s_{i_{j+1}}, 0 \leq j \leq m-1, \text{ with} \\
& b_1, \dots, b_k, c_1, \dots, c_m \in V_2\} \\
\cup\ & \{S \to E' s_f S E \mid s_f \in F\} \\
\cup\ & \{S \to s_0 \$ Z\}, \\
P_2 = \{ & S \to s s S Z \mid s \in Q\} \\
\cup\ & \{S \to \bar{b} S b C \mid b \in V_2\} \\
\cup\ & \{S \to E' S E,\ S \to \$\}.
\end{aligned}$$

It is easy to see that G_1 generates the strings in the upper strand of sequences which can be produced by γ using only the pairs from groups 1 and 3, plus the central substring $s_0\$Z$, whereas G_2 generates the strings in the lower strand of sequences produced by γ using only the pairs from groups 3, 4, 5, plus the central substring \$. By the intersection we check the complementarity relation ρ (which is the identity). Therefore, $L(G_1) \cap L(G_2) = L_n(\gamma)$, which completes the proof. □

4.6 More About Regular Sticker Systems

The regular sticker systems generate only regular languages, hence they cannot characterize RE by using AFL operations as squeezing mechanisms. On the other hand, mainly in the simple variant, such devices are attractive from a mathematical and a biochemical point of view. For instance, the use of couples of dominoes, essentially involved in the proof of Theorem 4.11, is not a very realistic assumption from a practical point of view. Using separated dominoes is much closer to the annealing operation in a test tube; in many places, "self-assembling" computations were reported or only proposed, which makes important the question of modifying the definition of simple sticker systems or of their language in such a way as to obtain characterizations of recursively enumerable languages for these sticker systems.

We consider here two restrictions on the language generated by a simple regular sticker system.

As we work here only with right-sided pairs of dominoes, we shall ignore the left hand member, the empty one. Moreover, we write separately the "upper dominoes" and the "lower dominoes". Thus, we write a simple regular sticker system in the form

$$\gamma = (V, \rho, A, D_l, D_u),$$

where V, ρ, A are as above, and $D_l \subseteq \binom{\lambda}{V^*}, D_u \subseteq \binom{V^*}{\lambda}$, with D_l and D_u finite sets. The languages $L_\alpha(\gamma)$ are defined as in the previous sections, $\alpha \in \{n, b, p\}$.

Consider two labelings of elements in D_l, D_u with elements in a set Lab, $e_\pi : D_\pi \longrightarrow Lab, \pi \in \{l, u\}$. For a complete computation $\sigma : x_1 \Longrightarrow x_2 \Longrightarrow \ldots \Longrightarrow x_k$ in γ we define the control words $e_\pi(\sigma), \pi \in \{l, u\}$, consisting of the labels of elements of D_l, D_u, respectively, used in σ. Formally, we denote

$$\delta(\pi, x \Longrightarrow y) = \begin{cases} e_\pi(\binom{w_1}{w_2}), & \text{if } x \Longrightarrow y \text{ uses the domino } \binom{w_1}{w_2} \in D_\pi, \\ \lambda, & \text{otherwise,} \end{cases}$$

for $\pi \in \{l, u\}$. Then,

$$e_\pi(\sigma) = \delta(\pi, x_1 \Longrightarrow x_2)\delta(\pi, x_2 \Longrightarrow x_3) \ldots \delta(\pi, x_{k-1} \Longrightarrow x_k), \ \pi \in \{l, u\}.$$

A complete computation $\sigma : x_1 \Longrightarrow^* x_k$, $x_1 \in A$, is said to be

– *fair*, if $|e_l(\sigma)| = |e_u(\sigma)|$,

– *coherent*, if $e_l(\sigma) = e_u(\sigma)$.

In a fair computation we use equally many upper blocks (elements of D_u) and lower blocks (elements of D_l); in a coherent computation we require that the sequence of labels associated to the upper blocks used in the computation is equal to the sequence of labels associated to the lower blocks used in the computation. Clearly, any coherent computation is also a fair one.

We denote by $L_f(\gamma), L_c(\gamma)$ the languages (of strings in the upper strand of molecules in $LM_n(\gamma)$) generated by γ using only fair computations or only coherent computations, respectively. The obtained families are denoted by $SRSL(f), SRSL(c)$. When the computations are also primitive, we replace f and c above with pf, pc, respectively.

The coherence condition leads again to a representation of recursively enumerable languages.

Theorem 4.12. *Each recursively enumerable language is the weak coding of a language in the family $SRSL(c)$ or $SRSL(pc)$.*

Proof. Consider a language $L \in RE, L \subseteq T^*$. According to Theorem 3.16, there are two alphabets V_1, V_2 with $T \subseteq V_2$, two λ-free morphisms $h_1, h_2 : V_1^* \longrightarrow V_2^*$, and a regular language $R \subseteq V_2^*$ such that $L = pr_T(h_1(EQ(h_1, h_2)) \cap R)$.

Assume that $V_1 = \{b_0, b_1, \ldots, b_{n-1}\}, n \geq 1$. Consider a deterministic finite automaton $M = (K, V_2, s_0, F, \delta)$ recognizing the language R; assume

$K = \{s_0, s_1, \ldots, s_{m-1}\}$, for some $m \geq 1$. We construct the simple sticker system

$$\gamma = (V, \rho, A, D_l, D_u),$$

where

$$\begin{aligned}
V &= V_2 \cup K \cup \{\bar{s} \mid s \in F\} \\
&\cup \{[s,j] \mid s \in K, 0 \leq j \leq m-1\}, \\
\rho &= \{(X,X) \mid X \in V_2\} \\
&\cup \{(s,s), ([s,j],s), (s,[s,k]), ([s,j],[s,k]) \mid s \in K, 0 \leq j,k \leq m-1\} \\
&\cup \{(\bar{s},\bar{s}) \mid s \in F\}, \\
A &= \{\begin{bmatrix} s_0 \\ s_0 \end{bmatrix} \binom{s_0}{\lambda}\}, \\
D_l &= \{\begin{pmatrix} \lambda \\ [s_{l_0},j]a_1 s_{l_1} s_{l_1} a_2 s_{l_2} s_{l_2} \ldots s_{l_{t_i-1}} s_{l_{t_i-1}} a_{t_i} s_{l_{t_i}} \end{pmatrix} \mid \\
&\quad a_1 a_2 \ldots a_{t_i} = h_2(b_i), 0 \leq i \leq n-1, 0 \leq j \leq m-1, \\
&\quad \delta(s_{l_k}, a_{k+1}) = s_{l_{k+1}}, 0 \leq l_k \leq m-1, 0 \leq k < t_i\} \\
&\cup \{\begin{pmatrix} \lambda \\ s_l \bar{s}_l \end{pmatrix} \mid s_l \in F\}, \\
D_u &= \{\begin{pmatrix} a_1[s_{l_1},j] s_{l_1} a_2 s_{l_2} s_{l_2} \ldots a_{t_i} s_{l_{t_i}} s_{l_{t_i}} \\ \lambda \end{pmatrix} \mid \\
&\quad a_1 a_2 \ldots a_{t_i} = h_1(b_i), 0 \leq i \leq n-1, 0 \leq j \leq m-1, \\
&\quad \delta(s_{l_k}, a_{k+1}) = s_{l_{k+1}}, 0 \leq l_k \leq m-1, 1 \leq k < t_i\} \\
&\cup \{\binom{\bar{s}_l}{\lambda} \mid s_l \in F\}.
\end{aligned}$$

We denote by $r_l(i,j,k)$, for $0 \leq i \leq n-1, 0 \leq j,k \leq m-1$, the elements in D_l associated as above with $h_2(b_i)$ and having the state s_j paired with the integer k; the sequence $\begin{pmatrix} \lambda \\ s_j \bar{s}_j \end{pmatrix}$ in D_l is denoted by $r_l(n,0,j)$. Similarly, we denote by $r_u(i,j,k)$, for $0 \leq i \leq n-1, 0 \leq j,k \leq m-1$, the sequence in D_u associated with $h_2(b_i)$ and having the state s_j paired with the integer k; the sequence $\binom{\bar{s}_j}{\lambda}$ in D_u is denoted by $r_u(n,j,0)$.

Clearly, $card(D_l) = card(D_u) = n \cdot m^2 + card(F)$. Define the labelings $e_l : D_l \longrightarrow \{1, 2, \ldots, card(D_l)\}$, $e_u : D_u \longrightarrow \{1, 2, \ldots, card(D_u)\}$ by

$$\begin{aligned}
e_l(r_d(i,j,k)) &= i \cdot m^2 + j \cdot m + k + 1, \\
e_u(r_u(i,j,k)) &= i \cdot m^2 + k \cdot m + j + 1.
\end{aligned}$$

By the construction above, it is clear that $u \Longrightarrow^* z$ is a complete computation in $\gamma, u \in A$, if and only if there is a sequence

$$s_{l_0} a_1 s_{l_1} a_2 \ldots s_{l_{t-1}} a_t s_{l_t},$$

such that (1) $s_{l_0} = s_0, s_{l_t} \in F, a_k \in V_2$, and $\delta(s_{l_{k-1}}, a_k) = s_{l_k}, 1 \leq k \leq t$, and (2) there is $x \in V^*$ such that $h_1(x) = h_2(x) = a_1 a_2 \ldots a_t$.

Define now the weak coding $h : WK_\rho(V) \to T^*$ by

$$h(\begin{bmatrix} X \\ X \end{bmatrix}) = \begin{cases} X, & \text{if } X \in T, \\ \lambda, & \text{otherwise,} \end{cases}$$

$$h(\begin{bmatrix} [s,j] \\ [s,k] \end{bmatrix}) = h(\begin{bmatrix} s \\ [s,k] \end{bmatrix}) = h(\begin{bmatrix} [s,j] \\ s \end{bmatrix}) = \lambda, \text{ for } s \in K, 0 \leq j,k \leq m-1.$$

We obtain the equality $h(L_c(\gamma)) = pr_T(h_1(EQ(h_1,h_2)) \cap R)$.

($\subseteq$). Consider a string $w \in h(L_c(\gamma))$. There is $z \in WK_\rho(V)$ such that $w = h(z)$ and there is a computation in γ of the form $\sigma : x \Longrightarrow^* z$, $x \in A$, $c_d(\sigma) = c_u(\sigma)$. By the construction of γ, there is a sequence

$$s_{l_0} a_1 s_{l_1} a_2 \ldots s_{l_{t-1}} a_t s_{l_t},$$

such that $s_{l_0} = s_0, s_{l_t} \in F, \delta(s_{l_{k-1}}, a_k) = s_{l_k}, 1 \leq k \leq t$. Consequently, $a_1 a_2 \ldots a_t \in R$. By the definition of D_l and D_u and the fact that $e_l(\sigma) = e_u(\sigma)$, it follows that $a_1 a_2 \ldots a_t = h_1(y) = h_2(y)$ for some $y \in V_1^*$. Then $a_1 a_2 \ldots a_t \in h_1(EQ(h_1,h_2)) \cap R$. Because $w = h(z) = pr_T(a_1 a_2 \ldots a_t)$, we obtain $w \in pr_T(h_1(EQ(h_1,h_2)) \cap R)$.

($\supseteq$). Let $w \in pr_T(h_1(EQ(h_1,h_2)) \cap R)$. There exist $x = b_{i_1} b_{i_2} \ldots b_{i_s} \in EQ(h_1,h_2)$ and $y = h_1(x) = h_2(x)$ such that $y \in R$ and $w = pr_T(y)$. Let $y = a_1 a_2 \ldots a_t, a_i \in V_2, 1 \leq i \leq t$. There is a sequence $s_{j_1}, s_{j_2}, \ldots, s_{j_{t+1}}$ of states in K such that $s_{j_1} = s_0, s_{j_{t+1}} \in F$, and $\delta(s_{j_k}, a_k) = s_{j_{k+1}}$, $1 \leq k \leq t$. Note that $h_1(x) = h_1(b_{i_1}) \ldots h_1(b_{i_s}) = a_1 a_2 \ldots a_t$. Let $h_1(b_{i_k}) = a_{q_k} \ldots a_{q_{k+1}-1}, 1 \leq k < s$, and $h_1(b_{i_s}) = a_{q_s} \ldots a_t$. Similarly, let $h_2(b_{i_k}) = a_{p_k} \ldots a_{p_{k+1}-1}, 1 \leq k < s$, and $h_2(b_{i_s}) = a_{p_s} \ldots a_t$. Then there is a computation σ using the following blocks from D_l:

$$r_l(i_1, j_{p_1}, j_{q_1+1}), \ldots, r_l(i_s, j_{p_s}, j_{q_s+1}), r_l(n, 0, i_t),$$

and the following blocks from D_u:

$$r_u(i_1, j_{q_1+1}, j_{p_1}), \ldots, r_u(i_s, j_{q_s+1}, j_{p_s}), r_u(n, i_t, 0).$$

Denote by z the result of this computation. By the definition of h we have $w = h(z)$. It is also easy to see that $e_l(\sigma) = e_u(\sigma)$, hence $w \in h(L_c(\gamma))$.

A complete computation in γ cannot be continued: there is no pair of blocks in D_l, D_u starting with two symbols which are complementary in the sense of the relation ρ. Therefore, $L_c(\gamma) = L_{pc}(\gamma)$, which completes the proof. □

In the case of fair computations we obtain non-regular languages, but not a characterization of recursively enumerable languages.

By a slight modification of the construction in the proof of Theorem 4.8. we get the following result.

Corollary 4.10. *Each regular language is the coding of a language in the family $SRSL(f)$.*

Proof. For a regular grammar $G = (N, T, S, P)$ we construct the simple regular sticker system

$$\gamma = (V, \rho, A, D_l, D_u),$$

with V and ρ as in the proof of Theorem 4.8, D_l containing the dominoes $\binom{\lambda}{y}$ for $(\binom{\lambda}{\lambda}, \binom{\lambda}{y}) \in D$ and D_u containing the dominoes $\binom{y}{\lambda}$ for $(\binom{\lambda}{\lambda}, \binom{y}{\lambda}) \in D$, for D as in the proof of Theorem 4.8, and with

$$\begin{aligned}
A = \{ & \begin{bmatrix}[S,a]_1\\ [S,a]_1\end{bmatrix} \binom{\lambda}{[X,b]_2} \mid S \to aX \in P, X \to bY \in P,\\
& X, Y \in N, a, b \in T\}\\
\cup \{ & \begin{bmatrix}[S,a]_1[X,b]_2\\ [S,a]_1[X,b]_2\end{bmatrix} \binom{\lambda}{[Y,c]_1} \mid S \to aX \in P, X \to bY \in P,\\
& Y \to cZ \in P, X, Y, Z \in N, a, b, c \in T\}\\
\cup \{ & \begin{bmatrix}[S,a]_1\\ [S,a]_1\end{bmatrix} \mid S \to a \in P, a \in T\}\\
\cup \{ & \begin{bmatrix}[S,a]_1[X,b]_2\\ [S,a]_1[X,b]_2\end{bmatrix} \mid S \to aX \in P, X \to b \in P, X \in N, a, b \in T\}\\
\cup \{ & \begin{bmatrix}[S,a]_1[X,b]_2[Y,c]_1\\ [S,a]_1[X,b]_2[Y,c]_1\end{bmatrix} \mid S \to aX \in P, X \to bY \in P,\\
& Y \to c \in P, X, Y \in N, a, b, c \in T\}.
\end{aligned}$$

It is easy to see that each string in $L(G)$ has a fair computation in γ: we can choose that axiom in A which ensures the use of an element of D_l in the last step, hence the number of blocks added in the upper strand is equal to the number of the blocks added in the lower strand. With the same coding h as in the proof of Theorem 4.8, we obtain $L(G) = h(L_f(\gamma))$. □

On the other hand, by imposing the fairness condition we can generate non-regular languages.

Theorem 4.13. $SRSL(f) - REG \neq \emptyset$.

Proof. Let us consider the sticker system

$$\begin{aligned}
\gamma &= (\{a, b\}, \{(a, a), (b, b)\}, A, D_l, D_u),\\
A &= \{\begin{bmatrix}a\\ a\end{bmatrix} \binom{a}{\lambda}\},\\
D_l &= \{\binom{\lambda}{a}, \binom{\lambda}{bb}\},\\
D_u &= \{\binom{aa}{\lambda}, \binom{b}{\lambda}\}.
\end{aligned}$$

There is only one axiom. If we use the domino $\binom{\lambda}{a}$, then we obtain a complete computation which is not fair. We can continue with any element of D_l and D_u. Due to the complementarity restriction, if a symbol b is introduced, then we have to continue by using dominoes $\binom{\lambda}{bb}, \binom{b}{\lambda}$, until obtaining again a complete computation.

Let us intersect the language $L_f(\gamma)$ with the regular language a^+b^+. We obtain a language consisting of strings of the form $a^{2n+2}b^m$, with $n \geq 0$, $m \geq 1$, produced by computations where:

- the first element of D_l is used $2n+1$ times,
- the second element of D_l is used $\frac{m}{2}$ times,
- the first element of D_u is used n times,
- the second element of D_u is used m times.

Because $\frac{m}{2}$ is an integer, we must have $m = 2k, k \geq 1$. Using the fairness, we obtain

$$2n + 1 + k = n + 2k,$$

which implies

$$n = k - 1.$$

The language

$$L_f(\gamma) \cap a^+b^+ = \{a^{2k}b^{2k} \mid k \geq 1\}$$

is not regular, hence $L_f(\gamma)$ is not regular either. □

Theorem 4.14. $SRSL(f) \subseteq MAT^{\lambda}$.

Proof. Consider a sticker system $\gamma = (V, \rho, A, D_l, D_u)$. Define

$$\begin{aligned}
V' &= \{a' \mid a \in V\},\\
L(A) &= \{[a_1, b_1'] \dots [a_k, b_k'] a_{k+1} \dots a_{k+r} \mid k \geq 1, r \geq 0,\\
&\qquad \begin{bmatrix} a_1 \dots a_k \\ b_1 \dots b_k \end{bmatrix} \begin{pmatrix} a_{k+1} \dots a_{k+r} \\ \lambda \end{pmatrix} \in A, a_i, b_i \in V \text{ for all } i\}\\
&\quad \cup \{[a_1, b_1'] \dots [a_k, b_k'] b_{k+1}' \dots b_{k+r}' \mid k \geq 1, r \geq 0,\\
&\qquad \begin{bmatrix} a_1 \dots a_k \\ b_1 \dots b_k \end{bmatrix} \begin{pmatrix} \lambda \\ b_{k+1} \dots b_{k+r} \end{pmatrix} \in A, a_i, b_i \in V \text{ for all } i\},\\
L(D_l) &= \{b_1' \dots b_k' \mid k \geq 1, \begin{pmatrix} \lambda \\ b_1 \dots b_k \end{pmatrix} \in D_l, b_i \in V, 1 \leq i \leq k\},\\
L(D_u) &= \{a_1 \dots a_k \mid k \geq 1, \begin{pmatrix} a_1 \dots a_k \\ \lambda \end{pmatrix} \in D_u, a_i \in V, 1 \leq i \leq k\}.
\end{aligned}$$

Consider the new symbols s, d, d' and construct the languages

$$\begin{aligned}
L_1 &= \{xd' \mid x \in L(D_l)\}^+, \\
L_2 &= \{xd \mid x \in L(D_u)\}^+, \\
L_1' &= L_1 \sqcup\!\sqcup\, c^+, \\
L_2' &= L_2 \sqcup\!\sqcup\, c^+, \\
L_3 &= (L(A)L_1' \sqcup\!\sqcup\, L_2') \cap \{[a, b'] \mid a, b \in V\}^*(VV' \cup \{cd', dc\})^*.
\end{aligned}$$

Clearly, L_1, L_2 are regular languages, hence also L_3 is regular: the family REG is closed under the shuffle operation and under intersection.

Consider the gsm g which:

- leaves unchanged the symbols $[a, b'], a, b \in V$,
- replaces each pair ab' by $[a, b'], a, b \in V$,
- replaces each pair cd' by $[c, d']$ and each pair dc by $[d, c]$.

The language $g(L_3)$ is also regular, over the alphabet

$$U = \{[a, b'] \mid a, b \in V\} \cup \{[c, d'], [d, c]\}.$$

Let $G = (N, U, S, P)$ be a regular grammar for $g(L_3)$ and construct the matrix grammar

$$G' = (N', V, S', M),$$

where

$$\begin{aligned}
N' &= N \cup U \cup \{S'\}, \\
M &= \{(S' \to S)\} \cup \{(r) \mid r \in P\} \\
&\cup \{([a, b'] \to a) \mid a, b \in V\} \\
&\cup \{([c, d'] \to \lambda, [d, c] \to \lambda)\}.
\end{aligned}$$

It is easy to see that $L(G')$ contains all the strings $w \in V^*$ such that $\binom{u_1}{u_2}\begin{bmatrix} x_1 \\ x_2 \end{bmatrix}\binom{v_1}{v_2} \Longrightarrow^* \begin{bmatrix} w \\ w' \end{bmatrix}$ in γ, $\binom{u_1}{u_2}\begin{bmatrix} x_1 \\ x_2 \end{bmatrix}\binom{v_1}{v_2} \in A$, and this is a fair derivation: the matrix $([c, d'] \to \lambda, [d, c] \to \lambda)$ checks whether or not the number of symbols d and d' is the same.

Therefore, $L_f(\gamma) \in MAT^\lambda$. □

Because the family MAT^λ is strictly included in RE and it is closed under arbitrary gsm mappings, we cannot obtain characterizations of RE starting from languages in the family $SRSL(f)$ and using codings, morphisms, or gsm mappings as squeezing mechanisms.

Open problem. Is the family $SRSL(f)$ included in CF (or in LIN)?

4.7 Bibliographical Notes

The sticker systems were introduced in [101] in the form of regular simple sticker systems. The primitive, balanced, coherent, and fair computations are also considered in [101]; the results in Sect. 4.6 are from this paper, as well as weaker counterparts of Theorem 4.1, Corollary 4.1, and Theorems 4.4 and 4.8. The sticker systems prolonging the sequences in both directions were introduced in [64], in the simple variant, under the name of bidirectional sticker systems. The bounded delay is also introduced in this paper, where the equalities $ASL(b) = LIN, OSL(b) = REG$ appear, as well. Theorem 4.11 and Corollary 4.9 are from [64], too.

Sticker systems in the general form (using dominoes of arbitrary shapes) are investigated in [165], where the results not mentioned above appear in the general framework used also in this chapter.

Chapter 5

Watson–Crick Automata

In this chapter we investigate the automata counterpart of the sticker systems studied in the previous chapter. We consider a new type of automata, working on tapes which are double stranded sequences of symbols related by a complementarity relation, similar to a DNA molecule (such a data structure is called a *Watson–Crick tape*). The automata scan separately each of the two strands, in a correlated manner. They can also have a finite number of states controlling the moves and/or they can have an auxiliary memory which is also a Watson–Crick tape, used in a FIFO-like manner. Combining such possibilities we obtain several types of automata. In most cases, these automata augmented with squeezing mechanisms, such as weak codings and deterministic sequential transducers, characterize the recursively enumerable languages.

We stress the essential difference between these automata and the customary ones, a difference based on the data structures they handle. While the customary automata operate on linear (one-dimensional) strings of symbols, our automata take double strands as their objects. Moreover, the double strands resemble DNA molecules in the following sense. The matching letters (nucleotides) are *complementary*, the relation of complementarity being defined for pairs of letters of the basic alphabet, similarly to the Watson–Crick complementarity of the pairs (A, T) and (C, G) of the DNA alphabet. Most importantly, we assume that such data structures, double strands satisfying the complementarity requirement mentioned, are freely available in the sense that we do not have to check in any way that the matching letters are indeed complementary.

Because of the complementarity, these automata are called *Watson–Crick automata*. Our main interest is in the basic variant, where the automaton scans separately each of the two strands in a correlated manner. However, we will also investigate other variants, such as transducers and automata with an auxiliary tape.

These automata make use of only one of the two essential features of the

DNA as a possible support for computations, the Watson–Crick complementarity (which renders the power of the twin-shuffle language available), but not of the second one, the multitude of DNA molecules which brings the *massive parallelism* to the computing scene. It remains an entirely open area to model this second feature, as well as to combine the two features into one model of DNA computing.

5.1 Watson–Crick Finite Automata

We are now going to define one of the classes of automata we have announced above. They are a counterpart of finite automata (they use states which control the transitions, as usual in automata theory), but work on Watson–Crick tapes, that is, on elements of $WK_\rho(V)$, for some alphabet V and its complementarity relation $\rho \subseteq V \times V$. (We use the notations established in the previous chapter.)

A *Watson–Crick finite automaton* is a construct

$$M = (V, \rho, K, s_0, F, \delta),$$

where V and K are disjoint alphabets, $\rho \subseteq V \times V$ is a symmetric relation, $s_0 \in K$, $F \subseteq K$, and $\delta : K \times \begin{pmatrix} V^* \\ V^* \end{pmatrix} \longrightarrow \mathcal{P}(K)$ is a mapping such that $\delta(s, \begin{pmatrix} x \\ y \end{pmatrix}) \neq \emptyset$ only for finitely many triples $(s, x, y) \in K \times V^* \times V^*$.

The elements of K are called *states*, V is the (input) alphabet, ρ is a complementarity relation on V, s_0 is the initial state, F is the set of final states, and δ is the transition mapping. The interpretation of $s' \in \delta(s, \begin{pmatrix} x_1 \\ x_2 \end{pmatrix})$ is: in state s, the automaton passes over x_1 in the upper level strand and over x_2 in the lower level strand of a double stranded sequence, and enters the state s'.

As in the case of finite automata, we can also write the transitions of M as rewriting rules of the form $s \begin{pmatrix} x_1 \\ x_2 \end{pmatrix} \rightarrow \begin{pmatrix} x_1 \\ x_2 \end{pmatrix} s'$; such a rule has the same meaning as $s' \in \delta(s, \begin{pmatrix} x_1 \\ x_2 \end{pmatrix})$.

Remark 5.1. In contrast to the case of finite automata, in Watson–Crick finite automata we have written first the alphabet V and the complementarity relation ρ, and after that the set K of states, in order to stress the fact that the pair (V, ρ) plays a fundamental role in our machines. Working with double stranded sequences is the crucial difference between traditional finite automata and Watson–Crick finite automata. □

A transition in a Watson–Crick finite automaton can be defined as follows: For $\binom{x_1}{x_2}, \binom{u_1}{u_2}, \binom{w_1}{w_2} \in \binom{V^*}{V^*}$ such that $\begin{bmatrix} x_1u_1w_1 \\ x_2u_2w_2 \end{bmatrix} \in WK_\rho(V)$ and $s, s' \in K$, we write

$$\binom{x_1}{x_2} s \binom{u_1}{u_2} \binom{w_1}{w_2} \Longrightarrow \binom{x_1}{x_2} \binom{u_1}{u_2} s' \binom{w_1}{w_2} \text{ iff } s' \in \delta(s, \binom{u_1}{u_2}).$$

We denote by $\Longrightarrow^*$ the reflexive and transitive closure of the relation $\Longrightarrow$.

As in the case of sticker systems, we investigate here the language of strings appearing in the upper strands of Watson–Crick tapes recognized by our automata, that is the language

$$L_u(M) = \{w_1 \in V^* \mid s_0 \begin{bmatrix} w_1 \\ w_2 \end{bmatrix} \Longrightarrow^* \begin{bmatrix} w_1 \\ w_2 \end{bmatrix} s_f, \text{ for } s_f \in F,$$
$$\text{and } w_2 \in V^*, \begin{bmatrix} w_1 \\ w_2 \end{bmatrix} \in WK_\rho(V)\}.$$

Remark 5.2. Of course, we can also consider the language of strings appearing in the lower strand, as well as the language of molecules, but we do not discuss such languages here (they are linked to $L_u(M)$ by the relation ρ; when ρ is injective, the three languages are isomorphic). □

Another important language associated to a Watson–Crick automaton can be defined taking into account the transitions, not the recognized sequence.

For a Watson–Crick finite automaton $M = (V, \rho, K, s_0, F, P)$ (hence with the transition rules written as rewriting rules) consider a labeling $e : P \longrightarrow Lab$, of rules in P with elements in a set Lab. For a computation $\sigma : s_0 w \Longrightarrow^* w s_f$, $w \in WK_\rho(V)$, $s_f \in F$, denote by $e(\sigma)$ the control word of σ, that is the sequence of labels of transition rules used in σ. In this way we obtain the language

$$L_{ctr}(M) = \{e(\sigma) \mid \sigma : s_0 w \Longrightarrow^* w s_f, w \in WK_\rho(V), s_f \in F\}.$$

Remark 5.3. The control word $e(\sigma)$ associated to a computation σ in a Watson–Crick finite automaton can be particularly useful in DNA computing, where we work with words over a prescribed reduced alphabet, hence we need codifications of symbols of larger alphabets arising from the problems we want to solve. Consider, for instance, the very first experiment in DNA computing, that was considered in Sect. 2.1. Associate a Watson–Crick automaton to a graph by using the codes of nodes in the upper level and the codes of the edges in the lower level when defining the transitions. Let each transition parse either a node or an edge. Label each transition with the name of the corresponding node or edge. Then the control word of a computation will be a shuffle of the description of the path associated to our computation, written as a sequence of nodes and simultaneously as a sequence of edges. By a weak coding, we can select from the control word the path description we want. Thus, in this case, the control word of a computation is more explicit than the recognized sequence. In particular, like in the Adleman's experiment, we can let the automaton work on nondeterministically chosen sequences, selecting the control words of interest. □

We say that the languages $L_\alpha(M), \alpha \in \{u, ctr\}$, are *recognized* by the Watson–Crick finite automaton M.

We note again that the work of Watson–Crick automata is defined for elements of $WK_\rho(V)$ only, that is, for double stranded sequences of elements in V paired according to the complementarity relation ρ. We can represent such a machine as consisting of a double tape on which an element of $WK_\rho(V)$ is written, a finite memory, able to store a state from a finite set of states, and two read only heads, one of them scanning the upper level and the other one scanning the lower level of the tape. Start with the two heads placed before the first symbol of each level, in state s_0. The two heads are moved to the right, according to the current state of the machine, as indicated by the transition mapping (the transition rules). Here a transition step means to move the two heads across blocks defined by a specific transition rule. Stop and accept the starting sequence when both heads reach the right hand end of the sequence written on the tape, entering a final state. Fig. 5.1 illustrates this representation.

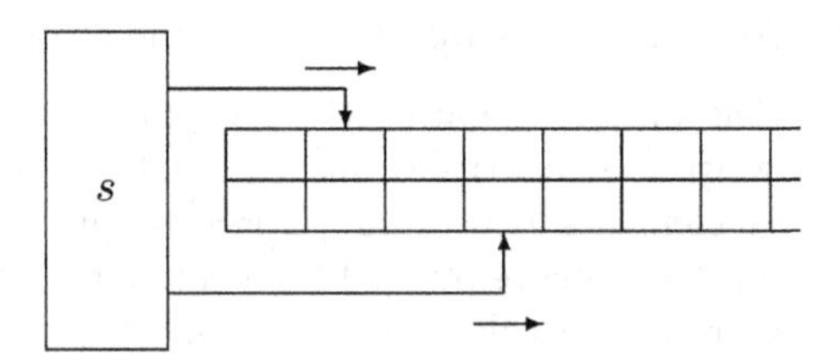

Figure 5.1: A Watson–Crick finite automaton

We consider also several variants of Watson–Crick finite automata. We say that $M = (V, \rho, K, s_0, F, P)$ is:

- *stateless*, if $K = F = \{s_0\}$;

- *all-final*, if $F = K$,

- *simple*, if for all $s \binom{x_1}{x_2} \to \binom{x_1}{x_2} s' \in P$ we have either $x_1 = \lambda$ or $x_2 = \lambda$,

- 1-*limited*, if for all $s \binom{x_1}{x_2} \to \binom{x_1}{x_2} s' \in P$ we have $|x_1 x_2| = 1$.

In the stateless automata, the components K, s_0, F can be omitted, and the transition rules can be written in the form $\begin{pmatrix} x_1 \\ x_2 \end{pmatrix}$. Then, the automaton is written in the form $M = (V, \rho, P)$.

We denote by $AWK(\alpha)$, $NWK(\alpha)$, $FWK(\alpha)$, $SWK(\alpha)$, $1WK(\alpha)$, $NSWK(\alpha)$, $N1WK(\alpha)$, $FSWK(\alpha)$, $F1WK(\alpha)$, the families of languages of the form $L_\alpha(M)$, $\alpha \in \{u, ctr\}$, recognized by Watson–Crick finite automata which are arbitrary (A), stateless (N, from "no state"), all-final (F), simple (S), 1-limited (1), stateless and simple (NS), stateless and 1-limited (N1), all-final and simple (FS), and all-final and 1-limited (F1), respectively. (The basic abbreviation, WK, is obtained by selecting the beginning and the end symbols of the single stranded sequence W A T S O N C R I C K.) We will use the generic term *WK families* to refer to all these language families.

5.2 Relationships Between the WK Families

In this section we investigate the relationships between the families of languages in the previous section, as well as the relationships of these families to the families in the Chomsky hierarchy.

Directly from the definitions we obtain:

Lemma 5.1. *$XWK(\alpha) \subseteq AWK(\alpha)$, $\alpha \in \{u, ctr\}, X \in \{N, F, S, 1, NS, N1, FS, F1\}$.*

Lemma 5.2. *$NWK(\alpha) \subseteq FWK(\alpha)$, $NSWK(\alpha) \subseteq FSWK(\alpha), N1WK(\alpha) \subseteq F1WK(\alpha)$, $\alpha \in \{u, ctr\}$.*

Lemma 5.3. *$XSWK(\alpha) \subseteq SWK(\alpha)$, $X1WK(\alpha) \subseteq 1WK(\alpha)$, $X1WK(\alpha) \subseteq XSWK(\alpha) \subseteq XWK(\alpha)$, $1WK(\alpha) \subseteq SWK(\alpha)$, $\alpha \in \{u, ctr\}$, $X \in \{N, F\}$.*

Moreover, it is easy to see that we also have the following relations:

Lemma 5.4. *$REG \subseteq 1WK(u)$.*

Lemma 5.5. *$AWK(\alpha) \subseteq CS$, $\alpha \in \{u, ctr\}$.*

Lemma 5.6. *Each language in a family $XWK(u)$ is a coding of a language in the family $XWK(ctr), X \in \{A, N, F, S, 1, NS, N1, FS, F1\}$.*

The use of states is powerful, in the sense that arbitrary transition rules can be replaced by simple transition rules without decreasing the power. The following lemma can also be viewed as a "normal form" result, customary in automata theory.

Lemma 5.7. *$AWK(u) \subseteq 1WK(u)$.*

Proof. Consider an unrestricted Watson–Crick finite automaton $M = (V, \rho, K, s_0, F, P)$ and construct the 1-limited Watson–Crick automaton

$$M' = (V, \rho, K', s_0, F, P'),$$

as follows.

For each transition rule

$$t : s \binom{a_1 a_2 \dots a_n}{b_1 b_2 \dots b_m} \rightarrow \binom{a_1 a_2 \dots a_n}{b_1 b_2 \dots b_m} s'$$

in P, $n \geq 0, m \geq 0, n + m \geq 2$, we introduce in P' the transitions

$$\begin{aligned}
& s \binom{a_1}{\lambda} \rightarrow \binom{a_1}{\lambda} s_{t,1}, \\
& s_{t,i} \binom{a_{i+1}}{\lambda} \rightarrow \binom{a_{i+1}}{\lambda} s_{t,i+1}, \ 1 \leq i \leq n-1, \\
& s_{t,n} \binom{\lambda}{b_1} \rightarrow \binom{\lambda}{b_1} s'_{t,1}, \\
& s'_{t,i} \binom{\lambda}{b_{i+1}} \rightarrow \binom{\lambda}{b_{i+1}} s'_{t,i+1}, \ 1 \leq i \leq m-2, \\
& s'_{t,m-1} \binom{\lambda}{b_m} \rightarrow \binom{\lambda}{b_m} s';
\end{aligned}$$

all states $s_{t,i}, s'_{t,i}$ are introduced in K', together with all states in K.

One can easily see that the obtained automaton is equivalent with M (the new states control the work of M' in a deterministic way) and 1-limited. □

Corollary 5.1. $1WK(u) = SWK(u) = AWK(u)$.

The construction above modifies the language $L_{ctr}(M)$; we do not know whether or not the inclusion in Lemma 5.7 also holds for families $1WK(ctr), SWK(ctr), AWK(ctr)$.

For an easy reference, we summarize the relations from the previous lemmas for families $XWK(ctr)$ in the diagram in Fig. 5.2; the arrows indicate inclusions which are not necessarily proper. The case of families $XWK(u)$ is postponed until new relations are established between them.

Remark 5.4. A notion which is related to the devices defined above is that of *two-head finite automata.*

A two-head finite automaton is a construct $M = (K, V, s_0, F, \delta)$, where K, V, s_0, F are as in a usual finite automaton and δ is the transition mapping, $\delta : K \times (V \cup \{\lambda\}) \times (V \cup \{\lambda\}) \longrightarrow \mathcal{P}(K)$. For $w_1, w_2, x_1, x_2 \in V^*$, $u_1, u_2 \in V \cup \{\lambda\}$, and $s, s' \in K$ we write

$$(w_1, w_2)s(u_1 x_1, u_2 x_2) \Longrightarrow (w_1 u_1, w_2 u_2)s'(x_1, x_2) \text{ iff } s' \in \delta(s, u_1, u_2).$$

The language recognized by M is defined by

$$L(M) = \{x \in V^* \mid s_0(x, x) \Longrightarrow^* (x, x)s_f, s_f \in F\}.$$

We denote by TH the family of languages recognized by such automata.

Some variants of two-head (or, more generally, multihead) finite automata were intensively investigated: *deterministic, simple* (one head reads the tape, the others can only distinguish the end markers of the input string), *sensing* (the heads can sense the case when two of them are placed in the same cell of the tape). Precise definitions, results and further references can be found in [49], [94], [97], [186].

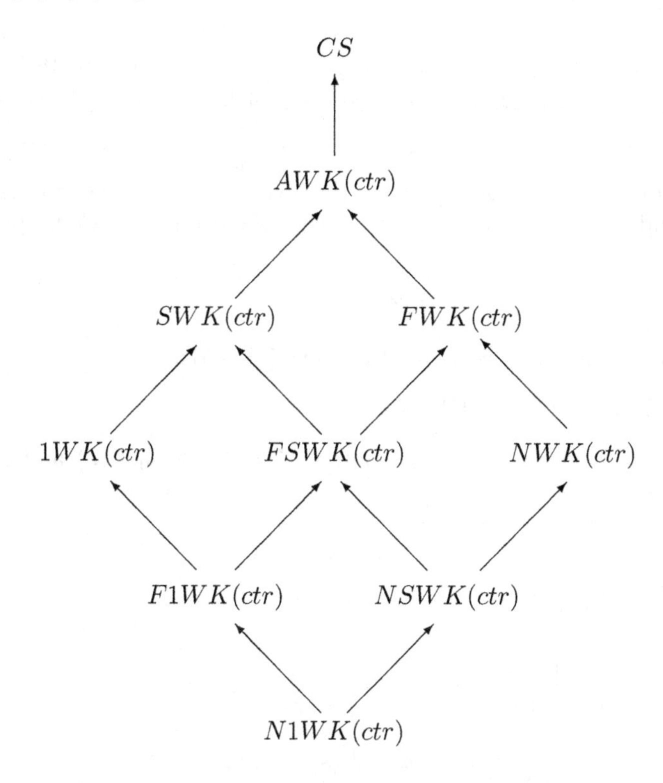

Figure 5.2: The hierarchy of *ctr* families

It is obvious that a two-head finite automaton is a particular case of a 1-limited Watson–Crick finite automaton: the complementarity relation is the identity, $(a, b) \in \rho$ if and only if $a = b$.

On the other hand, a 1-limited Watson–Crick finite automaton can be simulated by a two-head finite automaton: one head parses the input string acting as the upper head of the Watson–Crick automaton, the second one parses this string but acts as the lower head of the Watson–Crick automaton: it guesses a complement of the current symbol and it crosses a symbol a only if the lower head of the Watson–Crick automaton can cross – in the same state – a symbol b which is complementary to a.

Consequently, we get the following equality:

Lemma 5.8. *$TH = 1WK(u)$.*

A Watson–Crick finite automaton can be also viewed as a two-tape, two-head finite automaton, but of a very particular type: the two tapes are interrelated through the complementarity relation; if this relation is one-to-one, then one tape precisely identifies the other one – this is the case of the DNA molecules. □

In the simple stateless case, the parsing of a sequence in $WK_\rho(V)$ can be controlled by examining a subsequence of length at most the length of the longest string w_1, w_2 in transition rules $\binom{w_1}{w_2}$: because one of w_1, w_2 is always empty, we can continue with the level whose reading head is behind, thus bounding the distance (delay) between the two heads. Consequently, we obtain:

Lemma 5.9. *$NSWK(u) \subseteq REG$.*

The following strenghtening of Lemma 5.4 holds.

Lemma 5.10. *$REG \subseteq F1WK(u)$.*

Proof. Consider a finite automaton $M = (K, V, s_0, F, \delta)$ and construct the all-final 1-limited Watson–Crick finite automaton

$$M' = (V, \rho, K', s_0, K', \delta'),$$

with

$$\begin{aligned}
&\rho = \{(a, a)) \mid a \in V\},\\
&K' = K \cup \{s_f\}, \text{ (for } s_f \notin K),\\
&\delta'(s, \binom{a}{\lambda}) = \delta(s, a) \cup F(s, a), \;\; s \in K, a \in V,\\
&\qquad\qquad \text{where } F(s, a) = \begin{cases} \{s_f\}, & \text{if } \delta(s, a) \cap F \neq \emptyset,\\ \emptyset, & \text{otherwise,} \end{cases}\\
&\delta'(s_f, \binom{\lambda}{a}) = \{s_f\}, \; a \in V,\\
&\delta'(s, \binom{u}{v}) = \emptyset, \text{ in all other cases.}
\end{aligned}$$

The recognition of a sequence $\begin{bmatrix} w \\ w \end{bmatrix}$ proceeds as follows: one first parses the first strand, from left to right, exactly as in M, except for the last step, when M reaches a final state; then M' enters the state s_f. Then one can also parse the second strand, making possible the completion of recognition.

Therefore, $L_u(M') = L(M)$. □

The relations between families $XWK(u)$ can now be synthesized as in the diagram in Fig. 5.3; as usual, the arrows indicate inclusions which are not necessarily proper.

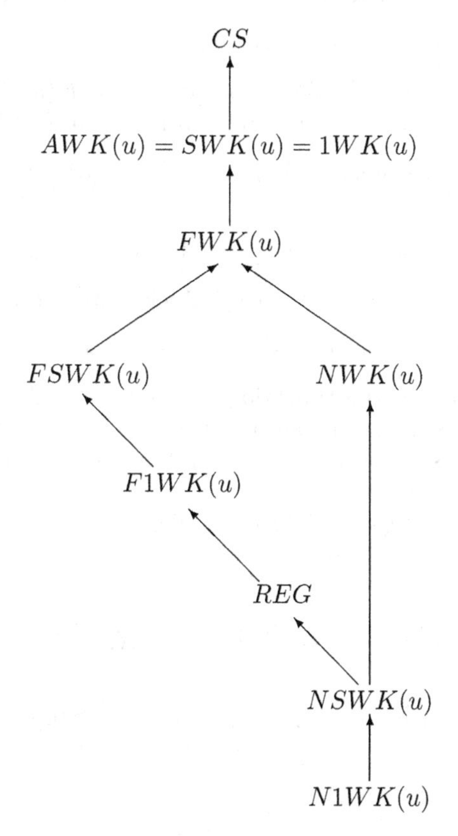

Figure 5.3: The hierarchy of u families

In some sense, the families above are "small": the languages in these families satisfy very strong conditions. The following lemmas provide two such necessary conditions.

Lemma 5.11. (i) *If* $L \in NWK(\alpha)$, *for* $\alpha \in \{u, ctr\}$, *then* $L = L^+$. (ii) *If* $L \in N1WK(u)$, *then there is an alphabet* V *such that* $L = V^+$.

Proof. (i) Consider a stateless Watson–Crick finite automaton $M = (V, \rho, P)$. If $w_1, w_2 \in WK_\rho(V)$ can be parsed by M, then $w_1 w_2$ can be parsed as well, using the same elements of P. Hence, $L_\alpha(M)^+ \subseteq L_\alpha(M), \alpha \in \{u, ctr\}$. The opposite inclusion, $L_\alpha(M) \subseteq L_\alpha(M)^+$, is obvious.

(ii) Obvious, because each string can be recognized by a stateless 1-limited Watson–Crick automaton. □

We shall see below (Lemma 5.14) that assertion (ii) is not true for the *ctr* case.

Corollary 5.2. $REG - XWK(\alpha) \neq \emptyset,\ \alpha \in \{u, ctr\}, X \in \{N, NS\}$.

Corollary 5.3. $F1WK(u) - NWK(u) \neq \emptyset$.

Proof. By Lemma 5.10, the language ab^* is in $F1WK(u)$. This language does not have the property stated in Lemma 5.11, so $ab^* \notin NWK(u)$. □

Corollary 5.4. *The inclusions* $NWK(u) \subset FWK(u)$ *and* $NSWK(u) \subset F1WK(u)$ *are proper.*

Lemma 5.12. *Every one-letter language in* $AWK(u)$ *is regular.*

Proof. Consider an unrestricted Watson–Crick finite automaton $M = (V, \rho, K, s_0, F, \delta)$. If there is a transition $s' \in \delta(s, \binom{a^i}{w})$ such that w contains a symbol b and $(a, b) \notin \rho$, then this transition can never be used when producing strings in $L_u(M)$. Thus, all such transitions can be ignored, that is we may assume that for all b as above we have $(a, b) \in \rho$. We construct the linear grammar

$$G = (K, \{a, b\}, s_0, P),$$

with

$$P = \{s \to a^i s' b^j \mid s' \in \delta(s, \binom{a^i}{w}), s, s' \in K, i \geq 0, j = |w|\}$$
$$\cup\ \{s \to a^i b^j \mid \delta(s, \binom{a^i}{w}) \cap F \neq \emptyset, s \in K, i \geq 0, j = |w|\}.$$

Consider also the linear language

$$L = \{a^n b^n \mid n \geq 1\}.$$

According to [77], Corollary 5.3.1, $L(G) \cap L$ is a linear language. (We have $L(G) \cap L = \{a^n b^m \mid (n, m) \in \Psi_{\{a,b\}}(L(G)) \cap \{(p, p) \mid p \in \mathbf{N}\}\}$; because the intersection of two semilinear sets is a semilinear set, it follows that $\Psi_{\{a,b\}}(L(G) \cap L)$ is a semilinear set. Together with the above mentioned result from [77], this implies that $L(G) \cap L$ is a linear language.) For the weak coding h defined by

$$h(a) = a,\ h(b) = \lambda,$$

we have

$$L_u(M) = h(L(G) \cap L),$$

which implies that $L_u(M)$ is regular. □

Corollary 5.5. $AWK(u) \subset CS$ *is a strict inclusion.*

However, the families discussed above (with the exception of $NSWK(u)$) contain languages of a very high complexity. We shall see in the next section that characterizations of recursively enumerable languages can be obtained starting from them and using AFL operations (in fact, weak codings and deterministic gsm's). We close this section with a result proving directly that stateless Watson–Crick finite automata can recognize complex languages. This is, of course, due to the free availability of double stranded sequences.

Lemma 5.13. $NWK(u) - MAT^{\lambda} \neq \emptyset$.

Proof. Consider the stateless Watson–Crick finite automaton

$$M = (\{a, b, c, d, e, f\}, \rho, P),$$

with

$$\rho = \{(a, a), (b, c), (c, b), (a, d), (d, a), (e, f), (f, e)\},$$
$$P = \{\binom{c}{\lambda}, \binom{d}{\lambda}, \binom{a}{a}, \binom{b}{b}, \binom{b}{c}, \binom{e}{c}, \binom{e}{d}, \binom{\lambda}{f}\}.$$

Consider also the regular language described by the following regular expression

$$R = c(dd^{+}b)(aa^{+}b)^{+}a^{+}e^{+},$$

and the weak coding h defined by

$$h(a) = a, \ h(b) = h(c) = h(d) = h(e) = \lambda.$$

The molecules recognized by M and having a string in R in their upper strand must be of the form

$$\begin{bmatrix} cd^{n_1}ba^{n_2}b \dots ba^{n_{m-1}}ba^{n_m}e^{n_{m+1}} \\ b\,x_1\,c\,x_2\,c \dots c\,x_{m-1}\,c\,x_m\,f^{n_{m+1}} \end{bmatrix},$$

with $m \geq 3, n_i \geq 2, 1 \leq i \leq m, n_{m+1} \geq 1$. Because of the complementarity, we also have $x_1 = a^{n_1}$, $x_i \in \{a, d\}^*$, $|x_i| = n_i$, for $2 \leq i \leq m$.

Each b in the upper strand is paired with an occurrence of b or c in the lower strand, because of the form of the pairs in P, as suggested by the subscripts of these symbols b, c in the following writing:

$$\begin{bmatrix} c_0d^{n_1}b_1a^{n_2}b_2 \dots b_{m-2}a^{n_{m-1}}b_{m-1}a^{n_m}e_me^{n_{m+1}-1} \\ b_1\,x_1\,c_2\,x_2\,c_3\,\dots\,c_{m-1}\,x_{m-1}\,c_m\,x_m\,f\,f^{n_{m+1}-1} \end{bmatrix}.$$

We have also indicated that the last occurrence of c in the lower strand is paired with the first occurrence of e in the upper strand.

Now, because of this precise pairing and because (1) the symbol a apears only in pairs $\binom{a}{a}$ of P, and (2) the symbol d can be introduced in the lower strand only by the pair $\binom{e}{d}$, it follows that:

1. $x_i \in a^+, 2 \leq i \leq m-1$,
2. $x_m = d^{n_{m+1}-1}$,
3. $|x_i| = n_{i+1}, 1 \leq i \leq m-1$.

From $|x_i| = n_i$ and $|x_i| = n_{i+1}$ we get $n_i = n_{i+1}, 1 \leq i \leq m-1$.

Therefore, the recognized molecule has the form

$$\begin{bmatrix} cd^n ba^n b \dots ba^n ba^n e^{n+1} \\ ba^n ca^n c \dots ca^n cd^n f^{n+1} \end{bmatrix},$$

with at least two blocks ba^n and $n \geq 2$.

Consequently, we obtain

$$\begin{aligned} h(L_u(M) \cap R) &= \{a^{nm} \mid n, m \geq 2\} \\ &= \{a^p \mid p \text{ is a composite number}\}. \end{aligned}$$

This language is not semilinear, hence is not regular; each one-letter language in MAT^λ is regular [85]; the family MAT^λ is closed under intersection with regular languages and arbitrary morphisms. Consequently, $L_u(M) \notin MAT^\lambda$. □

Corollary 5.6. *The inclusion* $NSWK(u) \subset NWK(u)$ *is proper.*

5.3 Characterizations of Recursively Enumerable Languages

We shall now give a series of representation results for recursively enumerable languages starting from languages recognized by Watson–Crick finite automata of various types. In fact, we have characterizations of RE, because RE is closed under the operations applied.

The proof of the following lemma, although technically simple, captures the essence of Watson–Crick tapes and the interconnection to the twin-shuffle language.

Lemma 5.14. *For every alphabet* V, *we have* $TS_V \in N1WK(ctr)$.

Proof. Consider the 1-limited stateless Watson–Crick finite automaton

$$M = (V, \rho, P),$$

with

$$\begin{aligned} \rho &= \{(a,a) \mid a \in V\}, \\ P &= \{\binom{a}{\lambda}, \binom{\lambda}{a} \mid a \in V\}, \end{aligned}$$

with the labeling

$$e(\binom{a}{\lambda}) = a, \ e(\binom{\lambda}{a}) = \bar{a}, \ a \in V.$$

If $\begin{bmatrix} x \\ x \end{bmatrix} \in WK_\rho(V)$ and $\sigma : s_0 \begin{bmatrix} x \\ x \end{bmatrix} \Longrightarrow^* \begin{bmatrix} x \\ x \end{bmatrix} s_0$ is a computation in M (the state s_0 is written only in order to make clear that this is a non-trivial computation), then $e(\sigma) \in x \sqcup\!\sqcup \bar{x}$, hence $e(\sigma) \in TS_V$.

As x above can be any string in V^* and each element of $x \sqcup\!\sqcup \bar{x}$ describes a correct computation in M, we have the equality $L_{ctr}(M) = TS_V$. □

Theorem 5.1. *Each language in the family RE is the image of a deterministic gsm mapping of a language in any family* $XWK(ctr), X \in \{A, N, F, S, 1, NS, N1, FS, F1\}$.

Proof. This is a direct consequence of the previous lemma and Theorem 3.17. □

Lemma 5.15. $SRSL(c) \subseteq FWK(u)$.

Proof. Consider a simple regular sticker system $\gamma = (V, \rho, A, D_l, D_u)$, with the elements of D_l, D_u labeled by a mapping $e : D_l \cup D_u \longrightarrow Lab$.

We construct the all-final Watson–Crick finite automaton

$$M = (V, \rho, \{s_0, s_1\}, s_0, \{s_0, s_1\}, P),$$

with the transition rules

$$s_0 \binom{u_1v_1}{u_2v_2} \rightarrow \binom{u_1v_2}{u_2v_2} s_1, \text{ for } \begin{bmatrix} u_1 \\ u_2 \end{bmatrix} \binom{v_1}{v_2} \in A, \begin{bmatrix} u_1 \\ u_2 \end{bmatrix} \in WK_\rho(V)$$

$$\binom{v_1}{v_2} \in \binom{V^*}{\lambda} \cup \binom{\lambda}{V^*},$$

$$s_1 \binom{x}{y} \rightarrow \binom{x}{y} s_1, \text{ for } \binom{\lambda}{y} \in D_l \text{ and } \binom{x}{\lambda} \in D_u$$

$$\text{such that } e(\binom{\lambda}{y}) = e(\binom{x}{\lambda}).$$

The coherent correct computations in γ can be simulated by correct computations in M and, conversely, the sequences recognized by M are also reached by correct (coherent) computations in γ. Indeed, the fact that we always have to start from an axiom is ensured by the initial state, s_0, while the coherence is ensured by the transition rules, which are defined only for pairs $\binom{x}{y}$ such that $\binom{x}{\lambda} \in D_u$ and $\binom{\lambda}{y} \in D_l$ and these blocks have the same label. Consequently, $L_c(\gamma) = L_u(M)$. □

Theorem 5.2. *Each recursively enumerable language is the weak coding of a language in any family* $XWK(u), X \in \{A, F, S, 1\}$.

Proof. The assertion follows from Lemma 5.15, the inclusions $FWK(u) \subseteq XWK(u), X \in \{A, S, 1\}$ (see Fig. 5.3), and Theorem 4.12. □

Lemma 5.16. *If $h_1, h_2 : V_1^* \longrightarrow V_2^*$ are two morphisms, then $h_1(EQ(h_1, h_2)) \in NWK(u)$.*

Proof. For h_1, h_2 given, we construct the stateless Watson–Crick finite automaton

$$M = (V_2, \rho, P),$$

with

$$\rho = \{(a, a) \mid a \in V_2\},$$
$$P = \{\begin{pmatrix} h_1(b) \\ h_2(b) \end{pmatrix} \mid b \in V_1\}.$$

A sequence $\begin{bmatrix} w_1 \\ w_2 \end{bmatrix} \in WK_\rho(V_2)$ is successfully parsed by M if and only if $w_1 = w_2$ (due to the relation ρ) and $w_1 = h_1(x), w_2 = h_2(x)$, for some $x \in V_1^*$ (due to the form of rules in P). Consequently, $x \in EQ(h_1, h_2)$ and $w_1 \in h_1(EQ(h_1, h_2))$, which implies $L_u(M) = h_1(EQ(h_1, h_2))$. □

Theorem 5.3. *Each language $L \in RE$ can be written in the form $L = h(L' \cap R)$, where $L' \in NWK(u), R \in REG$, and h is a projection.*

Proof. This is a direct consequence of the previous lemma and of Theorem 3.16. □

Theorem 5.3 can be modified to the following, perhaps more interesting, result: every recursively enumerable language is a projection of some language $L_u(M)$, where M is a Watson–Crick finite automaton.

Theorem 5.4. *Each language $L \in RE$ can be written in the form $L = h(L')$, where $L' \in AWK(u)$ and h is a projection.*

Proof. We use the representation of Theorem 3.16 and write $L = h(h_1(EQ(h_1, h_2)) \cap R_0)$, where h is a projection, $h_1, h_2 : V_1^* \longrightarrow V_2^*$ are λ-free morphisms, and $R_0 \subseteq V_2^*$ is a regular language. Let $M_0 = (K, V_2, s_0, F, \delta)$ be a deterministic finite automaton such that $R_0 = L(M_0)$. It suffices to construct a Watson–Crick finite automaton M with the property

$$L_u(M) = h_1(EQ(h_1, h_2)) \cap R_0.$$

Indeed, define a Watson–Crick finite automaton by

$$M = (V_2, \rho, K, s_0, F, \delta'),$$

with

$$\rho = \{(a, a) \mid a \in V_2\},$$

$$\delta'(s, \binom{h_1(a)}{h_2(a)}) = \{s'\}, \text{ whenever } s, s' \in K, a \in V_1,$$
$$\text{and } (s, h_1(a)) \vdash^* (s', \lambda),$$
$$\delta'(s, \binom{u}{v}) = \emptyset, \text{ otherwise.}$$

(Here $\vdash^*$ refers to the transition in the automaton M_0, as explained in Chap. 3.)

Analogously as in Lemma 5.16, it is now seen that M possesses the required property. □

Although a result like that in Lemma 5.16 seems not to be true for the family $FSWK(u)$, a result like that in Theorem 5.3 also holds true for this family.

Theorem 5.5. *Each language $L \in RE$ can be written in the form $L = h(L' \cap R)$, where $L' \in FSWK(u), R \in REG$, and h is a projection.*

Proof. Using Theorem 3.16, we can write L as $L = h(h_1(EQ(h_1, h_2)) \cap R_0)$, where h_1, h_2 are λ-free morphisms, h is a projection, and R_0 is a regular language. For $h_1, h_2 : V_1^* \longrightarrow V_2^*$, we construct the all-final simple Watson–Crick automaton

$$M = (V, \rho, K, s_0, K, \delta),$$

with

$$V = V_2 \cup \{c\},$$
$$\rho = \{(a, a) \mid a \in V_2\} \cup \{(c, c)\},$$
$$K = \{s_0, s_1\} \cup \{s_a \mid a \in V_1\},$$
$$\delta(s_0, \binom{h_1(a)}{\lambda}) = \{s_a\},$$
$$\delta(s_a, \binom{\lambda}{h_2(a)}) = \{s_0\}, \ a \in V_1,$$
$$\delta(s_0, \binom{c}{\lambda}) = \{s_1\},$$
$$\delta(s_1, \binom{\lambda}{c}) = \{s_1\},$$
$$\delta(s, \binom{u}{v}) = \emptyset \text{ in all other cases.}$$

For the regular language

$$R_1 = V_2^*\{c\}$$

we obviously have

$$L_u(M) \cap R_1 = h_1(EQ(h_1, h_2))\{c\}.$$

(The alternation of states s_0, s_a and the fact that after introducing s_1 – from s_0 – we have to remain in this state together ensure that the parsed molecule is of the form $\begin{bmatrix} h_1(x)c \\ h_2(x)c \end{bmatrix}$, for some $x \in V_1^*$.)

Considering $R = R_0\{c\}$ and extending the projection h by $h(c) = \lambda$, we obtain $L = h(L_u(M) \cap R)$. □

Of course, the projection h and the intersection with the regular language R in Theorems 5.3, 5.5 can be done by a deterministic gsm, hence a result like Theorem 5.1 also holds true for the families $NWK(u)$ and $FSWK(u)$.

Consequently, we may say that modulo a deterministic gsm – and in some cases modulo a weak coding only – all families $XWK(\alpha)$, $X \in \{A, N, S, 1, F, FS\}$, are equal to RE, and, in this sense, equal to each other. This also leads to the following estimation of the size of these families.

Corollary 5.7. *For each family of languages FL such that $FL \subset RE$ and FL is closed under intersection with regular languages and arbitrary morphisms, we have $XWK(u) - FL \neq \emptyset$, $X \in \{A, S, F, 1, N, FS\}$.*

Among the important language families FL satisfying the premise of the statement of Corollary 5.7 are MAT^λ and $ET0L$.

Note that in the previous results the final states play no role, they do not increase the power of Watson–Crick finite automata – modulo a deterministic gsm or, in certain cases, modulo a weak coding – and that simple Watson–Crick finite automata (with only three states) suffice to characterize RE modulo a weak coding. Automata without states, using transition pairs $\binom{u}{v}$ with non-restricted strings u, v, are also very powerful. These observations illustrate again the power of Watson–Crick complementarity, and they will be also confirmed by the machines considered in the subsequent sections.

It is also worth mentioning that in most of the constructions above (this is the case for the proofs of Theorem 5.5 and of Lemmas 5.14, 5.16) the complementarity relation is the identity relation, $(a, b) \in \rho$ if and only if $a = b$. This is not the case for the proof of Theorem 4.12, hence for Theorem 5.2. As we have mentioned several times already, in the DNA case the complementarity relation is one-to-one. When trying to be closer to the "DNA reality", we have to take a symmetric one-to-one relation as a basic complementarity relation for our models. This can raise some problems, because the above proof of Theorem 4.12 uses a relation which is not even injective: $(s, s) \in \rho$ and $(s, [s, k]) \in \rho$, too.

5.4 Watson–Crick Finite Transducers

An output can be associated to a Watson–Crick finite automaton in the same way as an output is associated to a finite automaton to form a gsm. This output is written on a normal tape rather than on a Watson–Crick double stranded tape (the other possibility will be considered in Sect. 5.6). Fig. 5.4 illustrates this idea.

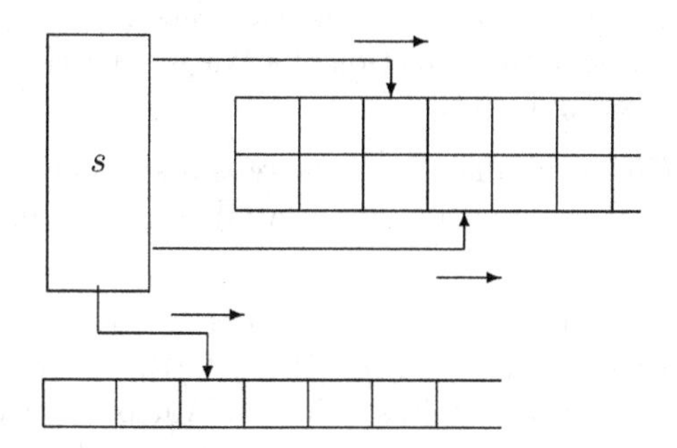

Figure 5.4: A Watson–Crick transducer

A *Watson–Crick gsm* is a construct

$$g = (V_I, \rho_I, V_O, K, s_0, F, \delta),$$

where V_I is the input alphabet, $\rho_I \subseteq V_I \times V_I$ is a symmetric relation (of complementarity), V_O is the output alphabet, K is the set of states, $s_0 \in K$ is the initial state, $F \subseteq K$ is the set of final states, and $\delta : K \times \begin{pmatrix} V_I^* \\ V_I^* \end{pmatrix} \longrightarrow \mathcal{P}_f(V_O^* \times K)$ is a mapping such that $\delta(s, \binom{u}{v}) \neq \emptyset$ only for a finite number of triples $(s, u, v) \in K \times V_I^* \times V_I^*$. The interpretation of $(x, s') \in \delta(s, \binom{u}{v})$ is: in state s, the transducer passes over u in the upper level and over v in the lower level of a double stranded sequence, produces the output x, and enters the state s'.

Formally, for $w_1, w_2, w_1', w_2' \in V_I^*, x, z \in V_O^*$, and $s, s' \in K$, we write

$$zs\begin{pmatrix} w_1 \\ w_2 \end{pmatrix} \Longrightarrow zxs'\begin{pmatrix} w_1' \\ w_2' \end{pmatrix} \quad \text{iff} \quad w_1 = x_1 w_1', w_2 = x_2 w_2', \text{ for}$$

$$(x, s') \in \delta(s, \begin{pmatrix} x_1 \\ x_2 \end{pmatrix}), x_1, x_2, w_1', w_2' \in V_I^*.$$

For a sequence $w = \begin{bmatrix} w_1 \\ w_2 \end{bmatrix} \in WK_{\rho_I}(V_I)$ we define

$$g(w) = \{z \in V_O^* \mid s_0 \begin{bmatrix} w_1 \\ w_2 \end{bmatrix} \Longrightarrow^* zs_f \begin{bmatrix} \lambda \\ \lambda \end{bmatrix}, s_f \in F\}.$$

We extend this definition to languages in $WK_{\rho_I}(V_I)$ in the usual way.

As for Watson–Crick finite automata, we consider stateless, simple, 1-limited, all-final, stateless simple, stateless 1-limited, all-final simple, and all-final 1-limited Watson–Crick gsm's.

As the labeling of transitions in a stateless Watson–Crick finite automaton can be expressed as the output function of a Watson–Crick gsm, from Lemma 5.14 we get the following result.

Corollary 5.8. *For every alphabet V, there is a stateless 1-limited Watson-Crick gsm g_V such that we have $TS_V = g_V(WK_\rho(V))$, for $\rho = \{(a,a) \mid a \in V\}$.*

This implies a result like that in Theorem 5.1: each language $L \in RE$ can be written in the form $L = g(g_V(WK_\rho(V)))$, where g is a usual gsm (deterministic) and g_V is a stateless 1-limited Watson–Crick gsm.

Both the intersection with a regular language and a morphism can be realized by a Watson–Crick gsm, hence from Theorem 5.5 we obtain:

Theorem 5.6. *For each language $L \in RE, L \subseteq V^*$, there is a simple all-final Watson–Crick gsm g_L such that $L = g_L(WK_\rho(V'))$, for some Watson–Crick domain $WK_\rho(V')$.*

Hence, Watson–Crick finite transducers not only do not preserve the families in the Chomsky hierarchy, but they map (very simple) regular languages in such a way that they cover the whole family RE.

In the same way as each two-head finite automaton can be considered a variant of a Watson–Crick finite automaton, a two-head finite transducer can be considered to be a special case of a Watson–Crick gsm.

5.5 Further Variants of Watson–Crick Finite Automata

Among the Watson–Crick finite automata considered in Sect. 5.1, the most intimately related to the DNA structure are the stateless automata which are using only Watson–Crick complementarity, and no additional automata theory-like features. These automata deserve further investigation.

Because $h(w_1w_2) = h(w_1)h(w_2)$, where h is a morphism, it follows from Lemma 5.11 that we cannot represent every regular language as the morphic image of a language in $NWK(\alpha), \alpha \in \{u, ctr\}$. Therefore, on the one hand, starting from $NSWK(u)$ we cannot even obtain representations of REG using weak codings (as in Theorem 5.2), while on the other hand, the results in Theorems 5.1 and 5.5 cannot be improved by replacing the deterministic gsm mapping by a morphism, or by not using an intersection with a regular language.

The "weak point" of stateless Watson–Crick finite automata is that they cannot control the first step of a computation. This suggests the following definition.

An *initial stateless Watson–Crick finite automaton* is a quadruple

$$M = (V, \rho, P_0, P),$$

where V is an alphabet, $\rho \subseteq V \times V$ is a symmetric relation, P_0 and P are finite subsets of $\binom{V^*}{V^*}$.

A double stranded sequence $\begin{bmatrix} w_1 \\ w_2 \end{bmatrix} \in WK_\rho(V)$ is recognized by M if and only if

$$\begin{aligned} & w_1 = x_0 x_1 \dots x_n,\ w_2 = y_0 y_1 \dots y_n,\ n \geq 0, \\ & \text{where } \binom{x_0}{y_0} \in P_0, \binom{x_i}{y_i} \in P, 1 \leq i \leq n. \end{aligned}$$

Therefore, at the beginning of the parsing of $\begin{bmatrix} w_1 \\ w_2 \end{bmatrix}$ we have to use an element of P_0 and after that no element of P_0 is used again – unless $P_0 \cap P \neq \emptyset$.

As usual, we denote by $L_\alpha(M), \alpha \in \{u, crt\}$, the languages associated to correct computations in M. By $INWK(\alpha), INSWK(\alpha)$ we denote the families of languages recognized by initial stateless and by initial stateless simple Watson–Crick finite automata.

As expected, controlling the first step of a recognition process increases the power of our machines. Still, one can easily see that a result like that in Lemma 5.9 is true:

Lemma 5.17. *$INSWK(u) \subseteq REG$.*

On the other hand, we have the following result, which does not hold for non-initial automata.

Theorem 5.7. *Each regular language is a coding of a language in the family $INSWK(u)$.*

Proof. Consider a deterministic finite automaton $M = (K, V, s_0, F, \delta)$ and construct the initial simple stateless Watson–Crick finite automaton

$$M' = (U, \rho, P_0, P),$$

where

$$\begin{aligned} U &= \{[s, a, s']_i \mid s, s' \in K, a \in V, 1 \leq i \leq 4\}, \\ \rho &= \{([s, a, s']_i, [s, a, s']_i) \mid s, s' \in K, a \in V, 1 \leq i \leq 4\}, \\ P_0 &= \{\binom{[s_0, a, s']_1}{\lambda} \mid s' = \delta(s_0, a), a \in V\}, \\ &\cup \{\binom{[s_0, a, s']_3}{\lambda} \mid s' = \delta(s_0, a) \in F, a \in V\}, \end{aligned}$$

$$
\begin{aligned}
P = \{ & \binom{[s, a_1, s']_2 [s', a_2, s'']_1}{\lambda} \mid s' = \delta(s, a_1), s'' = \delta(s', a_2), \\
& s, s', s'' \in K, a_1, a_2 \in V\} \\
\cup \{ & \binom{[s, a_1, s']_2 [s', a_2, s'']_3}{\lambda} \mid s' = \delta(s, a_1), s'' = \delta(s', a_2), \\
& s, s' \in K, s'' \in F, a_1, a_2 \in V\} \\
\cup \{ & \binom{[s, a, s']_4}{\lambda} \mid s' = \delta(s, a) \in F, s \in K, a \in V\} \\
\cup \{ & \binom{\lambda}{[s, a_1, s']_1 [s', a_2, s'']_2} \mid s' = \delta(s, a_1), s'' = \delta(s', a_2), \\
& s, s', s'' \in K, a_1, a_2 \in V\} \\
\cup \{ & \binom{\lambda}{[s, a_1, s']_1 [s', a_2, s'']_4} \mid s' = \delta(s, a_1), s'' = \delta(s', a_2) \in F, \\
& s, s' \in K, s'' \in F, a_1, a_2 \in V\} \\
\cup \{ & \binom{\lambda}{[s, a, s']_3} \mid s' = \delta(s, a) \in F, s \in K, a \in V\}.
\end{aligned}
$$

Each recognition of a sequence must start with $\binom{[s_0, a, s_1]_1}{\lambda}$ in P_0, with the exception of the recognition of sequences of length one, which starts with $\binom{[s_0, a, s_1]_3}{\lambda}$ in P_0, and ends immediately by using $\binom{\lambda}{[s_0, a, s_1]_3}$ in P. All elements of P of the form $\binom{x}{\lambda}$ have the strand x starting with a symbol $[s, a, s']_2$ or $[s, a, s']_4$; all elements of P of the form $\binom{\lambda}{y}$ have y starting with a symbol $[s, a, s']_1$ or $[s, a, s']_3$. Consequently, after completing the parsing of an element of $WK_\rho(U)$ no further steps can be taken, because the pairing imposed by ρ cannot be observed.

This means that we can successfully parse only sequences $w \in WK_\rho(U)$ composed of two identical strands of the form

$$[s_0, a_1, s_1]_1 [s_1, a_2, s_2]_2 [s_2, a_3, s_3]_1 [s_3, a_4, s_4]_2 \ldots [s_{k-1}, a_k, s_k]_i [s_k, a_{k+1}, s_{k+1}]_j,$$

with $i = 2, j = 3$ when k is even, and $i = 1, j = 4$ when k is odd; moreover, $s_0 s_1 \ldots s_{k+1}$ is a state sequence corresponding to the recognition of the string $a_1 a_2 \ldots a_{k+1}$ in M.

Let h be the coding that maps triples $[s, a, s']_i$ to a. Then we obtain $L(M) = h(L_u(M'))$. □

Because each usual stateless Watson–Crick finite automaton can be considered to be an initial one, by taking $P_0 = P$, we obtain the inclusions $XWK(\alpha) \subseteq IXWK(\alpha), X \in \{N, NS\}, \alpha \in \{u, ctr\}$. Consequently, the characterizations of the family RE in Theorems 5.1 and 5.3 hold also true for the corresponding families $IXWK(\alpha), \alpha \in \{u, ctr\}$.

As we know from Chap. 1, the two strands of a DNA molecule have opposite $5' - 3'$ orientations. This suggests considering a variant of Watson–Crick finite automata that parse the two strands of a Watson–Crick tape in opposite directions. Figure 5.5 illustrates the initial configuration of such an automaton.

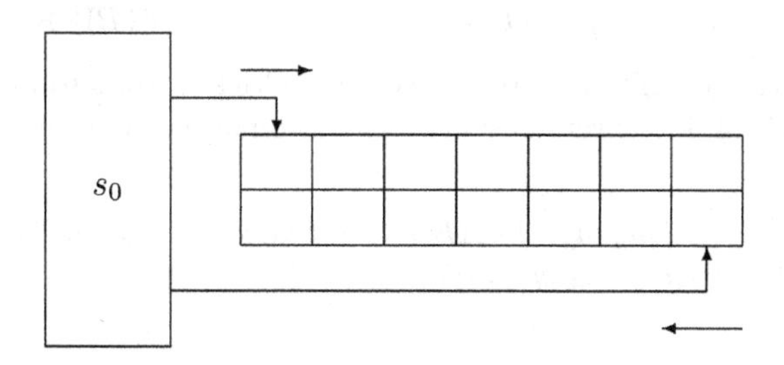

Figure 5.5: A reverse Watson–Crick finite automaton

Formally, a *reverse Watson–Crick finite automaton* is a construct

$$M = (V, \rho, K, s_0, F, \delta),$$

with the components defined exactly as for Watson–Crick finite automata, but with the relation $\Longrightarrow$ defined as follows:

For $w_1, w_2, w_1', w_2', x, y \in V^*, s, s' \in K$, we write

$$\binom{w_1}{w_2 y} s \binom{x w_1'}{w_2'} \Longrightarrow \binom{w_1 x}{w_2} s' \binom{w_1'}{y w_2'} \text{ iff } \begin{bmatrix} w_1 x w_1' \\ w_2 y w_2' \end{bmatrix} \in WK_\rho(V),\ s' \in \delta(s, \binom{x}{y}).$$

Then $\begin{bmatrix} w_1 \\ w_2 \end{bmatrix} \in WK_\rho(V)$ is recognized by M if and only if

$$\binom{\lambda}{w_2} s_0 \binom{w_1}{\lambda} \Longrightarrow^* \binom{w_1}{\lambda} s_f \binom{\lambda}{w_2}, \text{ for } s_f \in F.$$

As in Sect. 5.1 we can associate two languages $L_\alpha(M), \alpha \in \{u, ctr\}$, with a reverse Watson–Crick finite automaton M. These families of languages associated with reverse Watson–Crick finite automata corresponding to families $XWK(\alpha)$ are denoted by $XRWK(\alpha), X \in \{A, N, F, S, 1, NS, N1, FS, F1\}$.

Clearly, a diagram like that in Fig. 5.2 also holds true for these families.

In the simple stateless automata the direction of head movement is not crucial, but only the possibility of covering the double stranded input sequence with upper and lower level blocks. We therefore obtain:

Lemma 5.18. $NSWK(u) = NSRWK(u)$.

In view of Lemma 5.9, we obtain

Corollary 5.9. $NSRWK(u) \subseteq REG$.

A counterpart of Theorem 5.1 is true in the new framework as well.

Lemma 5.19. *For every alphabet* V, *we have* $RTS_V \in N1RWK(ctr)$.

Proof. For the 1-limited stateless Watson–Crick automaton M in the proof of Lemma 5.14, interpreted as a reverse automaton, we have $L_{ctr}(M) = RTS_V$. □

Theorem 5.8. *Each language in RE is the image by a deterministic gsm mapping of a language in any family* $XRWK(ctr), X \in \{A, N, F, S, 1, NS, N1, FS, F1\}$.

We do not know which of the other results in Sects. 5.2 and 5.3 are valid also for reverse Watson–Crick finite automata. Anyway, we have

Theorem 5.9. $NRWK(u) - CF \neq \emptyset$.

Proof. Consider the reverse stateless Watson–Crick finite automaton

$$M = (\{a, b, c\}, \rho, P),$$

with

$$\rho = \{(a, a), (b, b), (c, c)\},$$
$$P = \{\binom{a}{c}, \binom{b}{\lambda}, \binom{c}{b}, \binom{\lambda}{a}\},$$

and the regular language $R = a^+b^+c^+$.

We are interested in the intersection $L_u(M) \cap R$. Only sequences $\begin{bmatrix} w \\ w \end{bmatrix}$ with $w = a^n b^m c^p$, $n, m, p \geq 1$, can be considered as inputs for M which can lead to strings in $L_u(M) \cap R$ because of the form of ρ. The parsing of such a sequence $\begin{bmatrix} w \\ w \end{bmatrix}$ proceeds as follows: we first obtain

$$\begin{aligned} \binom{\lambda}{a^n b^m c^p} s_0 \binom{a^n b^m c^p}{\lambda} &\Longrightarrow^* \binom{a^n}{a^n b^m c^{p-n}} s_0 \binom{b^m c^p}{c^n} \\ &\Longrightarrow^* \binom{a^n b^m}{a^n b^m c^{p-n}} s_0 \binom{c^p}{c^n}, \text{ if } p \geq n, \end{aligned}$$

or

$$\binom{\lambda}{a^n b^m c^p} s_0 \binom{a^n b^m c^p}{\lambda} \Longrightarrow^* \binom{a^p}{a^n b^m} s_0 \binom{a^{n-p} b^m c^p}{c^p}, \text{ if } n > p.$$

(We use the only state s_0 of the system – useless in controlling the work of M – just to indicate the places of the two read heads in the two strands.)

No further step is possible in the second case and no further step is possible in the first case when $p > n$. If $p = n$, then we can continue with

$$\binom{a^n b^m}{a^n b^m} s_0 \binom{c^n}{c^n} \Longrightarrow^* \binom{a^n b^m c^n}{a^n b^{m-n}} s_0 \binom{\lambda}{b^n c^n}, \text{ if } m \geq n,$$

or

$$\binom{a^n b^m}{a^n b^m} s_0 \binom{c^n}{c^n} \Longrightarrow^* \binom{a^n b^m c^m}{a^n} s_0 \binom{c^{n-m}}{b^m c^n}$$
$$\Longrightarrow^* \binom{a^n b^m c^m}{\lambda} s_0 \binom{c^{n-m}}{a^n b^m c^n}, \text{ if } n > m.$$

No further step can be done in the second case and no further step is possible in the first case when $m > n$. If $n = m$, then we can continue with

$$\binom{a^n b^n c^n}{a^n} s_0 \binom{\lambda}{b^n c^n} \Longrightarrow^* \binom{a^n b^n c^n}{\lambda} s_0 \binom{\lambda}{a^n b^n c^n}.$$

Consequently,

$$L_u(M) \cap R = \{a^n b^n c^n \mid n \geq 1\},$$

which is not a context-free language. □

Theorem 5.10. $F1RWK(u) - CF \neq \emptyset$.

Proof. Consider the all-final 1-limited reverse Watson–Crick finite automaton

$$M = (\{a, b, c\}, \rho, K, s_0, K, \delta),$$

with

$$\begin{aligned}
&\rho = \{(a, a), (b, b), (c, c)\},\\
&K = \{s_0, s_1, s_2, s_3\},\\
&\delta(s_0, \binom{a}{\lambda}) = \{s_1\},\\
&\delta(s_1, \binom{\lambda}{c}) = \delta(s_0, \binom{b}{\lambda}) = \delta(s_2, \binom{\lambda}{b}) = \{s_0\},\\
&\delta(s_0, \binom{c}{\lambda}) = \{s_2\},\\
&\delta(s_0, \binom{\lambda}{a}) = \delta(s_3, \binom{\lambda}{a}) = \{s_3\},\\
&\delta(s, \binom{u}{v}) = \emptyset, \text{ otherwise.}
\end{aligned}$$

For the regular language $R = a^+ b^+ c^+$ we obtain

$$L_u(M) \cap R = \{a^n b^n c^n \mid n \geq 1\}$$

(the states in K control the work of M when parsing a sequence $\begin{bmatrix} w \\ w \end{bmatrix}$ with $w = a^n b^m c^p$ in the same way as the pairs $\binom{u}{v} \in P$ do in the proof of Theorem 5.9). □

Corollary 5.10. *The inclusions* $NSRWK(u) \subset FSRWK(u)$, $NSRWK(u) \subset NRWK(u)$ *are proper.*

A natural generalization of Watson–Crick finite automata, also suggested by the idea of reverse automata, are the Watson–Crick two-way finite automata, where one or both of the two read heads can move on the corresponding strand of a Watson–Crick tape in both directions.

We do not define here formally such machines (in Sect. 5.7 we shall do it for the case when only one of the heads, the lower one, works in the two-way manner). However, because two-way automata are generalizations of usual one-way automata, all characterizations of recursively enumerable languages in Sect. 5.3 also remain true for the corresponding variants of two-way Watson–Crick automata. Moreover, Lemma 5.9 also remains true: the work on the two strands is independent of each other in the case of simple stateless automata; in Chap. 3 we mentioned that two-way finite automata characterize regular languages; checking the correctness of the pairing of the symbols on the two strands according to the complementarity relation can be done by a literal shuffle followed by a gsm, hence this is an operation which preserves the regularity.

If we also supplement the model with end markers of the input, then a two-way Watson–Crick finite automaton (with states) can also simulate a reverse Watson–Crick finite automaton.

We do not consider here further study of these variants of Watson–Crick finite automata.

5.6 Watson–Crick Automata with a Watson–Crick Memory

The Watson–Crick finite transducers discussed in Sect. 5.4 are somewhat hybrid devices, as they use an input Watson–Crick tape and a usual output tape (single stranded). This suggests considering output tapes to be also Watson–Crick tapes, leading to the following device.

A *Watson–Crick prefix automaton* is a construct

$$M = (V_1, \rho_1, V_2, \rho_2, K, s_0, F, \delta),$$

where V_1, V_2 are alphabets, $\rho_1 \subseteq V_1 \times V_1, \rho_2 \subseteq V_2 \times V_2$ are symmetric relations on V_1 and V_2, respectively, K is a (finite) set of states, $s_0 \in K, F \subseteq K$, and

$$\delta : K \times \binom{V_1^*}{V_1^*} \longrightarrow \mathcal{P}_f(\binom{V_2^*}{V_2^*} \times K),$$

where $\delta(s, \binom{x_1}{x_2}) \neq \emptyset$ only for a finite number of triples $(s, x_1, x_2) \in K \times V_1^* \times V_1^*$.

The interpretations of these elements are as follows: V_1 is the alphabet of the first Watson–Crick tape of M, V_2 is the alphabet of the second Watson–Crick tape, ρ_1, ρ_2 are complementarity relations on V_1, V_2, K is the set of states, s_0 is the initial state, F is the set of final states, and δ is the transition mapping. The meaning of $(\binom{y_1}{y_2}, s') \in \delta(s, \binom{x_1}{x_2})$ is: in state s, one parses the strings x_1, x_2 in the two strands of the first tape of the automaton, one passes to state s', and one writes/parses the strings y_1, y_2 in the two strands of the second tape of the automaton. The automaton starts in state s_0, with the four heads placed at the left hand end of the four strands of the two tapes, and stops in a final state, with all the four heads placed at the right ends of the four strands. Fig. 5.6 illustrates this idea.

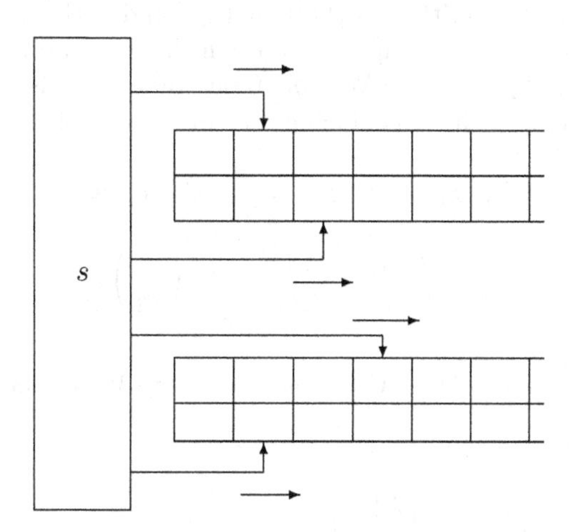

Figure 5.6: A Watson–Crick automaton with memory

We can interpret and use such a machinery in two ways: as a recognizing device, and as a transducer.

In the first interpretation, as a recognizer, the second tape is used as a control of the automaton, as a memory. This can be done in two modes: (1) start with the second tape empty and at each transition $(\binom{y_1}{y_2}, s') \in \delta(s, \binom{x_1}{x_2})$ the strings y_1, y_2 are written on the strands of the second tape, starting with the leftmost empty positions, or (2) start with the second tape containing a sequence $\begin{bmatrix} w_1 \\ w_2 \end{bmatrix} \in WK_{\rho_2}(V_2)$ and at each transition $(\binom{y_1}{y_2}, s') \in$

$\delta(s, \binom{x_1}{x_2})$ move the heads reading the second tape over y_1, y_2, respectively. In the first mode, the automaton stops correctly only if it reaches a sequence $\begin{bmatrix} w_1 \\ w_2 \end{bmatrix} \in WK_{\rho_2}(V_2)$ in the last configuration. Clearly, the two modes of using the second tape when M is a recognizing machine are equivalent.

In either case, we can also define accepting configurations to be ones with the second tape empty: in case (1) any column $\begin{bmatrix} a \\ b \end{bmatrix}$ with $(a, b) \in \rho_2$, appearing in the second tape is deleted immediately after being produced; in case (2) we delete any column $\begin{bmatrix} a \\ b \end{bmatrix}$ with $(a, b) \in \rho_2$, immediately after moving both the two heads to the right of it. Thus, in the first case we have a sort of FIFO (first-in-first-out) memory, which "melts" from the left as soon as completing the columns. This is why we call these automata Watson–Crick prefix automata.

We define formally the transition only for the case where we start with an element of $WK_{\rho_1}(V_1)$ written in the first tape and with the second tape empty; while exploring the first tape, we write in the second tape, completing an element of $WK_{\rho_2}(V_2)$, hence a Watson–Crick tape. In this way, we cover both the recognizing and the transducing interpretations of a Watson–Crick prefix automaton.

For $w_1, w_2, x_1, x_2 \in V_1^*, z_1, z_2, y_1, y_2 \in V_2^*, s, s' \in Q$, we write

$$s \binom{x_1 w_1}{x_2 w_2} / \binom{z_1}{z_2} \Longrightarrow s' \binom{w_1}{w_2} / \binom{z_1 y_1}{z_2 y_2} \text{ iff } (\binom{y_1}{y_2}, s') \in \delta(s, \binom{x_1}{x_2}).$$

A sequence $\begin{bmatrix} w_1 \\ w_2 \end{bmatrix} \in WK_{\rho_1}(V_1)$ is accepted by M if there is $\begin{bmatrix} z_1 \\ z_2 \end{bmatrix} \in WK_{\rho_2}(V_2)$ such that

$$s_0 \begin{bmatrix} w_1 \\ w_2 \end{bmatrix} / \begin{bmatrix} \lambda \\ \lambda \end{bmatrix} \Longrightarrow^* s_f \begin{bmatrix} \lambda \\ \lambda \end{bmatrix} / \begin{bmatrix} z_1 \\ z_2 \end{bmatrix}.$$

Therefore, a computation (recognition or translation) is correctly terminated if the contents of the first tape – which was a Watson–Crick tape – is exhausted and on the second tape one produces another Watson–Crick tape, a complete "molecule" observing the complementarity relation on the alphabet V_2.

Remark 5.5. The slash / in above notation indicates the fact that we have here two different double stranded sequences, written on different tapes, not two concatenated double stranded sequences. Note therefore the crucial difference between $\begin{bmatrix} x_1 \\ x_2 \end{bmatrix} \begin{bmatrix} y_1 \\ y_2 \end{bmatrix}$ and $\begin{bmatrix} x_1 \\ x_2 \end{bmatrix} / \begin{bmatrix} y_1 \\ y_2 \end{bmatrix}$; in the first case we can also write $\begin{bmatrix} x_1 y_1 \\ x_2 y_2 \end{bmatrix}$, which makes no sense in the second case. □

When interpreting a Watson–Crick prefix automaton M as a transducer, the result of translating a sequence $w = \begin{bmatrix} w_1 \\ w_2 \end{bmatrix} \in WK_{\rho_1}(V_1)$ is

$$M(w) = \{\begin{bmatrix} z_1 \\ z_2 \end{bmatrix} \in WK_{\rho_2}(V_2) \mid s_0 \begin{bmatrix} w_1 \\ w_2 \end{bmatrix} / \begin{bmatrix} \lambda \\ \lambda \end{bmatrix} \Longrightarrow^* s_f \begin{bmatrix} \lambda \\ \lambda \end{bmatrix} / \begin{bmatrix} z_1 \\ z_2 \end{bmatrix},$$
$$\text{for some } s_f \in F\}.$$

We emphasize the fact that here we only translate Watson–Crick tapes (over V_1) into Watson–Crick tapes (over V_2); if the output is not an element of $WK_{\rho_2}(V_2)$, then the translation fails.

We shall not investigate the Watson–Crick prefix automata used as transducers. Anyway, it is easy to see that the Watson–Crick gsm's in Sect. 5.4 can be simulated by Watson–Crick prefix transducers: take the identity relation as a complementarity relation on V_2 and $y_1 = y_2$ for each transition $(\begin{pmatrix} y_1 \\ y_2 \end{pmatrix}, s') \in \delta(s, \begin{pmatrix} x_1 \\ x_2 \end{pmatrix})$, etc. Thus, the assertion in Theorem 5.6 holds true also for (simple all-final) Watson–Crick prefix transducers.

As for Watson–Crick finite automata, we can associate two languages also to a Watson–Crick prefix automaton: $L_\alpha(M), \alpha \in \{u, ctr\}$. For instance,

$$L_u(M) = \{w \in V_1^* \mid s_0 \begin{bmatrix} w \\ w' \end{bmatrix} / \begin{bmatrix} \lambda \\ \lambda \end{bmatrix} \Longrightarrow^* s_f \begin{bmatrix} \lambda \\ \lambda \end{bmatrix} / \begin{bmatrix} z_1 \\ z_2 \end{bmatrix}, \text{ for some}$$
$$w' \in V_1^*, z_1, z_2 \in V_2^*, s_f \in F, \text{ such that}$$
$$\begin{bmatrix} w \\ w' \end{bmatrix} \in WK_{\rho_1}(V_1), \begin{bmatrix} z_1 \\ z_2 \end{bmatrix} \in WK_{\rho_2}(V_2)\}.$$

Just as for Watson–Crick finite automata, we can consider Watson–Crick prefix automata which are simple, 1-limited, stateless, all-final, or both simple and stateless, 1-limited and stateless, or both simple and all-final, or 1-limited and all-final. In the simple case, for each transition $(\begin{pmatrix} y_1 \\ y_2 \end{pmatrix}, s') \in \delta(s, \begin{pmatrix} x_1 \\ x_2 \end{pmatrix})$ we have both one of x_1, x_2 and one of y_1, y_2 empty, but not necessarily on the same strand, the upper or the lower one, in the two tapes; similarly, in the 1-limited case we have at the same time one of x_1, x_2 and one of y_1, y_2 empty and the others equal to symbols. We can also consider initial stateless Watson–Crick prefix automata, reverse, or two-way automata; we shall not investigate these cases here.

The stateless Watson–Crick prefix automata are presented in the form $M = (V_1, \rho_1, V_2, \rho_2, P)$, where $P \subseteq \begin{pmatrix} V_1^* \\ V_1^* \end{pmatrix} \times \begin{pmatrix} V_2^* \\ V_2^* \end{pmatrix}$; the rules $(\begin{pmatrix} x_1 \\ x_2 \end{pmatrix}, \begin{pmatrix} y_1 \\ y_2 \end{pmatrix})$ in P are usually written in the form $\begin{pmatrix} x_1 \\ x_2 \end{pmatrix} / \begin{pmatrix} y_1 \\ y_2 \end{pmatrix}$, where $\begin{pmatrix} x_1 \\ x_2 \end{pmatrix}$ indicates the strings to be parsed in the strands of the first tape, and $\begin{pmatrix} y_1 \\ y_2 \end{pmatrix}$ indicates the strings to be written (parsed) in the strands of the second tape (both actions are done within one transition step of the automaton).

We denote by $XPWK(\alpha)$ the families obtained in this way, for $X \in \{A, N, F, S, 1, NS, N1, FS, F1\}$ and $\alpha \in \{u, ctr\}$.

A Watson–Crick finite automaton can be simulated by a Watson–Crick prefix automaton with the two tapes identical, and the transitions operating in the same way on both of them (hence no further control is provided by the second tape). Consequently, we get

Lemma 5.20. *$XWK(\alpha) \subseteq XPWK(\alpha)$, for all X and α as above.*

In this way, all the representations of RE obtained for Watson–Crick finite automata hold true also for Watson–Crick prefix automata.

Also the relations between families $XPWK(ctr)$ corresponding to the diagram from Fig. 5.2 are true.

The use of the second tape increases the power of Watson–Crick automata, hence of prefix automata compared to finite automata.

Lemma 5.21. *If $h_1, h_2 : V^* \longrightarrow U^*$ are two morphisms, then $h_1(EQ(h_1, h_2)) \in NSPWK(u)$.*

Proof. For given morphisms h_1, h_2, we construct the simple stateless Watson–Crick prefix automaton

$$M = (V_1, \rho_1, V_2, \rho_2, P),$$

where

$$\begin{aligned}
V_1 &= U, \\
\rho_1 &= \{(a, a) \mid a \in U\}, \\
V_2 &= V, \\
\rho_2 &= \{(a, a) \mid a \in V\}, \\
P &= \{\begin{pmatrix} h_1(a) \\ \lambda \end{pmatrix} / \begin{pmatrix} a \\ \lambda \end{pmatrix}, \begin{pmatrix} \lambda \\ h_2(a) \end{pmatrix} / \begin{pmatrix} \lambda \\ a \end{pmatrix} \mid a \in V\}.
\end{aligned}$$

It is easy to see that

$$\begin{bmatrix} w_1 \\ w_2 \end{bmatrix} / \begin{bmatrix} \lambda \\ \lambda \end{bmatrix} \Longrightarrow^* \begin{bmatrix} \lambda \\ \lambda \end{bmatrix} / \begin{bmatrix} z_1 \\ z_2 \end{bmatrix}$$

if and only if $w_1 = w_2 \in U^*, z_1 = z_2 \in V^*, w_1 = h_1(z_1)$, and $w_2 = h_2(z_1)$. Therefore, $L_u(M) = h_1(EQ(h_1, h_2))$. □

Corollary 5.11. *If $h_1, h_2 : V^* \longrightarrow U^*$ are two morphisms, then $EQ(h_1, h_2) \in NSPWK(u)$.*

Proof. In the proof of Lemma 5.21, the second tape contains $EQ(h_1, h_2)$ in each of its strands; interchanging the two tapes, we get a Watson–Crick prefix automaton M' such that $L_u(M') = EQ(h_1, h_2)$. □

Theorem 5.11. *Each language $L \in RE$ can be written in the form $L = h(L' \cap R)$, where $L' \in NSPWK(u)$, $R \in REG$, and h is a projection.*

Proof. We use Lemma 5.21 in combination with Theorem 3.16, or Corollary 5.11 in combination with Theorem 3.15. □

Corollary 5.12. *For every family of languages FL such that $FL \subset RE$ and FL is closed under intersection with regular languages and arbitrary morphisms, we have $NSPWK(u) - FL \neq \emptyset$.*

It also follows from Theorem 5.11 that the inclusion $NSWK(u) \subset NSPWK(u)$ is proper: $NSWK(u)$ contains only regular languages.

A characterization of RE related to the one from Theorem 5.11 can also be obtained on the basis of the following result.

Lemma 5.22. *For every alphabet V we have $TS_V \in N1PWK(u)$.*

Proof. Consider the simple stateless Watson–Crick prefix automaton

$$M = (V_1, \rho_1, V_2, \rho_2, P),$$

with

$$\begin{aligned} V_1 &= V \cup \overline{V}, \\ \rho_1 &= \{(a,a), (\bar{a},\bar{a})) \mid a \in V\}, \\ V_2 &= V, \\ \rho_2 &= \{(a,a) \mid a \in V\}, \\ P &= \{\binom{\lambda}{a} / \binom{\lambda}{\lambda}, \binom{\lambda}{\bar{a}} / \binom{\lambda}{\lambda} \mid a \in V\} \\ &\cup \{\binom{a}{\lambda} / \binom{a}{\lambda}, \binom{\bar{a}}{\lambda} / \binom{\lambda}{a} \mid a \in V\}. \end{aligned}$$

Therefore, the correct recognitions in M are of the form

$$\begin{bmatrix} w \\ w \end{bmatrix} / \begin{bmatrix} \lambda \\ \lambda \end{bmatrix} \Longrightarrow^* \begin{bmatrix} \lambda \\ \lambda \end{bmatrix} / \begin{bmatrix} z \\ z \end{bmatrix},$$

for $w \in (V \cup \overline{V})^*, z \in V^*$, and $w \in z \sqcup\!\sqcup \bar{z}$: the second strand of the first tape is parsed without any control from the second tape; when parsing a symbol $a \in V$ in the first strand of the first tape one also parses an occurrence of a in the first strand of the second tape, and when parsing $\bar{a}$ in the first strand of the first tape one also parses an occurrence of a in the second strand of the second tape.

Consequently, $L_u(M) = TS_V$. □

Corollary 5.13. *Each language in RE is the image through a deterministic gsm mapping of a language in $N1PWK(u)$.*

5.7 Universality Results for Watson–Crick Automata

In view of the characterizations of recursively enumerable languages by means of Watson–Crick automata (Sect. 5.3), it is of interest to find universal Watson–Crick (finite) automata. Using, for instance, Theorem 5.2, such an automaton will be universal – modulo a weak coding – for the whole class of Turing machines.

We will postpone the discussion of this subject and remark first that a Watson–Crick finite automaton can be an elegant implementation of a "partially universal" finite automaton as constructed at the end of Sect. 3.3.

Let us proceed as follows: Let $M = (K, V, s_0, F, P)$ be a finite automaton and x be a string in V^*. Consider the string $w = code(M)^n$, where $code(M)$ is as specified in Sect. 3.3 and $n = |x|$. Consider also the string xc^m, for $m = |w| - |x|$ (that is, $|w| = |xc^m|$). Consider the relation $\rho = (K \cup V \cup \{c, c_1, c_2\}) \times (K \cup V \cup \{c, c_1, c_2\})$ (the total relation). Write the string xc^m in the upper strand and the string w in the lower strand of a Watson–Crick tape. The way of working of the universal finite automaton constructed at the end of Sect. 3.3 suggests a way of defining the transitions of a Watson–Crick automaton which works on the Watson–Crick tape mentioned above and recognizes the strings xc^m if and only if $x \in L(M)$: the lower head scans a copy of the code of M in the lower strand of the tape, nondeterministically chooses a move $sa \rightarrow as'$ of M, according to it the occurrence of the symbol a read by the upper head is scanned, and the state of the Watson–Crick automaton checks the correct linking of states of M, memorizing them. When reaching the first occurrence of the symbol c in the upper strand (there is at least one such an occurrence), a final state of M must be reached, in order to correctly finish the parsing. We leave the straightforward details of the construction to the reader (see also the construction below).

This time, the "program" of M is rather simple, a sequence of $|x|$ copies of a single string, the code of M, separated from the "input data" (the string x). However, there are two drawbacks: the length of this "program" is still rather large, while the string in the first strand, that recognized by our universal machine, ends with a tail of symbols c which is also rather long. Both these drawbacks can be eliminated if we allow the lower head of the Watson–Crick automaton to move in both directions. Then one copy of $code(M)$ is enough, hence the tape can be of the form $\begin{bmatrix} xc^m \\ code(M)c^p \end{bmatrix}$, where m and p are integers such that $m \geq 1$ and $|xc^m| = |code(M)c^p|$ (at least one occurrence of c is present in the upper strand, to mark the end of the tape, while the shorter strand, whichever it is, is completed with additional occurrences of c). Again the construction is similar in essence to that used at the end of Sect. 3.3, but we present it in full details because it can be of interest to see a concrete

Watson–Crick automaton which is universal in the sense considered above.

First, let us specify one further notation: one move in a Watson–Crick automaton with a two-way lower head is given in the form of two rewriting rules, one specifying the move of the upper head and one specifying the move of the lower head; to be more suggestive, we write these rules one above the other; the states involved in the two moves should always be identical (the state belongs to the automaton, not to each of the heads).

Let K and V be the set of states and the alphabet which are maximal for the finite automata we want to simulate. We construct the Watson–Crick automaton

$$M_u = (K_u, V \cup K \cup \{c, c_1, c_2\}, \rho_u, q_{0,u}, \{q_f\}, P_u),$$

where

$$\begin{aligned} \rho_u &= (V \cup K \cup \{c, c_1, c_2\}) \times (V \cup K \cup \{c, c_1, c_2\}), \\ K_u &= \{q_{0,u}, q'_{0,u}, q''_{0,u}, q_f\} \\ &\cup \{[s], (s), (s)', (s)'', (s)''', (s)^{iv}, \overline{[s]} \mid s \in K\} \\ &\cup \{[sa] \mid s \in K, a \in V\}, \end{aligned}$$

and the set P_u consists of the following transition rules (rules of the form $s\lambda \to s'\lambda$ in the upper positions mean that the upper strand head does not move):

1. $\begin{pmatrix} q_{0,u}\lambda \to q'_{0,u}\lambda \\ q_{0,u}s \to sq'_{0,u} \end{pmatrix}, \ s \in K,$

2. $\begin{pmatrix} q'_{0,u}\lambda \to q''_{0,u}\lambda \\ q'_{0,u}a \to aq''_{0,u} \end{pmatrix}, \ a \in V,$

3. $\begin{pmatrix} q''_{0,u}\lambda \to q_{0,u}\lambda \\ q''_{0,u}s \to sq_{0,u} \end{pmatrix}, \ s \in K,$

4. $\begin{pmatrix} q_{0,u}\lambda \to [s_0]\lambda \\ q_{0,u}s_0 \to s_0[s_0] \end{pmatrix},$

5. $\begin{pmatrix} [s]a \to a[sa] \\ [s]a \to a[sa] \end{pmatrix}, \ s \in K, a \in V,$

6. $\begin{pmatrix} [sa]\lambda \to (s')\lambda \\ [sa]s' \to s'(s') \end{pmatrix}, \ s, s' \in K, a \in V,$

7. $\begin{pmatrix} (s)\lambda \to (s)'\lambda \\ (s)s' \to s'(s)' \end{pmatrix}, \ s, s' \in K,$

8. $\begin{pmatrix} (s)'\lambda \to (s)''\lambda \\ (s)'a \to a(s)'' \end{pmatrix}, \ s \in K, a \in V,$

9. $\begin{pmatrix} (s)''\lambda \to (s)\lambda \\ (s)''s' \to s'(s) \end{pmatrix}, \ s, s' \in K,$

$$10.\quad \begin{pmatrix} (s)\lambda \to [s]\lambda \\ (s)s \to s[s] \end{pmatrix}, \quad s \in K,$$

$$11.\quad \begin{pmatrix} (s)\lambda \to (s)'''\lambda \\ s'(s) \to (s)'''s' \end{pmatrix}, \quad s, s' \in K,$$

$$12.\quad \begin{pmatrix} (s)'''\lambda \to (s)^{iv}\lambda \\ a(s)''' \to (s)^{iv}a \end{pmatrix}, \quad s \in K, a \in V,$$

$$13.\quad \begin{pmatrix} (s)^{iv}\lambda \to (s)\lambda \\ s'(s)^{iv} \to (s)s' \end{pmatrix}, \quad s, s' \in K,$$

$$14.\quad \begin{pmatrix} (s)c \to c\overline{[s]} \\ (s)c_1 \to c_1\overline{[s]} \end{pmatrix}, \quad s \in K,$$

$$15.\quad \begin{pmatrix} \overline{[s]}\lambda \to \overline{[s]}\lambda \\ \overline{[s]}s' \to s'\overline{[s]} \end{pmatrix}, s, s' \in K, s \neq s',$$

$$16.\quad \begin{pmatrix} \overline{[s]}\lambda \to q_f\lambda \\ \overline{[s]}s \to sq_f \end{pmatrix}, \quad s \in F,$$

$$17.\quad \begin{pmatrix} q_f\lambda \to q_f\lambda \\ q_fs \to sq_f \end{pmatrix}, \quad s \in K,$$

$$18.\quad \begin{pmatrix} q_fc \to cq_f \\ q_f\lambda \to q_f\lambda \end{pmatrix},$$

$$19.\quad \begin{pmatrix} q_f\lambda \to q_f\lambda \\ q_fc_2 \to c_2q_f \end{pmatrix},$$

$$20.\quad \begin{pmatrix} q_f\lambda \to q_f\lambda \\ q_fc \to cq_f \end{pmatrix}.$$

Assume now that we start with a finite automaton $M = (K', V', s_0, F, P)$, with $K' \subseteq K, V' \subseteq V$, and we write a string xc^m in the upper strand and $code(M)c^p$ in the lower strand of the input tape of M_u, with m, p as specified above.

While the upper head of M_u remains in the same place of its strand, the lower one looks for a transition of M which can parse the currently read symbol of x. At the beginning, this must be a transition of the form $s_0a \to as$. By rules 5, 6 one then simulates one step of the work of M. The state (s') can go to the right (rules 7, 8, 9) or to the left (rules 11, 12, 13), looking for a valid continuation (rule 10). When the string x is finished and one reaches c in the upper strand (we may assume that at the same moment we reach c_1 in the lower strand, because we can move freely the lower head), we pass to checking whether or not the current state is final with respect to M. The work of M_u stops correctly only in the affirmative case (rules 16). Therefore, $xc^m \in L(M_u)$ if and only if $x \in L(M)$. (Rules 17–20 are used for scanning the suffixes c^m, c_2c^p of the two strands.)

The complete proof of Theorem 5.2 proceeds along the following phases: (1) starting from a type-0 Chomsky grammar G, one constructs three morphisms h_1, h_2, h_3 such that $L(G) = h_3(h_1(EQ(h_1, h_2)) \cap R)$, where $EQ(h_1, h_2)$ is the equality set of h_1, h_2 (that is, the set of words w such that $h_1(w) = h_2(w)$), and R is a regular language (Theorem 3.16); (2) for h_1, h_2 as above, one constructs a sticker system (of a certain type: *coherent*) which generates a set of double stranded sequences having the language $h_1(EQ(h_1, h_2))$ in their upper strands (Theorem 4.12); finally, (3) a Watson–Crick automaton can be constructed starting from this sticker system, recognizing the language $h_1(EQ(h_1, h_2)) \cap R$ (Lemma 5.15); thus, a further morphism (h_3 above, which is in fact a weak coding) suffices to characterize RE. All the three steps are constructive. If we start from a universal type-0 grammar G_u instead of G, then we obtain a unique Watson–Crick automaton which should be universal in a natural sense. However, the above path from a universal type-0 grammar to a universal Watson–Crick automaton is too long and indirect, the result will be too complex (we do not even see an easy way to write the "program" to be run on such a universal machine). The task of finding *simple* Watson–Crick finite automata which are universal for the whole class of such automata, hence for the class of Turing machines, remains as a *research topic*. As in the case of finite automata, we consider here only the easier task of finding Watson–Crick finite automata which are universal for the class of automata with a bounded number of states and of input symbols.

For a Watson–Crick finite automaton in the 1-limited normal form (from Corollary 5.1 we know that such automata are equivalent with arbitrary Watson–Crick finite automata), $M = (K, V, \rho, s_0, F, P)$, we can consider a codification of it of the form

$$code(M) = [s_1 \begin{pmatrix} \alpha_1 \\ \beta_1 \end{pmatrix} s_1'] \ldots [s_n \begin{pmatrix} \alpha_n \\ \beta_n \end{pmatrix} s_n'][s_{f1}] \ldots [s_{fm}],$$

where each $[s_i \begin{pmatrix} \alpha_i \\ \beta_i \end{pmatrix} s_i']$ is a symbol corresponding to a move $s_i \begin{pmatrix} \alpha_i \\ \beta_i \end{pmatrix} \rightarrow \begin{pmatrix} \alpha_i \\ \beta_i \end{pmatrix} s_i'$ in P (hence one of α_i, β_i is a symbol, the other one is empty), and each $[s_{fi}]$ is a symbol associated to a final state in F.

When parsing a molecule $x = \begin{bmatrix} a_1 a_2 \ldots a_r \\ b_1 b_2 \ldots b_r \end{bmatrix}$, $a_i, b_i \in V, 1 \leq i \leq r, r \geq 1$, we scan one symbol at a time, in either of the two strands, hence we do $2r$ steps. The "speed" of the two heads is different, the distance between them can be arbitrarily large. Thus, we have to merge the code of M with the symbols in each strand of x, considering the molecule

$$\begin{aligned} w_0 &= \begin{bmatrix} code(M)a_1 \; code(M)a_2 \ldots code(M)a_r \; code(M)c \; code(M) \\ code(M)b_1 \; code(M)b_2 \ldots code(M)b_r \; code(M)c \; code(M) \end{bmatrix} \\ &= \begin{bmatrix} bls(code(M), a_1 a_2 \ldots a_r c) \\ bls(code(M), b_1 b_2 \ldots b_r c) \end{bmatrix}. \end{aligned}$$

(The complementarity relation contains all pairs in ρ plus all pairs (α, α), with α appearing in *code*(M) or $\alpha = c$.)

Now, a universal Watson–Crick finite automaton can be constructed following the same idea as in the case of usual finite automata (Sect. 3.3): scan an occurrence of *code*(M) in either of the two strands, choose a move of the form $s\binom{\alpha}{\beta} \to \binom{\alpha}{\beta}s'$ encoded by some symbol $[s\binom{\alpha}{\beta}s']$, simulate this move (clearly, when working in the upper strand, we must have $\beta = \lambda$, and when working in the lower strand we must have $\alpha = \lambda$); the states of the universal automaton will ensure the correct linking of states of M, thus controlling the parsing in the two strands exactly as the states of M do; we advance with different speeds along the two strands; when reaching the column $\binom{c}{c}$ we continue by checking whether or not the current state is final with respect to M (that is a state identified by a symbol $[s]$ in *code*(M); we end in a final state of the universal automaton only in the affirmative case.

Denoting by M_u the Watson–Crick automaton whose construction is sketched above, we obtain

$$w_0 \in L_u(M_u) \text{ iff } x \in L_u(M),$$

hence the universality property.

The "program" w_0 above (it also contains the data to be processed, the molecule x, intercalated with copies of *code*(M)) is rather complex (of a non-context-free type, because of the repeated copies of *code*(M) in each of the two strands). A way to reduce the complexity of the starting molecule of this universal Watson–Crick automaton is the same as in the case of normal finite automata (where we have passed to Watson–Crick automata): consider one more strand of the tape, that is, work with Watson–Crick automata with triple-stranded tapes and three heads scanning them, controlled by a common state.

Firstly, in such a case we can simplify the shape of the "program" w_0: write $a_1a_2\ldots a_r$ in the first strand, $b_1b_2\ldots b_r$ in the second strand, maybe followed by a number of occurrences of the symbol c such that these two strands are of the same length as the third one, where we write $2r$ copies of *code*(M). The head scanning the third strand can go from left to right in the usual way, identifying in each occurrence of *code*(M) a move which is simulated in one of the other two strands. Note that this time the "program" (*code*(M)) is separated from data (molecule x).

Secondly, if we allow the head in the third strand to work in a two-way manner (the other heads remain usual one-way heads), then only one copy of *code*(M) suffices (see the above construction of a two-way two-strands Watson–Crick finite automaton which is universal for the class of finite automata with a bounded number of states and of symbols).

We leave the technical details concerning universal Watson–Crick automata (with two or three strands in their tapes) to the interested reader.

Remark 5.6. We have investigated here only some basic questions about Watson–Crick automata, as recognizing devices essentially using the new data structure, the Watson–Crick tape, based on the complementarity relation specific to DNA sequences. We balance in this way the generative approach in the previous chapter. After having defined these new classes of machines, the usual program in automata theory can be followed in investigating them. We have considered here mostly only one class of problems, important from the DNA computing point of view: representations (hence characterizations) of recursively enumerable languages. The results are encouraging: most of our machines characterize RE modulo a simple squeezing device, a weak coding in some cases and a deterministic gsm in other cases. Such squeezing mechanisms are inherent to DNA computing because of the necessity of encoding the information we deal with, using the alphabet of the four DNA letters: A, C, G, T.

Many other (classes of) problems remain to be investigated. We mention only some of them:

1. Investigate the "pure" families of languages associated with Watson–Crick automata (where "pure" means: without using squeezing mechanisms). What are their mutual relationships and what are their relationships with the families in the Chomsky hierarchy or in any other standard hierarchy of languages?

2. Improve, if possible, the characterizations of RE presented here, by using simpler Watson–Crick automata and/or simpler squeezing mechanisms.

3. Find concrete Watson–Crick automata which are universal (modulo a weak coding or another squeezing mechanism) for the whole classs of Turing machines.

4. Consider deterministic Watson–Crick automata of various types. Are they strictly weaker than the non-deterministic ones? The determinism can be defined here also in a dynamic manner, as the possibility of branching during a computation (remember that we use transition rules of the form $s' \in \delta(s, \binom{x_1}{x_2})$ with x_1, x_2 strings, possibly empty, but not necessarily symbols).

5. Define and investigate complexity classes based on Watson–Crick automata. Because an essential part of the information necessary during a computation is embedded in the data structure we use, the Watson–Crick tape, and this is considered as provided for free, by DNA strands "automatically" checked for the Watson–Crick complementarity, it is expected that the usual complexity classes, based on Turing machines

and string-like data structures, will be different from the Watson–Crick complexity classes. This research topic (based on answers to (2)) is of a particular interest for the DNA computing area, because it can prove (or disprove) the usefulness of using DNA molecules – or associated theoretical models – for computing.

6. Define and investigate descriptional complexity measures for Watson–Crick automata and for their languages. The number of states is the basic measure to consider, as well as the number of transition rules (given as rewriting rules, as in Sect. 5.1). This latter measure is particularly interesting for stateless automata.

7. All of these classes of problems deal with the automata introduced in this chapter and, thus, are concerned with the Watson–Crick data structure. The problem area concerning the parallelism of possible DNA computing remains entirely open. In our estimation, the full usage of this important feature of DNA may render drastic changes to some basic ideas of complexity such as deterministic and nondeterministic polynomial time.

As a general "background" research topic there remains the question of implementability – if not implementation – of a Watson–Crick automaton of any of the variants considered here, or of other types which will be introduced. However, this is a challenge which should be approached in an interdisciplinary manner (team).

5.8 Bibliographical Notes

This chapter is mainly based on the paper [66], where the Watson–Crick automata (in all variants considered above, except the 1-limited one) were introduced. The Watson–Crick finite automata were also presented in [65]. The discussion about universal Watson–Crick automata follows [127], where the 1-limited variant is introduced.

Chapter 6

Insertion-Deletion Systems

6.1 Inserting-Deleting in the DNA Framework

In this chapter we consider computing models based on two operations which were already considered in formal language theory, mainly with motivation from linguistics. These operations – insertion and deletion, with context dependence – can also be encountered in the genetic area and they can be performed, at least theoretically, in the following ways.

Let us imagine that in a test tube we have a single stranded DNA sequence of the form $5' - x_1uvx_2z - 3'$, where all x_1, x_2, u, v, z are strings. Add to this test tube the single stranded DNA sequence $3' - \bar{u}\bar{y}\bar{v} - 5'$, where $\bar{u}, \bar{v}$ are the Watson–Crick complements of the strings u, v, and $\bar{y}$ is the complement of some new string y.

The two strings will anneal, $\bar{u}$ will stick to u and $\bar{v}$ to v, folding $\bar{y}$. We obtain the situation in Fig. 6.1(b). If we cut the double stranded subsequence uv (by a restriction enzyme), then we pass on to a structure like that in Fig. 6.1(c); adding $\bar{z}$, which acts as a primer (and also adding a polymerase), we shall obtain a complete double stranded sequence as in Fig. 6.1(d). Melting the solution, the two strands will be separated, hence we obtain two strings, one of them being x_1uyvx_2z. This means that the string y has been inserted between u and v.

By a similar mismatching annealing we can – theoretically – perform a deletion operation, also controlled by a context: take uyv in the starting string and add $\bar{u}\bar{v}$, then follow a similar procedure. Fig. 6.2 illustrates the operation (the passing from step (b) to step (c) is done by polymerisation and the removing of the loop y by a restriction enzyme).

Therefore, in the DNA framework we can perform insertion and deletion operations. Such operations are also present in the natural evolution process, under the form of (random) point mutations, where single symbols are inserted in or deleted from DNA sequences, in general without an explicit contextual control as in the case illustrated in Figs. 6.1 and 6.2. These operations are also present in the RNA editing, see e.g., [14], with an additional feature: inserting or deleting U (uracyl) is easier and more frequent than inserting or deleting A, C, G.

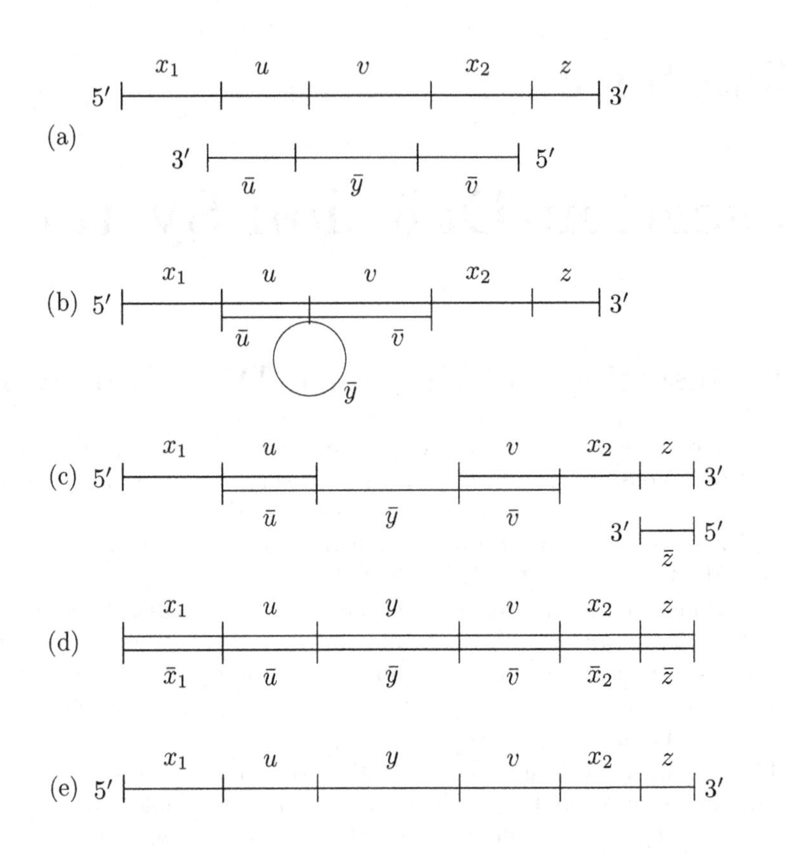

Figure 6.1: Inserting by mismatching annealing

As expected, insertion and deletion operations, working together, are very powerful, leading to characterizations of recursively enumerable languages. (Roughly speaking, in order to have a generative mechanism equal in power to type-0 Chomsky grammars it is necessary to have "enough" context-sensitivity embedded in the model and erasing possibilities. Clearly, insertion and deletion operations provide both these facilities.) We shall prove such characterizations below, for various restricted cases, formulated in terms of a computability model called an *insertion-deletion system* – in short, an *insdel system*.

6.2 Characterizations of Recursively Enumerable Languages

An *insdel system* is a construct

$$\gamma = (V, T, A, R),$$

where V is an alphabet, $T \subseteq V$, A is a finite language over V, and R is a finite set of triples of the form $(u, \alpha/\beta, v)$, where $u, v \in V^*$, $(\alpha, \beta) \in (V^+ \times \{\lambda\}) \cup (\{\lambda\} \times V^+)$. The elements of T are *terminal* symbols, those of A are *axioms*, the triples in R are *insertion-deletion rules.* The meaning of $(u, \lambda/\beta, v)$ is that β can be inserted in between u and v; the meaning of $(u, \alpha/\lambda, v)$ is that α can be deleted from the context (u, v). Stated otherwise, $(u, \lambda/\beta, v)$ corresponds to the rewriting rule $uv \to u\beta v$, and $(u, \alpha/\lambda, v)$ corresponds to the rewriting rule $u\alpha v \to uv$.

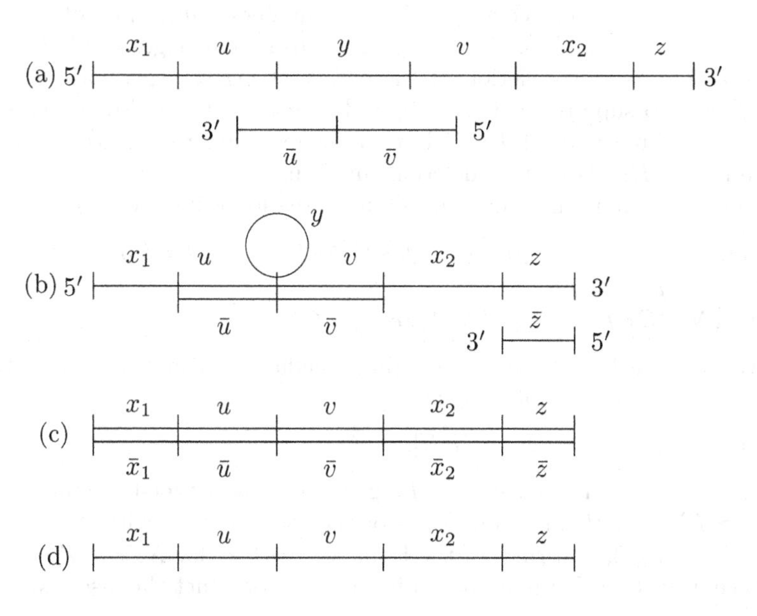

Figure 6.2: Deleting by mismatching annealing

Consequently, for $x, y \in V^*$ we can write $x \Longrightarrow y$ if y can be obtained from x by using an insertion or a deletion rule as above. Explicitly, this means that one of the following cases holds:

1) $x = x_1uvx_2, y = x_1u\beta vx_2$, for some $x_1, x_2 \in V^*$ and $(u, \lambda/\beta, v) \in R$,

2) $x = x_1u\alpha vx_2, y = x_1uvx_2$, for some $x_1, x_2 \in V^*$ and

$$(u, \alpha/\lambda, v) \in R.$$

The language generated by γ is defined by

$$L(\gamma) = \{w \in T^* \mid x \Longrightarrow^* w, \text{ for some } x \in A\},$$

where $\Longrightarrow^*$ is the reflexive and transitive closure of the relation $\Longrightarrow$.

We say that an insdel system $\gamma = (V, T, A, R)$ is *of weight* $(n, m; p, q)$ if

$$\begin{aligned} n &= \max\{|\beta| \mid (u, \lambda/\beta, v) \in R\}, \\ m &= \max\{|u| \mid (u, \lambda/\beta, v) \in R \text{ or } (v, \lambda/\beta, u) \in R\}, \\ p &= \max\{|\alpha| \mid (u, \alpha/\lambda, v) \in R\}, \\ q &= \max\{|u| \mid (u, \alpha/\lambda, v) \in R \text{ or } (v, \alpha/\lambda, u) \in R\}. \end{aligned}$$

We denote by $INS_n^m DEL_p^q$, for $n, m, p, q \geq 0$, the family of languages $L(\gamma)$ generated by insdel systems of weight $(n', m'; p', q')$ such that $n' \leq n$, $m' \leq m, p' \leq p, q' \leq q$. When one of the parameters n, m, p, q is not bounded, we replace it by $*$. Thus, the family of all insdel languages is $INS_*^* DEL_*^*$. Because the insertion-deletion of the empty string changes nothing, when $n = 0$ we also suppose that $m = 0$, and when $p = 0$ we also suppose that $q = 0$. The meaning of INS_0^0 is that no insertion rule is used, and the meaning of DEL_0^0 is that no deletion rule is used.

From the definitions, we obviously have the following inclusions.

Lemma 6.1. (i) *$INS_n^m DEL_p^q \subseteq INS_{n'}^{m'} DEL_{p'}^{q'}$, for all $0 \leq n \leq n'$, $0 \leq m \leq m'$, $0 \leq p \leq p'$, $0 \leq q \leq q'$.*

(ii) *$INS_*^* DEL_*^* \subseteq RE$, $INS_*^* DEL_0^0 \subseteq CS$.*

By using insdel systems of arbitrary weights, we can easily characterize the recursively enumerable languages.

Theorem 6.1. *$RE = INS_3^2 DEL_3^0$.*

Proof. Take a language $L \subseteq T^*$ generated by a type-0 grammar $G = (N, T, S, P)$ with P containing rules of the form $X \to x$ with $X \in N$, $x \in (N \cup T)^*, |x| \leq 2$, and rules of the form $XY \to UZ$, for $X, Y, U, Z \in N$ (for instance, take G in Kuroda normal form). We construct the insdel system

$$\begin{aligned} \gamma &= (N \cup T \cup \{E, K_1, K_2\}, T, \{SEE\}, R), \\ R &= \{(X, \lambda/K_1 x, \alpha_1\alpha_2) \mid X \to x \in P \text{ with } X \in N, x \in (N \cup T)^*, \\ &\quad\ |x| \leq 2, \text{ and } \alpha_1, \alpha_2 \in N \cup T \cup \{E\}\} \\ &\cup \{(XY, \lambda/K_2 UZ, \alpha_1\alpha_2) \mid XY \to UZ \in P \text{ with } X, Y, U, Z \in N, \\ &\quad\ \text{and } \alpha_1, \alpha_2 \in N \cup T \cup \{E\}\} \\ &\cup \{(\lambda, XK_1/\lambda, \lambda) \mid X \in N\} \\ &\cup \{(\lambda, XYK_2/\lambda, \lambda) \mid X, Y \in N\} \\ &\cup \{(\lambda, EE/\lambda, \lambda)\}. \end{aligned}$$

The symbol E is only used when checking the context $\alpha_1\alpha_2$ at the end of the strings. The symbols K_1, K_2 are "killers": K_1 removes one symbol, the one placed immediately to its left hand, and K_2 removes two symbols, those placed immediately to its left hand. Making use of these symbols, the insertion rules in R simulate the rules in P. Symbols already marked by the "killers" K_1, K_2 cannot be used as contexts of rules in R (α_1, α_2 in the above writing of rules in R cannot be equal to K_1, K_2). Therefore, we get $L(G) = L(\gamma)$. □

Natural from a mathematical point of view and also motivated from restrictions appearing in the DNA/RNA area, the problem arises whether or not the result in Theorem 6.1 can be strengthened, by considering shorter inserted or deleted strings, and shorter contexts controlling these operations. In particular, it is of interest to consider the case when only symbols are inserted or deleted: as we have mentioned in the previous section, such operations correspond to point mutations in genetic evolution.

Insdel systems of a very reduced weight characterize the recursively enumerable languages:

Theorem 6.2. $RE = INS_1^2DEL_1^1$.

Proof. Of course, we have to prove only the inclusion $RE \subseteq INS_1^2DEL_1^1$.

Consider a language $L \subseteq T^*, L \in RE$, generated by a type-0 grammar $G = (N, T, S, P)$ in the Penttonen normal form (Theorem 3.4), that is containing context-free rules $X \to x$ with $|x| \leq 2$, and non-context-free rules of the form $XY \to XZ$, for $X, Y, Z \in N$.

Without loss of generality we may assume that in each rule $X \to \alpha_1\alpha_2 \in P$ we have $X \neq \alpha_1$, $X \neq \alpha_2$, $\alpha_1 \neq \alpha_2$. (If necessary, we replace $X \to \alpha_1\alpha_2$ with $X \to X'$, $X' \to \alpha_1\alpha_2'$, $\alpha_2' \to \alpha_2$, where X', α_2' are new symbols.) Similarly, we may assume that for each rule $XY \to XZ \in P$ we have $X \neq Y$, $X \neq Z$, $Y \neq Z$. Moreover, by replacing each rule $X \to \alpha \in P$, $\alpha \in N \cup T$, by $X \to \alpha Z, Z \to \lambda$, we obtain an equivalent grammar. Hence, we may assume that the rules in P are of the following three forms:

1. $X \to \alpha_1\alpha_2$, for $\alpha_1, \alpha_2 \in N \cup T$ such that $X \neq \alpha_1, X \neq \alpha_2, \alpha_1 \neq \alpha_2$,
2. $X \to \lambda$,
3. $XY \to XZ$, for $X, Y, Z \in N$ such that $X \neq Y, X \neq Z, Y \neq Z$.

Moreover, we assume the rules of P are labeled in a one-to-one manner.

We construct the insdel system

$$\gamma = (V, T, A, R),$$

where

$$\begin{aligned} V &= N \cup T \cup \{[r], (r) \mid r \text{ is the label of a rule in } P\} \cup \{B, E\}, \\ A &= \{BSE\}, \end{aligned}$$

and the set R is constructed as follows.

1. For each rule $r : X \to \alpha_1\alpha_2 \in P$ of type 1, with $\alpha_1, \alpha_2 \in N \cup T$, we consider the following insertion-deletion rules:

$$
\begin{array}{ll}
(r.1) & (\beta_1, \lambda/[r], X\beta_2), \text{ for } \beta_1 \in N \cup T \cup \{B\}, \text{ and} \\
 & \beta_2 \in N \cup T \cup \{E\}, \\
(r.2) & ([r]X, \lambda/(r), \beta), \text{ for } \beta \in N \cup T \cup \{E\}, \\
(r.3) & ([r], X/\lambda, (r)), \\
(r.4) & ([r], \lambda/\alpha_1, (r)), \\
(r.5) & (\alpha_1, \lambda/\alpha_2, (r)), \\
(r.6) & (\lambda, [r]/\lambda, \alpha_1), \\
(r.7) & (\alpha_2, (r)/\lambda, \lambda).
\end{array}
$$

2. For each rule $r : X \to \lambda \in P$ of type 2, we introduce the deletion rule

$$
\begin{array}{ll}
(r.1) & (\beta_1, X/\lambda, \beta_2), \text{ for } \beta_1 \in N \cup T \cup \{B\} \text{ and} \\
 & \beta_2 \in N \cup T \cup \{E\}.
\end{array}
$$

3. For each rule $r : XY \to XZ \in P$ of type 3, with $X, Y, Z \in N$, we consider the following insertion-deletion rules:

$$
\begin{array}{ll}
(r.1) & (\beta_1 X, \lambda/[r], Y\beta_2), \text{ for } \beta_1 \in N \cup T \cup \{B\} \text{ and} \\
 & \beta_2 \in N \cup T \cup \{E\}, \\
(r.2) & ([r]Y, \lambda/(r), \beta), \text{ for } \beta \in N \cup T \cup \{E\}, \\
(r.3) & ([r], Y/\lambda, (r)), \\
(r.4) & ([r], \lambda/Z, (r)), \\
(r.5) & (X, [r]/\lambda, Z), \\
(r.6) & (Z, (r)/\lambda, \lambda).
\end{array}
$$

4. We also consider the deletion rules

$$
\begin{array}{l}
(\lambda, B/\lambda, \lambda), \\
(\lambda, E/\lambda, \lambda).
\end{array}
$$

We obtain the equality $L(G) = L(\gamma)$.

($\subseteq$) Each derivation step $w \Longrightarrow w'$ in G is simulated in γ by a derivation $BwE \Longrightarrow^* Bw'E$, using the rules $(r.i)$ associated as above with the rule from P used in $w \Longrightarrow w'$. For instance, assume that $w = w_1 X w_2$, $w' = w_1\alpha_1\alpha_2 w_2$, for $r : X \to \alpha_1\alpha_2 \in P$. Then we successively obtain:

$$\begin{array}{llr}
Bw_1Xw_2E & \Longrightarrow Bw_1[r]Xw_2E & \text{by the rule } (r.1) \\
& \Longrightarrow Bw_1[r]X(r)w_2E & \text{by the rule } (r.2) \\
& \Longrightarrow Bw_1[r](r)w_2E & \text{by the rule } (r.3) \\
& \Longrightarrow Bw_1[r]\alpha_1(r)w_2E & \text{by the rule } (r.4) \\
& \Longrightarrow Bw_1[r]\alpha_1\alpha_2(r)w_2E & \text{by the rule } (r.5) \\
& \Longrightarrow Bw_1\alpha_1\alpha_2(r)w_2E & \text{by the rule } (r.6) \\
& \Longrightarrow Bw_1\alpha_1\alpha_2w_2E & \text{by the rule } (r.7) \\
& = Bw'E. &
\end{array}$$

We proceed in a similar way when $w \Longrightarrow w'$ is done by using a rule $r : XY \to XZ$. The details are left to the reader.

We start from BSE; at any moment, the markers B, E can be removed. Thus, any terminal string generated by G is in $L(\gamma)$.

($\supseteq$) Consider a string BwE; initially we have $w = S$. We can apply to it a rule $(r.1)$ from group 1, or a deletion rule $(\beta_1, X/\lambda, \beta_2)$ associated with $X \to \lambda \in P$, or a rule $(r.1)$ from group 3, or a rule from group 4. Assume that we apply $(\beta_1, \lambda/[r], X\beta_2)$ for some $r : X \to \alpha_1\alpha_2 \in P$. We have

$$Bw_1Xw_2E \Longrightarrow Bw_1[r]Xw_2E.$$

Since the rules in P are labeled in a one-to-one way, $X \neq \alpha_1$, and rules of the form of $(r.1)$ in groups 1 and 3 have a left context checking the symbol placed immediately to the left of X (the same assertion holds for the deletion rules in group 2), the only rule which can use the symbol X is $(r.2)$. Eventually this rule must be applied, otherwise the derivation cannot lead to a terminal string. Thus, the substring $[r]X$ of $Bw_1[r]Xw_2$ leads to $[r]X(r)$. Again there is only one possible continuation, by the rule $(r.3)$, which erases the symbol X. Only after inserting α_1 between $[r]$ and (r) we can remove the symbol $[r]$. In the presence of α_1 and of (r) we can introduce α_2, too, by the rule $(r.5)$. As (r) is introduced after $[r]$, and $X \neq \alpha_1$, the symbol α_1 used by this rule $(r.5)$ as a left context should be introduced at a previous step, by the corresponding rule $(r.4)$. After introducing α_2, which is different from both α_1 and X, we can delete (r), by the rule $(r.7)$. Due to the contexts, no other rule can use the mentioned symbols as contexts or can delete any of them. Thus, after using $(r.1)$, we have to use all rules $(r.i), 2 \leq i \leq 7$, associated with $r : X \to \alpha_1\alpha_2$, simulating the use of $X \to \alpha_1\alpha_2$.

In the same way, after using a rule $(\beta_1X, \lambda/[r], Y\beta_2)$ associated with $r : XY \to XZ \in P$, we have to continue with $(r.i), 2 \leq i \leq 6$ (possibly not immediately or at consecutive steps, but using the same symbols of the current string), hence we have to simulate the rule $XY \to XZ$.

The deletion rules $(\beta_1, X/\lambda, \beta_2)$ directly correspond to erasing rules in P. The markers B, E can be deleted at any step. Consequently, γ can generate only strings in $L(G)$. □

Theorem 6.3. $RE = INS_1^1 DEL_2^0$.

Proof. Consider a language $L \in RE, L \subseteq T^*$, and take a grammar $G = (N, T, S, P)$ in the Geffert normal form given in Theorem 3.5(2) such that $L = L(G)$. Therefore, $P = P_1 \cup P_2$, where P_1 contains only context-free rules of the forms

$$S \to uSv, \text{ for } u, v \in (N \cup T - \{S\})^+,$$
$$S \to x, \text{ for } x \in (N \cup T - \{S\})^+,$$

and P_2 contains rules of the form $XY \to \lambda$ for $X, Y \in N$.

We construct the insdel system

$$\gamma = (V, T, A, R),$$

with

$$\begin{aligned} V &= N \cup T \cup \{c, K, K', F\} \\ &\cup \{[S, r] \mid r : S \to x \in P_1\} \\ &\cup \{[X, r, i] \mid r : S \to zXw \in P_1, z, w \in (N \cup T)^*, i = |z| + 1, \\ &\quad X \in N \cup T\}, \\ A &= \{Sc\}, \end{aligned}$$

and R contains the following insertion-deletion rules:

A. Replace each rule $S \to uSv$ in P by the rule $S \to uScv$. The rules $S \to x$ with $|x|_S = 0$ remain unchanged. Denote by P_1' the set obtained in this way. For each rule $r : S \to X_1 X_2 \dots X_k \in P_1'$ with $X_i \in N \cup T \cup \{c\}, 1 \leq i \leq k, k \geq 1$, we introduce the rules:

1. $(S, \lambda/[S, r], c)$,
2. $(S, \lambda/K, [S, r])$,
3. $(\lambda, SK/\lambda, \lambda)$,
4. $([S, r], \lambda/[X_1, r, 1], c)$,
5. $([X_i, r, i], \lambda/[X_{i+1}, r, i+1], c), 1 \leq i \leq k-1$,
6. $([S, r], \lambda/K, [X_1, r_1, 1])$,
7. $(\lambda, [S, r]K/\lambda, \lambda)$,
8. $([X_i, r, i], \lambda/X_i, [X_{i+1}, r, i+1]), 1 \leq i \leq k-1$,
9. $([X_i, r, i], \lambda/K, X_i), 1 \leq i \leq k-1$,
10. $(\lambda, [X_i, r, i]K/\lambda, \lambda), 1 \leq i \leq k$,
11. $([X_k, r, k], \lambda/F, c)$,
12. $([X_k, r, k], \lambda/X_k, F)$,
13. $([X_k, r, k], \lambda/K, X_k)$,
14. $(X_k, \lambda/K', F)$,
15. $(\lambda, K'F/\lambda, \lambda)$.

B. We also introduce in R the rules

$$(\lambda, XY/\lambda, \lambda), \text{ for } XY \to \lambda \in P_2,$$
$$(\lambda, c/\lambda, \lambda).$$

It is clear that if the insdel system γ simulates correctly the context-free rules in P_1, then it generates the language $L(G)$.

Consider a string $w \in (N \cup T \cup \{c\})^*$. The only rules in R applicable to such a string are those in group B, and a rule of type 1 in group A. Namely, a substring Sc of w can be replaced by $S[S,r]c$, for some $r \in P_1'$. Assume that $w = w_1Scw_2$ (initially, $w_1 = w_2 = \lambda$) and use a rule as above for $r : S \to X_1X_2 \dots X_k, k \geq 1$. We get $w_1S[S,r]cw_2$. From $[S,r]$ we will produce $X_1 \dots X_k$. Because $X_1 \neq c$ and $k \geq 1$, the occurrence of S in $w_1S[S,r]cw_2$ will never be followed by c again, hence it will not be used again by rules of type 1. We have to remove it using rules 2 and 3 (K is a "killer" of the symbol placed to its left hand):

$$w_1S[S,r]cw_2 \Longrightarrow w_1SK[S,r]cw_2 \Longrightarrow w_1[S,r]cw_2.$$

The only way to continue is by using rules of types 4, 5, 8, because the "killer" K able to remove the symbols $[S,r], [X_i,r,i], 1 \leq i \leq k-1$, can be introduced only when both to the left and to the right hand of it there appear symbols of the type $[S,r], [X_i,r,i]$, $1 \leq i \leq k-1$. Thus, eventually we have to perform the derivation

$$w_1[S,r]cw_2 \Longrightarrow w_1[S,r][X_1,r,1]cw_2 \Longrightarrow w_1[S,r][X_1,r,1][X_2,r,2]cw_2$$
$$\Longrightarrow \dots \Longrightarrow w_1[S,r][X_1,r,1] \dots [X_{k-1},r,k-1][X_k,r,k]cw_2.$$

After this derivation or intercalated with its steps, in between $[S,r]$ and $[X_1,r,1]$ we have to introduce the symbol K and in between each $[X_i,r,i]$ and $[X_{i+1},r,i+1]$, $1 \leq i \leq k-1$, we have to introduce X_i. We obtain the string

$$w_1[S_1,r]K[X_1,r,1]X_1[X_2,r,2]X_2 \dots [X_{k-1},r,k-1]X_{k-1}[X_k,r,k]cw_2.$$

The block $[S,r]K$ can be deleted (and this is the only way of removing $[S,r]$). In between $[X_i,r,i]$ and X_i we can introduce K, which will be removed together with $[X_i,r,i]$; this is the only way of removing the symbols $[X_i,r,i], 1 \leq i \leq k-1$. Thus, we get the string

$$w_1X_1X_2 \dots X_{k-1}[X_k,r,k]cw_2.$$

In order to remove $[X_k,r,k]$ we have to introduce first the symbol F, in the presence of which we can introduce X_k:

$$w_1X_1X_2 \dots X_{k-1}[X_k,r,k]cw_2 \Longrightarrow w_1X_1 \dots X_{k-1}[X_k,r,k]Fcw_2$$
$$\Longrightarrow w_1X_1 \dots X_{k-1}[X_k,r,k]X_kFcw_2.$$

Now, $[X_k, r, k]$ can be removed just as each symbol $[X_i, r, i]$ by introducing K in between $[X_k, r, k]$ and X_k, whereas F can be removed by introducing the symbol K', which is then removed together with F.

Note that if there is no occurrence of S in $X_1 \dots X_k$, then no further derivation step using rules in group A can be done; if there is one occurrence of S in $X_1 \dots X_k$, then it is followed by c, hence new applications of rules in group A are possible. The occurrence of c can be removed by the deletion rule $(\lambda, c/\lambda, \lambda)$ in group B. If this is done while a symbol S is still present, and it has not been used for simulating a rule in P_1, then this symbol cannot be removed, and the string will not produce a terminal string. Consequently, $L(\gamma) = L(G)$. □

It is not known how large the family $INS_1^1DEL_1^1$ is, but it seems that at least one of the parameters m (insertion contexts), p (deleted strings), q (deletion contexts) must be at least two in order to generate non-context-free languages. For instance, we have the following result, supporting this conjecture.

Theorem 6.4. $INS_*^1DEL_0^0 \subseteq CF$.

Proof. Take an insdel system $\gamma = (V, T, A, R)$ of weight $(n, 1; 0, 0)$, for some $n \geq 1$. Because no deletion is possible, we can remove from A all axioms not in T^* and from R all rules containing a symbol not in T. Thus, we may assume that $V = T$.

We construct the context-free grammar $G = (N, V, S, P)$, where

$$\begin{aligned}
N &= \{S\} \cup \{(\lambda, a), (a, b), (a, \lambda) \mid a, b \in V\}, \\
P &= \{S \to (\lambda, a_1)(a_1, a_2)(a_2, a_3) \dots (a_{k-1}, a_k)(a_k, \lambda) \mid a_1 a_2 \dots a_k \in A, \\
&\qquad k \geq 1, a_i \in V, 1 \leq i \leq k\} \\
&\quad \cup \{(a, b) \to (a, a_1)(a_1, a_2) \dots (a_k, b) \mid (a, \lambda/a_1 a_2 \dots a_k, b) \in R, \\
&\qquad \text{or } (a, \lambda/a_1 a_2 \dots a_k, \lambda) \in R, \text{ or } (\lambda, \lambda/a_1 a_2 \dots a_k, b) \in R, \\
&\qquad \text{or } (\lambda, \lambda/a_1 a_2 \dots a_k, \lambda) \in R, k \geq 1, a_i \in V, 1 \leq i \leq k, a, b \in V\} \\
&\quad \cup \{(\lambda, a) \to (\lambda, a_1)(a_1, a_2) \dots (a_{k-1}, a_k)(a_k, a) \mid (\lambda, \lambda/a_1 a_2 \dots a_k, a) \in R, \\
&\qquad \text{or } (\lambda, \lambda/a_1 a_2 \dots a_k, \lambda) \in R, k \geq 1, a_i \in V, 1 \leq i \leq k, a \in V\} \\
&\quad \cup \{(a, \lambda) \to (a, a_1)(a_1, a_2) \dots (a_{k-1}, a_k)(a_k, \lambda) \mid (a, \lambda/a_1 a_2 \dots a_k, \lambda) \in R, \\
&\qquad \text{or } (\lambda, \lambda/a_1 a_2 \dots a_k, \lambda) \in R, k \geq 1, a_i \in V, 1 \leq i \leq k, a \in V\} \\
&\quad \cup \{(\lambda, a) \to a, (a, \lambda) \to \lambda \mid a \in V\} \\
&\quad \cup \{(a, b) \to b \mid a, b \in V\}.
\end{aligned}$$

The rules $S \to x$, together with the terminal rules in P, introduce the strings of A; the rules in R are simulated by the corresponding rules in P. The symbols (a, b) keep track of pairs of symbols in the current string, whereas $(\lambda, a), (a, \lambda)$ make it possible to use rules $(u, \lambda/x, v)$ with $u = \lambda$ or $v = \lambda$, respectively, at the ends of the current string. Consequently, we have $L(\gamma) = L(G)$, and so $L(\gamma) \in CF$. □

This is the best result of this type, because we have

Theorem 6.5. *$INS_2^2DEL_0^0$ contains non-semilinear languages.*

Proof. We consider the following insdel system, the weight of which is equal to $(2,2;0,0)$:

$$\gamma = (\{a,b,c,d,f,g\},\{a,b,c,d,f,g\},\{fabcdf\},R),$$

with set R containing the following rules:

1) $(f,\lambda/ga,ab)$,
$(aa,\lambda/b,bc)$,
$(bb,\lambda/c,cd)$,
$(cc,\lambda/d,da)$,
$(dd,\lambda/a,ab)$,
$(cc,\lambda/d,df)$.

(Starting from the substring fab of the current string, these rules double each occurrence of the symbols a,b,c,d, step-by-step, from left to right. Note that, except for the first rule, each rule has the form $(u,\lambda/x,v)$ with $u=\alpha\alpha, \alpha \in \{a,b,c,d\}$, and v belongs to the set $\{ab,bc,cd,da\}$ – except the last rule above, for which $v=df$. The pairs ab,bc,cd,da are called *legal*; they are the only two-letter substrings of a string of the form $(abcd)^n$.

Clearly, starting from a string of the form $wf(abcd)^nf$ (initially we have $w=\lambda$ and $n=1$), we can move on to a string

$$wfg(aabbccdd)^m xy(abcd)^p f, \qquad (*)$$

with $m\geq 0, p\geq 0, m+p+1=n, y$ is a suffix of $abcd$, $abcd=zy$, and x is obtained by doubling each symbol in z. When $m=n-1$ and $y=\lambda$, we obtain the string $wfg(aabbccdd)^nf$, so the length of the string obtained between g and f is equal to $8n$, twice the length of the initial string $(abcd)^n$.)

2) $(g,\lambda/c,aa)$,
$(ca,\lambda/c,a)$,
$(ca,\lambda/d,bb)$,
$(db,\lambda/d,b)$,
$(db,\lambda/a,cc)$,
$(ac,\lambda/a,c)$,
$(ac,\lambda/b,dd)$,
$(bd,\lambda/b,d)$,
$(bd,\lambda/c,aa)$.

(Starting from the substring gaa, that is from the symbol g introduced by the rules of group 1, these rules replace each substring $\alpha\alpha, \alpha\in\{a,b,c,d\}$, by $\beta\alpha\beta\alpha, \beta\in\{a,b,c,d\}$, in such a way that all pairs $\beta\alpha,\alpha\beta$ are not legal. In view of the fact that, except the first rule, all the rules in group 2 are of the form $(u,\lambda/x,v)$ with u being a non-legal pair, it follows that these rules can be applied only in a step-by-step manner, from left to right. Since each rule $(u, \lambda/x, v)$ as above contains a pair $\alpha\alpha, \alpha \in \{a, b, c, d\}$, in the string uv, it follows that rules in group 2 can be applied only after the rules of group 1 have been applied. Consequently, from a string of the form $(*)$, using the rules of group 2, we can either move on to a string of the form

$$wfg(cacadbdbacacbdbd)^r uv(aabbccdd)^s xy(abcd)^p f, \qquad (**)$$

where $0 \leq r \leq m, r + s + 1 = m$, v is a suffix of $aabbccdd$, and u is obtained by "translating" the string z for which we have $zv = aabbccdd$ by means of the rules in group 2, or we get a string of the form

$$wfg(cacadbdbacacbdbd)^m x'y(abcd)^p f,$$

where x' is obtained from a prefix of x by "translating" it using the rules above.

Let us note that the rules of group 2 also double the number of symbols in the substring to which they are applied, so, when the string $(*)$ is of the form $wfg(aabbccdd)^n f$, we can obtain a string $wfg(cacadbdbacacbdbd)^n f$, that is with the substring bounded by g and f of length $16n$, twice the length of $(aabbccdd)^n$ and four times the length of the initial string $(abcd)^n$.)

3) $(b, \lambda/c, df)$,
$(d, \lambda/a, bc)$,
$(b, \lambda/c, da)$,
$(c, \lambda/a, bc)$,
$(c, \lambda/d, ab)$,
$(a, \lambda/b, cd)$,
$(a, \lambda/c, da)$,
$(b, \lambda/d, ab)$,
$(d, \lambda/b, cd)$,
$(gc, \lambda/f, ab)$.

(All the rules above are of the form $(u, \lambda/x, v)$ with v being a legal pair, or $v = df$ in the first rule. Moreover, with the exception of the last rule, each rule has $v = \alpha\beta$ with $\alpha, \beta \in \{a, b, c, d\}$, $u \in \{a, b, c, d\}$, and $u\alpha$ is a non-legal pair. Each rule introduces a symbol x between u and v in such a way that $x\beta$ is a legal pair. Consequently, the rules of group 3 can be applied only in the step-by-step manner, from right to left, starting either from the rightmost symbol f – by the first rule – or from the rightmost position where the rules of group 2 have been applied; indeed, only in that position does a three-letter substring $u\alpha\beta$ appear as above, with $u\alpha$ being a non-legal pair and $\alpha\beta$ a legal pair. Using the above rules we obtain only legal pairs, therefore we move on to a string containing substrings $abcd$.

As both groups of rules 1 and 2 need substrings $\gamma\gamma, \gamma \in \{a, b, c, d\}$, in order to be used, it follows that the rules of group 1 can be applied only after "legalizing" all pairs of symbols. So, the first rule in group 1 can be applied only after using the last rule of group 3, which introduces a new occurrence of the symbol f.

The application of rules in group 3 again doubles the length of the string. Consequently, a string of the form (**) is translated by rules in group 3 into

$$wfgcf(abcd)^{8r}u'v(aabbccdd)^s xy(abcd)^p f,$$

where u' is obtained from u in the above manner. When the string $wfg(aabbccdd)^n f$ has been transformed into $wfg(cacadbdbacacbdbd)^n f$ by means of rules in group 2, then the above group of rules provides the string $wfgcf(abcd)^{8n} f$.

Clearly, after using the rules of group 3 as many times as possible, the derivation can be reiterated, again using the rules of group 1.)

The above grammar generates a non-semilinear language. To this end, we use the following auxiliary result.

Assertion. *If $E \subseteq (\mathbf{N} - \{0\})^n$ is a semilinear set, then for each pair (i,j), $1 \leq i,j \leq n$, one of the following two properties holds:*

1. *There is a constant $k_{i,j}$ such that $\frac{u_i}{u_j} \leq k_{i,j}$ for all vectors $(u_1, \ldots, u_n) \in E$;*
2. *There are vectors $(u_1, \ldots, u_n) \in E$ with one of u_i, u_j fixed and the other one arbitrarily large.*

This assertion can be proved as follows. If $E = \bigcup_{t=1}^{k} E_t$, where $E_t \subseteq (\mathbf{N} - \{0\})^n$are linear sets, $1 \leq t \leq k$, and $E_t = \{v_{t_0} + \sum_{s=1}^{m_t} v_{t_s} x_s \mid x_1, \ldots, x_{m_t} \in \mathbf{N}\}$, for some vectors $v_{t_s}, 0 \leq s \leq m_t$, then $v_{t_0}(r) > 0$ for all $1 \leq r \leq n$ (we have no zero component in the vectors of E). Then

1. if $v_{t_s}(i) > 0, v_{t_s}(j) > 0$, for all $1 \leq s \leq m_t$, then for all $(u_1, \ldots, u_n) \in E_t$ we have $\frac{u_i}{u_j} \leq k_{i,j}^{(t)}$ for
$$k_{i,j}^{(t)} = \frac{\max\{v_{t_s}(i) \mid 0 \leq s \leq m_t\}}{\min\{v_{t_s}(j) \mid 0 \leq s \leq m_t\}};$$
2. if, say, $v_{t_s}(j) = 0, v_{t_s}(i) > 0$, for a given $s, 1 \leq s \leq m_t$, then the set $\{v_{t_0} + v_{t_s} x \mid x \in \mathbf{N}\}$ contains vectors with the ith component equal to $v_{t_0}(i) + x v_{t_s}(i)$, which is arbitrarily large, and the jth component equal to $v_{t_0}(j)$.

Now, if point 2 above holds for a linear set E_t, then it holds for E, otherwise property 1 holds for E, taking

$$k_{i,j} = \max\{k_{i,j}^{(t)} \mid 1 \leq t \leq k\}.$$

Returning to the proof of our theorem, consider the Parikh mapping, Ψ_V, associated to $V = \{g, a, b, c, d, f\}$ (please note the order). The above

Assertion is not true for the set $\Psi_V(L(\gamma))$. Indeed, consider the positions 1 and 2 (corresponding to symbols g and a) of 6-tuples in $\Psi_V(L(\gamma))$. From the above explanations, one can see that the rules in groups 1, 2, and 3 can be applied only in this order. One symbol g and some symbols a are introduced into each cycle of this form such that from a string x one goes to a string y with at most 8 times more occurrences of the symbol a. Consequently, each 6-tuple $(u_1, \ldots, u_6) \in \Psi_V(L(\gamma))$ has $u_1 \leq u_2 \leq 8^{u_1}$. As the ratio $\frac{8^{u_1}}{u_1}$ can be arbitrarily large, but for each given u_1 the component u_2 cannot have arbitrarily large values, it follows that the *Assertion* above is not fulfilled, so $\Psi_V(L(\gamma))$ is not semilinear. □

On the other hand, we have the following result, proving that insertion only is not "too strong".

Theorem 6.6. $LIN - INS_*^*DEL_0^0 \neq \emptyset$.

Proof. The language $L = \{a^n b a^n \mid n \geq 1\}$ is not in the family $INS_*^*DEL_0^0$: clearly, if an infinite language L_0 is in $INS_*^*DEL_0^0$, then an infinite number of strings $z \in L_0$ can be written in the form $z = uxv$ such that $x \neq \lambda$ and $uv \in L_0$. Such a property does not hold for the language L. □

6.3 One Symbol Insertion-Deletion Systems

Bounding the length of the insertion-deletion contexts at a low value (at most two, as in Theorems 6.2 and 6.3) is mathematically challenging, but not very important from a molecular computing point of view: the contexts (u, v) in Figs. 6.1 and 6.2 should be "large enough" in order to ensure the stability of the obtained structures. A restriction which is, however, asked for by biochemical constraints is that of computing by insertion-deletion of strings composed of one symbol only (such as U in the RNA case). Of course, we cannot work with only one symbol in our alphabet: on the one hand, codifying two-symbol strings over a one symbol alphabet leads to exponentially longer strings as compared with the starting ones. On the other hand, the operation cannot be controlled, since any context is present in any sufficiently long string.

We need at least one further symbol, but this symbol cannot be introduced or removed during the computation. Therefore, we have to provide enough copies of all symbols different from the insertion-deletion one, as a sort of workspace, and moreover, we have to accept an output modulo occurrences of such symbols, because they cannot be removed.

Moreover, because we have to manipulate only occurrences of one symbol, in order to generate a language over an arbitrary alphabet we have to codify the symbols of this general alphabet using the elements of our restricted alphabet and, after the computation, we have to decodify, returning to strings over the general alphabet. In this way, we are led to the following variant of insdel systems.

A *restricted insdel system* is a construct

$$\gamma = (V, \{a, c\}, A, R, h),$$

where V is an alphabet, a, c are specified symbols (not necessarily from V), A is a finite subset of $\{a, c\}^*$, R is a finite set of triples of the form $(u, \alpha/\beta, v)$, with $u, v \in \{a, c\}^*$ and $\alpha, \beta \in c^*$, one of α, β being empty, and $h : V^* \longrightarrow \{a, c\}^*$ is a morphism. Therefore, only substrings of the form $c^i, i \geq 1$, can be inserted or deleted; the contexts can contain occurrences of both symbols a and c. The relation $\Longrightarrow$ is defined in the usual way, over $\{a, c\}^*$. Then, the language generated by γ is

$$L(\gamma) = h^{-1}(\{w \in \{a, c\}^* \mid z(aca)^n \Longrightarrow^* (aca)^m w, \text{ for some } n, m \geq 0, z \in A\}).$$

In words, we start from an axiom $z \in A$, prolonged with an arbitrary number of "empty spaces" aca, we use an arbitrary number of insertion-deletion rules, we discard the "spaces" aca placed to the left hand end of the obtained string, and we map by h^{-1} the remaining string into a string in V^*. In this way, strings w for which $h^{-1}(w)$ is not defined are removed, hence we can ensure the termination of the derivation in the same way as when using a specified terminal alphabet.

We denote by $1INSDEL$ the family of languages generated by restricted insdel systems of arbitrary weight; because we work here with a coding of strings over V as strings over $\{a, c\}$, we cannot keep bounded (for instance, independent of the cardinality of V) the weight of the used systems.

Expected from the point of view of Theorems 6.2, 6.3 and encouraging from DNA/RNA computing point of view, we have the following result.

Theorem 6.7. $RE = 1INSDEL$.

Proof. Consider a language $L \subseteq T^*, L \in RE$, and consider a grammar $G = (N, T, S, P)$ in the Kuroda normal form generating L. Assume that

$$N \cup T = \{\alpha_1, \ldots, \alpha_n\},$$

with $\alpha_1 = S$.

The rules in P can be of the following forms:

1. $\alpha_i \rightarrow \alpha_j$,
2. $\alpha_i \rightarrow \lambda$,
3. $\alpha_i \rightarrow \alpha_j \alpha_k$,
4. $\alpha_i \alpha_j \rightarrow \alpha_k \alpha_p$.

Consider one new special symbol, $\alpha_0 = \#$, denoting an empty space, and assume that we start not from S but from $S\#^s$, for some $s \geq 0$. Then we may replace all rules of forms 1, 2, 3 in P with rules of the form 4, as follows:

$$\begin{aligned} &1. \quad \alpha_i\alpha_0 \to \alpha_j\alpha_0, \text{ for } \alpha_i \to \alpha_j \in P,\\ &2. \quad \alpha_i\alpha_0 \to \alpha_0\alpha_0, \text{ for } \alpha_i \to \lambda \in P,\\ &3. \quad \alpha_i\alpha_0 \to \alpha_j\alpha_k, \text{ for } \alpha_i \to \alpha_j\alpha_k \in P, \end{aligned}$$

providing that we also add the rules of the form

$$5. \quad \alpha_i\alpha_0 \to \alpha_0\alpha_i, \ 1 \leq i \leq n.$$

(They move the symbol $\alpha_0 = \#$ to the left.)

Let us denote by G' the grammar obtained in this way.

Then, for every derivation $S \Longrightarrow^* w$ in the grammar G we can find a derivation $S\#^s \Longrightarrow^* \#^t w$ in G' for some $t \geq 0$.

Starting from G', we construct a restricted insdel system $\gamma = (N \cup T \cup \{\#\}, \{a, c\}, A, R, h)$ as follows.

Consider the morphism g defined by $g(\alpha_i) = ac^{i+1}a, \ 0 \leq i \leq n$. (Hence the space $\#$ is encoded by aca.) Then

$$\begin{aligned} A &= \{ac^2a\},\\ h(\alpha_i) &= ac^{i+1}a, \text{ for } \alpha_i \text{ a terminal symbol of } G, \end{aligned}$$

and the set R contains the following rules:

For the q-th rule of G', $r_q : \alpha_i\alpha_j \to \alpha_k\alpha_p$, we consider the "codified" rule

$$ac^{i+1}aac^{j+1}a \to ac^{k+1}aac^{p+1}a$$

and we introduce in R the rules

$$\begin{aligned} &(r_q.1) \quad (ac^{i+1}aa, \lambda/c^{(2q-1)(n+1)}, c^{j+1}aac^{s+1}a), \ 0 \leq s \leq n,\\ &(r_q.2) \quad (ac^{i+1}, \lambda/c^{2(q-1)(n+1)}, aac^{(2q-1)(n+1)+j+1}a),\\ &(r_q.3) \quad (ac^{2(q-1)(n+1)+i+1}aa, c^{(2q-1)(n+1)+j-p}/\lambda, c^{p+1}a),\\ &(r_q.4) \quad (ac^{k+1}, c^{2(q-1)(n+1)+i-k}/\lambda, aac^{p+1}a). \end{aligned}$$

Note that:

- no rule in G is of the form $\#\alpha_j \to \alpha_k\alpha_p$, hence in the previous rules we have $i \geq 2$;
- in contrast to this observation, we have $j, k, p \geq 1$ (and they can be equal to 1).

We obtain $L(G) = L(\gamma)$. According to the previous discussion, it is sufficient to prove that

$$\{w \in T^* \mid S\#^s \Longrightarrow^* \#^t w \text{ in } G \text{ for some } s, t \geq 0\} = L(\gamma).$$

($\subseteq$) Consider a derivation step $x \Longrightarrow y$ in G, using a rule $r_q : \alpha_i\alpha_j \to \alpha_k\alpha_p$. Therefore,

$$x = x_1\alpha_i\alpha_j x_2 \Longrightarrow x_1\alpha_k\alpha_p x_2 = y.$$

The "codified" string corresponding to x is $g(x)$ such that

$$g(x) = g(x_1)ac^{i+1}aac^{j+1}ag(x_2).$$

Because we start from $S\#^s$ with large enough s, we may assume that $g(x_2) \neq \lambda$. Thus, we can use the associated rules $(r_q.1.) - (r_q.4.)$ and we successively get

$$\begin{aligned}
&g(x) = g(x_1)ac^{i+1}aac^{j+1}ag(x_2)\\
\Longrightarrow\ &g(x_1)ac^{i+1}aa^{(2q-1)(n+1)+j+1}ag(x_2)\\
\Longrightarrow\ &g(x_1)ac^{2(q-1)(n+1)+i+1}aac^{(2q-1)(n+1)+j+1}ag(x_2)\\
\Longrightarrow\ &g(x_1)ac^{2(q-1)(n+1)+i+1}aac^{p+1}ag(x_2)\\
\Longrightarrow\ &g(x_1)ac^{k+1}aac^{p+1}ag(x_2)\\
&= g(y).
\end{aligned}$$

Therefore, each derivation in G' can be simulated as above by a derivation in γ. Using rules of the form $\alpha_i\alpha_0 \to \alpha_0\alpha_i$, for each terminal derivation in G' which produces a string w, we can find a derivation in γ producing a string $\#^t g(w)$. Therefore, $h^{-1}(g(w)) = w$, that is, $L(G) \subseteq L(\gamma)$.

($\supseteq$) Consider a string

$$z = ac^{i_1}aac^{i_2}a \dots ac^{i_k}a,$$

where $1 \leq i_j \leq n+1, 1 \leq j \leq k$.

A block ac^ia, with $1 \leq i \leq n+1$, is said to be *low*; in contrast, ac^ia with $i > n+1$ is said to be *high*.

To a string z as above (with all blocks being *low*) we can only apply an insertion rule in R of the form $(r_q.1.)$, hence corresponding to a rule r_q in P. Let the "codified" rule associated to r_q be of the form

$$ac^{i_m}aac^{i_{m+1}}a \to ac^{p_1}aac^{p_2}a.$$

Using the rule (r_q.1.) we get

$$z \Longrightarrow ac^{i_1}a \ldots ac^{i_{m-1}}aac^{i_m}aac^{(2q-1)(n+1)+i_{m+1}}aac^{i_{m+2}}a \ldots ac^{i_k}a = z'.$$

Note that a *high* block has been introduced, $ac^{(2q-1)(n+1)+i_{m+1}}a$. When using a rule of type (r_q.1.), a *low* block, surrounded by two *low* blocks, is replaced by a *high* block. Specifically, $(2q-1)(n+1)$ new occurrences of the symbol c are introduced; because $2q-1$ is odd, we say that we have an *odd high* block.

No rule in R can use the *high* block in z' as a context, excepting the rule (r_q.2.) for the same q as above. The inverse morphism h^{-1} is not defined on *high* blocks, hence eventually the rule (r_q.2.) should be applied. This means that we replace the substring

$$ac^{i_m}aac^{(2q-1)(n+1)+i_{m+1}}a$$

of z' by

$$ac^{2(q-1)(n+1)+i_m}aac^{(2q-1)(n+1)+i_{m+1}}a. \qquad (*)$$

Therefore, the *low* block $ac^{i_m}a$ has been replaced by the *high* block $ac^{2(q-1)(n+1)+i_m}a$. As $2(q-1)$ is even, we say that we have an *even high* block.

Thus, we have obtained a string containing a pair of *high* blocks, one *even* and one *odd*, precisely identified by the index q of the rule r_q in G'.

None of the rules in R can use these high blocks, excepting (r_q.3.), which replaces the substring $(*)$ by

$$ac^{2(q-1)(n+1)+i_m}aac^{p_2}a.$$

Thus, the previous *odd high* block is replaced by a *low* block. The only possible way to replace the remaining *even high* block by a *low* one is by using the rule (r_q.4.), which leads to

$$ac^{p_1}aac^{p_2}a.$$

Thus, the rule $ac^{i_m}aac^{i_{m+1}}a \rightarrow ac^{p_1}aac^{p_2}a$ has been simulated and this is the only way to proceed towards a successful derivation (that is, a derivation which produces a string in $(aca)^*g(T^*)$). Consequently, when $ac^2a(aca)^s \Longrightarrow^* (aca)^t g(w)$, for some $w \in T^*$ (only for such strings w we have $g^{-1}(w)$ defined), we have $S\#^s \Longrightarrow^* \#^t w$ in G', $w \in T^*$. This implies $w \in L(G)$, completing the proof. □

From the proofs of Theorems 6.2, 6.3, and 6.7 we can obtain universality results: there are insdel systems $\gamma_u = (V_u, T, -, R_u)$ of weight (1, 2; 1, 1), or (1, 1; 2, 0), as well as restricted insdel systems such that for any insdel system $\gamma = (V_0, T, A_0, R_0)$, we can construct a set $A(\gamma)$ over V_u such that $L(\gamma'_u) = L(\gamma)$, for $\gamma'_u = (V_u, T, A(\gamma), R_u)$. Therefore, the universal insdel system simulates the given insdel system γ, the particular system γ can be run on γ_u as a program. This can be obtained as follows. Take a universal type-0 grammar $G_u = (N_u, T, P_u)$ and construct the equivalent insdel system as in the proof of Theorems 6.2, 6.3 (or 6.7). Since G_u has no axiom, the obtained system, γ_u, will have no axiom set. However, γ_u is the universal system we look for. If we consider an insdel system γ, with the same terminal alphabet, then there is a type-0 grammar, $G = (N, T, S, P)$, equivalent with γ. Consider (as at the end of Sect. 3.3, when we have constructed a universal type-0 grammar) the code $w(G)$ of this grammar. In the axiom strings BSE, Sc used in the proofs of Theorems 6.2, 6.3, respectively, we replace S by $w(G)S$. Then, because $L(G'_u) = L(G)$, for $G'_u = (N_u, T, w(G)S, P_u)$ and $L(G) = L(\gamma)$, $L(G'_u) = L(\gamma'_u)$, for γ'_u obtained from γ_u as above, we get $L(\gamma'_u) = L(\gamma)$, hence the universality property holds. One sees that, in fact, the "program" of γ to be run on γ_u consists of one string only. The same result holds true for restricted insdel systems, with the difference that the unique program-axiom should be supplemented with arbitrarily many empty spaces (aca, in the coding from the proof of Theorem 6.7). Thus, via the existence of universal Turing machines and of universal Chomsky type-0 grammars, we find a proof of the theoretical possibility of designing universal (programmable) molecular computers based on the insertion-deletion operations.

The proof of Theorem 6.7 suggests another interesting speculation, concerning the so-called "junk DNA." It is known that a large part of the human genome, about 97% of it, consists of short repeated sequences, thus unable to encode much information, and having no known function. A "computer science explanation" of this situation is given in [207]. We do not enter into details, but we only mention the fact that the basic assumption is that the higher life forms have to have complete computational power in order to possess an efficient immune system. But computational completeness is fragile, hence dangerous for life itself. Hence it must be kept under control. Thus, a "replicon police" ("replicon killers") should exist. Conclusion: "junk DNA may be in large part composed of the corpses of former replicons" [207].

The proof above provides a much more "peaceful" explanation: if we need a high computational complexity, then we need an arbitrarily large workspace, which at the beginning of the computation is given as a sequence of the encoding of the empty space (aca, here); during the computation the empty space is shuffled with the current meaningful string, the spaces are consumed and reintroduced in the string, such that at the end of the computation we again have a sequence of repeated strings, which can be arbitrarily long. In short, the junk DNA might be the working space of the cell "computation device". This explains both its abundance and the fact that it consists of repetitions of the same short string. In the terms of [207], these speculations show that computer science approaches to DNA/RNA might be at least as interesting to biology as for DNA/RNA computing.

6.4 Using Only Insertion

When designing an insdel "computer" it is natural to try to keep the underlying model as simple as possible. One idea is to use either only insertion operations or only deletion operations. As we have pointed out in Lemma 6.1(ii), using only insertion operations we generate only context-sensitive languages. Moreover, as we have seen in Sect. 6.2, the family $INS_*^*DEL_0^0$ has serious limitations (Theorem 6.6). However, supplementing the model with some squeezing mechanisms (direct and inverse morphisms, for instance), we can again characterize the recursively enumerable languages. It is not clear how to supplement a device based on deletion only with some additional mechanisms (other than insertion rules) in such a way to get all languages in RE. Therefore, we shall consider here only the families $INS_n^m DEL_0^0$, $n, m \geq 0$, including $n = *$ or $m = *$.

In order to have an image about the size and the properties of these families we mention a series of known results about them; for proofs we refer to [69], [146], [147], [209] (some of these results are proved in Sect. 6.2; full details can be found in [159]).

1. $FIN \subset INS_*^0 DEL_0^0 \subset INS_*^1 DEL_0^0 \subset \ldots \subset INS_*^* DEL_0^0 \subset CS$.
2. REG is incomparable with all families $INS_*^m DEL_0^0, m \geq 0$.
3. LIN and CF are incomparable with all families $INS_*^m DEL_0^0, m \geq 2$, and $INS_*^* DEL_0^0$.
4. $INS_2^2 DEL_0^0$ contains non-semilinear languages.
5. All families $INS_*^m DEL_0^0, m \geq 0$, are anti-AFL's.

Note a difference between points 2 and 3 above: REG is not incomparable with $INS_*^* DEL_0^0$, like LIN and CF. As we shall see below, using only insertion we can generate each regular language, hence we can compute at the level of finite automata. When arbitrary contexts can be used, this can be done without any additional help, when a morphism is added, then contexts of length 1 suffice.

Theorem 6.8. $REG \subset INS_*^* DEL_0^0$.

Proof. Let L be a regular language and let $M = (K, V, q_0, F, \delta)$ be the minimal deterministic finite automaton recognizing L.

For each $w \in V^*$, we define the mapping $\rho_w : K \longrightarrow K$ by

$$\rho_w(q) = q' \text{ iff } (q, w) \vdash^* (q', \lambda),\ q, q' \in K.$$

Obviously, if $x_1, x_2 \in V^*$ are such that $\rho_{x_1} = \rho_{x_2}$, then for every $u, v \in V^*$, ux_1v is in L if and only if ux_2v is in L.

The set of mappings from K to K is finite. Hence the set of mappings ρ_w as above is finite. Let n_0 be their number. We construct the insdel system $\gamma = (V, V, A, R)$ with

$$A = \{w \in L \mid |w| \leq n_0 - 1\},$$
$$R = \{(w, \lambda/v, \lambda) \mid |w| \leq n_0 - 1, 1 \leq |v| \leq n_0, |wv| \leq n_0, \text{ and } \rho_w = \rho_{wv}\}.$$

From the definition of mappings ρ_w and the definitions of A, R, it follows immediately that $L(\gamma) \subseteq L$.

Assume that the converse inclusion is not true and let $x \in L - L(\gamma)$ be a string of minimal length with this property. Thus $x \notin A$. Hence $|x| \geq n_0$. Let $x = zz'$ with $|z| = n_0$ and $z' \in V^*$. If $z = a_1a_2 \ldots a_{n_0}$, then it has $n_0 + 1$ prefixes, namely $\lambda, a_1, a_1a_2, \ldots, a_1 \ldots a_{n_0}$. There are only n_0 different mappings ρ_w. Therefore there are two prefixes u_1, u_2 of z such that $u_1 \neq u_2$ and $\rho_{u_1} = \rho_{u_2}$. With no loss in generality we may assume that $|u_1| < |u_2|$. By substituting u_2 by u_1 we obtain a string x' which is also in L. As $|x'| < |x|$ and x was of minimal length in $L - L(\gamma)$, we obtain $x' \in L(\gamma)$. However, $|u_2| - |u_1| \leq |u_2| \leq n_0$, so if $u_2 = u_1u_3$, then $(u_1, \lambda/u_3, \lambda)$ is an insertion rule in R. This implies that $x' \Longrightarrow_{in} x$, that is $x \in L(\gamma)$, a contradiction. In conclusion, $L \subseteq L(\gamma)$.

The strictness of the inclusion is obvious (see, for instance, Theorem 6.5 in Sect. 6.2). □

Theorem 6.9. *Each regular language is the coding of a language in the family* $INS_*^1DEL_0^0$.

Proof. Let $G = (N, T, S, P)$ be a regular grammar (hence with rules of the forms $X \to aY, X \to a$, for $X, Y \in N, a \in T$). We construct the regular grammar $G' = (N, N \times T, S, P')$, where

$$P' = \{X \to (X, a)Y \mid X \to aY \in P, \text{ for } a \in T, X, Y \in N\}$$
$$\cup \{X \to (X, a) \mid X \to a \in P, \text{ for } X \in N, a \in T\}.$$

Consider also the coding $h : (N \times T)^* \longrightarrow T^*$ defined by $h((X, a)) = a$, $X \in N, a \in T$. Clearly, $L(G) = h(L(G'))$, so it is sufficient to prove that $L(G') \in INS_*^1DEL_0^0$.

We consider the set

$$W = \{x \in (N \times T)^* \mid \text{if } x = x_1(X, a)x_2(Y, b)x_3, \text{ for }$$
$$x_1, x_2, x_3 \in (N \times T)^*, \text{ then } X \neq Y\}.$$

Clearly, for each $y \in W$ we have $|y| \leq card(N)$, so W is a finite set. We construct the insdel system $\gamma = (N \times T, N \times T, A, R)$, where

$$A = L(G') \cap (N \times T)W,$$
$$R = \{((X, a), \lambda/(X_1, a_1) \ldots (X_k, a_k), (Y, b)) \mid (X_1, a_1) \ldots (X_k, a_k) \in W,$$
$$b \in T, X \to aX_1 \in P, X_k \to a_kY \in P, X_i \to a_iX_{i+1} \in P,$$
$$\text{for all } 1 \leq i \leq k - 1\}.$$

The inclusion $L(\gamma) \subseteq L(G')$ is obvious. Conversely, let $x \in L(G')$ be an arbitrary string. If $x \in (N \times T)W$, then $x \in L(\gamma)$. If $x \notin W$, then $x = x_1(X,a)x_2(X,b)x_3$, $x_1, x_2, x_3 \in (N \cup T)^*$. Clearly, $y = x_1(X,b)x_3 \in L(G')$ and $|y| < |x|$. Let us take x_1, x_2, x_3 in such a way that $(X,a)x_2 \in W$. Then $y \Longrightarrow x$ is a correct derivation according to the rules in R. If $y \in (N \times T)W$, then $x \in L(\gamma)$. Otherwise, we repeat the procedure above until we obtain a string $z \in (N \times T)W$ such that $z \Longrightarrow \ldots \Longrightarrow y \Longrightarrow x$, so $x \in L(\gamma)$, which completes the proof. □

When both a direct and an inverse morphism are available and "not very short" contexts are used, then we reach the power of Turing machines.

Theorem 6.10. *Each language $L \in RE$ can be written in the form $L = g(h^{-1}(L'))$, where g is a weak coding, h is a morphism, and $L' \in INS_4^7DEL_0^0$.*

Proof. Consider a language $L \subseteq T^*, L \in RE$, generated by a type-0 Chomsky grammar $G = (N, T, S, P)$ in Kuroda normal form. Therefore, P contains rules of the following two types:

1. $X \to YZ, X \to a, X \to \lambda$, for $X, Y, Z \in N, a \in T$,
2. $XY \to UZ$, for $X, Y, U, Z \in N$.

From the form of these rules, we may assume that each string in $L(G)$ is generated by a derivation consisting of two phases, one when only nonterminal rules are used and one when only terminal rules are used. (If necessary, when symbols Q should be erased in order to prepare substrings XY for non-context-free rules in P, we replace Q by Q' and move Q' to an end of the string, where it will eventually be erased by a rule $Q' \to \lambda$.) Moreover, we may assume that during the second phase, the derivation is performed in the leftmost mode.

Consider the new symbols $\#, \$, c$ and construct the insdel system

$$\gamma = (N \cup T \cup \{\#, \$, c\}, N \cup T \cup \{\#, \$, c\}, \{c^4Sc^6\}, R),$$

with P' containing the following insertion rules:

(1) for each context-free rule $r : X \to x \in P$ we consider the rules:

$$\begin{aligned}(1.r) \ : \ & (\alpha_1\alpha_2\alpha_3\alpha_4X, \lambda/\#\$x, \alpha_5\alpha_6\alpha_7\alpha_8\alpha_9\alpha_{10}), \text{ for} \\ & \alpha_i \in N \cup \{\#, \$, c\}, 1 \leq i \leq 10, \\ & \alpha_3\alpha_4 \notin N\{\$\}, \ \alpha_2\alpha_3\alpha_4 \notin N\{\$\}N, \\ & \alpha_1\alpha_2\alpha_3\alpha_4 \notin N\{\$\}NN, \alpha_5 \notin \{\#, \$\}, \text{ and} \\ & \alpha_5\alpha_6\alpha_7\alpha_8\alpha_9 \notin N\{\#\$\}N\{\#\};\end{aligned}$$

(2) for each non-context-free rule $r : XY \to UZ \in P$ we consider the rules:

$$\begin{aligned}
(2.r.1) &: (\alpha_1\alpha_2\alpha_3 X, \lambda/\$UZ, Y\alpha_4), \text{ for} \\
&\quad \alpha_i \in N \cup \{\#, \$, c\}, 1 \leq i \leq 4, \text{ and} \\
&\quad \alpha_1\alpha_2\alpha_3 \notin N\{\$\}N, \ \alpha_2\alpha_3 \notin N\{\$\}, \ \alpha_4 \notin \{\#, \$\}, \\
(2.r.2) &: (X\$UZY, \lambda/\#\$, \alpha), \text{ for } \alpha \in N \cup \{c\}, \\
(2.r.3) &: (X, \lambda/\#, \$UZY\#\$);
\end{aligned}$$

(3) for each $X, Y \in N$ we consider the rules:

$$\begin{aligned}
(3.XY.1) &: (\alpha_1\alpha_2\alpha_3 XY\#\$, \lambda/X\#, \alpha_4\alpha_5\alpha_6), \text{ for} \\
&\quad \alpha_i \in N \cup \{\#, \$, c\}, 1 \leq i \leq 6, \\
&\quad \alpha_1\alpha_2\alpha_3 \notin N\{\$\}N, \text{ and if } \alpha_4\alpha_5 = X\#, \text{ then } \alpha_6 = \$; \\
(3.XY.2) &: (X, \lambda/\#\$, Y\#\$X\#\alpha), \text{ for } \alpha \in N \cup \{c\}, \\
(3.XY.3) &: (\$Y\#\$X\#, \lambda/\$X, \alpha), \text{ for } \alpha \in N \cup \{c\}.
\end{aligned}$$

We say that all rules $(1.r)$ are *of type* 1, all rules $(2.r.i)$, for r a non-context-free rule in P and $1 \leq i \leq 3$, are *of type* 2, and that all rules $(3.XY.i)$, for $X, Y \in N$ and $1 \leq i \leq 3$, are *of type* 3.

Denote by M the set of strings $\alpha\#\$$, for $\alpha \in N \cup T$. For each string $w \in M$ we consider a symbol b_w. Let W be the set of these symbols. We define the morphism

$$h : (W \cup T \cup \{c\})^* \longrightarrow (N \cup T \cup \{\#, \$, c\})^*,$$

by

$$\begin{aligned}
h(b_w) &= w, \ w \in M, \\
h(a) &= a, \ a \in T, \\
h(c) &= c.
\end{aligned}$$

Consider also the weak coding

$$g : (W \cup T \cup \{c\})^* \longrightarrow T^*,$$

defined by

$$\begin{aligned}
g(b_w) &= \lambda, \ w \in M, \\
g(c) &= \lambda, \\
g(a) &= a, \ a \in T.
\end{aligned}$$

We obtain

$$L(G) = g(h^{-1}(L(\gamma))).$$

The reasoning behind the construction above is the following.

The insertion rules of type 1 simulate the context-free rules of G, the rules of type 2 simulate the non-context-free rules of G. The rules of type 3 are used in order to prepare the current string for making possible the use of rules of type 2. This is done as follows:

The symbols $\#, \$$ are called *markers*. A nonterminal followed by $\#$ and then by a symbol different from $\$$ is said to be *#-marked*. A nonterminal followed by $\$$ is said to be *\$-marked*. A nonterminal followed by $\#\$$ is said to be *#\$-marked*. A nonterminal which is #-, \$-, or #\$-marked is said to be *marked*, otherwise it is called *unmarked*. A string consisting of unmarked symbols in $N \cup T \cup \{c\}$ and of blocks $\alpha\#\$$, for $\alpha \in N \cup T$, is said to be *legal*.

For example, c^4Sc^6 (the axiom of γ) is legal, $cX\#\$XaY\#\$c$ is also legal. The first occurrence of X and the occurrence of Y in this latter string are marked (#\$-marked), the second occurrence of X, as well as all occurrences of c and a are unmarked. However, $cX\$XaY\#\$c$ is not legal, because the first occurrence of X is \$-marked but not #\$-marked.

Now, the rules of type 3 are able to move an unmarked nonterminal X across a block $X\#\$$ placed immediately to the right of X. In this way, pairs XY can be created, which are needed for simulating the context-sensitive rules of G.

The marked symbols, plus the markers and the symbol c are considered "invisible garbage"; at each moment, the string of the unmarked symbols is intended to correspond to a sentential form of G. By the definitions of h and g, this "invisible garbage" is erased, indeed, from each legal string generated by γ. Because no unmarked nonterminal can be mapped by h^{-1}, what remains will be a terminal string.

In order to prove the equality $L(G) = g(h^{-1}(L(\gamma)))$ we shall first prove that rules in groups 1, 2, 3 in G' are doing what we have said that they are supposed to do (in this way we obtain the inclusion $\subseteq$), then we shall prove that they cannot do anything else (that is, also $\supseteq$ is true).

Claim 1. *When using a rule* $(\alpha_1\alpha_2\alpha_3\alpha_4X, \lambda/\#\$x, \alpha_5\alpha_6\alpha_7\alpha_8\alpha_9\alpha_{10})$ *of type 1, the occurrence of* X *in the derived string is unmarked, but it is #\$-marked in the resulting string, where also each symbol of* x *is unmarked.*

The fact that X is unmarked in the string to which the rule is applied is ensured by α_5, which is different from $\#$ and $\$$. As we obtain the substring $X\#\$x\alpha_5$, the other assertions are obvious.

Claim 2. *When using a group of rules* $(2.r.i), 1 \leq i \leq 3$, *associated with a rule* $r : XY \rightarrow UZ$ *in* P, *then the symbols* XY *are unmarked in the derived string, both of them will be #\$-marked in the resulting string, where* UZ *are unmarked.*

The substring of the string to which the rule $(2.r.1)$ is applied is $\alpha_1\alpha_2\alpha_3XY\alpha_4$, with $\alpha_4 \notin \{\#, \$\}$, hence X and Y are unmarked. We get the string $\alpha_1\alpha_2\alpha_3X\$UZY\alpha_4$, to which the rule $(2.r.2)$ is applied, leading to $\alpha_1\alpha_2\alpha_3X\$UZY\#\$\alpha_4$. Now, by the third rule, we get

$\alpha_1\alpha_2\alpha_3X\#\$UZY\#\$\alpha_4$. One sees how the third rule completes the #$-marking of X, whereas Y has been #$-marked by the second rule. Clearly, UZ are always unmarked. From a substring where the only unmarked block (not involving the substrings $\alpha_1\alpha_2\alpha_3$ and α_4) is XY we have obtained a substring where the only unmarked block (not involving the substrings $\alpha_1\alpha_2\alpha_3$ and α_4) is UZ.

Claim 3. *Starting from a legal string, the rules in a group* $(3.XY.i), 1 \leq i \leq 3$, *can replace a substring* $XY\#\$\alpha$ *(hence with an unmarked* X*) by a substring consisting of blocks in* $N\{\#\$\}$ *and ending with* $X\alpha$ *(hence with an unmarked* X*).*

The rule $(3.XY.1)$ can be applied to a string $x\alpha_1\alpha_2\alpha_3XY\#\$\alpha_4\alpha_5\alpha_6y$ and it produces the string $x\alpha_1\alpha_2\alpha_3XY\#\$X\#\alpha_4\alpha_5\alpha_6y$. The second rule is now applicable, leading to $x\alpha_1\alpha_2\alpha_3X\#\$Y\#\$X\#\alpha_4\alpha_5\alpha_6y$. Finally, the third rule produces $x\alpha_1\alpha_2\alpha_3X\#\$Y\#\$X\#\$X\alpha_4\alpha_5\alpha_6y$. Therefore, the substring $XY\#\$$ has been replaced by $X\#\$Y\#\$X\#\$X$, having an unmarked X in the rightmost position.

Thus, starting from a legal string (initially, we have c^4Sc^6), the rules of G' can simulate the rules of G, producing legal strings. Moreover, if we denote by $umk(x)$ the string of the unmarked symbols in a legal string x generated by G', then we have

Claim 4. (i) *If* $x \Longrightarrow^* y$ *by using a rule in group 1 or all three rules* $(2.r.i), 1 \leq i \leq 3$, *associated with a non-context-free rule* r *of* G, *then* $umk(x) \Longrightarrow umk(y)$ *by the coresponding rule in* G.

(ii) *If* $x \Longrightarrow^* y$ *by using the three rules in group 3 associated to the same* X, Y *in* N, *then* $umk(x) = umk(y)$.

Claim 5. *If* $x = g(h^{-1}(y))$, *for some* $y \in L(G')$, *then* y *is a legal string and* $x = umk(y), y \in T^*$. *Conversely, if* $y \in L(\gamma)$ *and* $umk(y) \in T^*$, *then* $umk(y) = g(h^{-1}(y))$.

This follows immediately from the definitions of the morphisms g and h. These claims prove the inclusion $L(G) \subseteq g(h^{-1}(L(\gamma)))$.

We shall now show that only derivations as above lead to legal strings.

Claim 6. *After using a rule* $(2.r.1)$, *no other rule but* $(2.r.2)$ *can be applied to the involved nonterminals* X, Y, U, Z. *Then, after* $(2.r.2)$, *only* $(2.r.3)$ *can be used.*

Indeed, let us consider only the subword $\alpha_1\alpha_2\alpha_3XY\alpha_4$ used by a rule $(2.r.1)$, for $r : XY \rightarrow UZ \in P$. After using $(2.r.1)$ we obtain $\alpha_1\alpha_2\alpha_3X\$UZY\alpha_4$. Now:

- No rule $(1.q)$ can be used to any of X, Y, U, Z, due to the symbols $\beta_i, 1 \leq i \leq 10$, in rules $(\beta_1\beta_2\beta_3\beta_4X, \lambda/\#\$x, \beta_5\beta_6\beta_7\beta_8\beta_9\beta_{10})$ of type $(1.q), q : X \rightarrow x \in P$. (For instance, $\beta_2\beta_3\beta_4 \notin N\{\$\}N$, hence Z above cannot be used by a rule $(1.q)$ corresponding to $q : Z \rightarrow x \in P$.)
- No rule $(2.q.1)$ can be used for a pair UZ or ZY, due to symbols $\beta_1\beta_2\beta_3$ in rules $(\beta_1\beta_2\beta_3X, \lambda/\$YZ, U\beta_4)$ of type $(2.q.1)$ for $q : XU \rightarrow YZ \in P$.

- No rule $(2.q.2), q \neq r$, can be used: this is obvious, because we need the subword $X\$UZY$, which identifies the rule r in P.

- No rule $(2.q.3)$ can be used, because we need a substring $X\$UZY\#\$$, and α_4 above is different from $\#$.

- No rule $(3.CD.1)$ can be used, because our string does not contain the substring $CD\#\$$; the same argument makes impossible the use of the rules $(3.CD.2)$ and $(3.CD.3)$, for all $C, D \in N$.

Using the rule $(2.r.2)$ we get the string $\alpha_1\alpha_2\alpha_3 X\$UZY\#\$\alpha_4$. Nothing has been changed to the left of $X\$UZY$ or inside this substring; moreover, Y is now $\#\$$-marked. As above, one can see that no rule can be applied to this string, excepting $(2.r.3)$. For instance:

- No rule $(3.ZY.1)$ can be used for the pair ZY (the only one which is followed by $\#\$$), because $\beta_1\beta_2\beta_3$ in a rule $(\beta_1\beta_2\beta_3 ZY\#\$, \lambda/Z\#, \beta_4\beta_5\beta_6)$ of this type cannot be $X\$U$.

- No rule $(3.CD.2)$ can be used, because there is no symbol C which is $\#$-marked in our string; the same reason makes impossible the use of a rule $(3.CD.3), C, D \in N$.

Claim 7. *After using a rule* $(3.XY.1)$, *no other rule but* $(3.XY.2)$ *can be applied to the involved nonterminals* X, Y. *Then, after using* $(3.XY.2)$, *no other rule than* $(3.XY.3)$ *can be used.*

The rule $(3.XY.1)$ replaces a substring $\alpha_1\alpha_2\alpha_3 XY\#\$\alpha_4\alpha_5\alpha_6$ by $w = \alpha_1\alpha_2\alpha_3\ XY\#\$X\#\alpha_4\alpha_5\alpha_6$. Now:

- No rule of type $(1.q) : (\beta_1\beta_2\beta_3\beta_4 X, \lambda/\#\$x, \beta_5\beta_6\beta_7\beta_8\beta_9\beta_{10})$ can be used (X is the only unmarked symbol in our string), because of $\beta_5\beta_6\beta_7\beta_8\beta_9\beta_{10}$ which cannot be equal to $Y\#\$X\#\alpha_4$.

- No rule of type $(2.q.1)$ can be used, because we do not have two unmarked symbols in w.

- No rule of types $(2.q.2), (2.q.3)$ can be used, because we do not have a $\$$-marked symbol in w.

- No rule $(3.CD.1) : (\beta_1\beta_2\beta_3 CD\#\$, \lambda/c\$, \beta_4\beta_5\beta_6)$ can be used; the only possibility is to use again $(3.XY.1)$ (no other symbols appear here), but $\beta_4\beta_5\beta_6$ prevents that.

- No rule $(3.CD.2)$ with $XY \neq CD$ can be used, just because we do not have the necessary occurrences of C and D.

- No rule $(3.CD.3)$ can be used, because we need a substring of the form $\$D\#\$C\#$, and such a substring does not appear in w.

Therefore, we have to continue with $(3.XY.2)$ and we get the string $\alpha_1\alpha_2\alpha_3X\#\$Y\#\$X\#\alpha_4\alpha_5\alpha_6$. There is no unmarked symbol here, hence rules of the forms $(1.r)$, $(2.q.1)$, $(2.q.2)$, $(2.q.3)$, $(3.CD.1)$, $(3.CD.2)$ cannot be used. A rule $(3.CD.3)$ can be used only if $XY = CD$, which concludes the proof of Claim 7.

Consequently, the rules in groups $(1.r)$, for r a context-free rule of P, and $(2.r.i)$, $1 \leq i \leq 3$, for r a non-context-free rule of P, and $(3.XY.i), 1 \leq i \leq 3$, for $X, Y \in N$, cannot be mixed; inside these groups, the rules have to be used in the order imposed by i, from 1 to 3, therefore, the system γ can only simulate derivations in G on unmarked symbols. This means that if h^{-1} is defined for $y \in L(G')$, then $c^4Sc^6 \Longrightarrow^* umk(y)$ in the grammar G and $g(h^{-1}(y)) \in L(G)$, proving the inclusion $g(h^{-1}(L(\gamma))) \subseteq L(G)$.

Note that the weight of γ is (4, 7; 0, 0) (4 is reached in rules of type $(1.r)$ and 7 is reached in rules of type $(3.XY.1)$). □

The proof of the previous theorem can be modified as follows:

– Write $L = \bigcup_{a\in T}(\partial_a^r(L)\{a\})$ and take a grammar $G_a = (N_a, T, S_a, P_a)$ for each language $\partial_a^r(L)$. Assume that alphabets $N_a, a \in T$, are mutually disjoint.

– Start from the axiom set $\{c^4S_aca \mid a \in T\}$.

– Together with all rules in the construction above associated with rules in $P_a, a \in T$, consider also the rules with the "witness" suffixes of the type $\alpha_1 \ldots \alpha_k$ ending with the symbol c. For instance, together with

$$(1.r) \; : \; (\alpha_1\alpha_2\alpha_3\alpha_4X, \lambda/\#\$x, \alpha_5\alpha_6\alpha_7\alpha_8\alpha_9\alpha_{10}),$$

consider also all rules with $\alpha_5\alpha_6\alpha_7\alpha_8\alpha_9\alpha_{10}$ replaced by:

$$\begin{aligned}
&\alpha_5\alpha_6\alpha_7\alpha_8\alpha_9c, \text{ for } \alpha_5 \in \{\#, \$\},\\
&\alpha_5\alpha_6\alpha_7\alpha_8c,\\
&\alpha_5\alpha_6\alpha_7c,\\
&\alpha_5\alpha_6c,\\
&\alpha_5c, \text{ for } \alpha_5, \alpha_6, \alpha_7, \alpha_8, \alpha_9 \in N \cup \{\$\},\\
&c.
\end{aligned}$$

Similarly for rules of all other types which involve suffixes of symbols α.

In this way, at the end of the current string we can use shortened rules and we can still prevent the derivations which can produce strings outside the languages $\partial_a^r(L)$.

– Also allow the terminal symbols to migrate to the right, by the rules in group 3, hence let X and Y in these rules be also terminals; moreover, let Y be also equal to c.

– Add the following rules:

$$(4.a.1) \; : \; (ac, \lambda/\#\$a\#, b), \; a, b \in T,$$

$$(4.a.2) : (a, \lambda/\#\$, c\#\$a\#b), \ a, b \in T,$$
$$(4.a.3) : (\$c\#\$a\#, \lambda/\$ca, b), \ a, b \in T.$$

Note the fact that the symbol c existing in the string is now #$-marked and that together with the unmarked occurrence of a moved to the right we introduce an unmarked occurrence of c. The derivation steps are

$$xacbx' \Longrightarrow xac\#\$a\#bx' \Longrightarrow xa\#\$c\#\$a\#bx' \Longrightarrow xa\#\$c\#\$a\#\$cabx',$$

hence the symbol a has been moved near the terminal b, across c.

– Add also the rule

$$(4.a.4) : (\$c\#\$a\#, \lambda/\$da, b), \ a, b \in T,$$

where d is a new symbol, which is introduced in the alphabet of G'.

As rule $(4.a.1)$ uses an unmarked occurrence of c, if we use rule $(4.a.4)$ instead of $(4.a.3)$, then we introduce no new unmarked occurrence of c, hence rules $(4.a.i)$ can no longer be applied. Therefore, if we consider the regular language

$$L_0 = \{\alpha\#\$ \mid \alpha \in (N \cup T \cup \{c\})^*\}\{d\},$$

then we obtain the equality

$$L = L_0 \backslash L(\gamma).$$

Indeed, the left quotient with respect to L_0 selects from $L(\gamma)$ those strings which contain the symbol d and which have in front of this symbol only #$-marked symbols. This means that all nonterminals were replaced by terminals and that all terminals were moved to the right, hence a copy of them is now present to the right of d. Consequently, we obtain

Corollary 6.1. *Each language $L \in RE$ can be written in the form $L = L_0 \backslash L'$, for L_0 a regular language and $L' \in INS_4^7DEL_0^0$.*

It is an *open problem* whether or not the parameters 4 and 7 appearing here can be replaced by smaller numbers. Anyway, from $INS_*^1DEL_0^0 \subseteq CF$ and the fact that CF is closed under inverse morphisms and arbitrary morphisms, the superscript 7 above cannot be replaced by 0 or by 1.

A quite interesting consequence about the size of families $INS_n^mDEL_0^0$ can be inferred:

Corollary 6.2. *Each family $INS_n^mDEL_0^0, n \geq 4, m \geq 7$, is incomparable with each family of languages FL such that $LIN \subseteq FL \subset RE$ and FL is closed under weak codings and inverse morphisms, or under left quotients with regular languages.*

Proof. Because $LIN - INS_*^*DEL_0^0 \neq \emptyset$, we get $FL - INS_*^*DEL_0^0 \neq \emptyset$. As the closure of FL under weak codings and inverse morphisms is strictly included in RE, we cannot have $INS_4^7DEL_0^0 \subseteq FL$ (then $RE \subseteq FL \subset RE$, a contradiction). □

As examples of families of languages having the properties of FL above we mention MAT^λ and $ET0L$.

6.5 Bibliographical Notes

Insertion systems (without using deletion operations) were first considered in [69]. Such systems were investigated in [146], [147], [209]. Insertion-deletion systems were considered in [104], following the systematic study of insertion-deletion operations in [99] (where, however, mainly combinatorial and decidability properties of several variants of these operations are considered, not generative mechanisms based on them). Theorem 6.1 is from [128] (a similar result appears in [104], with a much more complex proof). Theorems 6.2 and 6.7 are from [103]. Theorem 6.3 is new. Theorem 6.4 is based on a similar proof in [69]. The proof of Theorem 6.8 is based on a proof in [50] (dealing not with insertion systems, but with contextual grammars). Theorem 6.10 and its consequences are from [128]. Details about families $INS_n^mDEL_0^0$ can be found in the papers mentioned above, as well as in the monograph [159].

Discussions about the insertion-deletion operations appearing in the DNA and RNA framework can be found in [207], [14], respectively.

Chapter 7

Splicing Systems

Starting with this chapter we investigate computability models based on the splicing operation, a formal model of the recombinant behavior of DNA molecules under the influence of restriction enzymes and ligases. Informally speaking, splicing two strings means to cut them at points specified by given substrings (corresponding to the patterns recognized by restriction enzymes) and to concatenate the obtained fragments crosswise (this corresponds to a ligation reaction).

After briefly discussing the abstraction process leading from the recombination operation as it takes place *in vivo* to the language-theoretic operation of splicing, we start the mathematical study of this latter operation, both in the non-iterated and the iterated form of it. Then we define the fundamental notion of the following chapters, that of an *extended H system*, a language generating device using as a basic ingredient the splicing operation. This chapter discusses the splicing operation and the splicing systems, in the general and in the *simple* form, from a mathematical point of view. Because H systems with finite components generate only regular languages, additional control mechanisms are considered in the subsequent chapters, controlling the work of H systems. Various such mechanisms suggested by the regulated rewriting area in formal language theory (Chap. 8) and architectures suggested by grammar systems area (Chap. 10) are investigated. In most cases, characterizations of recursively enumerable languages are obtained, that is to say, computational completeness. From the corresponding proofs, *universal* H systems of the considered types are also obtained.

7.1 From DNA Recombination to the Splicing Operation

Let us start by an example, illustrating the cut and paste activity carried out *in vitro* on double stranded DNA sequences with restriction enzymes and ligases, resembling the recombination of DNA as it takes place *in vivo*.

Consider the following three DNA molecules:

5′ – CCCCCTCGACCCCC – 3′
3′ – GGGGGAGCTGGGGG – 5′

5′ – AAAAAGCGCAAAAA – 3′
3′ – TTTTTCGCGTTTTT – 5′

5′ – TTTTTGCGCTTTTT – 3′
3′ – AAAAACGCGAAAAA – 5′

The restriction enzymes (endonucleases) are able to recognize specific substrings of double stranded DNA molecules and to cut molecules at the middle of such substrings, either producing "blunt" ends or "sticky" ends. For instance, the sequences where the enzymes *Taq*I, *Sci*NI, and *Hha*I cut are, respectively:

T|C G A   G|C G C   G C G|C
A G C|T   C G C|G   C|G C G

We have also indicated the way of cutting the DNA molecules. Specifically, when acting on the three molecules mentioned above, the three enzymes *Taq*I, *Sci*NI, and *Hha*I will cut these molecules at the unique sites occurring in them and the following six fragments are produced:

5′ – CCCCCT   CGACCCCC – 3′
3′ – GGGGGAGC   TGGGGG – 5′

5′ – AAAAAG   CGCAAAAA – 3′
3′ – TTTTTCGC   GTTTTT – 5′

5′ – TTTTTGCG   CTTTTT – 3′
3′ – AAAAAC   GCGAAAAA – 5′

Note that in all cases we have obtained fragments with identical overhangs, CG when reading in the 5′ to 3′ direction, but there is a crucial difference between the case of *Taq*I, *Sci*NI, and that of *Hha*I: the free tips of the overhangs created by the first two enzymes are at 5′ ends, but the free tips of the two overhangs created by *Hha*I are at 3′ ends. This makes the ends of the first four fragments compatible. If a ligase is added, then the four fragments can be bound together, either restoring the initial molecules, or producing new molecules by recombination. The recombination of the first four fragments above gives the new molecules below:

5′ – CCCCCTCGCAAAAA – 3′
3′ – GGGGGAGCGTTTTT – 5′

$$5' - \text{AAAAAGCGACCCCC} - 3'$$
$$3' - \text{TTTTTCGCTGGGGG} - 5'$$

The formation of these recombinant molecules is possible because the overhangs match.

Thus the operation we have to model consists of two phases: (1) *cut* the sequences at well-specified sites, and (2) *paste* the fragments with matching ends.

Now, because of the precise Watson–Crick complementarity, we can consider the operation above as acting on single stranded sequences (hence on strings). As far as the DNA molecules are concerned, this abstraction step is obvious; for instance, the three molecules we have started with are precisely identified by the strings

$$\text{CCCCCTCGACCCCC},$$

$$\text{AAAAAGCGCAAAAA},$$

$$\text{TTTTTGCGCTTTTT},$$

respectively, with the convention that they represent a strand of a DNA molecule read in the $5'$ to $3'$ direction.

In what concerns the patterns of restriction enzymes, we have to keep not only the information about the involved nucleotides, but also about the type of tips created when cutting the molecules. The pattern is described by a triple (u, x, v), of strings over the alphabet {A, C, G, T}, with the meaning: (u, v) is the context where the cutting takes place and x is the overhanging sequence. In the case of the three enzymes above we have the triples:

$$(\text{T}, \text{CG}, \text{A}), \quad (\text{G}, \text{CG}, \text{C}), \quad (\text{G}, \text{CG}, \text{C}).$$

However, we know that the first two enzymes produce matching ends, whereas the third one does not (although its associated triple is identical to the triple describing the second enzyme). We simply distinguish the two possible classes, for instance, saying that the first two triples above are *of Class 1* and the last one is *of Class 2*. When recombining fragments of strings, we allow only the concatenation of fragments produced according to triples of the same class.

Formally, having two strings w_1, w_2 and two triples (u_1, x_1, v_1), (u_2, x_2, v_2), such that

$$w_1 = w_1' u_1 x_1 v_1 w_1'',$$
$$w_2 = w_2' u_2 x_2 v_2 w_2'',$$

we allow the recombination operation only when (u_1, x_1, v_1) and (u_2, x_2, v_2) are patterns of the same class and $x_1 = x_2$; the strings obtained by recombination are

$$z_1 = w_1' u_1 x v_2 w_2'',$$

$$z_2 = w_2' u_2 x v_1 w_1'',$$

where $x = x_1 = x_2$.

Tacitly, we have made here one more generalizing step, by working with strings over an unspecified, arbitrary alphabet.

In order to get the most general operation with strings, modeling the previously described one, we have to advance three more steps.

Firstly, instead of the relation of patterns (u, x, v) "to be of the same class", we can consider an arbitrary relation, by starting directly from pairs $((u_1, x, v_1), (u_2, x, v_2))$. The meaning of such a pair is that $(u_1, x, v_1), (u_2, x, v_2)$ are triples of the same class, they can produce matching ends, that is, the recombination of the fragments they produce is allowed.

Secondly, when having a pair $((u_1, x, v_1), (u_2, x, v_2))$ and two strings w_1, w_2 as above, $w_1 = w_1' u_1 x v_1 w_1''$ and $w_2 = w_2' u_2 x v_2 w_2''$, we can consider only the string $z_1 = w_1' u_1 x v_2 w_2''$ as a result of the recombination, because the string $z_2 = w_2' u_2 x v_1 w_1''$ is the result of the one-output-recombination with respect to the symmetric pair, $((u_2, x, v_2), (u_1, x, v_1))$.

Thirdly, instead of pairs of triples $((u_1, x, v_1), (u_2, x, v_2))$ as above, we can consider pairs of pairs: the passing from $w_1 = w_1' u_1 x v_1 w_1''$, $w_2 = w_2' u_2 x v_2 w_2''$ to $z_1 = w_1' u_1 x v_2 w_2''$ with respect to $((u_1, x, v_1), (u_2, x, v_2))$ is equivalent with the passing from $w_1 = w_1' u_1' v_1 w_1''$, $w_2 = w_2' u_2' v_2 w_2''$ to $z_1 = w_1' u_1' v_2 w_2''$ with respect to $((u_1', v_1), (u_2', v_2))$, where $u_1' = u_1 x$ and $u_2' = u_2 x$. Similarly, we can consider the quadruple $((u_1, x v_1), (u_2, x v_2))$.

Altogether, we are led to the following operation with strings over an alphabet V: a quadruple $(u_1, u_2; u_3, u_4)$, of strings over V, is called a *splicing rule*; with respect to such a rule r, for $x, y, z \in V^*$ we write

$$\begin{aligned}(x, y) \vdash_r z \quad \text{iff} \quad & x = x_1 u_1 u_2 x_2,\\ & y = y_1 u_3 u_4 y_2,\\ & z = x_1 u_1 u_4 y_2,\\ & \text{for some } x_1, x_2, y_1, y_2 \in V^*.\end{aligned}$$

We say that we *splice* x, y at the *sites* $u_1 u_2, u_3 u_4$, respectively, and the result is z. This is the basic operation we shall deal with in this chapter. When investigating it from a mathematical point of view, we shall consider it in this form. When we build computability models (in the subsequent chapters), in order to keep these models as close as possible to the reality, we shall consider the operation of the form

$$\begin{aligned}(x, y) \models_r (z, w) \quad \text{iff} \quad & x = x_1 u_1 u_2 x_2,\\ & y = y_1 u_3 u_4 y_2,\\ & z = x_1 u_1 u_4 y_2,\\ & w = y_1 u_3 u_2 x_2,\\ & \text{for some } x_1, x_2, y_1, y_2 \in V^*.\end{aligned}$$

We shall explicitly specify the variant in use when giving the relevant definitions. We shall say that $\vdash$ is a 1-splicing and $\models$ is a 2-splicing operation. When we say only "splicing", it will be clear from the context which type of splicing is meant. In this chapter we deal only with 1-splicing.

Of course, if we want to bring our models back to laboratory, we have to renounce all the aforementioned abstraction steps, going back to considering specified enzymes, with specified recognition patterns. For certain models we shall discuss some of the problems raised by such an attempt. In general, the power of our models will be essentially based on the control mechanisms imposed on the splicing operation, as mentioned at the beginning of this chapter. All these mechanisms look unrealistic for the present day laboratory techniques. Hence, the "computers" we shall discuss need significant progress in biochemical engineering, a task which is far beyond the scope of the present book.

7.2 Non-Iterated Splicing as an Operation with Languages

In all real circumstances, the sets of strings and the sets of splicing rules (of enzymes behind them) are finite. Because the strings can be arbitrarily long (there is no *a priori* bound on their length), it is just natural to also consider languages of arbitrary cardinality. This is not the case with the splicing rules: very few restriction enzymes have overhangs of length greater than six. This means that in the writing (u, x, v) as a representation of the recognized pattern, in most cases we have $|x| \leq 6$, that is, when considering splicing rules $(u_1, u_2; u_3, u_4)$, each string u_1, u_2, u_3, u_4 is of a rather limited length. In a mathematical set-up, such a limitation is not necessary. Moreover, as we shall see in Sect. 7.3, even in the iterated case, the splicing with respect to a finite set of rules preserves the regularity. From a computational point of view, this means that we can reach in this way at most the power of finite automata or Chomsky regular grammars. These observations suggest considering "arbitrarily long" splicing rules, i.e., infinite sets of splicing rules. In order to keep some control on such infinite sets, we shall codify the rules as strings; then their sets are languages and we can consider the type of these languages with respect to a specified classification, for example, the Chomsky hierarchy.

This is the style we shall adopt in this section.

Consider an alphabet V and two special symbols, $\#, \$$, not in V. A *splicing rule* (over V) is a string of the form

$$r = u_1 \# u_2 \$ u_3 \# u_4,$$

where $u_1, u_2, u_3, u_4 \in V^*$. (For a maximal generality, we place no restriction on the strings u_1, u_2, u_3, u_4. The cases when $u_1u_2 = \lambda$ or $u_3u_4 = \lambda$ could be ruled out as unrealistic.)

For a splicing rule $r = u_1\#u_2\$u_3\#u_4$ and strings $x, y, z \in V^*$ we write

$$\begin{aligned}(x, y) \vdash_r z \quad \text{iff} \quad & x = x_1u_1u_2x_2,\\ & y = y_1u_3u_4y_2,\\ & z = x_1u_1u_4y_2,\\ & \text{for some } x_1, x_2, y_1, y_2 \in V^*.\end{aligned}$$

(Therefore, a rule $u_1\#u_2\$u_3\#u_4$ corresponds to a rule $(u_1, u_2; u_3, u_4)$ as at the end of the previous section.)

The strings x, y are sometimes called the *terms* of the splicing; when understood from the context, we omit the specification of r and write $\vdash$ instead of $\vdash_r$.

The passing from x, y to z, via $\vdash_r$, can be represented as shown in Fig. 7.1.

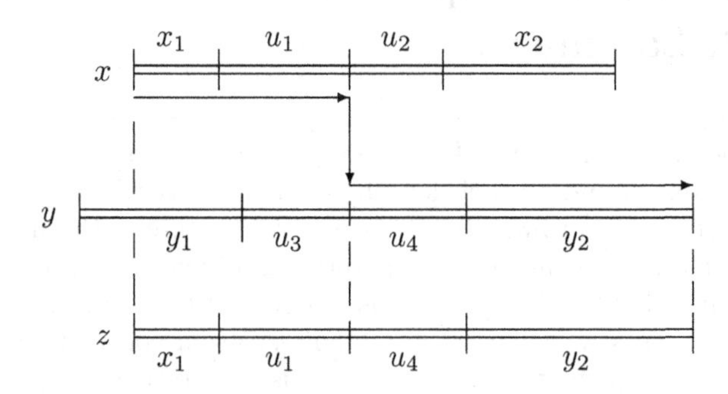

Figure 7.1: The splicing operation

Often, when the splicing of specific strings is presented, for better readability we shall indicate by a vertical bar the place where the terms of the splicing are cut, in the style:

$$(x_1u_1|u_2x_2,\ y_1u_3|u_4y_2) \vdash_r x_1u_1u_4y_2,$$

for $r = u_1\#u_2\$u_3\#u_4$.

The way of building the result of the splicing by concatenating a prefix of the first term of the splicing with a suffix of the second term is visible in this writing. It could be also useful to represent the splicing rules in a graphic way, as in Fig. 7.2(a), in order to make clearer the way of working of these rules: the "window" in Fig. 7.2(a) should identify simultaneously the sites u_1u_2 and u_3u_4 in two DNA molecules, as in Fig. 7.2(b).

An *H scheme* is a pair

$$\sigma = (V, R),$$

where V is an alphabet and $R \subseteq V^*\#V^*\$V^*\#V^*$ is a set of splicing rules.

Note that R can be infinite, and that we can consider its place in the Chomsky hierarchy, or in another classification of languages. In general, if $R \in FL$, for a given family of languages, FL, then we say that the H scheme σ is *of FL type.*

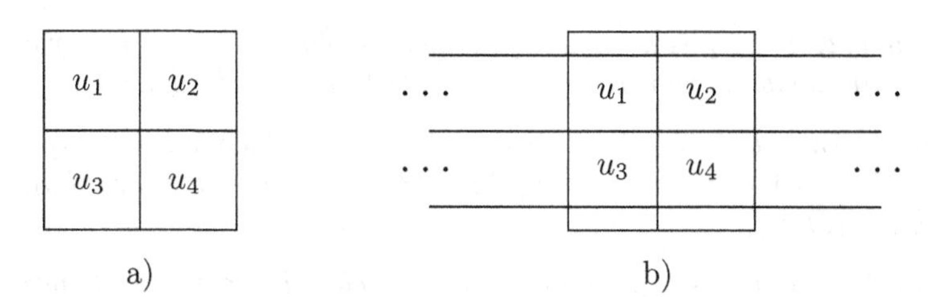

Figure 7.2: A graphical representation of a splicing rule

For a given H scheme $\sigma = (V, R)$ and a language $L \subseteq V^*$, we define

$$\sigma_1(L) = \{z \in V^* \mid (x, y) \vdash_r z, \text{ for some } x, y \in L, r \in R\}.$$

Thus, $\sigma_1(L)$ is the result of one step 1-splicing of strings in L with respect to the rules in R.

Sometimes, given an H scheme $\sigma = (V, R)$ and an ordered pair (x, y), $x, y \in V^*$, we also denote

$$\sigma_1(x, y) = \{z \in V^* \mid (x, y) \vdash_r z, \text{ for some } r \in R\}.$$

Note that $\sigma_1(x, y)$ is different from $\sigma_1(\{x, y\})$, which is the union of the four sets $\sigma_1(x, x), \sigma_1(x, y), \sigma_1(y, x), \sigma_1(y, y)$. We can write

$$\sigma_1(L) = \bigcup_{x,y \in L} \sigma_1(x, y).$$

Given two families FL_1, FL_2 of languages, we denote

$$S_1(FL_1, FL_2) = \{\sigma_1(L) \mid L \in FL_1 \text{ and } \sigma = (V, R) \text{ with } R \in FL_2\}.$$

(The subscript 1 in $S_1(\ldots, \ldots)$ reminds us that we are using here the 1-splicing operation.)

Therefore, the family FL_1 is closed under splicing of FL_2 type (we also say "FL_2 splicing") if $S_1(FL_1, FL_2) \subseteq FL_1$. In general, the power of FL_2 splicing is measured by investigating the families $S_1(FL_1, FL_2)$, for various FL_1.

We shall examine the families $S_1(FL_1, FL_2)$ for FL_1, FL_2 in the set $\{$*FIN, REG, LIN, CF, CS, RE*$\}$ (therefore, FL_2 is always assumed to contain at least all finite languages). We shall first establish a series of lemmas, connecting the splicing operation with other operations on languages, then collect the results about families $S_1(FL_1, FL_2)$ in a synthesis theorem.

Lemma 7.1. *For all families FL_1, FL_2, FL'_1, FL'_2, if $FL_1 \subseteq FL'_1$ and $FL_2 \subseteq FL'_2$, then $S_1(FL_1, FL_2) \subseteq S_1(FL'_1, FL'_2)$.*

Proof. Obvious, from the definitions. □

Lemma 7.2. *If FL_1 is a family of languages which is closed under concatenation with symbols, then $FL_1 \subseteq S_1(FL_1, FL_2)$, for all FL_2.*

Proof. Take $L \subseteq V^*, L \in FL_1$, and $c \notin V$. Then $L_0 = L\{c\} \in FL_1$. For the H scheme $\sigma = (V \cup \{c\}, \{\#c\$c\#\})$ we have $L = \sigma_1(L_0)$, hence $L \in S_1(FL_1, FL_2)$, for all FL_2. □

Lemma 7.3. *If FL is a family of languages closed under concatenation and arbitrary gsm mappings, then FL is closed under REG splicing.*

Proof. Take $L \subseteq V^*, L \in FL$, and an H scheme $\sigma = (V, R)$ with $R \subseteq V^*\#V^*\$V^*\#V^*$, $R \in REG$. Consider a new symbol, $c \notin V$, and a finite automaton $M = (K, V \cup \{\#, \$\}, s_0, F, \delta)$ recognizing the language R. By a standard construction, we can obtain a gsm g, associated with M, which transforms every string of the form

$$w = x_1u_1u_2x_2cy_1u_3u_4y_2,$$

for $x_1, x_2, y_1, y_2 \in V^*$, $u_1\#u_2\$u_3\#u_4 \in R$, to the string

$$g(w) = x_1u_1u_4y_2.$$

Consequently, $\sigma_1(L) = g(L\{c\}L)$. From the closure properties of FL, we obtain $\sigma_1(L) \in FL$. □

Lemma 7.4. *If FL is a family of languages closed under union, concatenation with symbols, and FIN splicing, then FL is closed under concatenation.*

Proof. Take two languages $L_1, L_2 \in FL, L_1, L_2 \subseteq V^*$, consider two new symbols, $c_1, c_2 \notin V$, and the H scheme

$$\sigma = (V \cup \{c_1, c_2\}, \{\#c_1\$c_2\#\}).$$

Obviously,

$$L_1L_2 = \sigma_1(L_1\{c_1\} \cup \{c_2\}L_2).$$

Hence, if FL has the mentioned properties, then $L_1L_2 \in FL$. □

Lemma 7.5. *If FL is a family of languages closed under concatenation with symbols and FIN splicing, then FL is closed under the operations Pref and Suf.*

Proof. For $L \subseteq V^*, L \in FL$, consider a new symbol, $c \notin V$, and the H schemes

$$\sigma = (V \cup \{c\}, \{\#a\$c\# \mid a \in V \cup \{c\}\}),$$
$$\sigma' = (V \cup \{c\}, \{\#c\$a\# \mid a \in V \cup \{c\}\}).$$

We have

$$Pref(L) = \sigma_1(L\{c\}), \quad Suf(L) = \sigma'_1(\{c\}L).$$

It is easy to see that using the rule $r = \#c\$c\#$ we obtain $(x|c, yc|) \vdash_r x$ and that for each $z \in Pref(x) - \{x\}, x \in L$, there is a rule of the form $r = \#a\$c\#$ in σ such that $(xc, yc) \vdash_r z$. Similarly for σ' and Suf. □

Lemma 7.6. *If FL is a family of languages which is closed under substitution with λ-free regular languages and arbitrary gsm mappings, then $S_1(REG, FL) \subseteq FL$.*

Proof. If FL has the above mentioned closure properties, then it is also closed under concatenation with symbols and intersection with regular languages (this follows directly from the closure under gsm mappings). Now, the closure under concatenation with symbols and under substitution with regular languages implies the closure under concatenation with regular languages. We shall use these properties below.

Take $L \subseteq V^*, L \in REG$, and an H scheme $\sigma = (V, R)$ with $R \in FL$. Consider the regular substitution $s : (V \cup \{\#, \$\})^* \longrightarrow \mathcal{P}((V \cup \{\#, \$\})^*)$ defined by

$$s(a) = \{a\}, \ a \in V,$$
$$s(\#) = \{\#\},$$
$$s(\$) = V^*\{\$\}V^*,$$

and construct the language

$$L_1 = V^* s(R) V^*.$$

Consider also the language

$$L_2 = (L \sqcup\!\sqcup \{\#\})\$(L \sqcup\!\sqcup \{\#\}).$$

As $L_1 \in FL$ and $L_2 \in REG$, we have $L_1 \cap L_2 \in FL$. The strings in $L_1 \cap L_2$ are of the form

$$w = x_1 u_1 \# u_2 x_2 \$ y_1 u_3 \# u_4 y_2,$$

for $x_1 u_1 u_2 x_2 \in L, y_1 u_3 u_4 y_2 \in L$, and $u_1 \# u_2 \$ u_3 \# u_4 \in R$.

If g is a gsm which erases the substring $\#z_2\$z_3\#$ from strings of the form $z_1\#z_2\$z_3\#z_4$ with $z_i \in V^*, 1 \leq i \leq 4$, then we get $\sigma_1(L) = g(L_1 \cap L_2)$, hence $\sigma_1(L) \in FL$. □

Lemma 7.7. *If FL is a family of languages which is closed under concatenation with symbols, then for all* $L_1, L_2 \in FL$ *we have* $L_1/L_2 \in S_1(FL, FL)$.

Proof. Take $L_1, L_2 \subseteq V^*, L_1, L_2 \in FL$, and $c \notin V$. For the H scheme

$$\sigma = (V \cup \{c\}, \{\#xc\$c\# \mid x \in L_2\}),$$

we obtain

$$L_1/L_2 = \sigma_1(L_1\{c\}).$$

Indeed, the only possible splicing of strings in $L_1\{c\}$ is of the form

$$(x_1|x_2c, yc|) \vdash_r x_1, \text{ for } x_1x_2 \in L_1, x_2 \in L_2, y \in L_1,$$

where $r = \#x_2c\$c\#$. □

Lemma 7.8. *If FL is a family of languages closed under concatenation with symbols, then for each* $L \in FL, L \subseteq V^*$, *and* $c \notin V$ *we have* $\{c\}L \in S_1(REG, FL)$.

Proof. For L, c as above, consider the H scheme

$$\sigma = (V \cup \{c, c'\}, \{cx\#c'\$c'\# \mid x \in L\}),$$

where c' is one further new symbol. Clearly, this is an H scheme of FL type. Then,

$$\{c\}L = \sigma_1(\{c\}V^*\{c'\}),$$

because the only splicings are of the form $(cx|c', cyc'|) \vdash_r cx$, for $r = cx\#c'\$c'\#, x \in L, y \in V^*$. □

Lemma 7.9. *If FL is a family of languages closed under concatenation with symbols and shuffle with symbols, then for each* $L \in FL$, $L \subseteq V^*$, *and* $c \notin V$, *we have* $\{c\}Pref(L) \in S_1(REG, FL)$.

Proof. For L, c as above, consider the H scheme of FL type

$$\sigma = (V \cup \{c, c'\}, \{cxc'\$c'\# \mid x \in L ⧢ \{\#\}\}),$$

where c' is one further new symbol. We have

$$\{c\}Pref(L) = \sigma_1(\{c\}V^*\{c'\}),$$

because the only possible splicings are of the form $(cx_1|x_2c', cyc'|) \vdash_r cx_1$, for rules $r = cx_1\#x_2c'\$c'\#,\ x_1x_2 \in L,\ y \in V^*$. □

We now synthesize the consequences of the previous lemmas for the families in the Chomsky hierarchy.

Theorem 7.1. *The relations in Table 7.1 hold, where at the intersection of the row marked with* FL_1 *with the column marked with* FL_2 *there appear either the family* $S_1(FL_1, FL_2)$, *or two families* FL_3, FL_4 *such that* $FL_3 \subset S_1(FL_1, FL_2) \subset FL_4$. *These families* FL_3, FL_4 *are the best possible estimations among the six families considered here.*

Table 7.1. The size of families $S_1(FL_1, FL_2)$

	FIN	REG	LIN	CF	CS	RE
FIN	FIN	FIN	FIN	FIN	FIN	FIN
REG	REG	REG	REG, LIN	REG, CF	REG, RE	REG, RE
LIN	LIN, CF	LIN, CF	RE	RE	RE	RE
CF	CF	CF	RE	RE	RE	RE
CS	RE	RE	RE	RE	RE	RE
RE	RE	RE	RE	RE	RE	RE

Proof. Clearly, $\sigma_1(L) \in FIN$ for all $L \in FIN$, whatever σ is. Together with Lemma 7.2, we have $S_1(FIN, FL) = FIN$ for all families FL.

Lemma 7.3 shows that $S_1(REG, REG) \subseteq REG$. Together with Lemma 7.2 we have $S_1(REG, FIN) = S_1(REG, REG) = REG$.

From Lemma 7.4 we get $S_1(LIN, FIN) - LIN \neq \emptyset$. From Lemma 7.3 we have $S_1(CF, REG) \subseteq CF$. Therefore, $LIN \subset S_1(LIN, FL) \subseteq CF = S_1(CF, FL)$ for $FL \in \{FIN, REG\}$.

Also the inclusions $S_1(LIN, FL) \subset CF, FL \in \{FIN, REG\}$, are proper. In order to see this, let us examine again the proof of Lemma 7.3. If $L \subseteq V^*, L \in LIN$, and $\sigma = (V, R)$ is an H scheme of REG type, then $\sigma_1(L) = g(L\{c\}L)$, where $c \notin V$ and g is a gsm. The language $L\{c\}L$ has a context-free index less than or equal to 2.

In Sect. 3.1 we have mentioned that the family of context-free languages of finite index is a full AFL, hence it is closed under arbitrary gsm mappings. Consequently, for each $L \in S_1(LIN, REG)$ we have $ind_{CF}(L) < \infty$. Since there are context-free languages of infinite index, it follows that $CF - S_1(LIN, REG) \neq \emptyset$.

From Theorem 3.12 we know that for every language $L \in RE, L \subseteq V^*$, there are $c_1, c_2 \notin V$ and a language $L' \subseteq L\{c_1\}\{c_2\}^*$ such that $L' \in CS$ and for each $w \in L$ there is $i \geq 0$ such that $wc_1c_2^i \in L'$. Take one further new symbol, c_3. The language $L'\{c_3\}$ is still in CS. For the H scheme

$$\sigma = (V \cup \{c_1, c_2, c_3\}, \{\#c_1\$c_3\#\}),$$

we have

$$\sigma_1(L'\{c_3\}) = \{w \mid wc_1c_2^ic_3 \in L'\{c_3\} \text{ for some } i \geq 0\} = L.$$

Consequently, $RE \subseteq S_1(CS, FIN)$. As $S_1(RE, RE) \subseteq RE$ (we can prove this in a straightforward way or we can invoke the Church–Turing Thesis), we get $S_1(CS, FL) = S_1(RE, FL) = RE$ for all FL.

According to Theorem 3.13, every language $L \in RE$ can be written as $L = L_1/L_2$, for $L_1, L_2 \in LIN$. By Lemma 7.7, each language L_1/L_2 with linear L_1, L_2 is in $S_1(LIN, LIN)$. Consequently, $S_1(LIN, FL) = S_1(CF, FL) = RE$, too, for all $FL \in \{LIN, CF, CS, RE\}$.

From Lemma 7.6 we have $S_1(REG, FL) \subseteq FL$ for $FL \in \{LIN, CF, RE\}$. All these inclusions are proper. More exactly, there are linear languages not in $S_1(REG, RE)$. Such an example is $L = \{a^n b^n \mid n \geq 1\}$.

Assume that $L = \sigma_1(L_0)$ for some $L_0 \in REG, L_0 \subseteq V^*$, and $\sigma = (V, R)$. Take a finite automaton for $L_0, M = (K, V, s_0, F, \delta)$, let $m = card(K)$, and consider the string $w = a^{m+1}b^{m+1}$ in L. Let $x, y \in L_0$ and $r \in R$ be such that $(x, y) \vdash_r w$, $x = x_1u_1u_2x_2$, $y = y_1u_3u_4y_2$, $w = x_1u_1u_4y_2$, for $r = u_1\#u_2\$u_3\#u_4$. We have either $x_1u_1 = a^{m+1}z$ or $u_4y_2 = z'b^{m+1}$, for some $z, z' \in \{a, b\}^*$. Assume that we have the first case; the second one is similar. Consequently, $x = a^{m+1}zu_2x_2$. When parsing the prefix a^{m+1}, the automaton M uses twice a state in K; the corresponding cycle can be iterated, hence L_0 contains strings of the form $x' = a^{m+1+ti}zu_2x_2$, for $t > 0$ and arbitrary $i \geq 0$. For such a string x' with $i \geq 1$ we have

$$(x', y) \vdash_r a^{m+1+ti}zu_4y_2 = a^{m+1+ti}b^{m+1}.$$

This string is not in L, a contradiction. The argument does not depend on the type of R. (Compare this with Lemma 7.8: $L \notin S_1(REG, RE)$, but $\{c\}L \in S_1(REG, LIN)$.)

According to Lemma 7.8, $S_1(REG, LIN) - REG \neq \emptyset$ and $S_1(REG, CF) - LIN \neq \emptyset$. From Lemma 7.9 we have $S_1(REG, CS) - CS \neq \emptyset$. (Consequently, $S_1(REG, CF)$ is incomparable with LIN and $S_1(REG, CS), S_1(REG, RE)$ are incomparable with LIN, CF, CS.)

All the assertions represented in the table are proved. □

Some remarks about the results in Table 7.1 are worth mentioning:

- All families $S_1(FL_1, FL_2)$ characterize families in the Chomsky hierarchy, with the exceptions of $S_1(REG, FL_2)$, with $FL_2 \in \{LIN, CF, CS, RE\}$, and $S_1(LIN, FL_2)$ with $FL_2 \in \{FIN, REG\}$, which are strictly intermediate between families in the Chomsky hierarchy. These six intermediate families need further investigation of their properties (for instance, closure under operations and decidability).
- A series of new characterizations of the family RE are obtained, starting, somewhat surprisingly, from "simple" pairs (FL_1, FL_2); especially interesting is the case (LIN, LIN), in view of the fact that it seems that the actual language of DNA sequences is not regular, or even context-free [23], [204]. Then, according to the previous results, it can be nothing else but recursively enumerable, of the highest complexity (in the Chomsky hierarchy).

We close this section by examining a possible hierarchy between LIN and CF, defined by subfamilies of $S_1(LIN, FIN)$.

For an H scheme $\sigma = (V, R)$ with a finite R, we define the *radius* of σ as

$$rad(\sigma) = \max\{|x| \mid x = u_i, 1 \leq i \leq 4, \text{ for some } u_1\#u_2\$u_3\#u_4 \in R\}.$$

Then, for $p \geq 1$, we denote by $S_1(FL, [p])$ the family of languages $\sigma_1(L)$, for $L \in FL$ and σ an H scheme of radius less than or equal to p.

Note that in the proof of Lemma 7.2 (and 7.4, 7.5), as well as in the proof of the inclusion $RE \subseteq S_1(CS, FIN)$ in Theorem 7.1, the schemes used are of radius 1. Hence for $FL \in \{FIN, REG, CF, CS, RE\}$ we have

$$S_1(FL, [1]) = S_1(FL, [p]), \text{ for all } p \geq 1,$$

that is, these hierarchies collapse. The same is true for $FL = LIN$. This follows from the next lemma.

Lemma 7.10. *If FL is a family of languages closed under λ-free gsm mappings, then* $S_1(FL, FIN) \subseteq S_1(FL, [1])$.

Proof. Take an H scheme $\sigma = (V, R)$ with finite R. Assume that the rules in R are labeled in a one-to-one manner, $R = \{r_1, \ldots, r_s\}$, $r_i = u_{i,1}\#u_{i,2}\$u_{i,3}\#u_{i,4}, 1 \leq i \leq s$. It is easy to construct a gsm g associated with R which transforms each string $w = x_1 u_{i,1} u_{i,2} x_2$ in $g(w) = x_1 u_{i,1} c_i u_{i,2} x_2$ and each string $w = y_1 u_{i,3} u_{i,4} y_2$ in $g(w) = y_1 u_{i,3} c_i' u_{i,4} y_2$, for $x_1, x_2, y_1, y_2 \in V^*$ and r_i as above, where c_i, c_i' are new symbols, associated with r_i. Consider now the H scheme $\sigma' = (V \cup \{c_i, c_i' \mid 1 \leq i \leq s\}, \{\#c_i\$c_i'\# \mid 1 \leq i \leq s\})$. We have $g(L) \in FL$ for each language $L \subseteq V^*$, $L \in FL$, and we obviously obtain $\sigma_1(L) = \sigma_1'(g(L))$. As $rad(\sigma') = 1$, the proof is complete. □

Theorem 7.2. $LIN \subset S_1(LIN, [p]) = S_1(LIN, FIN), p \geq 1$.

Proof. The inclusions $S_1(LIN, [p]) \subseteq S_1(LIN, [p+1]) \subseteq S_1(LIN, FIN)$, $p \geq 1$, follow by the definitions. From Lemma 7.10 we also get $S_1(LIN, FIN) \subseteq S(LIN, [1])$. The relation $LIN \subset S_1(LIN, FIN)$ is known from Theorem 7.1. □

7.3 Iterated Splicing as an Operation with Languages

When some restriction enzymes and a ligase are present in a test tube, they do not stop acting after one cut and paste operation, but they act iteratively.

For an H scheme $\sigma = (V, R)$ and a language $L \subseteq V^*$ we define

$$\sigma_1^0(L) = L,$$
$$\sigma_1^{i+1}(L) = \sigma_1^i(L) \cup \sigma_1(\sigma_1^i(L)), \ i \geq 0,$$

and

$$\sigma_1^*(L) = \bigcup_{i \geq 0} \sigma_1^i(L).$$

Consequently, $\sigma_1^*(L)$ is the closure of L under the splicing with respect to σ, i.e., the smallest language L' which contains L, and is closed under the splicing with respect to σ, that is to say, $\sigma_1(L') \subseteq L'$.

Note that $\sigma_1^1(L)$ is not equal to $\sigma_1(L)$, but to $L \cup \sigma_1(L)$.

For two families of languages, FL_1, FL_2, we define

$$H_1(FL_1, FL_2) = \{\sigma_1^*(L) \mid L \in FL_1 \text{ and } \sigma = (V, R) \text{ with } R \in FL_2\}.$$

Thus, the families $H_1(FL_1, FL_2)$ correspond to families $S_1(FL_1, FL_2)$ in the previous section. In the same way as in the case of the uniterated splicing, we can consider the hierarchies on the radius of finite H schemes, that is the families $H_1(FL, [p])$, of languages $\sigma_1^*(L)$ for $L \in FL$ and σ an H scheme of radius less than or equal to p.

Lemma 7.11. (i) *For all families FL_1, FL_1', FL_2, FL_2', if $FL_1 \subseteq FL_1'$ and $FL_2 \subseteq FL_2'$, then $H_1(FL_1, FL_2) \subseteq H_1(FL_1', FL_2')$.*

(ii) *$H_1(FL, [p]) \subseteq H_1(FL, [q])$, for all FL and $p \leq q$.*

Proof. Obvious from the definitions. □

Lemma 7.12. *$FL \subseteq H_1(FL, [1])$, for all families FL.*

Proof. Given $L \subseteq V^*, L \in FL$, consider a symbol $c \notin V$ and the H scheme

$$\sigma = (V \cup \{c\}, \{\#c\$c\#\}).$$

We clearly have $\sigma_1^i(L) = L$ for all $i \geq 0$, hence $\sigma_1^*(L) = \sigma_1^0(L) = L$. □

Lemma 7.13. *If FL_1, FL_2, FL_3 are families of languages such that both FL_1 and FL_2 are closed under shuffle with symbols and both FL_2 and FL_3 are closed under intersection with regular languages, then $H_1(FL_1, FL_2) \subseteq FL_3$ implies $S_1(FL_1, FL_2) \subseteq FL_3$.*

Proof. Take a language $L \subseteq V^*, L \in FL_1$, and an H scheme $\sigma = (V, R)$ with $R \in FL_2$. For $c \notin V$, consider the language

$$L' = L \sqcup\!\sqcup \{c\}$$

and the H scheme $\sigma' = (V \cup \{c\}, R')$ with

$$R' = (R \sqcup\!\sqcup \{cc\}) \cap V^*\#cV^*\$V^*c\#V^*.$$

From the properties of FL_1, FL_2 we have $L' \in FL_1, R' \in FL_2$. Moreover,

$$\sigma_1(L) = {\sigma_1'}^*(L') \cap V^*.$$

Indeed, ${\sigma_1'}^i(L') = \sigma_1'(L') \cup L'$ for all $i \geq 1$ (any splicing removes the symbol c from the strings of L', hence no further splicing is possible having as one of its terms the obtained string). Therefore, if ${\sigma_1'}^*(L') \in FL_3$, then $\sigma_1(L) \in FL_3$, too. □

Corollary 7.1. *In the conditions of Lemma 7.13, each language* $L \in S_1(FL_1, FL_2)$ *can be written as* $L = L' \cap V^*$, *for* $L' \in H_1(FL_1, FL_2)$.

We are now going to present two basic results in this area, of crucial importance for DNA computing based on splicing.

Lemma 7.14. (The Regularity Preserving Lemma) $H_1(REG, FIN) \subseteq REG$.

Proof. Let $L \subseteq V^*$ be a regular language recognized by a finite automaton $M = (K, V, s_0, F, \delta)$. Consider also an H scheme $\sigma = (V, R)$ with a finite set $R \subseteq V^*\#V^*\$V^*\#V^*$. Assume that $R = \{r_1, \ldots, r_n\}$ with $r_i = u_{i,1}\#u_{i,2}\$u_{i,3}\#u_{i,4}, 1 \leq i \leq n, n \geq 1$. Moreover, assume that $u_{i,1}u_{i,4} = a_{i,1}a_{i,2} \ldots a_{i,t_i}$, for $a_{i,j} \in V, 1 \leq j \leq t_i, t_i \geq 0$, $1 \leq i \leq n$. For each $i, 1 \leq i \leq n$, consider the new states $q_{i,1}, q_{i,2}, \ldots, q_{i,t_i}, q_{i,t_i+1}$. Denote their set by K' and consider the finite automaton

$$M_0 = (K \cup K', V, s_0, F, \delta_0),$$

where

$$\begin{aligned} &\delta_0(s, a) = \delta(s, a), \text{ for } s \in K, a \in V, \\ &\delta_0(q_{i,j}, a_{i,j}) = \{q_{i,j+1}\}, \ 1 \leq j \leq t_i, 1 \leq i \leq n. \end{aligned}$$

We construct a sequence of finite automata (with λ transitions) $M_k = (K \cup K', V, s_0, F, \delta_k)$, $k \geq 1$, starting from M_0, by passing from M_k to $M_{k+1}, k \geq 0$, in the following way.

Consider each splicing rule $r_i = u_{i,1}\#u_{i,2}\$u_{i,3}\#u_{i,4}$, $1 \leq i \leq n$.

If s is a state in $K \cup K'$ such that

1. $q_{i,1} \notin \delta_k(s, \lambda)$,

2. there is $s_1 \in K \cup K'$ and $x_1, x_2 \in V^*$ such that

$$\begin{aligned} &s \in \delta_k(s_0, x_1), \\ &s_1 \in \delta_k(s, u_{i,1}u_{i,2}), \\ &\delta_k(s_1, x_2) \cap F \neq \emptyset, \end{aligned}$$

(therefore, $x_1u_{i,1}u_{i,2}x_2 \in L(M_k)$), then we put

$$\delta_{k+1}(s, \lambda) = \{q_{i,1}\}.$$

We say that this is an *initial transition* of level $k + 1$.

Moreover, if s' is a state in $K \cup K'$ such that:

1. $s' \notin \delta_k(q_{i,t_i+1}, \lambda)$,

2. there is $s_1 \in K \cup K'$ and $y_1, y_2 \in V^*$ such that

$$s_1 \in \delta_k(s_0, y_1),$$
$$s' \in \delta_k(s_1, u_{i,3}u_{i,4}),$$
$$\delta_k(s', y_2) \cap F \neq \emptyset,$$

(therefore, $y_1 u_{i,3} u_{i,4} y_2 \in L(M_k)$), then we put

$$\delta_{k+1}(q_{i,t_i+1}, \lambda) = \{s'\}.$$

We say that this is a *final transition* of level $k+1$.

Then, δ_{k+1} is the extension of δ_k with the initial and final transitions of level $k+1$, with respect to all splicing rules in R and all states s, s' in $K \cup K'$.

As the set of states is fixed, the above procedure stops after at most $2 \cdot n \cdot card(K \cup K')$ steps, that is, there is an integer m such that $M_{m+1} = M_m$.

We shall prove that $L(M_m) = \sigma_1^*(L)$.

Since $\sigma_1^*(L)$ is the smallest language containing L and closed under the 1-splicing with respect to σ, it is enough to prove that

i) $L \subseteq L(M_m)$,

ii) $L(M_m)$ is closed under the 1-splicing with respect to σ,

iii) $L(M_m) \subseteq \sigma_1^*(L)$.

Point i) is obvious from the construction of the automaton M_m.

In order to prove point ii), let us consider a splicing rule $r_i = u_{i,1}\#u_{i,2}\$u_{i,3}\#u_{i,4}$ in R and two strings $x, y \in L(M_m)$ such that $x = x_1 u_{i,1} u_{i,2} x_2, y = y_1 u_{i,3} u_{i,4} y_2$. There are two states $s_1, s_2 \in K \cup K'$ such that

$$s_1 \in \delta_m(s_0, x_1),\ \delta_m(s_1, u_{i,1}u_{i,2}x_2) \cap F \neq \emptyset,$$
$$s_2 \in \delta_m(s_0, y_1 u_{i,3} u_{i,4}),\ \delta_m(s_2, y_2) \cap F \neq \emptyset.$$

From the construction of M_m we have

$$q_{i,1} \in \delta_m(s_1, \lambda) \text{ and } s_2 \in \delta_m(q_{i,t_i+1}, \lambda).$$

This implies that $x_1 u_{i,1} u_{i,4} y_2 \in L(M_m)$. The situation is illustrated in Fig. 7.3. Consequently, $\sigma_1(L(M_m)) \subseteq L(M_m)$.

In order to prove that each string recognized by M_m can be produced by iterated splicing with respect to σ starting from strings in L, we proceed by induction on the *level transition complexity* (abbreviated by *ltc*) of accepting paths (sequences of states) in M_m.

For a sequence π of states $s_0, s_1, \ldots, s_p$ in $K \cup K'$ such that $s_p \in F$, $s_{j+1} \in \delta_m(s_j, \alpha_j)$, for $\alpha_j \in V \cup \{\lambda\}$, $0 \leq j \leq p-1$, we denote by $ltc(\pi)$ the vector $(c_1, \ldots, c_m) \in \mathbf{N}^m$, where c_k is the number of initial or final transitions of level k in π, that is the number of subscripts l such that $s_{l+1} \in \delta_k(s_l, \lambda)$. We order the vectors in $\mathbf{N}^m$ in the *right-to-left lexicographic mode*: $(c_1, \ldots, c_m) < (d_1, \ldots, d_m)$ if there is $j, 1 \leq j \leq m$, such that $c_i = d_i$ for $i > j$ and $c_j < d_j$. Since this relation is a total order, we can use it as a basis for the induction arguments.

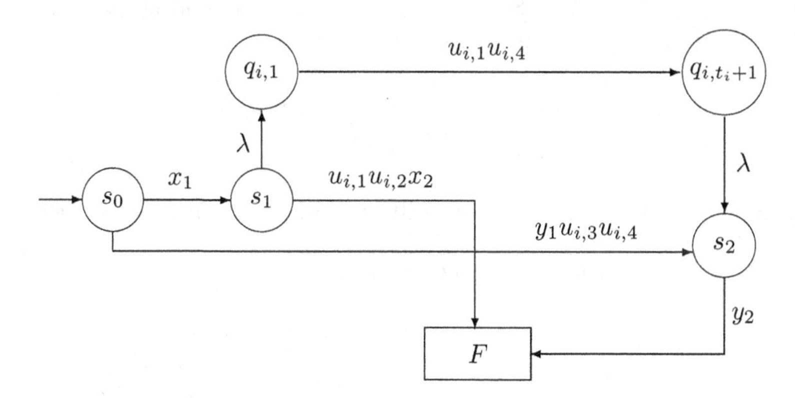

Figure 7.3: Simulating a splicing in M_m

For an accepting path π as above, we denote by $yield(\pi)$ the recognized string $\alpha_0\alpha_1 \ldots \alpha_{p-1}$.

If π is an accepting path such that $ltc(\pi) = (0, 0, \ldots, 0)$, then clearly $yield(\pi) \in L$, so $yield(\pi) \in \sigma_1^*(L)$.

Consider now that for some $(c_1, \ldots, c_m) \in \mathbf{N}^m$, $(c_1, \ldots, c_m) > (0, 0, \ldots, 0)$, all accepting paths π such that $ltc(\pi) < (c_1, \ldots, c_m)$ have $yield(\pi) \in \sigma_1^*(L)$. Consider an accepting path π in M_m such that $ltc(\pi) = (c_1, \ldots, c_m)$. (If no such a path exists, then the inductive step is fulfilled by default.) The path π should be of the form

$$\pi = s_0, s_1, \ldots, s_p, \; s_p \in F.$$

Since we start from $s_0 \in K$, we end with $s_p \in F \subseteq K$, and since $ltc(\pi) > (0, \ldots, 0)$, there are level transitions in the path; as from K to K' we can only go by initial level transitions and from K' to K we can only go by final level transitions, there are $s_{j_1}, s_{j_2}, 1 \leq j_1 < j_2 < p$, such that $s_{j_1} = q_{i,1}, s_{j_2} = q_{i,t_i+1}$, and all s_j with $j_1 \leq j \leq j_2$ are elements of K' (the parsing of $u_{i,1}u_{i,4}$ is uniquely determined: after reaching $q_{i,1}$, if the next level transition is a final one, we have to reach q_{i,t_i+1}). Therefore, there is a splicing rule $r = u_{i,1}\#u_{i,2}\$u_{i,3}\#u_{i,4}$ corresponding to which the above mentioned pair of level transitions have been introduced.

Assume that the transition corresponding to s_{j_1} is of level k and the transition corresponding to s_{j_2} is of level k'. Therefore, there is an accepting path $\pi_1 = s_0, \ldots, s_{j_1-1}, s'_{j_1}, \ldots, s'_{p'}$ in M_{k-1} such that $s'_{p'} \in F$ and

$$yield(s_0, \ldots, s_{j_1-1}) = x_1,$$
$$yield(s_{j_1-1}, s'_{j_1}, \ldots, s'_{p'}) = u_{i,1}u_{i,2}x_2,$$

for some $x_1, x_2 \in V^*$. Similarly, there is an accepting path $\pi_2 = s_0, s'_1, \ldots, s'_{j'_2}, s_{j_2+1}, \ldots, s_p$ in $M_{k'-1}$ such that

$$yield(s_0, s'_1, \ldots, s'_{j'_2}) = y_1u_{i,3}u_{i,4},$$
$$yield(s_{j_2+1}, \ldots, s_p) = y_2,$$

for some $y_1, y_2 \in V^*$. The situation is illustrated in Fig. 7.4. Consequently,

$$(yield(\pi_1), yield(\pi_2)) \vdash_r yield(\pi).$$

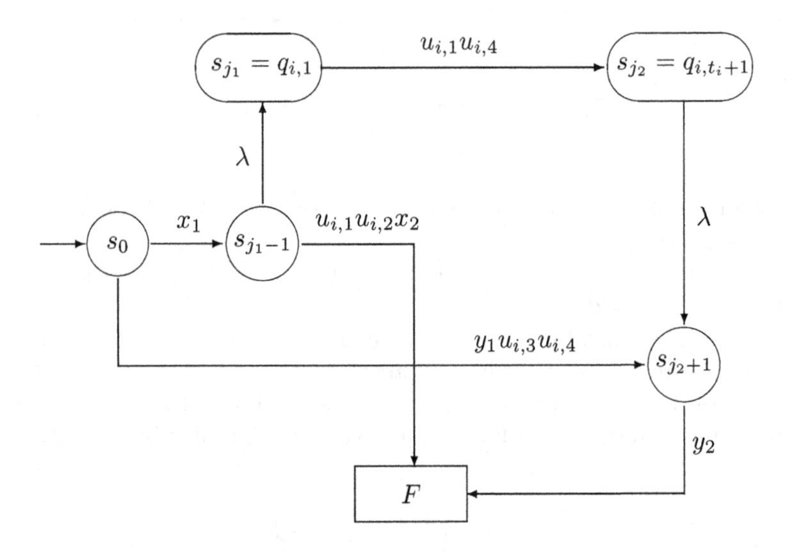

Figure 7.4: Finding a splicing in M_m

Examine now the level transition complexity of π_1; the case of π_2 is similar. The parsing of $u_{i,1}u_{i,2}x_2$ (passing from s_{j_1-1} to $s'_{p'} \in F$) is done in M_{k-1}, hence all the transitions here are of a level smaller than k. All the level transitions in the passing from s_0 to s_{j_1-1} are common to π and π_1. Thus, for $ltc(\pi_1) = (d_1, \ldots, d_m)$ we have $d_j \leq c_j$ for all $j \geq k$. Moreover, the path π contains at least one transition of level k which is not in π_1, that from s_{j_1-1} to s_{j_1}. This means that $d_k < c_k$; therefore, $(d_1, \ldots, d_m) < (c_1, \ldots, c_m)$, which concludes the inductive argument, hence the proof. □

A stronger result has been presented in [176]; a detailed proof of it can also be found in [90].

Lemma 7.15. *If FL is a full AFL, then $H_1(FL, FIN) \subseteq FL$.*

In fact, the proof in [90] gives a stronger result, because the assertion in Lemma 7.15 is shown to hold for any family FL of languages which contains the regular languages, is closed under right and left quotients by regular languages, and under substitution *into* regular languages. (If $s : V^* \longrightarrow 2^{U^*}$ is a substitution such that $s(a) \in FL$ for each $a \in V$, and $L \subseteq V^*$ is a regular language, then $s(L) \in FL$; each full AFL has this property, see Theorem 11.5 in [93].)

Lemma 7.16. (The Basic Universality Lemma) *Every language $L \in RE, L \subseteq T^*$, can be written in the form $L = L' \cap T^*$ for some $L' \in H_1(FIN, REG)$.*

Proof. Consider a type-0 grammar $G = (N, T, S, P)$, denote $U = N \cup T \cup \{B\}$, where B is a new symbol, and construct the H scheme

$$\sigma = (V, R),$$

where

$$V = N \cup T \cup \{X, X', B, Y, Z\} \cup \{Y_\alpha \mid \alpha \in U\},$$

and R contains the following groups of rules:

$$\begin{array}{rlll}
\textit{Simulate}: & 1. & Xw\#uY\$Z\#vY, & \text{for } u \rightarrow v \in P, w \in U^*, \\
\textit{Rotate}: & 2. & Xw\#\alpha Y\$Z\#Y_\alpha, & \text{for } \alpha \in U, w \in U^*, \\
 & 3. & X'\alpha\#Z\$X\#wY_\alpha, & \text{for } \alpha \in U, w \in U^*, \\
 & 4. & X'w\#Y_\alpha\$Z\#Y, & \text{for } \alpha \in U, w \in U^*, \\
 & 5. & X\#Z\$X'\#wY, & \text{for } w \in U^*, \\
\textit{Terminate}: & 6. & \#ZY\$XB\#wY, & \text{for } w \in T^*, \\
 & 7. & \#Y\$XZ\#. &
\end{array}$$

Consider also the language

$$\begin{aligned}
L_0 = & \{XBSY,\ ZY, XZ\} \\
& \cup \{ZvY \mid u \rightarrow v \in P\} \\
& \cup \{ZY_\alpha,\ X'\alpha Z \mid \alpha \in U\}.
\end{aligned}$$

We obtain $L = \sigma_1^*(L_0) \cap T^*$.

Indeed, let us examine the work of σ, namely the possibilities to obtain a string in T^*.

No string in L_0 is in T^*. All rules in R involve a string containing the symbol Z, but this symbol will not appear in the string produced by splicing. Therefore, at each step we have to use a string in L_0 and, excepting the case of using the string $XBSY$ in L_0, a string produced at a previous step.

The symbol B is a marker for the beginning of the sentential forms of G simulated by σ.

By rules in group 1 we can simulate the rules in P. Rules in groups 2 – 5 move symbols from the right hand end of the current string to the left hand end, thus making possible the simulation of rules in P at the right hand end of the string produced by σ. However, because B is always present and marks the place where the string of G begins, we know in each moment which is that string. Namely, if the current string in σ is of the form $\beta_1 w_1 B w_2 \beta_2$, for some β_1, β_2 markers of types X, X', Y, Y_α with $\alpha \in U$, and $w_1, w_2 \in (N \cup T)^*$, then $w_2 w_1$ is a sentential form of G.

We start from $XBSY$, hence from the axiom of G, marked to the left hand with B and bracketed by X, Y.

Let us see how the rules 2–5 work. Take a string $Xw\alpha Y$, for some $\alpha \in U$, $w \in U^*$. By a rule of type 2 we get

$$(Xw|\alpha Y, Z|Y_\alpha) \vdash XwY_\alpha.$$

The symbol Y_α memorizes the fact that α has been erased from the right hand end of $w\alpha$. No rule in R can be applied to XwY_α, excepting the rules of type 3:

$$(X'\alpha|Z, X|wY_\alpha) \vdash X'\alpha wY_\alpha.$$

Note that the same symbol α removed at the previous step is now added in the front of w. Again we have only one way to continue, namely by using a rule of type 4. We get

$$(X'\alpha w|Y_\alpha, Z|Y) \vdash X'\alpha wY.$$

If we use now a rule of type 7, removing Y, then X' (and B) can never be removed, the string cannot be turned to a terminal one. We have to use a rule of type 5:

$$(X|Z, X'|\alpha wY) \vdash X\alpha wY.$$

We have started from $Xw\alpha Y$ and have obtained $X\alpha wY$, a string with the same end markers. We can iterate these steps as long as we want, so any circular permutation of the string between X and Y can be produced. Moreover, what we obtain are exactly the circular permutations and nothing more (for instance, at every step we still have one and only one occurrence of B).

To every string XwY we can also apply a rule of type 1, providing w ends with the left hand member of a rule in P. Any rule of P can be simulated in this way, at any place we want in the corresponding sentential form of G, by preparing the string as above, using rules in groups 2–5.

Consequently, for every sentential form w of G there is a string $XBwY$, produced by σ, and, conversely, if Xw_1Bw_2Y is produced by σ, then w_2w_1 is a sentential form of G.

The only way to remove the symbols not in T from the strings produced by σ is by using rules in groups 6, 7. More precisely, the symbols XB can only be removed in the following conditions: (1) Y is present (hence the work is blocked if we use first rule 7, removing Y: the string cannot participate to any further splicing, and it is not terminal), (2) the current string bracketed by X, Y consists of terminal symbols only, and (3) the symbol B is in the left hand position. After removing X and B we can remove Y, too, and what we obtain is a string in T^*. From the previous discussion, it is clear that such a string is in $L(G)$, hence $\sigma_1^*(L_0) \cap T^* \subseteq L(G)$. Conversely, each string in $L(G)$ can be produced in this way, hence $L(G) \subseteq \sigma_1^*(L_0) \cap T^*$. We have the equality $L(G) = \sigma_1^*(L_0) \cap T^*$, which completes the proof. □

Many variants of the rotate-and-simulate procedure used in the previous proof will be presented in the following chapters.

The families $H_1(FL_1, FL_2)$ have serious limitations.

Lemma 7.17. *Let FL be a family of languages closed under intersection with regular languages and restricted morphisms. For every $L \subseteq V^*, L \notin FL$, and $c, d \notin V$, we have $L' \notin H_1(FL, RE)$, for*

$$L' = (dc)^* L (dc)^* \ \cup \ c(dc)^* L (dc)^* d.$$

Proof. For L, c, d as above, denote

$$L_1 = (dc)^* L (dc)^*,$$
$$L_2 = c(dc)^* L (dc)^* d.$$

As $L = L_1 \cap V^* = L' \cap V^*$ and $L = h(L_2 \cap cV^*d)$, where h is the morphism defined by $h(a) = a, a \in V$, and $h(c) = h(d) = \lambda$, it follows that $L_1 \notin FL, L_2 \notin FL$, and $L' = L_1 \cup L_2 \notin FL$.

Assume that $L' = \sigma_1^*(L_0)$, for some $L_0 \in FL, L_0 \subseteq L'$, and $\sigma = (V, R)$ with arbitrary R. As $L' \notin FL$, it follows that $L_0 \neq L'$ and we need effective splicing operations in order to produce L' from L_0. That is, splicings $(x, y) \vdash_r z$ with $x \neq z, y \neq z$ are necessary, $x, y \in L_0$. Write $x = x_1u_1u_2x_2$, $y = y_1u_3u_4y_2$, for some $x_1, x_2, y_1, y_2 \in (V \cup \{c, d\})^*$, and $r = u_1\#u_2\$u_3\#u_4 \in R$.

If $x \in L_1$, then $x' = cxd \in L_2, x' = cx_1u_1u_2x_2d$, hence we can perform $(x', y) \vdash_r z' = cx_1u_1u_4y_2 = cz$. If $z \in L'$, then $cz \notin L'$, a contradiction.

Therefore, x must be from L_2. Then $x' = dxc \in L_1$, $x' = dx_1u_1u_2x_2c$, hence we can perform $(x', y) \vdash_r z' = dx_1u_1u_4y_2 = dz$. Again we obtain a string not in L'. Since no splicing is possible without producing strings not in L', we must have $\sigma_1^*(L_0) = L_0$, which contradicts the relation $L' \neq L_0$.

As the type of the set R plays no role in the previous argument, we have $L' \notin H_1(FL, RE)$. □

Theorem 7.3. *The relations in Table 7.2 hold, where at the intersection of the row marked with FL_1 with the column marked with FL_2 there appear either the family $H_1(FL_1, FL_2)$, or two families FL_3, FL_4 such that $FL_3 \subset H_1(FL_1, FL_2) \subset FL_4$. These families FL_3, FL_4 are the best possible estimations among the six families considered here.*

Table 7.2. The size of families $H_1(FL_1, FL_2)$

	FIN	REG	LIN	CF	CS	RE
FIN	FIN, REG	FIN, RE	FIN, RE	FIN, RE	FIN, RE	FIN, RE
REG	REG	REG, RE	REG, RE	REG, RE	REG, RE	REG, RE
LIN	LIN, CF	LIN, RE	LIN, RE	LIN, RE	LIN, RE	LIN, RE
CF	CF	CF, RE	CF, RE	CF, RE	CF, RE	CF, RE
CS	CS, RE	CS, RE	CS, RE	CS, RE	CS, RE	CS, RE
RE	RE	RE	RE	RE	RE	RE

Proof. From Lemma 7.12 we have the inclusions $FL_1 \subseteq H_1(FL_1, FL_2)$, for all values of FL_1, FL_2. On the other hand, $H_1(FL_1, FL_2) \subseteq RE$ for all FL_1, FL_2. With the exception of the families $H_1(RE, FL_2)$, which are equal to RE, all inclusions $H_1(FL_1, FL_2) \subseteq RE$ are proper: from Lemma 7.17, we see that all the following differences are non-empty $REG - H_1(FIN, RE)$, $LIN - H_1(REG, RE)$, $CF - H_1(LIN, RE)$, $CS - H_1(CF, RE)$, $RE - H_1(CS, RE)$.

Lemma 7.14 and Lemma 7.12 together imply that $H_1(REG, FIN) = REG$. Hence we have $H_1(FIN, FIN) \subseteq REG$. This inclusion is strict by Lemma 7.17.

From Lemma 7.13 (and the results in Sect. 7.2) we obtain the strictness of the inclusions $LIN \subset H_1(LIN, FIN)$, and $CS \subset H_1(CS, FIN)$. The same result is obtained if FIN is replaced by any family FL_2.

Lemma 7.15, together with Lemma 7.12, implies $H_1(CF, FIN) = CF$. We also have $H_1(LIN, FIN) \subseteq CF$. The inclusion is proper by Lemma 7.17.

From Lemma 7.17 we see that RE is the best estimation for $H_1(FL_1, FL_2)$, $FL_2 \neq FIN$ (we have $H_1(FIN, REG) - FL \neq \emptyset$ for all families $FL \subset RE$ which are closed under intersection with regular languages).

The only assertion which remains to be proved is the fact that $H_1(FIN, FIN)$ contains infinite languages. This is true even for $H_1(FIN, [1])$: for $\sigma = (\{a\}, \{a\#\$\#a\})$ we have $\sigma_1^*(\{a\}) = a^+$. Thus, the proof is complete. □

Many of the relations in Table 7.2 are of interest:

- The iterated splicing with respect to regular sets of rules leads from the regular languages (even from the finite ones) to non-regular (even non-recursive) languages (this is not true for the "weaker" case of uniterated splicing); therefore, the result in Lemma 7.14 cannot be improved, by replacing FIN with a family of languages which is larger than REG.
- The iterated splicing with respect to (at least) regular sets of rules already leads from the finite languages to non-context-sensitive languages. In fact, for all FL_2 containing the regular languages, the intersections of the languages in $H_1(FL_1, FL_2)$ with regular languages of the form V^* characterize the family of recursively enumerable languages.

– However, all the families $H_1(FL_1, FL_2), FL_1 \neq RE$, have surprising limitations. When FL'_1 is the smallest family among those considered here which strictly includes FL_1, there are languages in FL'_1 which are not in $H_1(FL_1, FL_2)$, for all FL_2, including $FL_2 = RE$.

In view of the equalities $H_1(FL, FIN) = FL$, for $FL \in \{REG, CF, RE\}$, the hierarchies on the radius of H schemes collapse in these cases. The problem is still *open* for $FL \in \{LIN, CS\}$, but for FIN we have

Theorem 7.4. *$FIN \subset H_1(FIN, [1]) \subset H_1(FIN, [2]) \subset \ldots \subset H_1(FIN, FIN) \subset REG$.*

Proof. The inclusions follow from the definitions and from Lemmas 7.12 and 7.14; the strictness of the first and last inclusions is already known.

For $k \geq 1$, consider the language

$$L_k = \{a^{2k}b^{2k}a^n b^{2k}a^{2k} \mid n \geq 2k+1\}.$$

It belongs to $H_1(FIN, [k+1])$, because $L_k = \sigma_1^*(L'_k)$, for

$$L'_k = \{a^{2k}b^{2k}a^{2k+2}b^{2k}a^{2k}\},$$
$$\sigma = (\{a, b\}, \{a^{k+1}\#a^k\$a^{k+1}\#a^k\}).$$

Indeed, the splicing rule can only be used with the sites $u_1u_2 = a^{2k+1}$ and $u_3u_4 = a^{2k+1}$ in the central substring, $a^{2k+i}, i \geq 1$, of strings in L_k. Hence we can obtain strings with a^{2k+i+1} as a central substring, for all $i \geq 0$ by splicings of the forms

$$(a^{2k}b^{2k}a^{k+1}|a^{k+1}b^{2k}a^{2k}, a^{2k}b^{2k}aa^{k+1}|a^k b^{2k}a^{2k}) \vdash a^{2k}b^{2k}a^{k+1}a^k b^{2k}a^{2k},$$
$$(a^{2k}b^{2k}a^j a^{k+1}|a^k b^{2k}a^{2k}, a^{2k}b^{2k}a^{k+1}|a^{k+1}b^{2k}a^{2k}) \vdash a^{2k}b^{2k}a^j a^{k+1}a^{k+1}b^{2k}a^{2k},$$

for $j \geq 1$.

Assume that $L_k = \sigma'^*_1(L''_k)$, for some finite language L''_k and an H scheme $\sigma' = (V, R)$ with $rad(\sigma') \leq k$. Take a rule $r = u_1\#u_2\$u_3\#u_4 \in R$ and two strings $x, y \in L_k$ to which this rule can be applied, $x = a^{2k}b^{2k}a^n b^{2k}a^{2k}, n \geq 2k+1$, $y = y_1u_3u_4y_2$. As $|u_1u_2| \leq 2k$, if $u_1u_2 \in a^*$, then u_1u_2 is a substring of both the prefix a^{2k} and of the suffix a^{2k}, as well as of the central subword a^n of x. Similarly, if $u_1u_2 \in b^*$, then u_1u_2 is a substring of both substrings b^{2k} of x. If $u_1u_2 \in a^+b^+$, then u_1u_2 is a substring of both the prefix $a^{2k}b^{2k}$ and of the subword $a^n b^{2k}$ of x; if $u_1u_2 \in b^+a^+$, then u_1u_2 is a substring of both the suffix $b^{2k}a^{2k}$ and of the subword $b^{2k}a^n$ of x. In all cases, splicing x, y according to the rule r we find at least one string which is not in L_k, hence the equality $L_k = {\sigma'_1}^*(L''_k)$ is not possible. Therefore, $H_1(FIN, [k+1]) - H_1(FIN, [k]) \neq \emptyset$, for all $k \geq 1$. □

Although the families $H_1(FIN, [1])$ and REG seem to be very different (situated at the ends of an infinite hierarchy), they are still equal modulo a coding.

Theorem 7.5. *Every regular language is a coding of a language in the family* $H_1(FIN, [1])$.

Proof. Let $L \in REG$ be generated by a regular grammar $G = (N, T, S, P)$; hence the rules in P have the forms $X \to aY, X \to a$, for $X, Y \in N, a \in T$. Consider the alphabet

$$\begin{aligned} V = & \{[X, a, Y] \mid X \to aY \in P, \text{ with } X, Y \in N, a \in T\} \\ & \cup \{[X, a, *] \mid X \to a \in P, \text{ with } X \in N, a \in T\}, \end{aligned}$$

the H scheme

$$\begin{aligned} \sigma = (V, & \{[X, a, Y]\#\$\#[Y, b, Z] \mid [X, a, Y], [Y, b, Z] \in V\} \\ & \cup \{[X, a, Y]\#\$\#[Y, b, *] \mid [X, a, Y], [Y, b, *] \in V\}), \end{aligned}$$

and the finite language

$$\begin{aligned} L_0 = & \{[S, a, *] \mid S \to a \in P, a \in T\} \\ & \cup \{[X_1, a_1, X_2][X_2, a_2, X_3] \ldots [X_k, a_k, X_{k+1}][X_{k+1}, a_{k+1}, *] \mid \\ & k \geq 1, \ X_1 = S, \ X_i \to a_i X_{i+1} \in P, \ 1 \leq i \leq k, \\ & X_{k+1} \to a_{k+1} \in P, \text{ and for no } 1 \leq i_1 < i_2 < i_3 \leq k \text{ we have} \\ & [X_{i_1}, a_{i_1}, X_{i_1+1}] = [X_{i_2}, a_{i_2}, X_{i_2+1}] = [X_{i_3}, a_{i_3}, X_{i_3+1}]\} \end{aligned}$$

(we can have at most pairs of equal symbols of V in a string of L_0, but not triples of equal symbols). Consider also the coding $h : V \longrightarrow T$ defined by

$$h([X, a, Y]) = h([X, a, *]) = a, \ X, Y \in N, a \in T.$$

We have the relation

$$L = h(\sigma_1^*(L_0)).$$

Indeed, each string in L_0 corresponds to a derivation in G and if x, y are strings in $\sigma_1^*(L_0)$ describing derivations in G, $x = x_1[X, a, Y][Y, a', Z']x_2$ and $y = y_1[X', b', Y]\ [Y, b, Z]y_2$, then $z = x_1[X, a, Y][Y, b, Z]y_2 \in \sigma_1(x, y)$ and obviously z corresponds to a derivation in G, too. The coding h associates to such a string w describing a derivation in G the string $h(w)$ generated by this derivation. Consequently, $h(\sigma_1^*(L_0)) \subseteq L$.

Conversely, consider the strings in V^* describing derivations in G. Such strings w of length less than or equal to two are in L_0 hence in $\sigma_1^*(L_0)$. Assume that all such strings of length less than or equal to some $n \geq 2$ are in $\sigma_1^*(L_0)$ and consider a string w of the smallest length greater than n for which a derivation in G can be found. As $|w| > n \geq 2$, it follows that $w \notin L_0$, hence w contains a symbol $[X, a, Y]$ in three different positions:

$$w = w_1[X, a, Y]w_2[X, a, Y]w_3[X, a, Y]w_4.$$

Then

$$w' = w_1[X, a, Y]w_2[X, a, Y]w_4,$$
$$w'' = w_1[X, a, Y]w_3[X, a, Y]w_4$$

describe correct derivations in G and $|w'| < |w|, |w''| < |w|$, hence $w', w'' \in \sigma_1^*(L_0)$ by the induction hypothesis. From the form of w', w'' and of the splicing rules of σ we have $w \in \sigma_1(w', w'')$, hence $w \in \sigma_1^*(L_0)$, too.

For each derivation in G we find a string $w \in \sigma_1^*(L_0)$ such that $h(w)$ is exactly the string generated by this derivation. In conclusion, we have $L \subseteq h(\sigma_1^*(L_0))$. □

7.4 Extended H Systems; Generative Power

We now introduce the basic computability model that we shall investigate in this chapter and in the following ones. We consider here its general (unrestricted) form.

In the previous section we already have a generative mechanism based on the splicing operation: a pair (σ, L), where $\sigma = (V, R)$ is an H scheme and $L \subseteq V^*$ is a given language, identifies the language $\sigma_1^*(L)$. We can write the pair (σ, L) in a more explicit way, as a triple $\gamma = (V, L, R)$, identifying the language $L(\gamma) = \sigma_1^*(L)$. Such a triple is called an *H system*. Note that we allow the components L, R to be infinite, which contrasts with the usual custom when defining a grammar: a finite mechanism generating a possibly infinite language. In this section we shall continue in this manner, on the one hand, for the sake of mathematical completeness, and on the other because of results like those in Lemmas 7.14, 7.15: using a finite set of splicing rules we cannot overpass the regularity barrier when starting from regular languages. We look for computationally complete mechanisms; a way towards this goal is suggested by Lemma 7.16: using regular sets of splicing rules. This lemma also suggests a generalization in the definition of H systems: to consider a terminal alphabet (as in Chomsky grammars and in extended Lindenmayer systems) and to accept only the strings over this alphabet which are produced by iterated splicing. We get in this way the concept of *extended H systems*, the fundamental notion investigated in this chapter.

An *extended H system* is a quadruple

$$\gamma = (V, T, A, R),$$

where V is an alphabet, $T \subseteq V$, $A \subseteq V^*$, and $R \subseteq V^*\#V^*\$V^*\#V^*$, where $\#, \$$ are special symbols not in V.

We call V the alphabet of γ, T is the *terminal* alphabet, A is the set of *axioms*, and R the set of splicing rules. Therefore, we have an *underlying H scheme*, $\sigma = (V, R)$, augmented with a given subset of V and a set of axioms.

When $T = V$ we say that γ is a *non-extended* H system; below we shall only incidentally consider such systems.

The *language generated* by γ is defined by

$$L(\gamma) = \sigma_1^*(A) \cap T^*,$$

where σ is the underlying H scheme of γ.

For two families of languages, FL_1, FL_2, we denote by $EH_1(FL_1, FL_2)$ the family of languages $L(\gamma)$ generated by extended H systems $\gamma = (V, T, A, R)$, with $A \in FL_1, R \in FL_2$. A number of the results in Sect. 7.3 can be reformulated in terms of extended H systems. Moreover, we have the following inclusion.

Lemma 7.18. $REG \subseteq EH_1(FIN, FIN)$.

Proof. Take a language $L \in REG, L \subseteq T^*$, generated by a regular grammar $G = (N, T, S, P)$.

We construct the H system

$$\gamma = (N \cup T \cup \{Z\}, T, A_1 \cup A_2 \cup A_3, R_1 \cup R_2),$$

with

$$\begin{aligned}
A_1 &= \{S\},\\
A_2 &= \{ZaY \mid X \to aY \in P, X, Y \in N, a \in T\},\\
A_3 &= \{ZZa \mid X \to a \in P, X \in N, a \in T\},\\
R_1 &= \{\#X\$Z\#aY \mid X \to aY \in P, X, Y \in N, a \in T\},\\
R_2 &= \{\#X\$ZZ\#a \mid X \to a \in P, X \in N, a \in T\}.
\end{aligned}$$

If we splice a string ZxX, possibly from A_2 (for $x = c \in T$ and $U \to cX \in P$) using a rule in R_1, then we get a string of the form $ZxaY$. The symbol Z cannot be eliminated, hence no terminal string can be obtained if we continue to use the resulting string as the first term of a splicing. On the other hand, a string ZxX with $|x| \geq 2$ cannot be used as the second term of a splicing. Consequently, the only way to obtain a terminal string is to start from S, to use splicings with respect to rules in R_1 an arbitrary number of times, and to end with a rule in R_2. Always the first term of a splicing is that obtained by a previous splicing and the second one is from A_2 or from A_3 (at the last step). This corresponds to a derivation in G, hence we have $L(\gamma) = L(G) = L$. □

Theorem 7.6. *The relations in Table 7.3 hold, where at the intersection of the row marked with* FL_1 *with the column marked with* FL_2 *there appear either the family* $EH_1(FL_1, FL_2)$*, or two families* FL_3, FL_4 *such that*

$FL_3 \subset EH_1(FL_1, FL_2) \subseteq FL_4$. These families FL_3, FL_4 are the best possible estimations among the six families considered here.

Table 7.3. The generative power of extended H systems

	FIN	*REG*	*LIN*	*CF*	*CS*	*RE*
FIN	*REG*	*RE*	*RE*	*RE*	*RE*	*RE*
REG	*REG*	*RE*	*RE*	*RE*	*RE*	*RE*
LIN	*LIN*, *CF*	*RE*	*RE*	*RE*	*RE*	*RE*
CF	*CF*	*RE*	*RE*	*RE*	*RE*	*RE*
CS	*RE*	*RE*	*RE*	*RE*	*RE*	*RE*
RE	*RE*	*RE*	*RE*	*RE*	*RE*	*RE*

Proof. Clearly, $FL \subseteq EH_1(FL, FIN)$ for all FL. From Lemma 7.18 we also have $REG \subseteq EH_1(FIN, FL)$ for all FL. From Lemmas 7.14 and 7.15 and the closure of REG, CF under intersection with regular languages we obtain $EH_1(REG, FIN) \subseteq REG$, $EH_1(CF, FIN) \subseteq CF$. If in the proof of the relation $RE \subseteq S_1(CS, FIN)$ in Theorem 7.1 we take c_1, c_2, c_3 as nonterminal symbols and V as a terminal alphabet, then we obtain $RE \subseteq EH_1(CS, FIN)$. Thus, the first column of Table 7.3 is obtained.

From the proof of Lemma 7.16 we obtain $RE \subseteq EH_1(FIN, REG)$. As $EH_1(FL_1, FL_2) \subseteq RE$ for all families FL_1, FL_2 (this can be directly proved in a straightforward way or we can invoke the Church–Turing Thesis), the proof is complete. □

The only family which is not equal to a family in the Chomsky hierarchy is $EH_1(LIN, FIN)$.

Two of the relations summarized in Table 7.3 are central for the DNA computability based on splicing:

1. $EH_1(FIN, FIN) = REG$,
2. $EH_1(FIN, REG) = RE$.

When using a finite extended H system, that is a system with a finite set of axioms and a finite set of splicing rules, we only obtain a characterization of regular languages. The power of such devices stops at the level of finite automata (Chomsky regular grammars). Increasing the set of axioms does not help very much: we need a context-sensitive set of axioms in order to characterize RE using a finite set of splicing rules. However, making the smallest step (in our framework) in generalizing the set of splicing rules, that is considering a regular set of splicing rules, leads to the jump to the full power of Turing machines (Chomsky type-0 grammars). An infinite set of splicing rules, even forming a regular language, is not of much practical interest, it is not realistic to deal with "infinite computers". Thus, we have to choose: either we are satisfied with "DNA computers" based on splicing able to compute only at the level of finite automata, or we supplement the model with further features, in the hope of still working with finite sets of splicing rules but preserving the power of extended H systems with regular sets of rules. Moreover, as we have mentioned in Chap. 3, there is no finite automaton which is universal, in a natural way, for all finite automata, hence we cannot hope to devise universal – hence programmable – DNA computers as extended H systems with finite components (and no additional control on the splicing operation or another feature able to increase the power). The choice is somewhat forced: the only way to obtain Turing universal programmable DNA computers as extended H systems with finite components is to regulate the work of these systems, adding a supplementary control on the splicing operation. This will be the goal of the subsequent chapters.

In Sect. 3.3 we have constructed a finite automaton M_u which is universal for the class of finite automata with a bounded number of states and symbols. From the proof of Lemma 7.18, starting from M_u we can construct an extended H system γ_u which has similar universality properties. However, in this way we obtain a splicing system producing strings of the form $bls(code(M), x)$ (remember the notation from Sect. 3.3). It is possible to improve this result, in the sense that we can construct a "partially universal" extended H system with finite components and producing exactly the strings x recognized by the automaton M, whose "program" is introduced in the axiom set of the universal system. To this aim we do not start from $bls(code(M), x)$ as in Sect. 3.3, but from a string containing one more copy of x. Using $bls(code(M), x)$ we check whether or not $x \in L(M)$ and only in the affirmative case do we remove all auxiliary symbols, producing the terminal string x. Here is such a universal H system associated with M_u:

$$\gamma_u = (W, V, A_u, R_u),$$

with

$$\begin{aligned}
W &= V \cup K \cup K_u \cup \{B_0, B, E, Z, c_1, c_2\},\\
A_u &= \{BqZ \mid q \in K_u\} \cup \{ZZ\},\\
R_u &= \{Bq_{0,u}\#Z\$B_0\#\}\\
&\quad \cup \{Bq'\#Z\$Bq\alpha\# \mid q\alpha \to \alpha q' \in P_u, \text{ for } q, q' \in K_u,\\
&\quad\quad \alpha \in V \cup K \cup \{c_1, c_2\}\}\\
&\quad \cup \{\#ZZ\$BqE\# \mid q \in K\}.
\end{aligned}$$

If we add to A_u the axiom

$$w_0 = B_0 \; bls(code(M), x)Ex,$$

then we get an extended H system $\gamma_u' = (W, V, A_u', R_u)$ such that $L(\gamma_u') = L(M)$. Indeed, the only way to obtain a string in V^*, that is without symbols in $K \cup K_u \cup \{B_0, B, E, Z, c_1, c_2\}$, is to simulate the rules in P_u on the prefix

B_0 $bls(code(M), x)E$ of the axiom w_0, step by step, from the left (this is ensured by the fact that all splicing rules require the presence of B_0 or of B), until reducing this string to BqE, for some $q \in K$ (in fact, by the mode of work of M_u we have $q \in F_u$); then the block BqE can also be removed by the splicing rule $\#ZZ\$BqE\#$.

We formulate this important (for DNA computing) conclusion of the discussion above in the form of a theorem.

Theorem 7.7. *There are extended H systems with finite sets of rules which are universal for the class of finite automata with a bounded number of states and a bounded number of input symbols.*

Note that M is "run" on the "computer" γ_u via the "program" w_0, which is a unique string (associated with both M and x, hence it contains both the "algorithm" and the "input data"). This "program" is not very simple (short), it even has a non-context-free character, because of the presence of copies of the code of M. However, the string $w_0 = B_0\ bls(code(M), x)Ex$ can be generated from simpler strings by splicing: construct the string $z = code(M)$, produce copies of it (by amplification), then produce arbitrarily many strings of the form $X_i z X_i', i \geq 1$ (in fact, we need exactly $n+2$ of these copies, where $n = |x|$). If $x = a_1 a_2 \dots a_n, a_i \in V, 1 \leq i \leq n$, consider also the strings $Y_i a_i Y_i', 1 \leq i \leq n$. Finally, consider the strings $B_0 Z_1, Z_2 Ex$. It is now a simple task to devise splicing rules which can build the string $w_0 = B_0\ bls(code(M), x)Ex$ starting from the blocks mentioned above; the symbols $X_i, X_i', Y_i, Y_i', Z_1, Z_2$ can control the operations in such a way that when none of them is present in a string, then that string is equal to w_0. We leave this task to the reader.

Remark 7.1. In mathematical terms, REG in $EH_1(FIN, REG) = RE$ can be substituted with $EH_1(FIN, FIN)$ and we get $EH_1(FIN, EH_1(FIN, FIN)) = RE$. At first sight, we have an answer to the above mentioned problem of characterizing RE by using extended H systems with finite components. However, this equality makes no sense from a biochemical point of view: REG from $EH_1(FIN, REG) = RE$ refers to languages of splicing rules (associated with *restriction enzymes*), whereas REG from $EH_1(FIN, FIN) = REG$ refers to languages of *DNA molecules*. From a practical point of view, they are completely different objects. □

Let us return to the equality $RE = EH_1(FIN, REG)$, and to the proof of the inclusion $RE \subseteq EH_1(FIN, REG)$, as given by Lemma 7.16. The H system provided by the construction in the proof of this lemma is $\gamma = (V, T, L_0, R)$, with

$$card(L_0) = card(P) + 2 \cdot card(N \cup T) + 5$$

(the notations are those from the proof of Lemma 7.16). A natural question arises: can we decrease the number of axioms? The answer is somewhat unexpected: one axiom suffices. However, decreasing the number of axioms can increase their length. Let us prove this trade-off result in a more general framework.

We define the following two complexity measures for an extended H system $\gamma = (V, T, A, R)$ with a finite set A of axioms:

$$\begin{aligned} nrax(\gamma) &= card(A), \\ lmax(\gamma) &= \max\{|x| \mid x \in A\} \end{aligned}$$

(the number of axioms, and the maximal length of an axiom, respectively). For such a measure $\mu \in \{nrax, lmax\}$ and a language $L \in EH_1(FIN, REG)$, we define

$$\mu(L) = \min\{\mu(\gamma) \mid L = L(\gamma)\}$$

and then we consider the families

$$\mu^{-1}(k) = \{L \in EH_1(FIN, REG) \mid \mu(L) \leq k\},$$

for $k \geq 1$.

The following relations are direct consequences of the definitions.

Lemma 7.19. *For* $\mu \in \{nrax, lmax\}$ *we have* $\mu^{-1}(1) \subseteq \mu^{-1}(2) \subseteq \ldots \subseteq EH_1(FIN, REG)$.

Both these hierarchies collapse. (In terms of descriptional complexity, the measures *nrax, lmax* are *trivial*, in the sense defined in Chap. 3.)

In fact, from the proof of Lemma 7.16, we can already obtain the inclusion $EH_1(FIN, REG) \subseteq lmax^{-1}(4)$: if we start with a grammar $G = (N, T, S, P)$ in Kuroda normal form, hence with rules of the forms $u \to v$, $|u| \leq 2$, $|v| \leq 2$, then the H system γ constructed in the mentioned proof has $lmax(\gamma) = 4$.

A stronger result is true:

Theorem 7.8. $lmax^{-1}(1) \subset lmax^{-1}(2) = EH_1(FIN, REG)$.

Proof. Take an H system $\gamma = (V, T, A, R)$ with a finite set A of axioms, namely $A = \{w_1, w_2, \ldots, w_n\}$. Assume $lmax(\gamma) = k$, $s = \max\{|uv| \mid u\#v\$u'\#v' \in R$, or $u'\#v'\$u\#v \in R\}$, and $r = \max(k, s)$. Consider the new symbols, $c_i, c'_i, 1 \leq i \leq n$, associated to the axioms of γ and construct

$$\gamma' = (V \cup \{c_i, c'_i \mid 1 \leq i \leq n\} \cup \{c, e\}, T, A', R'),$$

where

$$\begin{aligned} A' &= \{c_i a \mid 1 \leq i \leq n, a \in V\} \cup \{cc'_i,\ c_i e,\ ec'_i \mid 1 \leq i \leq n\} \\ &\cup \{\lambda \mid \text{if } \lambda \in L(\gamma)\}, \\ R' &= R'' \cup \bigcup_{i=1}^{n} R_i, \end{aligned}$$

$$
\begin{aligned}
R'' &= \{u_1'u_1\#u_2u_2'\$u_3'u_3\#u_4u_4' \mid u_1\#u_2\$u_3\#u_4 \in R,\\
&\quad u_1', u_3' \in e^*V^*, u_2', u_4' \in V^*e^*, |u_1'u_1u_2u_2'| = 2r, |u_3'u_3u_4u_4'| = 2r\},\\
R_i &= \{c_ix\#\$c_i\#a \mid x \in Pref(w_i), a \in V, xa \in Pref(w_i)\}\\
&\cup \{c_iw_i\#\$c\#c_i',\ \#cc_i'\$ce^r\#,\ \#e^rc_i'\$cc_i'\#\}\\
&\cup \{c_ie\#\$c_i\#e^jw_ic_i',\ c_ie^rw_ie^j\#c_i'\$\#ec_i' \mid 0 \le j \le r-1\}, 1 \le i \le n.
\end{aligned}
$$

We have $L(\gamma) = L(\gamma')$.

The inclusion $\subseteq$ is rather obvious: Starting from $c_ia, a \in V$, we can add symbol by symbol, reconstructing w_i from the left to the right; when we obtain c_iw_i we can also introduce the symbol c_i'; in the presence of c_i' we can add r occurrences of e in the left hand side of the string. After completing this operation, we can also add r occurrences of e in the right hand end of the string. Strings of the form $c_ie^rwe^rc_i'$ can be spliced by using the rules of R'', and this corresponds to using the rules from R. At any moment, the prefix c_ie^r and the suffix e^rc_i' can be removed, hence every string of $L(\gamma)$ can be generated by γ'.

Conversely, no unintended string can be produced in γ'. Indeed, all axioms of γ' (except λ, providing that the empty string is in $L(\gamma)$) contain non-terminal symbols. The non-terminals c_i, c_i' can be removed only in the presence of r occurrences of e, and such occurrences are introduced only after completing the reconstruction of the axiom w_i of γ. No "incomplete" string c_ix, c_ixc_i' with $x \in Pref(w_i)$, or $c_ie^jx, c_ie^jxc_i'$ with $j < r, x \in Pref(w_i)$, or $j = r, x \in Pref(w_i) - \{w_i\}$, can be spliced by the rules of R'', because the length of such a string is smaller than the length of the sites necessary for applying the rules from R'' (note also that the rules of R'' do not contain the symbols c_i, c_i'). If a string $c_ie^rxc_i'$ is spliced, then x should be an axiom of γ (of the length equal to k), hence the splicing corresponds to a splicing in γ.

Consequently, the work of γ' consists precisely of producing the axioms of γ, bounded by c_ie^r to the left and possibly also by e^rc_i' to the right, then applying the rules in R'', which corresponds to applying the rules from R to the strings without the prefixes c_ie^r and the suffixes e^rc_i'; at any moment, c_ie^r and e^rc_i' can be removed. This means that $L(\gamma') \subseteq L(\gamma)$, hence we have the equality $L(\gamma) = L(\gamma')$ and the inclusion $EH_1(FIN, REG) \subseteq lmax^{-1}(2)$.

This bound cannot be improved, the inclusion $lmax^{-1}(1) \subset lmax^{-1}(2)$ is proper: Take the language $L = \{aa\}$ and assume that $L = L(\gamma)$, for some $\gamma = (V, \{a\}, A, R)$ with $A \subseteq V$. Since all axioms are of length 1, the symbol a must appear in at least one axiom, hence $a \in A$. This means $a \in L(\gamma)$, contradicting the equality $L = L(\gamma)$. □

Also the hierarchy on the number of axioms collapses (this time to one level only).

Theorem 7.9. $nrax^{-1}(1) = EH_1(FIN, REG)$.

Proof. For a given system $\gamma = (V, T, A, R)$ with $A = \{w_1, w_2, \ldots, w_n\}$,

$n \geq 2$, construct

$$\gamma' = (V \cup \{c, c'\}, T, \{w\}, R'),$$

where c, c' are new symbols, and

$$w = c'cw_1cw_2 \dots cw_ncc',$$
$$R' = R \cup \{\#c'c\$c\#c', \ \#c\$c'\#, \ \#c'\$c\#\}.$$

We have $L(\gamma) = L(\gamma')$.

The inclusion $\subseteq$ is easy to see: using the rule $\#c'c\$c\#c'$ we can obtain $(w, w) \vdash c'$; using the rule $\#c\$c'\#$, from each string of the form xc (and from w, too) we can remove the rightmost c (the suffix cc', respectively) by a splicing with c'; using the rule $\#c'\$c\#$, from each string of the form x_1cx_2 we can separate the suffix x_2. Therefore, all axioms of γ can be separated from w. Using the rules of R we then get every string of $L(\gamma)$.

Conversely, as long as occurrences of c, c' are present, the string is not terminal (this is the case with w). The use of rules in $R' - R$ cannot mix symbols of V, but only cut down strings of the form $c'cx_1cx_2c \dots cx_r$ (maybe also ending with c or with cc') near the symbols c. If we have two strings x, y of the forms $x = x_1cx_2cx_3$, $y = y_1cy_2cy_3$, with $x_2, y_2 \in V^*$, and we use a splicing rule in R on x_2, y_2, then we obtain a string $z = x_1cz_1cy_3$ such that $(x_2, y_2) \vdash z_1$ for $z_1 \in V^*$. We start from w containing exactly the axioms of γ. If we use now rules in $R' - R$ and we separate z_1 from z, this is a string which can be also produced directly by γ. Therefore, mixing up the new rules of R' with the rules of R does not lead to unintended strings. This implies $L(\gamma') \subseteq L(\gamma)$. □

Corollary 7.2. $nrax^{-1}(1) = RE$.

In the proofs of the previous theorems, when decreasing the length of axioms it was necessary to increase the number of axioms, and conversely. As expected, in general the two measures cannot be simultaneously improved.

For a measure $\mu : GM \longrightarrow \mathbf{N}$ of complexity of generative mechanisms in a given class GM, denote $\mu_{GM}(L) = \min\{\mu(G) \mid L = L(G), G \in GM\}$. Then, for a language L, define

$$\mu^{-1}(L) = \{G \in GM \mid L = L(G), \mu(G) = \mu_{GM}(L)\}$$

(the set of optimal generative devices in GM producing L). Two measures μ_1, μ_2 are said to be *incompatible* (on GM) if there is a language L in the family generated by GM, such that

$$\mu_1^{-1}(L) \cap \mu_2^{-1}(L) = \emptyset.$$

(The two measures cannot be simultaneously minimized for the elements in GM generating the language L.)

Theorem 7.10. *The measures nrax, lmax are incompatible.*

Proof. Consider any language $L \subseteq a^+ \cup b^+$ such that $alph(L) = \{a, b\}$. According to Theorem 7.9, $nrax(L) = 1$. Take $\gamma = (V, \{a, b\}, A, R)$ such that $L = L(\gamma)$ and $A = \{w\}, w \in V^*$. We must have $|w| \geq 3$. Indeed, both symbols a and b must be present in w, otherwise they cannot appear in the strings of $L(\gamma)$. However, neither ab nor ba can be the axiom, at least one symbol in $V - \{a, b\}$ appears in w, hence $|w| \geq 3$.

On the other hand, from Theorem 7.8 we know that $lmax(L) = 2$.

Consequently, for no language L, as above, can we find γ with both $nrax(\gamma) = 1$ and $lmax(\gamma) = 2$. Languages L of the considered form appear in $EH_1(FIN, REG)$, hence the proof is complete. □

7.5 Simple H Systems

In the previous sections of this chapter we have looked for variants of generative mechanisms based on the splicing operation which are as powerful as possible. Here we follow the opposite approach, considering a rather particular type of splicing systems. As expected, in such a particular case many questions can be solved in a nice way, which makes these devices attractive from a mathematical point of view.

A *simple H system* is a triple

$$\gamma = (V, A, Q),$$

where V is an alphabet, A is a finite language over V, and $Q \subseteq V$. The elements of A are called *axioms*, those of Q are called *markers*.

For $x, y, z \in V^*$ and $a \in Q$ we write

$$(x, y) \vdash_a z \quad \text{iff} \quad x = x_1 a x_2, y = y_1 a y_2, z = x_1 a y_2, \text{ for some } x_1, x_2, y_1, y_2 \in V^*.$$

Consequently, for each marker $a \in Q$ we can imagine that we have the splicing rule $r_a = a\#\$a\#$ (or $r'_a = \#a\$\#a$). Denote

$$R_Q = \{r_a \mid a \in Q\}$$

and consider the H scheme $\sigma_Q = (V, R_Q)$. Then the language generated by γ, denoted by $L(\gamma)$, is defined as being equal to $\sigma_Q^*(A)$. (In this section we omit the subscript 1 in $\sigma_1(L), \sigma_1^*(L)$, avoiding heavy notations of the type $(\sigma_Q)_1^*(L)$.)

Here is an *example*: consider the simple H system

$$\gamma = (\{a, b, c\}, \{abaca, acaba\}, \{b, c\}).$$

We obtain

$$L(\gamma) = (abac)^+ a \cup (abac)^* aba \cup (acab)^+ a \cup (acab)^* aca.$$

Indeed,

$$((abac)^n a, acaba) \vdash_c (abac)^{n-1}abacaba = (abac)^n aba,$$
$$((abac)^n aba, abaca) \vdash_b (abac)^n abaca = (abac)^{n+1}a,$$

and similarly for strings $(acab)^n a, (acab)^n aca$, hence we have the inclusion $\supseteq$. Conversely, when splicing two strings of one of the forms $(abac)^n a, (acab)^n a$ (initially we have $n = 1$) or $(abac)^n aba, (acab)^n aca$, we identify either a substring aba or a substring aca of them, hence the obtained strings are of the same form.

Let us denote by SH the family of languages generated by simple H systems.

The following necessary conditions for a language to be in the family SH can be easily proved.

Lemma 7.20. (i) *If $L \in SH$ is an infinite language, then there is $Q \subseteq alph(L), Q \neq \emptyset$, such that $\sigma_Q(L) \subseteq L$.*

(ii) *If $L \in SH, L \subseteq V^*$, and $a^+ \subseteq L$ for some $a \in V$, then $\sigma_{\{a\}}(L) \subseteq L$.*

(iii) *Take $w \in V^+$. We have $w^* \in SH$ if and only if there is a symbol $a \in V$ such that $|w|_a = 1$.*

Using these conditions we can show that the following languages are not in the family SH:

$$L_1 = a^+b^+a^+b^+,$$
$$L_2 = a^+b \cup b^+,$$
$$L_3 = (aabb)^+,$$

and that a language $L \subseteq a^*$ is in the family SH if and only if it is either finite or equal to one of a^*, a^+. Moreover, we get

Corollary 7.3. *The family SH is an anti-AFL.*

Because each simple H system is a (non-extended) finite H system of a particular type, from Lemma 7.14 we obtain $SH \subseteq REG$. We shall see below that this result can be obtained in a much easier way as a consequence of a representation theorem for languages in SH. In view of the previous counterexamples, the inclusion $SH \subset REG$ is proper. However, SH and REG are "equal modulo a coding":

Lemma 7.21. *Every regular language is the projection of a language in the family SH.*

Proof. Let $M = (K, V, s_0, F, \delta)$ be a deterministic finite automaton. We construct the simple H system $\gamma = (K \cup V, A, K)$, with

$$\begin{aligned} A = \{ & s_0a_1s_1a_2s_2 \dots s_ra_{r+1}s_{r+1} \mid r \geq 0, s_i \in K, 0 \leq i \leq r+1, \\ & s_{r+1} \in F, s_{i+1} = \delta(s_i, a_{i+1}), 0 \leq i \leq r, \\ & \text{each state } s_i, 0 \leq i \leq r+1, \text{ appears at most twice}\}. \end{aligned}$$

Clearly, A is a finite set. Consider also the projection h defined by $h(a) = a$, for $a \in V$, and $h(s) = \lambda$, for $s \in K$.

The inclusion $h(L(\gamma)) \subseteq L(M)$ follows from the construction of γ and the definition of h. The reverse inclusion can be easily proved by induction on the length of strings in $L(M)$. □

Lemma 7.22. *For every language $L \in SH$ there are five finite languages L_1, L_2, L_3, L_4, L_5 and a projection h such that $L = h(L_1 L_2^* L_3 \cap L_4^*) \cup L_5$.*

Proof. Let $\gamma = (V, A, Q)$ be a simple H system. For each $a \in V$ consider a new symbol, a'; denote $V' = \{a' \mid a \in V\}$.

Define

$$\begin{aligned}
L_1 &= \{xa \mid xay \in A, x, y \in V^*, a \in Q\},\\
L_2 &= \{a'xb \mid yaxbz \in A, x, y, z \in V^*, a, b \in Q\},\\
L_3 &= \{a'x \mid yax \in A, x, y \in V^*, a \in Q\},\\
L_4 &= V \cup \{aa' \mid a \in V\},\\
L_5 &= \{x \in A \mid |x|_a = 0 \text{ for all } a \in Q\},\\
h &: (V \cup V')^* \longrightarrow V^*, \ h(a) = a, a \in V, \text{ and } h(a') = \lambda, a \in V.
\end{aligned}$$

Then we claim that

$$L(\gamma) = h(L_1 L_2^* L_3 \cap L_4^*) \cup L_5.$$

Let us denote by B the right hand member of this equality.

(1) $L(\gamma) \subseteq B$. Clearly, from the definitions, it is enough to prove that (i) B includes the set A, and (ii) $\sigma_Q(B) \subseteq B$.

(i) If $x \in A$ and $|x|_a = 0$ for all $a \in Q$, then $x \in L_5 \subseteq B$. If $x \in A$ and $x = x_1 a x_2, a \in Q$, then $x_1 a \in L_1, a' x_2 \in L_3$, hence $x_1 a a' x_2 \in L_1 L_3$. Clearly, $x_1 a a' x_2 \in L_4^*$, too. As $h(x_1 a a' x_2) = x_1 a x_2 = x$, we have $x \in B$. Consequently, $A \subseteq B$.

(ii) Take two strings $x, y \in B$. If one of them is in L_5, then $\sigma_Q(x, y) = \{x, y\} \subseteq B$.

Take $x', y' \in L_1 L_2^* L_3 \cap L_4^*$ such that $x = h(x'), y = h(y')$, and take $z \in \sigma_Q(x, y)$, $(x, y) \vdash_a z$ for some $a \in Q$. Write

$$x = z_1 a z_2', \ y = z_1' a z_2, \ \text{for } z = z_1 a z_2,$$

and

$$\begin{aligned}
x' &= x_1 a_1 a_1' x_2 \dots x_k a_k a_k' x_{k+1}, \ k \geq 1,\\
y' &= y_1 b_1 b_1' y_2 \dots y_s b_s b_s' y_{s+1}, \ s \geq 1,
\end{aligned}$$

for $a_i, b_i \in Q, x_i, y_i \in V^*, a'_{i-1}x_ia_i, b'_{i-1}y_ib_i \in L_2$, for all i and x_1a_1, $y_1b_1 \in L_1$, $a'_kx_{k+1}, b'_sy_{s+1} \in L_2$. Then

$$x = x_1a_1x_2 \dots x_ka_kx_{k+1},\ y = y_1b_1y_2 \dots y_sb_sy_{s+1}.$$

Identify the marker a in x, respectively in y, as used in $(x, y) \vdash_a z$.

If $a = a_i$, then $z_1 = x_1a_1 \dots x_{i-1}a_{i-1}x_i$, $z'_2 = x_{i+1}a_{i+1} \dots a_kx_{k+1}$.

If $a_i \neq a$ for all $1 \leq i \leq k$, then there is $x_i = x'_iax''_i$. For $i = 1$ we have $x'_1a \in L_1$ and $a'x''_1a_1 \in L_2$. For $1 < i < k+1$ we have $a'_{i-1}x'_ia \in L_2, a'x''_ia_{i+1} \in L_2$. For $i = k+1$ we have $a'_kx'_{k+1}a \in L_2$ and $a'x''_{k+1} \in L_3$. In all cases we can find a string of the form $x'' = w_1aa'w_2 \in L_1L_2^*L_3 \cap L_4^*$ such that $x = h(x'')$.

Similarly, we can find $y'' = w'_1aa'w'_2 \in L_1L_2^*L_3 \cap L_4^*$ such that $y = h(y'')$. For the string $z' = w_1aa'w'_2$ we clearly have $z' \in L_1L_2^*L_3 \cap L_4^*$ and $z = h(z')$. Consequently, $z \in B$, which completes the proof of the property (ii), hence of the inclusion $L(\gamma) \subseteq B$.

(2) $B \subseteq L(\gamma)$. Take $x \in B$. If $x \in L_5$, then $x \in A \subseteq L(\gamma)$.

If $x = h(x'), x' = x_1a_1a'_1x_2a_2a'_2 \dots x'_ka_ka'_kx_{k+1}, k \geq 1$, with $x_1a_1 \in L_1$, $a'_{i-1}x_ia_i \in L_2, 2 \leq i \leq k$, $a'_kx_{k+1} \in L_3$, then from the definitions of L_1, L_2, L_3 there are the strings $x_1a_1x'_1, y_ia_{i-1}x_ia_iy'_i, 2 \leq i \leq k$, $z_{k+1}a_kx_{k+1}$, all of them in A. Then

$$(x_1a_1x'_1, y_2a_1x_2a_2y'_2) \vdash_{a_1} x_1a_1x_2a_2y'_2 = w_2,$$
$$(w_2, y_3a_2x_3a_3y'_3) \vdash_{a_2} x_1a_1x_2a_2x_3a_3y'_3 = w_3,$$
$$\dots\dots\dots\dots\dots$$
$$(w_k, z_{k+1}a_kx_{k+1}) \vdash_{a_k} x_1a_1x_2a_2 \dots x_ka_kx_{k+1} = x.$$

Consequently, $x \in L(\gamma)$. □

This representation is not a characterization of languages in SH. In fact, a similar result holds true for all regular languages: just combine Lemmas 7.21 and 7.22. However, this representation has a series of interesting consequences, one of them referring exactly to the regularity of simple splicing languages.

Corollary 7.4. *$SH \subseteq REG$.*

Moreover, from Lemma 7.22, we also obtain the following useful necessary condition for a language to be in SH.

Corollary 7.5. *If $\gamma = (V, A, Q)$ is a simple H system, then for every $x \in Sub(L(\gamma)) \cap (V - Q)^*$ we have $|x| \leq \max\{|w| \mid w \in A\}$.*

Making use of this property, we get

Theorem 7.11. *It is decidable whether or not a regular language is a simple H language.*

Proof. Let $L \subseteq V^*$ be a regular language, given, for instance, by a regular grammar or a finite automaton. For any subset Q of V, denote

$$\begin{aligned} R_Q = &(V - Q)^* \\ &\cup \{x_1a_1x_2a_2 \dots x_ka_kx_{k+1} \mid 1 \leq k \leq 2 \cdot card(Q), \\ &\quad x_i \in (V - Q)^*, 1 \leq i \leq k,\ a_i \in Q, 1 \leq i \leq k, \text{ and} \\ &\quad \text{there are no } 1 \leq i < j < l \leq k+1 \text{ such that } a_i = a_j = a_l\}. \end{aligned}$$

(Therefore, R_Q contains all strings x over V such that each symbol of Q appears at most twice in x.)

(1) If $L \cap R_Q$ is an infinite set, then there is no H system $G = (V, A, Q)$ such that $L = L(G)$.

Indeed, $L \cap R_Q$ being infinite means that there is $x \in Sub(L) \cap (V - Q)^*$ of arbitrary length, contradicting the previous corollary.

(2) If $L \cap R_Q$ is a finite set, then we consider all H systems $\gamma = (V, A, Q)$ with $A \subseteq L \cap R_Q$. Then there is an H system $\gamma' = (V, A', Q)$ such that $L = L(\gamma')$ if and only if $L = L(\gamma)$ for a system γ constructed above.

(*if*): trivial.

(*only if*): Take $\gamma' = (V, A', Q)$ such that $L(\gamma') = L$ and A' is not a subset of $L \cap R_Q$. This means that A' contains a string of the form

$$z = x_1ax_2ax_3ax_4,$$

for $x_1, x_2, x_3, x_4 \in V^*, a \in Q$. Consider the strings

$$z_1 = x_1ax_2ax_4, \quad z_2 = x_1ax_3ax_4.$$

Both of them are in $\sigma_Q(\{z\})$, hence in $L(\gamma')$. Moreover, $(z_1, z_2) \vdash_a z$. Therefore, replacing A' by

$$A'' = (A' - \{z\}) \cup \{z_1, z_2\},$$

we get a system $\gamma'' = (V, A'', Q)$ such that $L(\gamma') = L(\gamma'')$. Continuing this procedure (for a finite number of times, because A' is finite and $|z_1| < |z|, |z_2| < |z|$) we eventually find a system $\gamma''' = (V, A''', Q)$ with $A''' \subseteq L \cap R_Q$.

Now, $L \in SH$ if and only if $L = L(\gamma)$ for some $\gamma = (V, A, Q)$ with $Q \subseteq V$. There are only finitely many such sets Q. Proceed as above with each of them. We have $L \in SH$ if and only if there is such a set Q_0 for which $L \cap R_{Q_0}$ is finite and there is $A_0 \subseteq L \cap R_{Q_0}$ (finitely many possibilities) such that $L = L(\gamma_0)$ for $\gamma_0 = (V, A_0, Q_0)$. The equality $L = L(\gamma_0)$ can be checked algorithmically. In conclusion, the question whether or not $L \in SH$ can be decided algorithmically. □

This result cannot be extended to context-free (not even to linear) languages.

Theorem 7.12. *The problem of whether or not a linear language is a simple H language is not decidable.*

Proof. Take an arbitrary linear language $L_1 \subseteq \{a,b\}^*$, as well as the language $L_2 = c^+d^+c^+d^+$, which is not in the family SH. Construct the language

$$L = L_1\{c,d\}^* \cup \{a,b\}^* L_2.$$

This is a linear language.

If $L_1 = \{a,b\}^*$, then $L = \{a,b\}^*\{c,d\}^*$ and this is a simple H language: for $\gamma = (\{a,b,c,d\}, \{xy \mid x \in \{a,b\}^*, y \in \{c,d\}^*, |x| \in \{0,2\}, |y| \in \{0,2\}\}, \{a,b,c,d\})$ we have $L(\gamma) = \{a,b\}^*\{c,d\}^*$.

If $L_1 \neq \{a,b\}^*$, then $\{a,b\}^* - L_1 \neq \emptyset$. Take $w \in \{a,b\}^* - L_1$ and consider the string $w' = wcdcd$. It is in L, and $(w', w') \vdash_e wcdcdcd$ for $e \in \{c,d\}$. This string is not in L, therefore none of c, d can be a marker in a simple H system for the language L. But L contains all string in $wc^+d^+c^+d^+$, hence $L \in SH$ would contradict point (i) in Lemma 6.20.

Consequently, $L \in SH$ if and only if $L_1 = \{a,b\}^*$, which is undecidable for linear languages. □

Because $SH \subset REG$, it is of interest to investigate the relationships between SH and other subfamilies of REG. We consider only one (important) such sub-regular family, that of strictly locally testable languages.

A language $L \subseteq V^*$ is *p-strictly locally testable*, for some $p \geq 1$, if we can write it in the form

$$\begin{aligned} L = & \{x \in L \mid |x| < 2p\} \\ & \cup (Pref(L) \cap V^p)V^*(Suf(L) \cap V^p) - V^*(V^p - Sub(L))V^*. \end{aligned}$$

A language is strictly locally testable if it is p-strictly locally testable for some $p \geq 1$. We denote by SLT the family of such languages.

Clearly, $SLT \subset REG$. In fact, SLT is contained in the family of *extended star-free languages*, the smallest family of languages containing the finite languages and closed under boolean operations and under concatenation.

Theorem 7.13. $SH \subset SLT$.

Proof. We shall use the characterization given in [45] for strictly locally testable languages.

According to [203], a string $x \in V^*$ is called *constant* with respect to a language $L \subseteq V^*$ if whenever $uxv \in L$ and $u'xv' \in L$, then also $uxv' \in L$ and $u'xv \in L$. In [45] it is proved that a language $L \subseteq V^*$ is strictly locally testable if and only if there is an integer k such that all strings in V^k are constants with respect to L.

Consider now a language $L \in SH, L = L(\gamma)$, for some $\gamma = (V, A, Q)$. Take the integer

$$k = \max\{|x| \mid x \in A\} + 1.$$

Every string in V^k is a constant with respect to L. Indeed, take such a string x and two strings $uxv, u'xv'$ in L. Because $|x| = k$, according to Corollary 7.5, we have $|x|_Q > 0$. Take $a \in Q$ such that $x = x_1 a x_2$. Therefore $uxv = ux_1ax_2v, u'xv' = u'x_1ax_2v'$, hence $(uxv, u'xv') \vdash_a ux_1ax_2v' = uxv'$ and $(u'xv', uxv) \vdash_a u'x_1ax_2v = u'xv$. In conclusion, $L \in SLT$ and $SH \subseteq SLT$.

The inclusion is proper: for $w = aabb$ we have $w^* \in SLT$ (obvious), but $w^* \notin SH$ (Lemma 7.20.(iii)). □

7.6 Bibliographical Notes

The splicing operation was introduced by T. Head in [86] in the form considered at the beginning of Sect. 7.1: one gives triples (u, x, v) describing the patterns recognized by the restriction enzymes and one builds splicing rules $((u_1, x, v_1), (u_2, x, v_2))$ using triples which produce matching ends. The association of x to u_1, u_2 (or to v_1, v_2) leading to contexts $(u_1', v_1), (u_2', v_2)$ describing the cutting places is introduced in [46]. The coding of splicing rules as strings, $u_1\#u_2\$u_3\#u_4$, is introduced in [149], where splicing schemes with infinite sets of rules are also considered.

Sect. 7.1 above is based on [86] and on the Introduction and the Appendix of [90]. Sect. 7.2 is based on [149], with some modifications as in [90].

One of the main problems raised in [86] concerns the power of H schemes with finite sets of rules, when these rules are iterated on finite sets of starting strings. The question is answered in [35], where it is proved that such an operation preserves regular languages. The proof uses complex arguments formulated in terms of the semigroup of dominoes. The proof presented here for the Regularity Preserving Lemma (Lemma 7.14) is based on [175]. The same idea is used in [177] for proving that the following generalization of the definition of the splicing operation preserves the regularity: consider splicing rules as triples $r : (v_1, v_2; v_3)$; if $x = x_1v_1x_2$ and $y = y_1v_2y_2$, then the effect of applying the rule r to x, y is the string $x_1v_3y_2$. This corresponds to writing a splicing rule $u_1\#u_2\$u_3\#u_4$ as used here in the form $(u_1u_2, u_3u_4; u_1u_4)$. In this way, the proof in [177] implies Lemma 7.14. But note that starting from a regular set of splicing rules written in the form $u_1\#u_2\$u_3\#u_4$ and passing to the corresponding set of rules of the form $(u_1u_2, u_3u_4; u_1u_4)$ (with a suitable encoding of "," and ";") we do not necessarily obtain a regular language, because of the repetitions of u_1 and u_4. However, if the number of strings u_1, u_4 is finite, then the regularity is preserved, even using an infinite regular language of splicing rules $u_1\#u_2\$u_3\#u_4$. The general result in terms of AFL's (Lemma 7.15) has been reported in [176], it appears in [177] for the case of rules of the form $(v_1, v_2; v_3)$, and, with full details, in [90]. A recent more general formulation appears in [178].

The Basic Universality Lemma is proved in [153], directly for extended H systems. Extended H systems were introduced in [167]. Morphic characterizations of regular languages (as Theorem 7.5) appear in [71], [73], [149].

Lemma 7.18 appears in [167]. Theorem 7.8 – 7.10 are from [153]. Theorem 7.7 is from [127].

Simple H systems are introduced and investigated in [130], where several results not mentioned in Sect. 7.5 can be found (about descriptional complexity, algebraic characterizations, comparison with other sub-regular families, etc).

A generalization of SH systems is considered in [88], namely with string markers: a *k-simple H system* is a triple $\gamma = (V, A, Q)$, where V is an alphabet, A is a finite subset of V^*, and Q is a finite set of strings over V, with the length less than or equal to k, $k \geq 1$. The strings in Q are used in the same way as the markers of a simple H system: $(uxv, u'x'v') \vdash_x uxv'$, $x \in Q$. Denote by S_kH the family of languages generated by k'-simple H systems, where $k' \leq k$. In [88] one proves that $SH = S_1H \subset S_2H \subset \ldots \subset \bigcup_{k \geq 1} S_kH = SLT$. Moreover, it is decidable whether or not a regular language belongs to a family $S_kH, k \geq 1$; in the affirmative case, the minimal value of k can be effectively found.

Chapter 8

Universality by Finite H Systems

As we have seen in the previous chapter, extended H systems with finite sets of axioms and splicing rules are able to generate only regular languages. As we are looking for generative (computability) models having the power of Turing machines, we have to consider features that can increase the power of H systems. This has been successfully done for Chomsky grammars and other generative mechanisms in the regulated rewriting area and the grammar systems area. Following suggestions from these areas, as well as suggestions offered by the proof of the Basic Universality Lemma (Lemma 7.16), in this chapter we shall consider a series of controlled H systems with finite components which characterize the recursively enumerable languages, hence are *computationally complete.* From the proofs, we shall also obtain *universal* computing devices, hence models of "programmable DNA computers based on splicing".

8.1 Using 2-Splicing Instead of 1-Splicing

The extended H systems we shall consider in this chapter (with regulated splicing or involving other features able to ensure computational completeness even when using finite sets of axioms and of splicing rules) are intended to be theoretical models of DNA computers based on splicing. That is why from now on we shall work with the more realistic 2-splicing operation defined by taking into account both the two possible strings obtained by recombination:

$$
\begin{aligned}
(x, y) \models_r (z, w) \quad \text{iff} \quad & x = x_1u_1u_2x_2, \\
& y = y_1u_3u_4y_2, \\
& z = x_1u_1u_4y_2,
\end{aligned}
$$

$$w = y_1 u_3 u_2 x_2,$$
$$\text{for some } x_1, x_2, y_1, y_2 \in V^*,$$

where $r = u_1\#u_2\$u_3\#u_4$ is a splicing rule.

For an H scheme $\sigma = (V, R)$ and a language $L \subseteq V^*$ we define

$$\sigma_2(L) = \{z \in V^* \mid (x, y) \models_r (z, w) \text{ or } (x, y) \models_r (w, z), \text{ for some } x, y \in L \text{ and } r \in R\}.$$

Then we can define $\sigma_2^i(L), i \geq 0$, and $\sigma_2^*(L)$ in the same way as we have done at the beginning of Sect. 7.3 of the previous chapter with $\sigma_1^i(L), i \geq 0$, and $\sigma_1^*(L)$, respectively. Thus, we can repeat all the considerations in Sects. 7.2–7.5 for the operation $\models$ instead of $\vdash$. We shall denote by $S_2(FL_1, FL_2), H_2(FL_1, FL_2), EH_2(FL_1, FL_2)$ the families of languages corresponding to $S_1(FL_1, FL_2), H_1(FL_1, FL_2)$ and $EH_1(FL_1, FL_2)$, respectively.

We say that a family FL of languages is closed under *marked circular permutation* if for each language $L \subseteq V^*\{c\}V^*$, where $c \notin V$, $L \in FL$, the language

$$perm_c(L) = \{vcu \mid ucv \in L, u, v \in V^*\}$$

is also an element of FL.

Lemma 8.1. *For every family FL_1 and for FL_2 closed under union and under marked circular permutation, we have $X_2(FL_1, FL_2) \subseteq X_1(FL_1, FL_2)$, $X \in \{S, H, EH\}$.*

Proof. Consider an H scheme $\sigma = (V, R)$ with $R \in FL_2$. For every language $L \in FL_1$ we have $\sigma_2(L) = \sigma_1'(L)$, where $\sigma' = (V, R \cup perm_\$(R))$. The equality is obvious and $R \cup perm_\$(R) \in FL_2$ by the closure properties of FL_2. This proves all the inclusions in the statement of the lemma. □

Most of the results in Chap. 7 about families $X_1(FL_1, FL_2), X \in \{S, H, EH\}$, are true also for the families $X_2(FL_1, FL_2)$. This does not happen, however, with Lemmas 7.2, 7.4, 7.5, but, if the families FL_1, FL_2 from the statement of these lemmas are also closed under intersection with regular languages, then these statements also hold for families $X_2(\ldots, \ldots)$. The gsm's in the statements and the proofs of Lemmas 7.3 and 7.6 can simulate such an intersection with a regular language, hence these lemmas also remain true for the case of the 2-splicing. We shall not pause to check each of the results in Chap. 7 from this point of view, because we are not interested in the *mathematical* properties of the operation $\models$ but in its *computational* properties. From this point of view, the following facts are important:

1. Lemma 7.14 combined with Lemma 8.1 implies that $H_2(REG, FIN) \subseteq REG$. Similarly, from Lemma 7.15, we have $H_2(CF, FIN) \subseteq CF$. Moreover, Lemma 7.18 remains valid for the 2-splicing operation,

$REG \subseteq EH_2(FIN, FIN)$. This can be easily seen from the proof of Lemma 7.18, by examining the second string produced by the 2-splicing operations in the extended H system γ: all such strings contain nonterminal symbols in N as well as occurrences of the symbol Z, and even by entering new splicings, these strings cannot produce terminal strings not in L. Consequently, $EH_2(FIN, FIN) = REG$.

2. Lemma 7.16 remains true when replacing H_1 by H_2, that is every language $L \in RE, L \subseteq V^*$, can be written in the form $L = L' \cap V^*$ for some $L' \in H_2(FIN, REG)$. This can be checked in the proof of Lemma 7.16 (the second string produced by 2-splicing operations cannot lead to strings in V^* which are not in L), but we do not emphasize this here; in some of the following proofs we shall consider this aspect in full detail.

3. In view of Lemma 8.1, the statement in Lemma 7.17 also remains true when replacing H_1 with H_2.

Therefore, the relations in Table 7.3 are valid for families $EH_2(FL_1, FL_2)$. In particular, we have

1. $EH_2(FIN, FIN) = REG$,
2. $EH_2(FIN, REG) = RE$.

Thus, the discussion after Table 7.3 is also valid for extended H systems based on 2-splicing: in order to get computational completeness for systems with finite components we have to add further features to our models.

8.2 Permitting and Forbidding Contexts

Examining the H system in the proof of Lemma 7.16, one can see that the set of splicing rules is infinite because of the appearance of substrings w in rules of types 1, 2, 3, 4, 5, 6. However, these substrings contain no information (except the case of rules of type 6, where $w \in T^*$), they are arbitrary strings over the alphabet $N \cup T \cup \{B\}$. The role of these substrings w is to allow information to be obtained about the symbol appearing behind them, namely X, X' in the left hand end of the first term of the splicing and $Y, Y_\alpha, \alpha \in N \cup T \cup \{B\}$, in the right hand end of the second term of the splicing. Otherwise stated, we have in fact finite splicing rules, applied only to strings containing (at their ends) certain symbols, from well specified sets. This suggests considering the following type of H systems with controlled splicing.

An extended H system *with permitting contexts* is a quadruple

$$\gamma = (V, T, A, R),$$

where V is an alphabet, $T \subseteq V$, A is a finite language over V, and R is a finite set of triples of the form $p = (r; C_1, C_2)$, with $r = u_1\#u_2\$u_3\#u_4$ being a splicing rule over V and C_1, C_2 being finite subsets of V^*.

Note that here we only consider systems with finite components.

For $x, y, z, w \in V^*$ and $p \in R, p = (r; C_1, C_2)$, we define $(x, y) \models_p (z, w)$ if and only if $(x, y) \models_r (z, w)$, every element of C_1 appears as a substring in x and every element of C_2 appears as a substring in y; when $C_1 = \emptyset$ or $C_2 = \emptyset$, then no condition on x, respectively y, is imposed.

The pair $\sigma = (V, R)$ is called (the underlying) H scheme with permitting context rules. The language generated by γ is defined in the natural way:

$$L(\gamma) = \sigma_2^*(A) \cap T^*.$$

We denote by $EH_2([n], p[m]), n, m \geq 1$, the family of languages $L(\gamma)$ generated by extended H systems with permitting contexts, $\gamma = (V, T, A, R)$, with $card(A) \leq n$ and $rad(R) \leq m$, where $rad(R)$ is the maximal radius of splicing rules r in triples $(r; C_1, C_2)$ from R. When no restriction on the number of axioms or on the maximal radius is considered (but, of course, these numbers are still finite), we replace $[n]$ or $[m]$, respectively, by FIN.

The proof of the next lemma is given with full details, both because the result is important for our purposes and because we want to exhibit the method of working of an H system based on the operation $\models$.

Lemma 8.2. $RE \subseteq EH_2(FIN, pFIN)$.

Proof. Consider a type-0 Chomsky grammar $G = (N, T, S, P)$. Let us denote $U = N \cup T \cup \{B\}$, where B is a new symbol. We construct the extended H system with permitting contexts

$$\gamma = (V, T, A, R),$$

where

$$\begin{aligned} V &= N \cup T \cup \{B, X, X', Y, Z, Z', Z''\} \\ &\cup \{Y_\alpha \mid \alpha \in U\}, \\ A &= \{XBSY, XZ, Z', Z'', ZY\} \\ &\cup \{ZY_\alpha, X'\alpha Z \mid \alpha \in U\} \\ &\cup \{ZvY \mid u \to v \in P\} \end{aligned}$$

and R contains the following rules with permitting contexts:

Simulate :	1.	$(\#uY\$Z\#vY; \{X\}, \emptyset)$,	for $u \to v \in P$,
Rotate :	2.	$(\#\alpha Y\$Z\#Y_\alpha; \{X\}, \emptyset)$,	for $\alpha \in U$,
	3.	$(X\#\$X'\alpha\#Z; \{Y_\alpha\}, \emptyset)$,	for $\alpha \in U$,
	4.	$(\#Y_\alpha\$Z\#Y; \{X'\}, \emptyset)$,	for $\alpha \in U$,
	5.	$(X'\#\$X\#Z; \{Y\}, \emptyset)$,	
Terminate :	6.	$(XB\#\$\#Z'; \{Y\}, \emptyset)$,	
	7.	$(\#Y\$Z''\#; \emptyset, \emptyset)$.	

This is precisely the construction from the proof of the Basic Universality Lemma, that is, the rotate-and-simulate procedure, written for splicing rules with permitting contexts.

The rules from group 1 allow us to simulate rules from P on a suffix of the first term of the splicing.

We must be able to simulate the application of a rule from P in an arbitrary position of the underlying sentential form, not only in the right-hand end of the word. To this aim, the rules in groups 2, 3, 4, and 5 allow us to "rotate" the word. A rule in group 2 cuts a symbol α from the right-hand end of the word, Y_α memorizes this symbol, in its presence a rule from group 3 will introduce α in the left hand end (together with X'), then Y_α is again replaced by Y (by using the appropriate rule from group 4), and X' is again replaced by X (by using the rule from group 5). Any circular permutation can be obtained in this way.

The rules from groups 6, 7 finally allow us to remove the markers X and Y (the former one only when B is adjacent to it).

Let us look in some detail at how the ideas mentioned above work.

When simulating derivation steps in G, we start from $XBSY$, and at every step the markers X and its variant X' as well as Y and its variants Y_β, $\beta \in U$, are present to indicate the ends of the word. Moreover, at any moment the symbol B tells us where the beginning of the word is, whose permutation we consider.

All the splicing rules with permitting contexts contained in R require an occurrence of the symbols Z, Z', Z'' in the second term of the splicing; in fact, these words are meant to be taken from A. If we start with rule 1 applied to $XBSY$ and ZvY, for some $S \to v \in P$, this starts the simulation of a derivation in G. In general, having a word Xx_1Bx_2uY and $u \to v \in P$, we can obtain Xx_1Bx_2vY by using the associated rules in group 1:

$$(Xw|uY, Z|vY) \models_p (XwvY, ZuY),$$

for $p = (\#uY\$Z\#vY; \{X\}, \emptyset)$, where $u \to v \in P$ and $w \in (N \cup T)^*\{B\}(N \cup T)^*$.

This corresponds to a derivation step $x_2ux_1 \Longrightarrow x_2vx_1$ in G.

As additional results (that are not used at a subsequent step of the simulation of a derivation in G) of the splicings of the form above we obtain the words ZuY for $u \to v \in P$.

To each word $Xw\alpha Y$, $\alpha \in N \cup T$, $w \in (N \cup T)^*\{B\}(N \cup T)^*$, respectively $Xw\alpha Y$, $\alpha = B$, $w \in (N \cup T)^*$, we can also apply the appropriate rule from group 2 and then proceed with applying the appropriate rules from groups 3 and 4; finally, by using the rule in group 5, we obtain the word $X\alpha wY$. The symbol α has been moved from the right-hand end to the left-hand end of the word, which is exactly what we need for rotating the underlying sentential form:

– $(Xw|\alpha Y, Z|Y_\alpha) \models_p (XwY_\alpha, Z\alpha Y)$, for $p = (\#\alpha Y\$Z\#Y_\alpha; \{X\}, \emptyset)$, where $w\alpha \in (N \cup T)^*\{B\}(N \cup T)^*$, $\alpha \in N \cup T \cup \{B\}$;

- $(X|wY_\alpha, X'\alpha|Z) \models_p (XZ, X'\alpha wY_\alpha)$, for $p = (X\#\$X'\alpha\#Z; \{Y_\alpha\}, \emptyset)$,
  where $w\alpha \in (N \cup T)^*\{B\}(N \cup T)^*$, $\alpha \in N \cup T \cup \{B\}$;

- $(X'\alpha w|Y_\alpha, Z|Y) \models_p (X'\alpha wY, ZY_\alpha)$, for $p = (\#Y_\alpha\$Z\#Y; \{X'\}, \emptyset)$,
  where $w\alpha \in (N \cup T)^*\{B\}(N \cup T)^*$, $\alpha \in N \cup T \cup \{B\}$;

- $(X'|\alpha wY, X|Z) \models_p (X'Z, X\alpha wY)$, for $p = (X'\#\$X\#Z; \{Y\}, \emptyset)$,
  where $w\alpha \in (N \cup T)^*\{B\}(N \cup T)^*$, $\alpha \in N \cup T \cup \{B\}$.

As additional results (that cannot already be found in the set of axioms A) of the splicings listed above, we obtain the words $Z\alpha Y$, for $\alpha \in U$, and, if $v \neq \lambda$ for all $u \to v \in P$, then $X'Z$ again, too.

Notice that every word obtained from $XBSY$ so far, not containing the symbol Z, is of the form $\alpha_1 x_1 B x_2 \alpha_2$, with (α_1, α_2) being one of the pairs (X, Y), (X, Y_α), (X', Y_α), (X', Y), $\alpha \in U$. Hence these symbols appearing as permitting contexts in the splicing rules of R precisely control the work of γ.

In order to obtain a terminal word we have to use rules from groups 6, 7; Y must be present when using rule 6 and B must be adjacent to X:

- $(XB|wY, |Z') \models_p (XBZ', wY)$, for $p = (XB\#\$\#Z'; \{Y\}, \emptyset)$,

- $(w|Y, |Z'') \models_p (w, Z''Y)$, for $p = (F'\#\$\#Z''; \{F'\}, \emptyset)$,

where $w \in (N \cup T)^*$.

As additional results of the splicings above we obtain the words XBZ', $Z''Y$.

Altogether, in γ we can produce every terminal word that can be produced by G, i.e., $L(\gamma) \supseteq L(G)$.

Conversely, no unintended terminal words can be generated in γ, i.e., $L(\gamma) \subseteq L(G)$.

Indeed, words of the form ZuY obtained after using a rule from group 1 associated with $u \to v \in P$ can be spliced by a rule of type 1 only when $ZuY \in A$, hence the operation is already discussed. Moreover, we can also perform a splicing

$$(Zu|Y, Z''|) \models_p (Zu, Z''Y), \text{ for } p = (\#Y\$Z''\#; \emptyset, \emptyset).$$

The string Zu cannot enter new splicings, while $Z''Y$ can enter splicing by the same rule, hence nothing new is produced:

$$(Xw|Y, Z''|Y) \models_p (XwY, Z''Y),$$
$$(Z''|Y, Z''|) \models_p (Z'', Z''Y).$$

Consider now the strings $Z\alpha Y, \alpha \in U$, obtained during the rotating phase. They cannot be used as the first term of a splicing using rules of types 1 – 6. If used in a splicing

$$(Z\alpha|Y, Z''|) \models_p (Z\alpha, Z''Y), \text{ for } p = (\#Y\$Z''\#; \emptyset, \emptyset),$$

then the string $Z\alpha$ cannot enter new splicings and $Z''Y$ cannot lead to terminal strings (see the previous paragraph). The string $Z\alpha Y$ can be used as the second term of a splicing only by a rule of type 1, providing that $Z\alpha Y \in A$.

Finally, $X'Z$ can enter no splicing; XBZ' can be spliced by the rule of type 6,

$$(XB|w, XB|Z') \models (XBZ', XBw), \ w \in (N \cup T)^*,$$

which produces nothing new, while $Z''Y$ can be spliced by using the rule of type 7 but again no terminal string can be produced.

In conclusion, we obtain $L(\gamma) = L(G)$. □

Remark 8.1. Note that in the rules $(p; C_1, C_2)$ of the H system with permitting contexts constructed in the proof of Lemma 8.2 (we can modify the rule of type 7 to $(\#Y\$Z''\#; \{Y\}, \emptyset))$, the pairs (C_1, C_2) of permitting contexts are of the special form $(\{D\}, \emptyset)$ for some nonterminal D, i.e. we only check the occurrence of one nonterminal in the first term of the splicing. This can be viewed as a normal form result for our systems. □

Remark 8.2. A permitting context splicing rule as in the proof of Lemma 8.2, i.e., with only one checked symbol which should appear at an end of the string, might be – theoretically – implemented in the following way.

As we have seen in Chap. 1, the restriction enzymes work only on double stranded sequences. We melt the solution in order to obtain single stranded sequences, and add a primer which contains the complement of the permitting symbol. This primer will only be attached to single stranded sequences containing the designated symbol (at an end of it). Only these single stranded sequences will enter the polymerization reaction, leading to double stranded sequences, hence the enzyme will only act on them. □

Theorem 8.1. $EH_2(FIN, pFIN) = RE$.

Proof. By a direct proof or by the Church–Turing Thesis, we have $EH_2(FIN, pFIN) \subseteq RE$; the converse inclusion is given in Lemma 8.2. □

In the proof of Lemma 8.2 we have paid no attention to the radius of the used splicing rules. Starting the proof of Lemma 8.2 from a grammar G in Kuroda normal form, we obtain an extended H system with permitting contexts of radius 3. This value is reached in splicing rules of type 1 ($\#uY\$Z\#vY$, where $|uY|$ and $|vY|$ can be equal to 3). This result can be strengthened:

Lemma 8.3. $RE \subseteq EH_2(FIN, p[2])$.

Proof. Consider a type-0 grammar $G = (N, T, S, P)$ in Kuroda normal form. Denote by P_1 the set of context-free rules in P and by P_2 the set of non-context-free rules in P.

We construct the extended H system with permitting contexts $\gamma = (V, T, A, R)$, where

$$\begin{aligned} V &= N \cup T \cup \{B, X, X', Z, Z', Z'', Y\} \\ &\cup \{Y_r \mid r \in P_2\} \cup \{Z_r \mid r \in P\} \\ &\cup \{Y_\alpha \mid \alpha \in N \cup T \cup \{B\}\}, \\ A &= \{XBSY, ZY, XZ, Z', Z''\} \\ &\cup \{ZY_\alpha, X'\alpha Z \mid \alpha \in N \cup T \cup \{B\}\} \\ &\cup \{Z_r xY \mid r : C \to x \in P_1\} \\ &\cup \{ZY_r, Z_r EFY \mid r : CD \to EF \in P_2\}, \end{aligned}$$

and R contains the following productions:

$$\begin{aligned} \textit{Simulate}: \quad & 1. \quad (\#CY\$Z_r\#x; \{X\}, \emptyset), \text{ for } r : C \to x \in P_1, \\ & 2. \quad (C\#DY\$Z\#Y_r; \{X\}, \emptyset), \\ & 3. \quad (\#CY_r\$Z_r\#EF; \{X\}, \emptyset), \text{ for } r : CD \to EF \in P_2, \\ \textit{Rotate}: \quad & 4. \quad (\#\alpha Y\$Z\#Y_\alpha; \{X\}, \emptyset), \\ & 5. \quad (X\#\$X'\alpha\#Z; \{Y_\alpha\}, \emptyset), \\ & 6. \quad (\#Y_\alpha\$Z\#Y; \{X'\}, \emptyset), \\ & 7. \quad (X'\#\$X\#Z; \{Y\}, \emptyset), \text{ for } \alpha \in N \cup T \cup \{B\}, \\ \textit{Terminate}: \quad & 8. \quad (XB\#\$\#Z'; \{Y\}, \emptyset), \\ & 9. \quad (\#Y\$Z''\#; \emptyset, \emptyset). \end{aligned}$$

The rules of type 1 simulate the rules in P_1, the rules of types 2, 3 simulate the rules in P_2. (Initially we have $XBSY$, hence at the first step we have to use a rule of type 1. Each splicing must involve an axiom and a string produced at a previous step and it produces a string of the form XwY, with $w \in (N \cup T \cup \{B\})^*$ and X, Y possibly replaced by variants of them, primed or having subscripts.) For instance, consider a string Xw_1Bw_2CDY and a rule $r : CD \to EF \in P_2$. We get

$$\begin{aligned} &(Xw_1Bw_2C|DY, Z|Y_r) \models_2 (Xw_1Bw_2CY_r, ZDY), \\ &(Xw_1Bw_2|CY_r, Z_r|EFY) \models_3 (Xw_1Bw_2EFY, Z_rCY_r). \end{aligned}$$

The rules of types 4, 5, 6, 7 are used for "rotating" the string and the rules of types 8 – 9 finish the work of γ in the same way as in the proof of Lemma 8.2. Thus, we have $L(G) = L(\gamma)$.

Clearly, $rad(\gamma) = 2$. □

Therefore, Theorem 8.1 can be written in the form

Corollary 8.1. $RE = EH_2(FIN, p[2])$.

Three *open* problems remain here: (1) Can the assertion in Lemma 8.3 be strengthened to $RE \subseteq EH_2(FIN, p[1])$? We *conjecture* that $EH_2(FIN, p[1]) \subseteq CF$, i.e., that the answer is negative. (2) Can the number of axioms be bounded in advance without loosing the computational completeness? We *conjecture* that the answer is affirmative, more precisely, that $RE = EH_2([1], pFIN)$, that is, one axiom suffices. (3) Can we simultaneously decrease both the number of axioms and the radius of splicing rules? As is customary in the area of descriptional complexity (we saw this at the end of Sect. 7.4, too), it is possible that this question has a negative answer, that is, a trade-off between these two complexity criteria, $card(A)$ and $rad(R)$, is highly probable. However, as we shall see below, in many cases of controlled H systems the two parameters can simultaneously be bounded by rather low thresholds.

In what concerns the above conjectured inclusion $EH_2(FIN, p[1]) \subseteq CF$, if this would be confirmed, then an entirely new characterization of context-free languages is obtained, because we have the next result.

Lemma 8.4. $CF \subseteq EH_2(FIN, p[1])$.

Proof. Consider a context-free grammar $G = (N, T, S, P)$ in the strong Chomsky normal form, that is, with the rules in P of the forms $X \to a$, $X \to YZ$, for $X, Y, Z \in N, a \in T$, and with the additional restrictions specified in Theorem 3.2:

1. if $X \to YZ$ is in P, then $Y \neq Z$,

2. if $X \to YZ$ is in P, then for each rule $X \to Y'Z'$ in P we have $Z' \neq Y$ and $Y' \neq Z$.

We construct the permitting context H system $\gamma = (V, T, A, R)$, where

$$V = T \cup \{X_l, X_r, X_r' \mid X \in N\} \cup \{D, E\},$$
$$A = \{X_l a X_r \mid X \to a \in P\} \cup \{X_l D, DX_r, DX_r' \mid X \in N\} \cup \{E\},$$

and R consists of the following splicing rules:

1) $(a\#Y_r\$Z_l\#b; \{Y_l\}, \{Z_r\})$, for $X \to YZ \in P, a, b \in T$,
2) $(a\#Z_r\$D\#X_r'; \{Y_l\}, \emptyset)$, for $X \to YZ \in P, a \in T$,
3) $(X_l\#D\$Y_l\#a; \emptyset, \{X_r'\})$, for $X, Y \in N, a \in T$,
4) $(a\#X_r'\$D\#X_r; \{X_l\}, \emptyset)$, for $X \in N, a \in T$,
5) $(a\#S_r\$E\#; \{S_l\}, \emptyset)$, for $a \in T$,
6) $(\#E\$S_l\#a; \emptyset, \emptyset)$ for $a \in T$.

The basic idea of this construction is that a string $X_l w X_r$, with $w \in T^+, X \in N$, is generated in γ if and only if $X \Longrightarrow^* w$ in the grammar G (the subscripts l, r in X_l, X_r stand for "left" and "right", respectively). When $X = S$, then w is an element of $L(G)$.

The rules $X \to a$ are directly simulated by the axioms, because the strings $X_l a X_r$ are introduced in A.

Consider two strings $Y_l w_1 Y_r, Z_l w_2 Z_r$ produced by γ; the axioms in A corresponding to terminal rules in P are of this form, all other axioms are of a different form. If there is a rule $X \to YZ$ in P, then a splicing rule of type 1 exists in R, hence we can perform

$$(Y_l w_1 | Y_r, Z_l | w_2 Z_r) \models_1 (Y_l w_1 w_2 Z_r, Z_l Y_r).$$

The second string cannot enter new splicings in γ (either a terminal symbol or the control symbol D must always be present in the terms of a splicing). The first string can be processed as follows:

$$(Y_l w_1 w_2 | Z_r, D | X_r') \models_2 (Y_l w_1 w_2 X_r', D Z_r),$$
$$(X_l | D, Y_l | w_1 w_2 X_r') \models_3 (X_l w_1 w_2 X_r', Y_l D),$$
$$(X_l w_1 w_2 | X_r', D | X_r) \models_4 (X_l w_1 w_2 X_r, D X_r').$$

These steps are possible only if the rule $X \to YZ$ is in P. Therefore, we obtain the string $X_l w_1 w_2 X_r$ which corresponds to a derivation $X \Longrightarrow^* w_1 w_2$ in G. The strings $D Z_r, Y_l D, D X_r'$ are axioms.

Thus, indeed, a string $X_l z X_r$ is produced in γ if and only if $X \Longrightarrow^* z$ in G. If we obtain a string $S_l x S_r$, that is with $x \in L(G)$, then we can use rules 5, 6 in R:

$$(S_l x | S_r, E |) \models_5 (S_l x, E S_r),$$
$$(| E, S_l | x) \models_6 (x, S_l E).$$

We obtain the string x, as well as strings which can only enter splicings of the same forms but produce nothing (for instance, $(S_l x | S_r, E | S_r) \models_5 (S_l x S_r, E S_r)$). Rule 6 cannot be used before using rule 5, because the symbol S_l must be present in the string when rule 5 is applied. This concludes the proof of the equality $L(G) = L(\gamma)$. □

Consider now again the proof of Lemma 8.2. The symbols whose presence is checked are elements of the set of control symbols

$$Q = \{X, X', Y\} \cup \{Y_\alpha \mid \alpha \in N \cup T \cup \{B\}\}.$$

The *presence* of a symbol is equivalent with the *absence* of all other symbols, because the control symbols are always present at the ends of the strings – except when finishing the generation of a terminal string. Thus, we can consider a dual variant of extended H systems with permitting contexts, that is, systems with forbidding contexts.

An extended H system *with forbidding contexts* is a quadruple $\gamma = (V, T, A, R)$, where V is an alphabet, $T \subseteq V$ (the terminal alphabet), A

is a finite language over V (axioms), and R is a finite set of triples (we call them rules with forbidding contexts) of the form $p = (r; D_1, D_2)$, where $r = u_1\#u_2\$u_3\#u_4$ is a splicing rule over V and D_1, D_2 are finite subsets of V^*.

For $x, y, z, w \in V^*$ and $p \in R, p = (r; D_1, D_2)$, we define $(x, y) \models_p (z, w)$ if and only if $(x, y) \models_r (z, w)$, no element of D_1 appears as a substring of x and no element of D_2 appears as a substring of y; when $D_1 = \emptyset$ or $D_2 = \emptyset$, then no condition on x, respectively y, is imposed.

From a biochemical point of view, the permitting contexts can be interpreted as catalysts or promoters, favoring the splicing by the associated splicing rule, while the forbidding contexts can be interpreted as inhibitors, suppressing the associated splicing rule.

The pair $\sigma = (V, R)$ is called an (underlying) H scheme with forbidding contexts rules. The language generated by γ is defined in the usual way:

$$L(\gamma) = \sigma_2^*(A) \cap T^*.$$

We denote by $EH_2([n], f[m]), n, m \geq 1$, the family of languages $L(\gamma)$ generated by extended H systems with forbidding contexts, $\gamma = (V, T, A, R)$, with $card(A) \leq n$ and $rad(R) \leq m$, where $rad(R)$ is the maximal radius of splicing rules r in triples $(r; D_1, D_2)$ in R. When no restriction on the number of axioms or on the maximal radius is imposed (except that these numbers are still finite), we replace $[n]$ or $[m]$ by FIN.

As expected from the previous discussion, we have the equality $EH_2(FIN, f[2]) = RE$. Actually, a stronger result is true:

Theorem 8.2. $EH_2([1], f[2]) = RE$.

The proof of this theorem is based on two lemmas corresponding to Lemmas 8.2 (combined with Lemma 8.3) and to the conjectured inclusion $RE \subseteq EH_2([1], pFIN)$. For the sake of completeness and because of the strong form of Theorem 8.2, we present the core construction of the proofs of both these lemmas.

Lemma 8.5. $RE \subseteq EH_2(FIN, f[2])$.

Proof. Consider a type-0 grammar $G = (N, T, S, P)$ in the Kuroda normal form, denote by P_1 the set of context-free rules in P and by P_2 the set of non-context-free rules in P. We construct an H system with forbidding contexts $\gamma = (V, T, A, R)$, where V, T, A are the same as in the proof of Lemma 8.3, $U = N \cup T \cup \{B\}$, and R contains the following rules with forbidding contexts:

Simulate : 1. $(\#CY\$Z_r\#x; \{X'\}, \emptyset)$, for $r : C \to x \in P_1$,
2. $(C\#DY\$Z\#Y_r; \{X'\}, \emptyset)$,
3. $(\#CY_r\$Z_r\#EF; \{X'\}, \emptyset)$, for $r : CD \to EF \in P_2$,

Rotate : 4. $(\#\alpha Y\$Z\#Y_\alpha; \{X'\}, \emptyset)$,
5. $(X\#\$X'\alpha\#Z; \{Y\} \cup \{Y_\beta \mid \beta \in U,$

$$\begin{aligned}
& \qquad\quad \beta \neq \alpha\} \cup \{Y_r \mid r \in P_2\}, \emptyset), \\
& \quad 6.\ \ (\#Y_\alpha\$Z\#Y; \{X\}, \emptyset), \text{ for } \alpha \in U, \\
& \quad 7.\ \ (X'\#\$X\#Z; \{Y_\beta \mid \beta \in U\} \cup \{Y_r \mid r \in P_2\}, \emptyset), \\
Terminate: & \quad 8.\ \ (XB\#\$\#Z'; \{Y_\beta \mid \beta \in U\} \cup \{Y_r \mid r \in P_2\}, \emptyset), \\
& \quad 9.\ \ (\#Y\$Z''\#; \{X, B\}, \emptyset).
\end{aligned}$$

The equality $L(\gamma) = L(G)$ can be checked in an easy way (repeating arguments like those in the proofs of Lemmas 8.2 and 8.3). □

Lemma 8.6. $EH_2(FIN, f[2]) \subseteq EH_2([1], f[2])$.

Proof. Consider the extended H system with forbidding contexts $\gamma = (V, T, A, R)$ with $A = \{w_1, w_2, \ldots, w_n\}, n \geq 2$, given by the construction in the previous proof, that is, with rules of the form $(r; D_1, \emptyset)$. Moreover, at each splicing step, the second term of the operation is an axiom. We construct the extended H system with forbidding contexts

$$\gamma' = (V \cup \{c\}, T, \{w\}, R'),$$

where

$$w = cw_1cw_2 \ldots cw_nc,$$

and

$$\begin{aligned}
R' &= \{(r; D_1 \cup \{c\}, \{c\}) \mid (r; D_1, \emptyset) \in R\} \\
&\cup \ \{(\#c\$c\#; \emptyset, \emptyset)\}.
\end{aligned}$$

By applying the splicing rule $\#c\$c\#$ to two copies of w we can cut one w at the symbol c in front of some axiom w_i, obtaining $w_ic \ldots w_nc$. Applying the same rule $\#c\$c\#$ to this word and to another copy of w we can produce the axiom w_i as a separate string.

Because the rules in R now have the forbidding contexts $D_1 \cup \{c\}, \{c\}$, the two terms of the splicing cannot contain the symbol c. Removing c amounts to separating the axioms of γ from w by the rule $\#c\$c\#$. Consequently, we have $L(\gamma) = L(\gamma')$. □

Note that in the case of forbidding contexts we do not obtain a normal form as in the case of permitting contexts, that is with condition sets of the form (C_1, C_2), with C_1 containing one symbol only and C_2 empty. On the other hand, in the forbidding case we obtain a result which is not known for permitting contexts: low bounds are found both for the number of axioms and for the radius.

The control through forbidding symbols as in the proof of Lemma 8.5, that is, with the checked symbols always appearing at the ends of the first term of the splicing, can be easily implemented by considering a *priority relation* on the set of splicing rules.

Specifically, an *ordered* extended H system is a construct $\gamma = (V, T, A, R, >)$, where V is an alphabet, $T \subseteq V$ (the terminal alphabet), A is a finite language over V (axioms), R is a finite set of splicing rules over V, and $>$ is a partial order relation on R.

For $x, y, z, w \in V^*$ and $r \in R$ we allow the relation $(x, y) \models_r (z, w)$ only if we do not have $(x, y') \models_{r'} (z', w')$ or $(y', x) \models_{r'} (z', w')$, for some $y', z', w' \in V^*$ and $r' \in R$ such that $r' > r$. (A splicing is performed by a rule which is maximal among all splicing rules which can be applied to the first string and any other string.)

We denote by $EH_2([n], ord[m]), n, m \geq 1$, the family of languages generated by ordered extended H systems with at most n axioms and of radius at most m.

Theorem 8.3. $RE = EH_2([1], ord[2])$.

Proof. Consider an H system with forbidding contexts as given by the constructions in the proof of Lemmas 8.5 and 8.6, that is, of the form $\gamma = (V, T, \{w\}, R)$, with $rad(\gamma) = 2$ and with the rules in R of two types: a rule $\#c\$c\#$ which cuts w in parts not containing the symbol c, and rules which always use these parts as the second term of the splicing. We modify γ as follows. Add the symbol Z_0 to the alphabet V and replace the axiom w by wcZ_0Z_0. Replace each rule with forbidding contexts of the form

$$(u_1\#u_2\$u_3\#u_4; D_1, D_2)$$

with $D_1 \cup D_2 \neq \emptyset$ by the set of rules

$$\begin{aligned} r &= u_1\#u_2\$u_3\#u_4, \\ r(\alpha) &= \alpha\#\$Z_0\#Z_0, \ \text{for } \alpha \in D_1, \\ r(\alpha) &= Z_0\#Z_0\$\alpha\#, \ \text{for } \alpha \in D_2. \end{aligned}$$

Consider the relation $>$ defined by

$$r(\alpha) > r, \ \text{for all } \alpha \in D_1 \cup D_2.$$

Consider also the rule

$$r_0 = Z_0\#\$Z_0\#Z_0,$$

with $r_0 > r$ for all rules $r = u_1\#u_2\$u_3\#u_4$ corresponding to a starting rule $(u_1\#u_2\$u_3\#u_4; D_1, D_2)$.

Denote by γ' the ordered H system obtained in this way.

By using the rule $\#c\$c\#$ as in the proof of Lemma 8.6, we can separate from wcZ_0Z_0 the blocks not containing the symbol c, as well as the string Z_0Z_0. Now, if for some $(r; D_1, D_2)$ a symbol $\alpha \in D_1 \cup D_2$ appears in a string x, then in order to splice (x, y) in γ', for some $y \in V^*$, we cannot use the rule r, but one of the rules $r(\alpha), r'(\alpha)$. However, the use of $r(\alpha)$ and $r'(\alpha)$ introduces the symbol Z_0 in both strings produced by splicing and this symbol cannot be eliminated (the rule r_0 has to be used for the first term of the splicing, the second one should be an axiom). Therefore, the priority restriction forces the use of splicing rules r in a forbidding manner. Consequently, $L(\gamma) = L(\gamma')$. □

The order relation in systems above can be interpreted as modeling the difference between the reactivity of the enzymes involved by the splicing rules: when two different enzymes can cut the same string, the more reactive one will actually work.

8.3 Target Languages

While extended H systems with permitting or forbidding contexts correspond to the biochemical activity of catalysts (promoters) and inhibitors, which control *in vivo* or *in vitro* reactions by their presence in or absence from the molecules entering the reaction, regulating the splicing by *target languages* corresponds to another biochemical aspect, encountered *in vivo*: nature selects the offsprings of the evolutionary process in a rather dramatic manner, not allowing the perpetuation of "unsuitable" forms of life. Formulated in "Lamarckian terms", we may say that evolution has a sense, that the mutations and recombinations are made "toward improvement". Such speculations can be easily modeled in our framework by considering a sort of *hypothesis language*: when splicing two strings, the resulting strings should be members of a given language. This corresponds to conditional grammars in the area of regulated rewriting and to grammar systems with hypothesis languages in grammar systems theory. As in these areas, too, the power of H systems is increased by considering such regulating mechanisms: we can again characterize the family RE by systems with finite sets of splicing rules.

An extended H system *with local targets* is a construct $\gamma = (V, T, A, R)$, where V is an alphabet, $T \subseteq V$ (the terminal alphabet), A is a finite language over V (axioms), and R is a finite set of pairs $p = (r, Q_p)$, where $r = u_1\#u_2\$u_3\#u_4$ is a splicing rule over V and Q_p is a regular language over V. For $x, y, z, w \in V^*$ and $p = (r, Q_p)$ in R we write $(x, y) \models_p (z, w)$ if and only if $(x, y) \models_r (z, w)$ and $z, w \in Q_p$ (the results of the splicing with respect to r belong to Q_p).

If, for such an extended H system with local targets, $\gamma = (V, T, A, R)$, we have $Q_{p_1} = Q_{p_2}$ for all $p_1 = (r_1, Q_{p_1}), p_2 = (r_2, Q_{p_2})$ in R, then we say that γ is a system *with a global target*. If Q is the common target language of rules in R, then we write the system in the form $\gamma = (V, T, A, R', Q)$, with R' consisting of the splicing rules in R.

In the customary style, we denote by $EH_2([n], lt[m]), n, m \geq 1$, the family of languages generated by extended H systems with local targets having at most n axioms and splicing rules of radius at most m; in the case of global targets we replace lt by gt; when no bound on the number of axioms or on the radius is imposed, we replace $[n], [m]$ by FIN.

From the definitions we have

Lemma 8.7. *$EH_2([n], gt[m]) \subseteq EH_2([n], lt[m])$, for all $n, m \geq 1$.*

Lemma 8.8. *$RE \subseteq EH_2(FIN, gt[2])$.*

Proof. Consider a type-0 grammar in Kuroda normal form, $G = (N, T, S, P)$, denote by P_1 the set of context-free rules in P and by P_2 the set of non-context-free rules in P. Assume the rules in P labeled in a one-to-one manner. Denote $U = N \cup T \cup \{B\}$, where B is a new symbol. We construct the extended H system with a global target

$$\gamma = (V, T, A, R, Q),$$

where

$$\begin{aligned} V &= N \cup T \cup \{B, X, X', X'', Y, Y', Z, Z'\} \\ &\cup \{Y_\alpha \mid \alpha \in U\} \cup \{Z_r \mid r \in P\} \\ &\cup \{Y_r \mid r : CD \to EF \in P_2\}, \\ A &= \{XBSY, ZY, XZ, ZY', X''Z, Z'\} \\ &\cup \{ZY_\alpha, X'\alpha Z \mid \alpha \in U\} \\ &\cup \{Z_r xY \mid r : C \to x \in P_1\} \\ &\cup \{ZY_r, Z_r EFY \mid r : CD \to EF \in P_2\}, \end{aligned}$$

and the following groups of splicing rules; we associate with them target languages in a local manner in order to make more explicit the work of γ, but Q is the union of all these local target languages:

$$\begin{array}{rll} \textit{Simulate}: & 1.\ \ \#CY\$Z_r\#x, & Q_{1,r} = XU^*Y \cup \{Z_rCY\}, \\ & & \text{for } r : C \to x \in P_1, \\ & 2.\ \ C\#DY\$Z\#Y_r, & Q_{2,r} = XU^*Y_r \cup \{ZDY\}, \\ & 3.\ \ \#CY_r\$Z_r\#E, & Q_{3,r} = XU^*Y \cup \{Z_rCY_r\}, \\ & & \text{for } r : CD \to EF \in P_2, \\ \textit{Rotate}: & 4.\ \ \#\alpha Y\$Z\#Y_\alpha, & Q_{4,\alpha} = XU^*Y_\alpha \cup \{Z\alpha Y\}, \\ & 5.\ \ X'\alpha\#Z\$X\#, & Q_{5,\alpha} = X'U^*Y_\alpha \cup \{XZ\}, \\ & 6.\ \ \#Y_\alpha\$Z\#Y', & Q_{6,\alpha} = X'U^*Y' \cup \{ZY_\alpha\}, \\ & & \text{for } \alpha \in U, \\ & 7.\ \ X''\#Z\$X'\#, & Q_7 = X''U^*Y' \cup \{X'Z\}, \\ & 8.\ \ \#Y'\$Z\#Y, & Q_8 = X''U^*Y \cup \{ZY'\}, \\ & 9.\ \ X\#Z\$X''\#, & Q_9 = XU^*Y \cup \{X''Z\}, \\ \textit{Terminate}: & 10.\ \ \#Z'\$XB\#, & Q_{10} = T^*Y \cup \{XBZ'\}, \\ & 11.\ \ \#Y\$Z'\#, & Q_{11} = T^* \cup \{Z'Y\}. \end{array}$$

Then,

$$Q = \bigcup_{r \in P_1} Q_{1,r} \cup \bigcup_{r \in P_2} (Q_{2,r} \cup Q_{3,r})$$
$$\cup \bigcup_{\alpha \in U} (Q_{4,\alpha} \cup Q_{5,\alpha} \cup Q_{6,\alpha}) \cup \bigcup_{i=7}^{11} Q_i.$$

The work of γ is based on the same rotate-and-simulate idea as in several of the proofs of similar relations above, with some further precautions in the rotation phase. As above, we start from the axiom $XBSY$ and at every splicing step we splice a string of the form XwY, maybe with X replaced with X' or X'' and with Y replaced with $Y', Y_\alpha, \alpha \in U$, or $Y_r, r \in P_2$, with an axiom (each splicing rule contains an occurrence of Z, Z', or Z_r, for $r \in P$). Together with the target language, these control symbols control the work of γ in a way that makes possible the simulation in γ of all correct derivations in G and, conversely, prevents the generation of terminal strings not in $L(G)$.

Let us examine, for instance, a rotation step. Starting from a string $Xw\alpha Y$, with $w \in U^*, \alpha \in U$, by using the rules in group $4-9$ we successively get:

$$\begin{aligned}
&(Xw|\alpha Y, Z|Y_\alpha) \models_{4,\alpha} (XwY_\alpha, Z\alpha Y),\\
&(X'\alpha|Z, X|wY_\alpha) \models_{5,\alpha} (X'\alpha wY_\alpha, XZ),\\
&(X'\alpha w|Y_\alpha, Z|Y') \models_{6,\alpha} (X'\alpha wY', ZY_\alpha),\\
&(X''|Z, X'|\alpha wY') \models_{7} (X''\alpha wY', X'Z),\\
&(X''\alpha w|Y', Z|Y) \models_{8} (X''\alpha wY, ZY'),\\
&(X|Z, X''|\alpha wY) \models_{9} (X\alpha wY, X''Z).
\end{aligned}$$

We have obtained the string $X\alpha wY$ which is a circular permutation of the starting string $Xw\alpha Y$. The target language does not contain strings of the form $X''zY_r$, for $r \in P_2$, hence $X''\alpha wY$ cannot enter splicings with respect to rules in group 2. Using a rule in group 1 does not lead to unintended strings. None of the other strings on the first position of the resulting pairs above can enter splicings with respect to simulating rules in R. Similarly, these strings cannot be spliced according to terminating rules: the resulting strings cannot be in $T^*Y \cup T^* \cup \{XBZ', Z'Y\}$. In what concerns the "by-product" strings above, $Z\alpha Y, XZ, ZY_\alpha, X'Z, ZY', X''Z$, part of them are axioms, the others either cannot enter splicings because of the target restriction, or they can enter splicings which do not lead to illegal terminal strings. For instance, $X'Z$, which is not an axiom, can be spliced with $X''Z$,

$$(X''|Z, X'|Z) \models_7 (X''Z, X'Z),$$

but the two strings are reproduced (note that both strings $X''Z, X'Z$ are in Q).

Thus, we can conclude that $L(\gamma) = L(G)$. □

Lemma 8.9. $EH_2(FIN, gt[m]) \subseteq EH_2([1], gt[m]), m \geq 1$.

Proof. Let $\gamma = (V, T, A, R, Q)$ be an extended H system with a global target, with $A = \{w_1, \ldots, w_n\}, n \geq 2$. We construct the extended H system with a global target

$$\gamma' = (V \cup \{c\}, T, \{w\}, R', Q'),$$

where

$$\begin{aligned} w &= cw_1cw_2 \ldots cw_nc, \\ R' &= R \cup \{\#c\$c\#\}, \\ Q' &= Q \cup \{w_i,\ w_ic \ldots cw_nc,\ cw_1c \ldots cw_{i-1}cw, \\ &\qquad wcw_{i+1}c \ldots cw_n \mid 1 \leq i \leq n\}. \end{aligned}$$

We obtain $L(\gamma) = L(\gamma')$: using the rule $\#c\$c\#$ we can separate each axiom w_i of γ from w:

$$\begin{aligned} &(|w, cw_1c \ldots cw_{i-1}c|w_ic \ldots cw_nc) \models (w_ic \ldots cw_nc, cw_1c \ldots cw_{i-1}cw), \\ &(w_i|cw_{i+1}c \ldots cw_nc, w|) \models (w_i, wcw_{i+1}c \ldots cw_nc). \end{aligned}$$

If a splicing rule in R is applied to strings containing the symbol c, then the resulting strings should be in $Q' - Q$, hence they must be either axioms in A or strings containing further occurrences of c, hence composed of blocks $w_i, 1 \leq i \leq n$, bounded by occurrences of c. This ensures the inclusion $L(\gamma') \subseteq L(\gamma)$. The reverse inclusion is obvious.

Note that the radius of rules in $R' - R$ is one, hence we have $rad(R') = rad(R)$. □

Theorem 8.4. $RE = EH_2([1], gt[2]) = EH_2([1], lt[2])$.

Proof. The inclusion $RE \subseteq EH_2([1], gt[2])$ follows from Lemmas 8.8, 8.9. The inclusion $EH_2([1], gt[2]) \subseteq EH_2([1], lt[2])$ is pointed out in Lemma 8.7, whereas the inclusion $EH_2([1], lt[2]) \subseteq RE$ can be proved by a straightforward construction of a type-0 grammar simulating an extended H system with local targets (or can be obtained from the Church–Turing Thesis). □

By using target languages, we have removed the infinity of the set of splicing rules, but we have reintroduced the infinity in the target languages. However, it is enough to use the information provided by the first and the last symbols of the strings produced by splicing, returning again to the style of permitting-forbidding context conditions in the previous section, but formulated for the strings obtained by splicing and not for the strings entering the splicing.

We can relate this also to the style of genetic algorithms area, formulating these conditions by means of *fitness mappings*: consider a mapping assessing the quality (fitness, reactivity) of the strings and let us control the process by asking that strings with a low degree of fitness are not used in further splicings.

Here we consider a boolean fitness mapping (a predicate).

An extended H system *with a fitness mapping* is a construct $\gamma = (V, T, A, R, f)$, where V is an alphabet, $T \subseteq V$ (the terminal alphabet), A is a finite subset of V^* (axioms), R is a finite set of splicing rules over V, and $f : V^* \longrightarrow \{0, 1\}$. The splicing of two strings $x, y \in V^*$ with respect to a rule in R is defined only when $f(x) = 1, f(y) = 1$. The language generated by γ is defined in the usual way. By $EH_2([n], fit[m]), n, m \geq 1$, we denote the family of languages generated by such systems with at most n axioms and the radius at most m.

By simply taking $f : V^* \longrightarrow \{0, 1\}$ as being the membership mapping of $A \cup Q$ in an extended H system with a global target, $\gamma = (V, T, A, R, Q)$, we get an extended H system with a fitness mapping $\gamma' = (V, T, A, R, f)$ such that $L(\gamma) = L(\gamma')$: the result of a splicing does not enter new splicings if it is not in $A \cup Q$. Thus we can write

Corollary 8.2. $RE = EH_2([1], fit[2])$.

If the fitness mapping is not restricted, it can introduce artificially complex features in the generated language, just by starting from an H system with a complex fitness mapping. Thus, it is important to define particular classes of such mappings. A natural idea is to consider a *local* definition of the fitness mapping: we say that $f : V^* \longrightarrow \{0, 1\}$ is *locally defined* if $f(\alpha x \beta) = f(\alpha x' \beta)$ for all $\alpha, \beta \in V, x, x' \in V^*$ (that is, the value of $f(\alpha x \beta)$ does not depend on x, but only on the symbols α, β bounding it). We denote by $EH_2([n], fit_l[m]), n, m \geq 1$, the family of languages generated by extended H systems with a locally defined fitness mapping, with at most n axioms and of radius at most m; $[n], [m]$ are replaced by FIN when no bound on the number of axioms or on the radius is imposed. Now, from the proof of Corollary 8.1, by a suitable definition of a mapping f capturing the restrictions imposed by the permitting context conditions (checked only at the ends of the spliced strings), we get

Corollary 8.3. $RE = EH_2([1], fit_l FIN) = EH_2(FIN, fit_l[2])$.

Note that, in fact, we have a particular form of a locally defined fitness mapping, that is depending only on the leftmost symbol of a string.

Although Corollaries 8.2 and 8.3 are obtained as reformulations of other results (involving reformulations of other notions), we have mentioned them because they look more appropriate from a "practical" point of view: in order to implement H systems with a fitness mapping it is enough to devise a mechanism able to remove or inhibit the non-fitting strings obtained by non-deterministic, unrestricted splicing; moreover, this mechanism has to check only the ends of the strings in order to evaluate their fitness.

8.4 Programmed and Evolving Systems

The ways of controlling the splicing operation in previous sections were based on contextual conditions restricting the use of the splicing rules: any rule in the set of splicing rules can be used at any time, but only for splicing strings fulfilling certain conditions. However, the splicing rules themselves can change from one step to the next one, influenced by the very strings to which they are applied or, more generally, by the currently available set of strings. Both the strings (DNA molecules) and the splicing rules (restriction enzymes and ligases) correspond to chemical complexes placed together in a given space (a cell, a test tube), hence their interaction takes place in both directions: not only do the splicing rules act on the strings, but the strings also influence the splicing rules.

In a general set-up, for a language $L \subseteq V^*$ and a set R of splicing rules over V, we can define:

1. A language $String(R, L)$, of all strings obtained by one-step splicing of the strings in L with respect to the rules in R, perhaps applied in a restricted mode;

2. A set $Rule(R, L)$, of splicing rules over V, obtained from the rules in R under the influence of the strings in L.

Then, starting from a language $L_0 \subseteq V^*$ and a set R_0 of splicing rules over V, we can define the sequence

$$(R_i, L_i) = (Rule(R_{i-1}, L_{i-1}), String(R_{i-1}, L_{i-1})), i \geq 1.$$

For an extended H system (maybe with a control *ctr* on the application of its rules), $\gamma = (V, T, A, R, ctr)$, we can define the sequence $(R_i, L_i), i \geq 0$, starting from $L_0 = A, R_0 = R$ and with the mapping $String$ depending on *ctr*. Then the language generated by γ can be defined by

$$L(\gamma) = (\bigcup_{i \geq 0} L_i) \cap T^*.$$

Such a system γ is said to be an *evolving* one.

The mapping $String$ can be defined as in the previous sections, using free splicing or splicing restricted in various ways. Permitting or forbidding contexts, target languages, or fitness mappings can be some variants. In this section we discuss the possibilities offered by the *Rule* mapping.

One variant is to define the set $Rule(R_{i-1}, L_{i-1})$ not depending on the whole language L_{i-1} but only on the strings used when passing from L_{i-2} to $L_{i-1} = String(R_{i-2}, L_{i-2})$, starting from the assumption that the strings which are "close" to the rules can modify them – and can be modified by them. Then, in this framework, we can define $Rule(R_{i-1}, L_{i-1})$ in a permitting or forbidding way, depending on the presence or absence of certain symbols in the spliced strings. We can cover in this way the permitting or forbidding ways of controlling the splicing.

A general class of evolving H systems can be based on mappings $Rule(R, L)$ depending on R only, hence independent of the current set of strings. This immediately suggests considering H systems corresponding to time-varying grammars or to programmed grammars in the regulated rewriting area.

We postpone the study of time-varying H systems for a subsequent chapter, because this type of generative mechanism has a distributed architecture which deserves to be emphasized and compared with other distributed H systems. In the programmed case we can once again characterize the family RE.

A *programmed* extended H system is a construct $\gamma = (V, T, A, R, next)$, where V is an alphabet, $T \subseteq V$ (the terminal alphabet), A is a finite language over V (axioms), R is a finite set of splicing rules over V, and $next : R \longrightarrow \mathcal{P}(R)$.

The language generated by γ is defined by

$$L(\gamma) = (A \cup \sigma_2(A) \cup \rho(A)) \cap T^*,$$

where $\sigma = (V, R)$ is the underlying H scheme of γ and

$$\begin{aligned} \rho(A) = \{ w \in V^* \mid & \text{ there is a sequence of splicings of the form} \\ & (x_1, y_1) \models_{r_1} (x_2, y'_2), (x_2, y_2) \models_{r_2} (x_3, y'_3), \ldots \\ & (x_k, y_k) \models_{r_k} (x_{k+1}, y'_{k+1}), \text{ such that } k \geq 2, \\ & r_i \in next(r_{i-1}), 2 \leq i \leq k, y'_i \in V^*, 2 \leq i \leq k+1, \\ & x_1, y_1 \in A, y_i \in A, 2 \leq i \leq k, \text{ and } w = x_{k+1} \}. \end{aligned}$$

In words, the language $L(\gamma)$ contains all terminal strings in A, as well as those strings which can be obtained by one splicing starting from strings in A (the set $\sigma_2(A)$), or by several splicings with the following properties: one starts from two axioms; at each subsequent step, one splices the first of the two strings produced at the previous step with one axiom; the rules used at consecutive steps are related by the *next* mapping.

The condition to use at each step – except the first one – a string produced at the previous step and an axiom could seem artificial and restrictive, but most of the H systems in the proofs considered in the previous sections work in this way when following "correct" paths of splicing, i.e., paths towards strings in the generated language. However, in these systems other splicings are also possible. Here, in order to take advantage of the programmed type of restriction, we have to impose this condition in an explicit manner on the terms of the splicing operations.

In the usual style, we denote by $EH_2([n], pr[m]), n, m \geq 1$, the family of languages generated by programmed extended H systems with at most

n axioms and with splicing rules of radius at most m. When n, m are not bounded, we replace $[n], [m]$ by FIN.

Theorem 8.5. $RE = EH_2(FIN, pr[2])$.

Proof. We only have to prove the inclusion $RE \subseteq EH_2(FIN, pr[2])$.

Consider a type-0 grammar $G = (N, T, S, P)$ in Kuroda normal form, denote by P_1 the set of context-free rules in P and by P_2 the set of non-context-free rules in P. Assume the rules in P are labeled in a one-to-one manner. We construct the programmed H system

$$\gamma = (V, T, A, R, next),$$

where

$$\begin{aligned} V &= N \cup T \cup \{X, X', Y, B\} \\ &\cup \{Y_p \mid p \in P_2\} \cup \{Z_p \mid p \in P\}, \\ A &= \{X'BSY, ZY, XZ\} \\ &\cup \{Z_p xY \mid p : C \to x \in P_1\} \\ &\cup \{ZY_p, Z_p EFY \mid p : CD \to EF \in P_2\} \\ &\cup \{X\alpha Z \mid \alpha \in N \cup T \cup \{B\}\}, \end{aligned}$$

R contains the following splicing rules

$$\begin{aligned} Start: \quad & s_0 = X\#Z\$X'\#B, \\ Simulate: \quad & s_p = \#CY\$Z_p\#x, && \text{for } p : C \to x \in P_1, \\ & s_{1,p} = C\#DY\$Z\#Y_p, \\ & s_{2,p} = \#CY_p\$Z_p\#E, && \text{for } p : CD \to EF \in P_2, \\ Rotate: \quad & r_{1,\alpha} = \#\alpha Y\$Z\#Y, \\ & r_{2,\alpha} = X\alpha\#Z\$X\#, && \text{for } \alpha \in N \cup T \cup \{B\}, \\ Terminate: \quad & t_1 = \#ZY\$XB\#, \\ & t_2 = \#Y\$ZY\#, \end{aligned}$$

and the mapping $next$ is defined as follows:

$$\begin{aligned} next(s_0) &= \{s_p \mid p \in P_1\}, \\ next(s_p) &= \{s_{p'} \mid p' \in P_1\} \\ &\cup \{s_{1,p'} \mid p' \in P_2\} \\ &\cup \{r_{1,\alpha} \mid \alpha \in N \cup T \cup \{B\}\} \\ &\cup \{t_1\}, \quad \text{for } p \in P_1, \\ next(s_{1,p}) &= \{s_{2,p}\}, \\ next(s_{2,p}) &= \{s_{p'} \mid p' \in P_1\} \\ &\cup \{s_{1,p'} \mid p' \in P_2\} \\ &\cup \{r_{1,\alpha} \mid \alpha \in N \cup T \cup \{B\}\} \end{aligned}$$

$$
\begin{aligned}
& \cup \{t_1\}, \quad \text{for } p \in P_2, \\
next(r_{1,\alpha}) &= \{r_{2,\alpha}\}, \\
next(r_{2,\alpha}) &= \{s_p \mid p \in P_1\} \\
& \cup \{s_{1,p} \mid p \in P_2\} \\
& \cup \{r_{1,\beta} \mid \beta \in N \cup T \cup \{B\}\} \\
& \cup \{t_1\}, \quad \text{for } \alpha \in N \cup T \cup \{B\}, \\
next(t_1) &= \{t_2\}, \\
next(t_2) &= \emptyset.
\end{aligned}
$$

The mapping *next* controls the work of γ in such a way that, after starting the simulation of a rule $p \in P_2$ by using the splicing rule $s_{1,p}$, we have to continue with the splicing rule $s_{2,p}$, and after removing a symbol α from the right hand end of w in a string XwY, we have to continue by introducing α in the left hand end of w.

Because rules $r_{2,\alpha}, \alpha \in N \cup T \cup \{B\}$, and t_1 cannot be used in the presence of X' and X' can be removed only by s_0, which can be used only in the first splicing step, we have to start with s_0. This prevents starting with rules $r_{2,\alpha}$, which would produce illegal strings.

In this way, each sequence of splicings in γ precisely corresponds to a derivation in G. Consequently, $L(G) = L(\gamma)$. □

Although elegant from a mathematical point of view, the notion of a programmed H system contains the non-biochemical ingredient of the *next* mapping, defined independently of the current strings, in a sort of a "total" manner: from one step to another the whole set of available splicing rules is changed, simply by replacing them with new rules. In the spirit of evolving H systems with the activity defined by two mappings *String* and *Rule*, we now look for a local manner of changing the rules from one step to the next. One way of doing this is to consider *point mutations*, insertion and deletion operations of single symbols, or substitutions of a symbol by another one. We consider such operations with context-dependence: a symbol is inserted in or deleted from a specified context. To this aim we consider insertion-deletion rules as in Chap. 6, of the form $(u, \alpha/\beta, v)$, with u, v strings and α, β symbols or λ, telling us that α can be replaced by β in the context (u, v); α/λ means deletion, λ/β means insertion, a/b means changing a for b. As we shall see below (and as expected from the results in Chap. 6), insertion and deletion rules suffice; still, we define our locally evolving H systems in the general way.

An *extended H system with locally evolving splicing rules* is a construct

$$\gamma = (V, T, A_0, A_c, E, C_0, P),$$

where

(i) V is the total alphabet of γ,

(ii) $T \subseteq V$ is the terminal alphabet,

(iii) $A_0 \subseteq V^*$ is the finite set of starting axioms,

(iv) $A_c \subseteq V^*$ is the finite set of current axioms,

(v) E is an alphabet, $E \cap V \neq \emptyset$,

(vi) C_0 is an initial sequence of splicing rules, $C_0 = (r_1, \ldots, r_k)$, $r_i \in E^*\#E^*\$E^*\#E^*$, $1 \leq i \leq k$,

(vii) P is a finite set of editing rules of the form $(u, \alpha/\beta, v)$, with $u, v \in (E \cup \{\#, \$\})^*$, and $\alpha, \beta \in E \cup \{\lambda\}, \alpha \neq \beta$.

The rules in P are used for editing the splicing rules, starting from the "template" rules in C_0. When $\alpha \in E, \beta \in E$, we have a substitution, when $\alpha \in E, \beta = \lambda$, we have a deletion, and when $\alpha = \lambda, \beta \in E$, we have an insertion. Note that the special symbols $\#, \$$ cannot be edited.

The idea is to use the components E, C_0, P in order to produce splicing rules. We pass from one step to the next by using one rule in P for each currently available splicing rule. If any of the splicing rules present at one time can be applied to a string produced at a previous step (initially, a string in A_0) and to a current axiom (a string in A_c), then it has to be used; otherwise, the strings present in the tube are not modified. When a splicing rule can be applied to a couple of strings, we assume that all copies of those strings are used and consumed, hence they are no longer present for the next steps – with the exception of the current axioms, which are supposed to be unexhaustible (new copies of them are added whenever they are necessary).

Formally, we are led to the following definition.

Denote by $\Longrightarrow_P$ the usual derivation relation with respect to rules in P, written as rewriting rules $u\alpha v \to u\beta v$, for $(u, \alpha/\beta, v)$ in P. For a splicing rule $r \in E^*\#E^*\$E^*\#E^*$ we define

$$P(r) = \{r' \mid r \Longrightarrow_P r'\}.$$

We extend the relation $\Longrightarrow_P$ to k-tuples of splicing rules by

$$(r_1, \ldots, r_k) \Longrightarrow_P (r'_1, \ldots, r'_k) \text{ iff } r'_j \in P(r_j), 1 \leq j \leq k.$$

Starting from C_0, at the time $i \geq 1$ we can obtain in this manner a sequence $C_i = (r_{i,1}, \ldots, r_{i,k})$; we associate to it the set of splicing rules

$$R_i = \{r \mid r = r_{i,j} \text{ for some } 1 \leq j \leq k\}.$$

Note that the set R_i contains exactly one descendant of every rule in C_0; out of the possible variants which can be obtained due to the possible nondeterminism of using the rules in P, only one is actually chosen.

Consider now the "genome evolution." We define the sets $A_i, i \geq 0$, as follows. The initial set A_0 is given.

For $x \in V^*$ and a given set R of splicing rules and for $i \geq 0$, we define

$$\delta_i(x, R) = \begin{cases} 1, & \text{if there is } y \in A_i \cup A_c \text{ and } r \in R \text{ such that} \\ & (x, y) \models_r (w, z) \text{ or } (y, x) \models_r (w, z), \text{ for some } w, z \in V^*, \\ 0, & \text{otherwise.} \end{cases}$$

Moreover, we define

$$R_i(A_i) = \{w \in V^* \mid (x, y) \models_r (w, z) \text{ or } (x, y) \models_r (z, w), \text{ for } r \in R_i, x, y \in A_i \cup A_c, \{x, y\} \cap A_i \neq \emptyset\}, \ i \geq 0.$$

Then

$$A_i = \{x \in A_{i-1} \mid \delta_{i-1}(x, R_{i-1}) = 0\} \cup R_{i-1}(A_{i-1}), \ i \geq 1.$$

(As at the beginning of this section, we can say that $R_i = Rule(R_{i-1}, A_{i-1})$ and $A_i = String(R_{i-1}, A_{i-1})$.)

In words, A_i consists of all strings in A_{i-1} which cannot enter a splicing with another string in A_{i-1} or in A_c according to a rule in the current set R_{i-1}, plus all the strings obtained by such splicing operations. Note that a string already used in a splicing operation does not survive, it is no longer present in the next set A_i. When no currently available splicing rule can be applied to the strings in the current set A_i (and to the axioms in A_c), then all strings in A_i will pass unchanged to the next step, hence $A_{i+1} = A_i$.

The language generated by γ is defined by

$$L(\gamma) = (\bigcup_{i \geq 0} A_i) \cap T^*.$$

In Chap. 6 (Theorem 6.2) we have seen that we can characterize RE by using insertion rules of the form $(u, \lambda/\alpha, v)$ with $|u|, |v| \leq 2$, and deletion rules of the form $(u, \lambda/\alpha, v)$ with $|u|, |v| \leq 1$, α being a symbol. Combining this result with that in Lemma 7.16 (carefully arranging the construction in such a way to meet the conditions in the work of an extended H system with locally evolving splicing rules as above), we get a characterization of RE in the new framework. In this way we use splicing rules of an arbitrary length, which is bad from a practical point of view.

Fortunately, a characterization of RE holds also true for a rather particular type of H systems with locally evolving rules.

An extended H system with locally evolving splicing rules $\gamma = (V, T, A_0, A_c, E, C_0, P)$ is said to be *restricted* if $card(A_0) = 1, card(C_0) = 1$.

Thus, we have exactly one splicing rule at each time.

Let us denote by $EH_2(FIN, rle[m]), m \geq 1$, the family of languages $L(\gamma)$, generated by restricted locally evolving H systems as above with splicing rules of radius at most $m, m \geq 1$.

Theorem 8.6. $EH_2(FIN, rle[4]) = RE$.

Proof. We have to prove only the inclusion $RE \subseteq EH_2(FIN, rle[4])$.

Consider a type-0 grammar $G = (N, T, S, P_0)$ in Kuroda normal form, that is with rules of the forms

1. $AB \to CD$, for $A, B, C, D \in N$,
2. $A \to BC$, for $A, B, C \in N$,
3. $A \to a$, for $A \in N, a \in T$,
4. $A \to \lambda$, for $A \in N$.

As usual, we assume all the rules in P_0 labeled in a one-to-one manner.

We construct the restricted extended H system with locally evolving splicing rules

$$\gamma = (V, T, A_0, A_c, E, C_0, P),$$

with

$$\begin{aligned} V &= N \cup T \cup \{X, Y, Z, B_0\}, \\ A_0 &= \{XB_0SY\}, \\ A_c &= \{ZY\} \cup \{ZvY \mid u \to v \in P_0\} \\ &\cup \{X\alpha Z \mid \alpha \in N \cup T \cup \{B_0\}\}, \\ E &= N \cup T \\ &\cup \{X, Y, Z, B_0, c_1, c_2, d_1, d_2, d_3, e_1, e_2, f_1, f_2, g_1, g_2, h_1, h_2\} \\ &\cup \{[r,1], [r,2] \mid r \in P_0\}, \\ C_0 &= (c_1\#Y\$Z\#), \end{aligned}$$

and the set P containing the following point mutation rules; in order to check the correctness of the construction, we present these rules together with the current configuration set R_i, consisting of exactly one splicing rule (which however is not always deterministically produced).

1. Producing splicing rules for *simulating* rewriting rules in P_0:

A. For each rule $r : AB \to CD \in P_0$ we consider the following insertion-deletion rules:

0.	$-------$	$c_1\#Y\$Z\#$,
1.	$(c_1\#, \lambda/[r,1], Y)$,	$c_1\#[r,1]Y\$Z\#$,
2.	$(\lambda, c_1/\lambda, \#[r,1])$,	$\#[r,1]Y\$Z\#$,
3.	$(\#, \lambda/A, [r,1])$,	$\#A[r,1]Y\$Z\#$,
4.	$(\#A, \lambda/B, [r,1])$,	$\#AB[r,1]Y\$Z\#$,
5.	$(AB[r,1]Y\$Z\#, \lambda/[r,2], \lambda)$,	$\#AB[r,1]Y\$Z\#[r,2]$,
6.	$(\lambda, [r,1]/\lambda, Y\$Z\#[r,2])$,	$\#ABY\$Z\#[r,2]$,
7.	$(\#, \lambda/C, [r,2])$,	$\#ABY\$Z\#C[r,2]$,
8.	$(\#C, \lambda/D, [r,2])$,	$\#ABY\$Z\#CD[r,2]$,
9.	$(CD, \lambda/Y, [r,2])$,	$\#ABY\$Z\#CDY[r,2]$,
10.	$(Y, [r,2]/\lambda, \lambda)$,	$\#ABY\$Z\#CDY$.

The splicing rule obtained is the first one in the sequence above that can be applied to strings of the form XwY and to axioms in A_c (the symbols $c_1, [r,1], [r,2]$ do not appear in V, hence in strings produced by splicing from A_0 and A_c). Note that the insertion rule 5 can be repeated, introducing several copies of the symbol $[r,2]$, hence producing strings of the form $\#AB[r,1]Y\$\#[r,2]^k$, with $k \geq 2$, but only the leftmost occurrence of the symbol $[r,2]$ will introduce symbols C, D and Y; then, in the presence of Y, all symbols $[r,2]$ are removed (otherwise the currently produced splicing rule cannot be applied).

B. For each rule $r : A \to BC \in P_0$ we consider the following mutation rules:

$$
\begin{array}{lll}
0. & - - - - - - - - & c_1\#Y\$Z\#, \\
1. & (c_1\#, \lambda/[r,1], Y), & c_1\#[r,1]Y\$Z\#, \\
2. & (\lambda, c_1/\lambda, \#[r,1]), & \#[r,1]Y\$Z\#, \\
3. & (\#, \lambda/A, [r,1]), & \#A[r,1]Y\$Z\#, \\
4. & (A[r,1]Y\$Z\#, \lambda/[r,2], \lambda), & \#A[r,1]Y\$Z\#[r,2], \\
5. & (\lambda, [r,1]/\lambda, Y\$Z\#[r,2]), & \#AY\$Z\#[r,2], \\
6. & (\#, \lambda/B, [r,2]), & \#AY\$Z\#B[r,2], \\
7. & (\#B, \lambda/C, [r,2]), & \#AY\$Z\#BC[r,2], \\
8. & (BC, \lambda/Y, [r,2]), & \#AY\$Z\#BCY[r,2], \\
9. & (Y, [r,2]/\lambda, \lambda), & \#AY\$Z\#BCY.
\end{array}
$$

C. For any rule of the type $r : A \to a \in P_0$ we introduce the mutation rules $1-5$ in group B above, as well as the following three insertion-deletion rules:

$$
\begin{array}{lll}
6. & (\#, \lambda/a, [r,2]), & \#AY\$Z\#a[r,2], \\
7. & (a, \lambda/Y, [r,2]), & \#AY\$Z\#aY[r,2], \\
8. & (Y, [r,2]/\lambda, \lambda), & \#AY\$Z\#aY.
\end{array}
$$

D. Finally, for any rule $r : A \to \lambda \in P_0$ we again introduce mutation rules $1-5$ from group B above, and we continue with the following mutation rules:

$$
\begin{array}{lll}
6. & (\#, \lambda/Y, [r,2]), & \#AY\$Z\#Y[r,2], \\
7. & (Y, [r,2]/\lambda, \lambda), & \#AY\$Z\#Y.
\end{array}
$$

In all cases, only in the last step (after using the rule B9, C8, and D7, respectively) can we obtain a splicing rule containing no symbol $c_1, [r,1], [r,2]$, hence applicable to strings obtained from A_0, A_c by splicing.

Denote, in general, the corresponding rule in P_0 by $u \to v$. The obtained splicing rule is $\#uY\$Z\#vY$, hence we can splice

$$(Xw|uY, Z|vY) \models (XwvY, ZuY).$$

The string $XwvY$ (corresponding to the simulation of the rule $u \to v$) will be used at a subsequent splicing step, the first one when a splicing rule is applicable to it. If ZuY is also used, then it will never produce a terminal string, because the symbol Z cannot be eliminated (we will see this below).

After using a rule $\#uY\$Z\#vY$, we continue to edit it, returning to the template rule $c_1\#Y\$Z\#$. This is done by using the following point mutation rules:

$$
\begin{array}{lll}
A'. & 0. & -------- \qquad \#ABY\$Z\#CDY, \\
 & 1. & (\$Z\#CD, \lambda/e_1,)Y, \quad \#ABY\$Z\#CDe_1Y, \\
 & 2. & (e_1, Y/\lambda, \lambda), \quad \#ABY\$Z\#CDe_1, \\
 & 3. & (\lambda, \alpha/\lambda, e_1), \quad \#ABY\$Z\#e_1, \ \alpha \in N \cup T, \\
 & 4. & (\lambda, \lambda/e_2, Y\$Z\#e_1), \quad \#ABe_2Y\$Z\#e_1, \\
 & 5. & (e_2Y\$Z\#, e_1/\lambda, \lambda), \quad \#ABe_2Y\$Z\#, \\
 & 6. & (\lambda, \alpha/\lambda, e_2), \quad \#e_2Y\$Z\#, \ \alpha \in N, \\
 & 7. & (\lambda, \lambda/c_1, \#e_2), \quad c_1\#e_2Y\$Z\#, \\
 & 8. & (c_1\#, e_2/\lambda, \lambda), \quad c_1\#Y\$Z\#.
\end{array}
$$

We have returned to the splicing rule in C_0. The same happens in the case B, without any modification (with CD being the right hand member of the corresponding rule in P_0; rule 6 is used only once). For the case C we have to replace rule 1 above with

$$1'. \ (\$Z\#a, \lambda/e_1, Y), \ \#AY\$Z\#ae_1Y,$$

whereas in the case D we replace rule 1 above with

$$1''. \ (\$Z\#, \lambda/e_1, Y), \ \#AY\$Z\#e_1Y.$$

In all cases, the subsequent mutation rules 2 – 8 work in the same way, reproducing the splicing rule $c_1\#Y\$Z\#$; no intermediate splicing rule can be applied to the string produced at a previous step and to an axiom in A_c, due to the control symbols e_1, e_2.

2. Producing splicing rules for *rotating* the current string:

E. For each symbol $\alpha \in N \cup T \cup \{B_0\}$ we consider the following mutation rules (for producing a splicing rule which cuts the symbol α from the right end of the string):

$$
\begin{array}{lll}
0. & -------- & c_1\#Y\$Z\#, \\
1. & (c_1\#, \lambda/d_1, Y), & c_1\#d_1Y\$Z\#, \\
2. & (\lambda, c_1/\lambda, \#d_1), & \#d_1Y\$Z\#, \\
3. & (\#, \lambda/\alpha, d_1), & \#\alpha d_1Y\$Z\#, \\
4. & (\alpha d_1Y\$Z\#, \lambda/c_2, \lambda), & \#\alpha d_1Y\$Z\#c_2, \\
5. & (\lambda, d_1/\lambda, Y\$Z\#c_2), & \#\alpha Y\$Z\#c_2, \\
6. & (\#, \lambda/Y, c_2), & \#\alpha Y\$Z\#Yc_2, \\
7. & (Y, c_2/\lambda, \lambda), & \#\alpha Y\$Z\#Y.
\end{array}
$$

The obtained splicing rule contains no control symbol c_1, d_1, c_2, hence it can be used for cutting α from the right hand end of the current string:

$$(Xw|\alpha Y, Z|Y) \models (XwY, Z\alpha Y).$$

The string XwY will be spliced at the first subsequent step when a splicing rule can be applied, while $Z\alpha Y$ will never lead to a terminal string.

F. For each symbol $\alpha \in N \cup T \cup \{B_0\}$ we also consider the following mutation rules (for producing a splicing rule which introduces the symbol α in the left hand end of the string):

0.	$- - - - - - - -$	$\#\alpha Y\$Z\#Y,$
1.	$(Z, \lambda/d_2, \#Y),$	$\#\alpha Y\$Zd_2\#Y,$
2.	$(d_2\#, Y/\lambda, \lambda),$	$\#\alpha Y\$Zd_2\#,$
3.	$(\lambda, Z/\lambda, d_2),$	$\#\alpha Y\$d_2\#,$
4.	$(\lambda, \lambda/d_3, Y\$d_2),$	$\#\alpha d_3Y\$d_2\#,$
5.	$(d_3Y\$, \lambda/X, d_2),$	$\#\alpha d_3Y\$Xd_2\#,$
6.	$(\$X, d_2/\lambda, \lambda),$	$\#\alpha d_3Y\$X\#,$
7.	$(d_3Y, \lambda/Z, \$),$	$\#\alpha d_3YZ\$X\#,$
8.	$(d_3, Y/\lambda, Z\$),$	$\#\alpha d_3Z\$X\#,$
9.	$(\lambda, \lambda/X, \#\alpha d_3),$	$X\#\alpha d_3Z\$X\#,$
10.	$(X, \lambda/\alpha, \#\alpha d_3),$	$X\alpha\#\alpha d_3Z\$X\#,$
11.	$(X\alpha\#, \alpha/\lambda, d_3),$	$X\alpha\#d_3Z\$X\#,$
12.	$(X\alpha\#, d_3/\lambda, Z),$	$X\alpha\#Z\$X\#.$

The obtained splicing rule can be applied:

$$(X\alpha|Z, X|wY) \models (X\alpha wY, XZ),$$

hence the string has been circularly permuted with one symbol. (Note that because of the way γ works, transforming the splicing rule from a step to the next one, we no longer need control symbols like $X', Y_\alpha, \alpha \in N \cup T \cup \{B_0\}$, as in the proof of Lemma 8.2.)

G. (Return to the template splicing rule $c_1\#Y\$Z\#$; here $\alpha \in N \cup T \cup \{B_0\}$):

0.	$- - - - - - - -$	$X\alpha\#Z\$X\#,$
1.	$(\alpha\#Z\$, \lambda/f_1, \lambda),$	$X\alpha\#Z\$f_1X\#,$
2.	$(f_1, X/\lambda, \#),$	$X\alpha\#Z\$f_1\#,$
3.	$(f_1, \lambda/Z, \#),$	$X\alpha\#Z\$f_1Z\#,$
4.	$(Z, \lambda/f_2, \$f_1Z),$	$X\alpha\#Zf_2\$f_1Z\#,$
5.	$(f_2\$, f_1/\lambda, \lambda),$	$X\alpha\#Zf_2\$Z\#,$
6.	$(\#, Z/\lambda, f_2),$	$X\alpha\#f_2\$Z\#,$
7.	$(\#f_2, \lambda/Y, \$),$	$X\alpha\#f_2Y\$Z\#,$
8.	$(\lambda, \lambda/c_1, X\alpha\#f_2Y),$	$c_1X\alpha\#f_2Y\$Z\#,$
9.	$(c_1X, \alpha/\lambda, \#f_2),$	$c_1X\#f_2Y\$Z\#,$
10.	$(c_1, X/\lambda, \#f_2),$	$c_1\#f_2Y\$Z\#,$
11.	$(c_1\#, f_2/\lambda, \lambda),$	$c_1\#Y\$Z\#.$

We have not only returned to $c_1\#Y\$Z\#$, but this is also the first time when the rules A1, B1, C1, D1, E1 (having the first hand member $c_1\#Y$)

can be applied again. Thus, we can continue by any of these rules, either simulating again a rule in P_0 or moving one more symbol α from the right hand end of the string bounded by X, Y to its left hand end.

3. *Finishing* the work of the system:

H. (removing the prefix XB_0):

$$
\begin{array}{rll}
0. & -------- & c_1\#Y\$Z\#, \\
1. & (c_1\#, \lambda/g_1, Y), & c_1\#g_1Y\$Z\#, \\
2. & (c_1, \lambda/X, \#g_1), & c_1X\#g_1Y\$Z\#, \\
3. & (\lambda, c_1/\lambda, X\#g_1), & X\#g_1Y\$Z\#, \\
4. & (X, \lambda/B_0, \#g_1), & XB_0\#g_1Y\$Z\#, \\
5. & (g_1, Y/\lambda, \$), & XB_0\#g_1\$Z\#, \\
6. & (g_1\$Z, \lambda/g_2, \#), & XB_0\#g_1\$Zg_2\#, \\
7. & (\lambda, g_1/\lambda, \$Zg_2), & XB_0\#\$Zg_2\#, \\
8. & (\$, Z/\lambda, g_2\#), & XB_0\#\$g_2\#, \\
9. & (\$g_2\#, \lambda/Y, \lambda), & XB_0\#\$g_2\#Y, \\
10. & (g_2\#, \lambda/Z, Y), & XB_0\#\$g_2\#ZY, \\
11. & (\lambda, g_2/\lambda, \#ZY), & XB_0\#\$\#ZY.
\end{array}
$$

Only the last splicing rule can be applied:

$$(XB_0|wY, |ZY) \models (XB_0ZY, wY).$$

J. (removing the end marker Y):

$$
\begin{array}{rll}
0. & -------- & XB_0\#\$\#ZY, \\
1. & (\$, \lambda/h_1, \#ZY), & XB_0\#\$h_1\#ZY, \\
2. & (h_1\#Z, Y/\lambda, \lambda), & XB_0\#\$h_1\#Z, \\
3. & (h_1\#, Z/\lambda, \lambda), & XB_0\#\$h_1\#, \\
4. & (h_1, \lambda/Y, \#), & XB_0\#\$h_1Y\#, \\
5. & (h_1, \lambda/Z, Y\#), & XB_0\#\$h_1ZY\#, \\
6. & (\lambda, \lambda/h_2, \#\$h_1ZY), & XB_0h_2\#\$h_1ZY\#, \\
7. & (h_2\#\$, h_1/\lambda, \lambda), & XB_0h_2\#\$ZY\#, \\
8. & (\lambda, B_0/\lambda, h_2), & Xh_2\#\$ZY\#, \\
9. & (\lambda, X/\lambda, h_2), & h_2\#\$ZY\#, \\
10. & (h_2\#, \lambda/Y, \$), & h_2\#Y\$ZY\#, \\
11. & (\lambda, h_2/\lambda, \#Y\$), & \#Y\$ZY\#.
\end{array}
$$

Thus we can splice

$$(w|Y, ZY|) \models (w, ZYY).$$

If the string w is terminal, then it is in $L(G)$; if not, then no further splicing can be applied to it, because the end markers X, Y are no longer present and no further splicing rules will be produced from now on. It is easy to see that using the mutation rules in "wrong" ways (for instance, using rule 11 in the previous group before using rules 8, 9, hence producing a splicing rule $\alpha\#Y\$ZY\#$, with $\alpha \in \{X, B_0, XB_0\}$), will not produce a terminal string (in the example above, the current string either contains no occurrence of X and B_0, because rule $XB_0\#\$\#ZY$ has been applied, or, if such a symbol appears, then no further splicing is done, hence the string is either XB_0Y, so we get λ, or X remains unchanged, so no terminal string is obtained). The reader can easily check such variants. The rule $\#Y\$ZY\#$ cannot be modified, the work of γ stops here.

Consequently, we get $L(\gamma) = L(G)$.

The longest strings u_1, u_2, u_3, u_4 in splicing rules $u_1\#u_2\$u_3\#u_4$ produced as above appear in rules obtained at steps A4, A5, A9, B8, A′4, A′5, and this length is four. In conclusion, $RE \subseteq EH_2(FIN, rle[4])$. □

Note in the previous construction that we have used only insertion and deletion rules (always of one symbol only) and no rule of the form $(u, \alpha/\beta, v)$, for $\alpha \neq \lambda \neq \beta$, is involved. Using the latter rules in addition, the construction can be slightly simplified.

It is also highly probable that the radius of the system can be decreased, to at most two, as it happens in all previous cases. This, however, will make the construction above still more complex, hence we do not continue here in this direction.

No attention has been paid in this construction to the length of contexts in insertion-deletion rules. For instance, the longest rules here are those in group A5, of the form $(u, \lambda/\alpha, v)$ with $v = \lambda$ and $|u| = 7$. Of course, this parameter can be improved; it is highly probable that rules of the form $(u, \alpha/\beta, v)$, with $|u| \leq 2, |v| \leq 2$, are sufficient.

On the other hand, the fact that the strings in A_c are always present can be arranged in an "internal" way, at least in the case of the system in the proof of the theorem above: Each string in A_c is of one of the forms Zx, xZ. Considering the splicing rules $Z\#x\$Z\#x$ and $x\#Z\$x\#Z$, these strings are passed from one step to the next unmodified, reproduced by splicings. Now, splicing rules as above can be permanently produced as follows. Consider the case of $Z\#x\$Z\#x$. If we have in C_0 the rules

$$Z\#x\$Z\#x, \; Z\#x\$c_3Z\#x,$$

and we also consider the mutation rules

$$(\$, \lambda/c_3, Z), \; (\$, c_3/\lambda, Z),$$

then at each step the two splicing rules above are reproduced, hence $Z\#x\$Z\#x$ is always present. The same can be done for rules of the form $x\#Z\$x\#Z$.

Considering further variants of evolving H systems, with the rules modification also depending on the currently available (or spliced) strings, remains a *research topic*.

8.5 H Systems Based on Double Splicing

We now consider a class of H systems which can be viewed as a counterpart of the matrix grammars in the regulated rewriting area. However, we do not have here sequences of splicing rules specified in advance, but we only ask that the work of an H system proceeds in a couple of steps: the two strings obtained after a splicing immediately enter a second splicing. The rules used in the two steps are not prescribed or dependent in any way to each other; also, the two output strings of a double splicing step are not related to the two input strings of a later double splicing step.

Consider a usual extended H system $\gamma = (V, T, A, R)$ with finite sets A and R. For $x, y, w, z \in V^*$ and $r_1, r_2 \in R$ we write

$$(x, y) \models_{r_1, r_2} (w, z) \text{ iff } (x, y) \models_{r_1} (u, v) \text{ and } (u, v) \models_{r_2} (w, z), \text{ for } u, v \in V^*.$$

For a language $L \subseteq V^*$ we define

$$\begin{aligned}\sigma_d(L) &= \{w \mid (x, y) \models_{r_1, r_2} (w, z) \text{ or } (x, y) \models_{r_1, r_2} (z, w),\\ &\qquad \text{for } x, y \in L, r_1, r_2 \in R\},\\ \sigma_d^*(L) &= \bigcup_{i \geq 0} \sigma_d^i(L), \text{ where}\\ \sigma_d^0(L) &= L,\\ \sigma_d^{i+1}(L) &= \sigma_d^i(L) \cup \sigma_d(\sigma_d^i(L)),\ i \geq 0.\end{aligned}$$

Then, we associate with γ the language

$$L_d(\gamma) = \sigma_d^*(A) \cap T^*.$$

By $EH_2(FIN, d[k])$ we denote the family of languages $L_d(\gamma)$ generated as above by extended H systems $\gamma = (V, T, A, R)$ of radius at most $k, k \geq 1$.

Let us examine an **example**: consider the extended H system

$$\gamma = (\{a, b, c, d, e\}, \{a, b, c, d\}, \{cabd, caebd\}, R),$$

with R containing the splicing rules

$$r_1 = c\#a\$ca\#ebd,\ r_2 = ce\#bd\$b\#d.$$

Take a string of the form $ca^nb^nd, n \geq 1$; one of the axioms is of this form, with $n = 1$. The only possible splicing involving this string is

$$(c|a^nb^nd, ca|ebd) \models_{r_1} (cebd, ca^{n+1}b^nd).$$

In the sense of the double splicing operation, we have to continue; the only possibility is

$$(ce|bd, ca^{n+1}b^n|d) \models_{r_2} (ced, ca^{n+1}b^{n+1}d).$$

Consequently, we have

$$(ca^n b^n d, caebd) \models_{r_1, r_2} (ced, ca^{n+1} b^{n+1} d).$$

The operation can be iterated.

Another possibility is to start with two copies of the axiom *caebd*:

$$(c|aebd, ca|ebd) \models_{r_1} (ce|bd, caaeb|d) \models_{r_2} (ced, caaebbd).$$

We can continue, but the symbol e will be present in all obtained strings; these strings cannot enter splicings with strings of the form $ca^n b^n d$, hence they do not lead to terminal strings.

In conclusion, we obtain

$$L_d(\gamma) = \{ca^n b^n d \mid n \geq 1\},$$

which is not a regular language. Consequently, the double splicing is strictly more powerful than the simple one. This assertion will be stressed below in the strongest possible way: extended H systems using the double splicing operation are equal in power to type-0 grammars.

Theorem 8.7. $RE = EH_2(FIN, d[2])$.

Proof. We prove only the inclusion $\subseteq$. The reverse inclusion can be proved by a straightforward construction of a type-0 grammar simulating an extended H system based on the double splicing operation (or we can invoke the Church–Turing Thesis).

The proof consists of two phases.

(1) Consider a grammar $G = (\{S, B_1, B_2, B_3, B_4\}, T, S, P \cup \{B_1 B_2 \to \lambda, B_3 B_4 \to \lambda\})$ in the Geffert normal form given in Theorem 3.5(2), that is, with P containing rules of the forms $S \to uSv, S \to x$, with $u, v, x \in (T \cup \{B_1, B_2, B_3, B_4\})^+$. We construct the extended H system $\gamma = (V, T, A, R)$ with:

$$\begin{aligned}
V &= T \cup \{S, B_1, B_2, B_3, B_4, X, Y, Z, Z'\}, \\
A &= \{SxS \mid S \to x \in P, x \in (T \cup \{B_1, B_2, B_3, B_4\})^*\} \\
&\quad \cup \{SuZvS \mid S \to uSv \in P\} \\
&\quad \cup \{Z', XY\}, \\
R &= \{S\#\$Su\#ZvS,\ SZ\#vS\$\#S \mid S \to uSv \in P\} \\
&\quad \cup \{S\#\$\#Z',\ SZ'\#\$\#S\} \\
&\quad \cup \{B_1\#B_2\$X\#Y,\ \#B_1Y\$XB_2\#\} \\
&\quad \cup \{B_3\#B_4\$X\#Y,\ \#B_3Y\$XB_4\#\}.
\end{aligned}$$

The idea of this construction is as follows. The splicing rules of the forms $S\#\$Su\#ZvS$, $SZ\#vS\$\#S$ simulate the context-free rules $S \to uSv$ in P, while the splicing rules $B_1\#B_2\$X\#Y$, $\#B_1Y\$XB_2\#$, $B_3\#B_4\$X\#Y$,

$\#B_3Y\$XB_4\#$ simulate the rules $B_1B_2 \to \lambda, B_3B_4 \to \lambda$, respectively; the terminal rules of G are simulated by the axioms SxS in A. The context-free derivations in G are simulated in γ in the reverse order, starting from the center of the produced string (from the substring introduced by a rule $S \to x$) towards the ends.

For instance, assume that we have a string of the form SwS with $w \in (T \cup \{B_1, B_2, B_3, B_4\})^*$; the axioms SxS are of this type. If we apply a splicing rule $r_1 = S\#\$Su\#ZvS$, associated with some rule $S \to uSv \in P$, then we get

$$(S|wS, Su|ZvS) \models_{r_1} (SZvS, SuwS).$$

We have to continue; because no symbol X, Y, Z' is present, the only possibility is to use the rule $r_2 = SZ\#vS\$\#S$ associated with the same rule $S \to uSv \in P$:

$$(SZ|vS, Suw|S) \models_{r_2} (SZS, SuwvS).$$

The double splicing

$$(SwS, SuZvS) \models_{r_1,r_2} (SZS, SuwvS)$$

has simulated the use of the rule $S \to uSv$ in the reverse order.

(The reader might check that starting with $(SwS|, Su|ZvS) \models_{r_1} (SwSZ|vS, |Su) \models_{r_2} (SwSZSu, vS)$ does not lead to terminal strings.)

If to a string SwS we apply the rule $r_1 = S\#\$\#Z'$, then we have to continue with the rule $r_2 = SZ'\#\$\#S$:

$$(S|wS, |Z') \models_{r_1} (SZ'|, w|S) \models_{r_2} (SZ'S, w).$$

The occurrences of S from the ends of the string are removed (this means that from now on no further rule of the form $S \to uSv \in P$ can be simulated in γ starting from the string w).

If to a string w, bounded or not by occurrences of S, we apply the splicing rule $r_1 = B_1\#B_2\$X\#Y$ (providing that a substring B_1B_2 appears in w, that is, $w = xB_1B_2y$), then we have to continue with the rule $r_2 = \#B_1Y\$XB_2\#$ (no other rule is applicable to the intermediate strings), hence we get:

$$(xB_1|B_2y, X|Y) \models_{r_1} (x|B_1Y, XB_2|y) \models_{r_2} (xy, XB_2B_1Y).$$

The occurrence of B_1B_2 specified above is removed from the input string.

The same assertions are true if we first apply the rule $B_3\#B_4\$X\#Y$; an occurrence of the substring B_3B_4 is removed.

The strings $SZS, SZ'S$ cannot enter splicings leading to terminal strings and this can be easily seen. If a string XB_2B_1Y, XB_4B_3Y enters new splicings, they produce nothing new. For instance, for $r = \#B_1Y\$XB_2\#$ we get:

$$(XB_2|B_1Y, XB_2|B_1Y) \models_r (XB_2|B_1Y, XB_2|B_1Y) \models_r (XB_2B_1Y, XB_2B_1Y).$$

No double splicing of a type different from those discussed above can lead to terminal strings. Consequently, the double splicing operations in γ correspond to using context-free rules in P, to removing the two occurrences of S from the ends of a string, or to using the erasing rules $B_1B_2 \rightarrow \lambda$, $B_3B_4 \rightarrow \lambda$. The order of using these rules is irrelevant. Consequently, $L(G) = L_d(\gamma)$.

(2) In the previous construction we can modify the "linear" rules $S \rightarrow uSv$ of P, replacing them by rules of the forms $D \rightarrow \alpha E\beta$, where $\alpha, \beta \in T \cup \{B_1, B_2, B_3, B_4\}$ and $|\alpha\beta| = 1$, in such a way that we obtain a grammar which is equivalent with G, but contains only rules with the right hand side of length two; moreover, we may assume that all rules $D \rightarrow \alpha E\beta$ have $D \neq E$; the nonterminal alphabet is now bigger, new symbols are used.

A linear grammar with several nonterminal symbols can be simulated by an extended H system using double splicing operations in a way similar to the way we have simulated the context-free rules of the grammar G in the previous construction.

Specifically, consider a linear grammar $G = (N, T, S, P)$ and construct the extended H system $\gamma = (V, T, A, R)$ with

$$\begin{aligned}
V &= N \cup T \cup \{Z, Z'\}, \\
A &= \{DxD \mid D \rightarrow x \in P, x \in T^*\} \\
&\quad \cup \{D\alpha Z\beta D \mid D \rightarrow \alpha E\beta \in P, \text{ where } D, E \in N, \alpha, \beta \in T \cup \{\lambda\}\} \\
&\quad \cup \{Z'\}, \\
R &= \{E\#\$D\alpha\#Z\beta,\ EZ\#\beta D\$\#E \mid D \rightarrow \alpha E\beta \in P, \\
&\qquad D, E \in N, \alpha, \beta \in T \cup \{\lambda\}\} \\
&\quad \cup \{S\#\$\#Z',\ SZ'\#\$\#S\}.
\end{aligned}$$

The reader can easily check that the derivations in G are simulated in γ in the reverse order, starting from strings DxD associated to terminal rules $D \rightarrow x$ and going back to a string of the form SzS, when the symbols S can be eliminated. Therefore, $L(G) = L_d(\gamma)$. Clearly, $rad(\gamma) = 2$.

Combining this idea with the manner of simulating erasing rules of the form $B_iB_j \rightarrow \lambda$ (note that the splicing rules associated with these rules are of radius one), we get an extended H system of radius two. □

8.6 Multisets

In the definition of splicing operations (of both types $\vdash$ and $\models$) used in the previous sections, after splicing two strings x, y and obtaining (in the case of $\models$) two possibly new strings z, w, we may use again x or y as a term of a splicing, these strings are not consumed by splicing; moreover, we may splice x or y with z or w, hence we may splice strings from one "generation" with strings from another "generation". Also the new strings, z and w, are supposed to appear in an arbitrary number of copies each.

This assumption, that if a string is available then arbitrarily many copies of it are available, is realistic in the sense that, usually, a large number of copies of each string are used whenever a string is used. Moreover, producing a large number of copies of a DNA sequence is easily feasible by amplification through polymerase chain reaction (PCR) techniques. This also reduces complexity: the computation can run in parallel, on a large number of string-processors.

However, the existence of several copies of each string raises the difficult problem of controlling the splicing so as to prevent "wrong" operations. For instance, after cutting several copies of a string x into fragments x_1, x_2 and modifying (part of the copies of) x_1, x_2 to some x_1', x_2', the test tube will contain strings of all four forms, x_1, x_2, x_1', x_2'; it might be possible to recombine x_1 with x_2' or x_1' with x_2 in such a way as to obtain illegal strings which "look like" legal strings x_1x_2 or $x_1'x_2'$.

A possibility to avoid this difficulty is to use at least some of the strings in a specified number of copies, and to keep track of these numbers during the work of the system. This leads us to consider *multisets*, i.e., sets with multiplicities associated with their elements.

Formally, a multiset over a set X of abstract elements is a mapping $M : X \longrightarrow \mathbf{N} \cup \{\infty\}$; $M(x)$ is the number of copies of $x \in X$ in the multiset M. When $M(x) = \infty$, then x is present in arbitrarily many copies. The set $\{x \in X \mid M(x) > 0\}$ is called the *support* of M and it is denoted by $supp(M)$. A usual set $S \subseteq X$ is interpreted as the multiset defined by $S(x) = 1$ for $x \in S$, and $S(x) = 0$ for $x \notin S$.

For two multisets M_1, M_2 over X we define their *union* by $(M_1 \cup M_2)(x) = M_1(x) + M_2(x)$, and their *difference* by $(M_1 - M_2)(x) = M_1(x) - M_2(x)$ for $x \in X$ such that $M_1(x) \geq M_2(x)$ and both $M_1(x), M_2(x)$ are finite, and $(M_1 - M_2)(x) = \infty$ for $x \in X$ such that $M_1(x) = \infty$; for other strings $x \in X$ the difference $M_1 - M_2$ is not defined. Usually, a multiset with a finite support, M, is presented as a set of pairs $(x, M(x))$, for $x \in supp(M)$.

For instance, $M_1 = \{(ab, 3), (abb, 1), (aa, \infty)\}$ is a multiset over $\{a, b\}^*$ with the support consisting of three words, ab, abb, aa; the first one appears in three copies, the second one appears in only one copy, whereas aa appears in an arbitrary number of copies. If we also take $M_2 = \{(ab, 1), (abb, 1), (aa, 17)\}$, then the difference $M_1 - M_2$ is defined and it is equal to $\{(ab, 2), (aa, \infty)\}$.

An *extended μH system* is a quadruple

$$\gamma = (V, T, A, R),$$

where V is an alphabet, $T \subseteq V$ (the terminal alphabet), A is a multiset over V^+ with $supp(A)$ finite (axioms), and R is a finite set of splicing rules over V.

For such a μH system and two multisets M_1, M_2 over V^* we define

$$\begin{array}{lll} M_1 \Longrightarrow_\gamma M_2 & \text{iff} & \text{there are } x, y, z, w \in V^* \text{ such that} \\ & \text{(i)} & M_1(x) \geq 1,\ (M_1 - \{(x,1)\})(y) \geq 1, \\ & \text{(ii)} & x = x_1u_1u_2x_2,\ y = y_1u_3u_4y_2, \\ & & z = x_1u_1u_4y_2,\ w = y_1u_3u_2x_2, \\ & & \text{for } x_1, x_2, y_1, y_2 \in V^*,\ u_1\#u_2\$u_3\#u_4 \in R, \\ & \text{(iii)} & M_2 = (((M_1 - \{(x,1)\}) - \{(y,1)\}) \\ & & \cup\{(z,1)\}) \cup \{(w,1)\}. \end{array}$$

At point (iii) we have operations with multisets. The writing above is meant to also cover the case when $x = y$ (then we must have $M_1(x) \geq 2$ and we must subtract 2 from $M_1(x)$), or $z = w$ (then we must add 2 to $M_2(z)$). When γ is understood, we write $\Longrightarrow$ instead of $\Longrightarrow_\gamma$.

In plain words, when passing from a multiset M_1 to a multiset M_2, according to γ, the multiplicity of two elements of M_1, x and y, is diminished by one, and the multiplicity of the resulting words, z and w, is augmented by one. The multiplicity of all other elements in $supp(M_1)$ is not changed. The obtained multiset is M_2.

The language generated by an extended μH system γ consists of all words containing only terminal symbols and whose multiplicity is at least once greater than or equal to one during the work of γ. Formally, we define this language by

$$L(\gamma) = \{w \in T^* \mid w \in supp(M) \text{ for some } M \text{ such that } A \Longrightarrow_\gamma^* M\}.$$

An extended H system $\gamma = (V, T, A, R)$, as defined in Sect. 7.4, can be interpreted as an extended μH system with $A(x) = \infty$ for all $x \in A$ and with $M(x) = \infty$ for all multisets M whose support is composed of strings x derived from A. Such multisets (with $M(x) = \infty$, if and only if $M(x) > 0$) are called ω-multisets, hence the corresponding H systems can be called ωH systems.

The family of languages generated by extended μH systems $\gamma = (V, T, A, R)$ with $card(supp(A)) \leq n$ and $rad(R) \leq m$, $n, m \geq 1$, is denoted by $EH_2(\mu[n], [m])$; when n or m are not bounded, then we replace $[n], [m]$ by FIN.

Similarly, we may write the families $EH_2(FL_1, FL_2)$ as $EH_2(\omega FL_1, FL_2)$ in order to stress the fact that we work with ω-multisets.

It is important to point out here the fact that writing $M(x) = \infty$ for a string in $supp(M)$ does not necessarily mean that we actually dispose of infinitely many copies of x. It only means that we do not count the number of copies of x: at any moment when we need a copy of x we have it. In the DNA framework, this means that when we need further copies of a given sequence, we can produce them (for instance, by amplification).

Using multisets, hence counting the number of occurrences (of some) of the strings used, provides once again the tools for controlling the work of H systems in such a way as to characterize the family RE.

We separate the proof of this assertion in several lemmas; the first one establishes the most important part of this result, the simulation of a type-0 grammar by an extended μH system.

Lemma 8.10. $RE \subseteq EH_2(\mu FIN, [5])$.

Proof. Consider a type-0 Chomsky grammar $G = (N, T, S, P)$, with the rules in P of the form $u \to v$ with $1 \leq |u| \leq 2$, $0 \leq |v| \leq 2$, $u \neq v$ (for instance, we can take G in the Kuroda normal form). Also assume that the rules in P are labeled in a one-to-one manner. By U we denote the set $N \cup T$ and we construct the extended μH system

$$\gamma = (V, T, A, R),$$

where

$$V = N \cup T \cup \{X_1, X_2, Y, Z_1, Z_2\} \cup \{(r), [r] \mid r \in P\},$$

the multiset A contains the word

$$w_0 = X_1^2 Y S X_2^2,$$

with the multiplicity $A(w_0) = 1$, and the following words with infinite multiplicity:

$$\begin{aligned} w_r &= (r)v[r], && \text{for } r : u \to v \in P, \\ w_\alpha &= Z_1 \alpha Y Z_2, && \text{for } \alpha \in U, \\ w'_\alpha &= Z_1 Y \alpha Z_2, && \text{for } \alpha \in U, \\ w_t &= YY. \end{aligned}$$

The set R contains the following splicing rules:

1. $\delta_1\delta_2 Y u\#\beta_1\beta_2\$(r)v\#[r]$, for $r : u \to v \in P$, $\beta_1, \beta_2 \in U \cup \{X_2\}$, $\delta_1, \delta_2 \in U \cup \{X_1\}$,
2. $Y\#u[r]\$(r)\#v\alpha$, for $r : u \to v \in P$, $\alpha \in U \cup \{X_2\}$,
3. $\delta_1\delta_2 Y\alpha\#\beta_1\beta_2\$Z_1\alpha Y\#Z_2$, for $\alpha \in U$, $\beta_1, \beta_2 \in U \cup \{X_2\}$, $\delta_1, \delta_2 \in U \cup \{X_1\}$,
4. $\delta\#Y\alpha Z_2\$Z_1\#\alpha Y\beta$, for $\alpha \in U$, $\delta \in U \cup \{X_1\}$, $\beta \in U \cup \{X_2\}$,
5. $\delta\alpha Y\#\beta_1\beta_2\beta_3\$Z_1 Y\alpha\#Z_2$, for $\alpha \in U$, $\beta_1 \in U$, $\beta_2, \beta_3 \in U \cup \{X_2\}$, $\delta \in U \cup \{X_1\}$,
6. $\delta\#\alpha Y Z_2\$Z_1\#Y\alpha\beta$, for $\alpha \in U$, $\delta \in U \cup \{X_1\}$, $\beta \in U \cup \{X_2\}$,
7. $\#YY\$X_1^2Y\#w$, for $w \in \{X_2^2\} \cup T\{X_2^2\} \cup T^2\{X_2\} \cup T^3$,
8. $\#X_2^2\$Y^3\#$.

The idea behind this construction is as follows. The rules in groups 1 and 2 simulate rules in P, in the presence of the symbol Y. The rules in groups 3 and 4 move the symbol Y to the right, the rules in groups 5 and 6 move the symbol Y to the left. The "main axiom" is w_0. All rules in groups 1 – 6 involve a word derived from w_0 and containing such a symbol Y introduced by this axiom, in the sense that they can use only one axiom different from w_0. In any one moment, we have two occurrences of X_1 at the beginning of a word and two occurrences of X_2 at the end of a word (maybe the same word). The rules in groups 1, 3, and 5 separate words of the form $X_1^2 z X_2^2$ into two words $X_1^2 z_1$, $z_2 X_2^2$, each one with multiplicity one; the rules in groups 2 and 4, 6 bring together these words, leading to a word of the form $X_1^2 z' X_2^2$. The rules in groups 7 and 8 remove the auxiliary symbols X_1, X_2, Y. If the remaining word is terminal, then it is an element of $L(G)$. The symbols $(r), [r]$ are associated with rules in P, while Z_1 and Z_2 are associated with *moving* operations.

Using these explanations, the reader can easily verify that each derivation in G can be simulated in γ, hence we have $L(G) \subseteq L(\gamma)$. (An induction argument on the length of the derivation can be used, but the details are straightforward and tedious, and we shall not adopt that strategy here. Moreover, the discussion below implicitly shows how to simulate a terminal derivation in G by splicing operations in γ.)

Let us consider in some detail the opposite inclusion. We claim that if $A \Longrightarrow_\gamma^* M$ and $w \in T^*, M(w) > 0$, then $w \in L(G)$.

As we have pointed out above, by a direct check we can see that we cannot splice two of the axioms $w_r, w_\alpha, w'_\alpha, w_t$ (for instance, the symbols δ, β in rules in group 4 and 6 prevent the splicing of $w_\alpha, w'_\alpha, \alpha \in U$). In the first step, we have to start with w_0, $w_0 = X_1^2 Y S X_2^2$, $A(w_0) = 1$. Now, assume that we have a word $X_1^2 w_1 Y w_2 X_2^2$ with multiplicity 1 (w_0 is of this form). If w_2 starts with the left hand member of a rule in P, then we can apply to it a rule of type 1. Assume that this is the case, the word is $X_1^2 w_1 Y u w_3 X_2^2$ for some $r : u \to v \in P$. Using the axiom $(r)v[r]$ from A we obtain

$$(X_1^2 w_1 Y u | w_3 X_2^2, (r)v | [r]) \models (X_1^2 w_1 Y u[r], \ (r) v w_3 X_2^2).$$

No rule from groups 1 and 3 – 8 can be applied to the obtained words. From group 2, the rule $Y\#u[r]\$(r)\#v\alpha$ can be applied involving both these words, which leads to

$$(X_1^2 w_1 Y | u[r], (r) | v w_3 X_2^2) \models (X_1^2 w_1 Y v w_3 X_2^2, \ (r) u[r]).$$

The word $(r)u[r]$ can never enter a new splicing, because in the rule $r : u \to v$ from P we have assumed $u \neq v$. The multiplicity of $X_1^2 w_1 Y u[r]$ and $(r) v w_3 X_1^2$ has been reduced to 0 again (hence these words are no longer available), the multiplicity of $X_1^2 w_1 Y v w_3 X_2^2$ is one. In this way, we have passed from $X_1^2 w_1 Y u w_3 X_2^2$ to $X_1^2 w_1 Y v w_3 X_2^2$, both having the multiplicity one, which corresponds to using the rule $r : u \to v$ in P. Moreover we see that at each moment there is only one word containing X_1^2 and only one word (maybe the same) containing X_2^2 in the current multiset.

If to a word $X_1^2 w_1 Y \alpha w_3 X_2^2$ we apply a rule of type 3, then we get

$$(X_1^2 w_1 Y \alpha | w_3 X_2^2, Z_1 \alpha Y | Z_2) \models (X_1^2 w_1 Y \alpha Z_2,\ Z_1 \alpha Y w_3 X_2^2).$$

No rule from groups 1 – 3 and 5 – 8 can be applied to the obtained words. By using a rule from group 4 we obtain

$$(X_1^2 w_1 | Y \alpha Z_2, Z_1 | \alpha Y w_3 X_2^2) \models (X_1^2 w_1 \alpha Y w_3 X_2^2,\ Z_1 Y \alpha Z_2).$$

The first of the obtained words has replaced $X_1^2 w_1 Y \alpha w_3 X_2^2$, which now has the multiplicity 0 (hence we have interchanged Y with α), the second one is an axiom.

In the same way, one can see that the use of a rule from group 5 must be followed by using the corresponding rule of type 6, which results in interchanging Y with its left hand neighbour.

Consequently, in each moment we have a multiset with either one word $X_1^2 w_1 Y w_2 X_2^2$ or two words $X_1^2 z_1$, $z_2 X_2^2$, each one with multiplicity 1. Only in the first case, provided $w_1 = \lambda$, we can remove $X_1^2 Y$ by using a rule from group 7; then we can also remove X_2^2 by using the rule in group 8. This is the only way to remove these nonterminal symbols. If the word obtained is not terminal, then it cannot be further processed, because it does not contain the symbol Y. In conclusion, we can only simulate derivations in G and move Y freely in the word of multiplicity one, hence $L(\gamma) \subseteq L(G)$. One sees that the radius of γ is five, reached by rules in group 1, where $|\delta_1 \delta_2 Y u| = 5$ when $|u| = 2$. □

Remark 8.3. Let us estimate the number of copies necessary for each axiom. We have said that $A(w_0) = 1$ (and this is essential for the correctness of the simulation of G by γ above). For all w of type $w_r, w_\alpha, w'_\alpha, w_t$ we have said that $A(w) = \infty$. Actually, one sees that for each $r \in P$ we need as many copies of w_r as many times the rule r is used in a derivation in G. Then, w_α and w'_α are necessary for each operation of moving Y to the left or to the right. The word w_t is used only once, by a rule of type 7, at the end of the work of γ. Thus, we might take $A(w_t) = 1$, too.

Moreover, we have seen above that in each moment there are exactly one or exactly two words whose multiplicity is controlled, namely equal to one. Thus, we do not have to "count", say, distinguishing between n and $n+1$ copies of a given word, for large n. It is enough to distinguish between 0 and 1, and that for at most two words; this distinction is made automatically, by the way the system above works, our only concern is to prevent making copies of these distinguished words.

This fact, plus the possibility of obtaining new copies of certain words, via PCR techniques, makes the construction above realistic – from these points of view. □

Lemma 8.11. *$EH_2(\mu FIN, [m]) \subseteq EH_2(\mu[2], [m])$, for all $m \geq 1$.*

Proof. Take an extended μH system $\gamma = (V, T, A, R)$, with finite *supp*(A). Let $w_1, w_2, \ldots, w_n$ be the words of *supp*(A) such that $A(w_i) < \infty$, $0 \leq i \leq n$, and let $z_1, \ldots, z_m$ be the words in *supp*(A) with $A(z_i) = \infty$, $0 \leq i \leq m$. We construct the extended μH system

$$\gamma' = (V \cup \{c, d_1, d_2\}, T, A', R'),$$

where A' contains the word

$$w = (w_1c)^{A(w_1)}(w_2c)^{A(w_2)} \ldots (w_nc)^{A(w_n)},$$

with multiplicity 1, and the word

$$z = d_1cz_1cz_2c \ldots cz_mcd_2,$$

with infinite multiplicity. If $n = 0$, then w does not appear, if $m = 0$, then $z = d_1cd_2$. Moreover

$$R' = R \cup \{\#c\$d_2\#,\ \#d_1\$c\#,\ c\#\$\#d_1\}.$$

The word z can be used for cutting each w_i and each z_j from w and z, respectively. For instance, in order to obtain z_j we splice z with z using $c\#\$\#d_1$ for the occurrence of c to the left hand of z_j, that is

$$(d_1cz_1c \ldots z_{j-1}c|z_jc \ldots cz_mcd_2,\ |z) \models (d_1cz_1c \ldots cz_{j-1}cz,\ z_jc \ldots cz_mcd_2),$$

then we splice the second word with z again using $\#c\$d_2\#$, and we get

$$(z_j|c \ldots cz_mcd_2,\ z|) \models (z_j,\ zcz_{j+1} \ldots cz_mcd_2).$$

Arbitrarily many words z_j can be produced, because $A'(z) = \infty$.

In order to produce the words $w_i, 1 \leq i \leq n$, we start from the left hand end of the string w, by applying $\#c\$d_2\#$ to w and z; we get w_1 and $zc(w_1c)^{A(w_1)-1}(w_2c)^{A(w_2)} \ldots (w_nc)^{A(w_n)}$, both of them with multiplicity 1. Using the rule $\#d_1\$c\#$ for z and the second word above, we obtain zcz and $(w_1c)^{A(w_1)-1}(w_2c)^{A(w_2)} \ldots (w_nc)^{A(w_n)}$, again both with multiplicity 1. From the first word we can separate axioms $z_j, 1 \leq j \leq m$, but this is not important, because these axioms appear with infinite multiplicity in A. From the second word we can continue as above, cutting again a prefix w_1. In this way, exactly $A(w_1)$ copies of w_1 will be produced; in a similar way we can proceed for the other axioms $w_2, \ldots, w_n$ in order to obtain exactly $A(w_i)$ copies of w_i, $i = 2, \ldots, n$.

The use of the nonterminals c, d_1, and d_2 guarantees that only the axioms of γ with infinite multiplicity can be generated in an arbitrary number of copies by the splicing rules in $R' - R$, whereas for each axiom w_i of γ with finite multiplicity $A(w_i)$ we can only obtain $A(w_i)$ copies of w_i. If a rule of R

is used for splicing words of the form x_1cx_2, i.e. containing the nonterminal c, then we finally will have to cut such a word by using the rules in $R'-R$ in order to obtain a terminal word. As we start from the axioms of γ, separated by occurrences of the symbol c, and with the correct multiplicities (guaranteed by the mode of constructing the words w and z), this also corresponds to a correct splicing in γ. Consequently, $L(\gamma') = L(\gamma)$. □

Lemma 8.12. $EH_2(\mu[1], FIN) \subseteq REG$.

Proof. Take an extended μH system $\gamma = (V, T, A, R)$ with $supp(A) = \{w\}$. If $A(w) < \infty$, then $L(\gamma)$ is obviously a finite language (every word in $L(\gamma)$ has a length not greater than $|w| \cdot A(w)$).

If $A(w) = \infty$, then $L(\gamma) \in EH_2([1], FIN) \subseteq EH_2(FIN, FIN) = REG$. Hence we conclude that $EH_2(\mu[1], FIN) \subseteq REG$. □

Lemma 8.13. $REG \subseteq EH_2(\omega[1], [2])$.

Proof. In Lemma 7.18 we have proved that $REG \subseteq EH_1(FIN, FIN)$. It is easy to see that, in fact, we also get $REG \subseteq EH_2(FIN, [2])$ (the system in the proof of Lemma 7.18 has radius two). Let $\gamma = (V, T, A, R)$ be the obtained H system.

Now, using the same construction as in the proof of Lemma 8.11 (the radius is not changed), we can combine all axioms in A with infinite multiplicity into one axiom, w, hence we obtain $REG \subseteq EH_2(\omega[1], [2])$. □

Theorem 8.8. *$REG = EH_2(\mu[1], [2]) = EH_2(\mu[1], FIN) \subset EH_2(\mu[2], FIN) = EH_2(\mu[2], [m]) = RE$, for all $m \geq 5$.*

Proof. For the reader's convenience, let us recall the relations proved in the four lemmas above:

Lemma 8.10: $RE \subseteq EH_2(\mu FIN, [5])$,

Lemma 8.11: $EH_2(\mu FIN, [m]) \subseteq EH_2(\mu[2], [m])$, for all $m \geq 1$,

Lemma 8.12: $EH_2(\mu[1], FIN) \subseteq REG$,

Lemma 8.13: $REG \subseteq EH_2(\omega[1], [2])$.

Now, from the definitions we have

$$EH_2(\omega[n], [m]) \subseteq EH_2(\mu[n], [m]),$$

for all $n, m \geq 1$, and also for $[n], [m]$ replaced by FIN. Thus, from Lemmas 8.12 and 8.13 we obtain

$$REG = EH_2(\mu[1], [2]) = EH_2(\mu[1], FIN).$$

Lemma 8.10 and Lemma 8.11 imply

$$RE \subseteq EH_2(\mu FIN, [m]) \subseteq EH_2(\mu[2], [m]), \ m \geq 5.$$

By a direct proof or from the Church–Turing Thesis we also have

$$EH_2(\mu[2],[m]) \subseteq RE,$$

for all $m \geq 5$, which completes the proof. □

In Sects. 8.1 – 8.5 we have characterized the family RE by imposing certain restrictions on the splicing operation in extended H systems, mainly restrictions inspired from the regulated rewriting area. These restrictions are of a non-biochemical nature, hence they raise serious difficulties for present day laboratory techniques if we want to implement them. More precisely, such restrictions can be (probably) implemented by manually controlling the splicing (e.g., by changing the temperature, acidity or other reaction conditions, in a way to favor or inhibit certain enzymes, by the primer technique described in Remark 8.2, etc.). However, this approach removes some of the central attractive features of DNA computing: the speed, the energy efficiency, the non-determinism (of parallel reactions). In particular, the speed of the process is dramatically decreased. The hope here is that control of the process can be carried out by intrinsic biochemical means. This requires significant progress in biochemical engineering.

Unfortunately, the multiset approach also has a serious drawback: having two strings, each one with multiplicity one, and splicing them is an event with a very low probability. In order to enter a ligation reaction, two strings must be close enough to each other. How to ensure this in a realistic way and in a short interval of time (not to speak about an *efficient* time) is an open problem. For instance, we can bind the two strings to a solid support (these techniques are well understood, see, e.g., [95], [114], [125]), in order to keep them closer and to increase the probability of splicing, but the extent to which this operation is feasible and efficient for strings of large length is a matter of bioengineering out of the scope of this book.

8.7 Universality Results

In the previous five sections we have proved that the family of recursively enumerable languages can be characterized by extended H systems with the work controlled by:

1. permitting contexts,
2. forbidding contexts,
3. local or global target languages,
4. fitness mappings,
5. next-rule mappings,

6. point mutations which edit the currently available splicing rules,
7. double splicing,
8. multisets.

This means that the extended H systems of these types are *computationally complete*, in the sense that they are equal in power to Turing machines (Chomsky type-0 grammars).

However, such results are not enough in order to provide *programmable* computability models ("computers") based on the splicing operation. To this aim, *universal* H systems of the considered types should be found, systems with all components fixed and able to simulate any particular H system in the corresponding class when adding a *code* of the particular system to the universal one. It is natural to add this code as a further axiom to the axiom set of the universal system. Thus, we are led to the following general definition.

Consider an alphabet T and a class $\mathcal{H}$ of extended H systems (for instance, the class of extended H systems with permitting contexts, or with multisets, and so on). An element of $\mathcal{H}$ of the form

$$\gamma_u = (V_u, T, A_u, R_u),$$

where V_u is an alphabet such that $T \subseteq V_u$, $A_u \subseteq V_u^*$, and R_u is a set of splicing rules over V_u, is said to be *universal* for the class $\mathcal{H}$ if for every $\gamma \in \mathcal{H}$ there is a string $w_\gamma \in V_u^*$ such that $L(\gamma) = L(\gamma_u')$, where $\gamma_u' = (V_u, T, A_u \cup \{w_\gamma\}, R_u)$.

Thus, w_γ, the code of γ, is a "program" which can be executed by γ_u in such a way that the work of γ is simulated by γ_u. The axioms in A_u can be viewed as constituting the "operating system" of the "computer" γ_u.

The restriction to a given terminal alphabet can be avoided by accepting a coding of T by elements of a fixed alphabet, for instance, consisting of only two symbols, a, b. Denoting by h this coding, $h : T^* \longrightarrow \{a, b\}^*$, we can then say that γ_u is universal if $L(\gamma) = h(L(\gamma_u'))$, for any given γ. We do not consider this case here, because we already have a restriction to the four letters of the DNA alphabet (hence we already need a coding in order to deal with arbitrary alphabets).

Starting the proofs of Lemmas 8.2, 8.3, 8.5, 8.8, Corollary 8.3, Theorems 8.5, 8.6, 8.7, and Lemma 8.10 from universal type-0 Chomsky grammars, we obtain H systems of the types used in these results whose components depend on the universal grammars, hence are fixed; moreover, these H systems have the universality property as defined above. Since this result is the most important one from the DNA computing point of view, we shall prove it in some detail. Moreover, the number of auxiliary symbols used when passing from a (universal) type-0 grammar to an extended H system of the mentioned types can be significantly decreased: two such symbols are enough. Specifically, the following general result is true.

Lemma 8.14. *Given an extended H system γ of one of the eight types 1–8 listed above, with the total alphabet V and the terminal alphabet T (and finite sets of axioms and of splicing rules), we can construct an extended H system γ' of the same type with γ, with the total alphabet $T \cup \{c_1, c_2\}$ and the terminal alphabet T, such that $L(\gamma) = L(\gamma')$.*

Proof. If $V - T = \{Z_1, \ldots, Z_n\}$, then we consider the morphism $h : V^* \longrightarrow (T \cup \{c_1, c_2\})^*$ defined by

$$h(Z_i) = c_1 c_2^i c_1, \text{ for } 1 \leq i \leq n,$$
$$h(a) = a, \text{ for } a \in T.$$

We construct the system γ' with the total alphabet $T \cup \{c_1, c_2\}$ and the other components obtained by applying the morphism h, in the usual way, to the components of γ. (For instance, for each axiom x of γ we introduce $h(x)$ as an axiom of γ', and for each splicing rule $u_1\#u_2\$u_3\#u_4$ of γ we introduce $h(u_1)\#h(u_2)\$h(u_3)\#h(u_4)$ as a splicing rule of γ'. We proceed in a similar way for the permitting or forbidding conditions, target languages, point mutation rules in evolving H systems – in this latter case the old markers c_1, c_2 in the proof of Theorem 8.6 should not be confused with the new symbols c_1, c_2.)

The equality $L(\gamma) = L(\gamma')$ follows from the fact that in all components of γ', whatever its type is, the blocks $c_1 c_2^i c_1, 1 \leq i \leq n$, are never broken by the splicing operations (or editing operations, in the case of evolving systems), they behave in the same way as the corresponding symbols Z_i. □

Note that in the proof of this lemma we pass from symbols Z_i to strings $c_1 c_2^i c_1$, hence we obtain a system γ' with a radius larger than that of γ (but with the same number of axioms). Moreover, in the case of permitting or forbidding contexts, the contexts are no longer symbols, but strings (of a bounded length). This was the reason for defining the extended H systems with permitting or forbidding contexts in the general manner, dealing with string conditions rather than symbols, as it is enough for the proofs of Theorems 8.1, 8.2. This does not introduce a significant additional difficulty in checking such conditions in the way described in Remark 8.2. This is true at least for short strings. On the other hand, we can work not with two auxiliary symbols but with several symbols; in this way the length of the encodings of the nonterminals in γ is decreased. The balance of the number of nonterminals and of the length of the mentioned codings is a matter of practical interest, hence it should be investigated under specific circumstances.

We are now ready to present one of the main results of this chapter, from the point of view of DNA computability. For precise references and details we formulate it explicitly for μH systems, but similar results hold true, *mutatis mutandis*, for all types of H systems investigated above.

Theorem 8.9. *For every given alphabet T there exists an extended μH system of type $(\mu[1], FIN)$, with only two auxiliary symbols, which is universal for the class of extended μH systems of type $(\mu FIN, FIN)$ with the terminal alphabet T.*

Proof. Consider an alphabet T and two different symbols c_1, c_2 not in T.

In Chap. 3 we have mentioned that for the class of type-0 Chomsky grammars with a given terminal alphabet there are universal grammars, i.e. constructs $G_u = (N_u, T, P_u)$ such that for any given grammar $G = (N, T, S, P)$ there is a word $w(G) \in (N_u \cup T)^+$ (the "code" of G) such that $L(G'_u) = L(G)$ for $G'_u = (N_u, T, w(G), P_u)$. (The language $L(G'_u)$ consists of all terminal words z such that $w(G) \Longrightarrow^* z$ using the rules in P_u.)

For a given universal type-0 grammar $G_u = (N_u, T, P_u)$, we follow the construction in the proof of Lemma 8.10, obtaining an extended μH system $\gamma_1 = (V_1, T, A_1, R_1)$, where the axiom (with multiplicity 1) $w_0 = X_1^2 Y S X_2^2$ is not considered. Notice that all the other axioms in A_1 (all having infinite multiplicity) and the rules in R_1 depend on N_u, T, and P_u only, hence they are fixed.

As in the proof of Lemma 8.11, we now pass from γ_1 to $\gamma_2 = (V_2, T, A_2, R_2)$. As A_1 contains only axioms with infinite multiplicity, A_2 consists of only one word (that one denoted by z in the proof of Lemma 8.11), which has infinite multiplicity.

We now follow the proof of Lemma 8.14, codifying all symbols in $V_2 - T$ by words over $\{c_1, c_2\}$; the obtained system,

$$\gamma_u = (\{c_1, c_2\} \cup T, T, A_u, R_u)$$

is the universal μH system we are looking for.

Indeed, take an arbitrary extended μH system $\gamma_0 = (V, T, A, R)$. Since $L(\gamma_0) \in RE$, there is a type-0 grammar $G_0 = (N_0, T, S_0, P_0)$ such that $L(\gamma_0) = L(G_0)$. Construct the code of G_0, $w(G_0)$, as imposed by the definition of universal type-0 grammars one uses, consider the word

$$w'_0 = X_1^2 Y w(G_0) X_2^2,$$

corresponding to the axiom w_0 in the proof of Lemma 8.10, then codify w'_0 over $\{c_1, c_2\} \cup T$ as we have done above with the axioms of γ_2. Denote the obtained word by $w(\gamma_0)$. Then $L(\gamma'_u) = L(\gamma_0)$, for $\gamma'_u = (\{c_1, c_2\} \cup T, T, \{(w(\gamma_0), 1)\} \cup A_u, R_u)$.

This can be seen easily. In the proof of Lemma 8.10, the system γ simulates the work of G, starting from the axiom S of G, bracketed as in $X_1^2 Y S X_2^2$. If instead of S we put an arbitrary word x over the alphabet of G, then in γ we obtain exactly the language of terminal words y such that $x \Longrightarrow^* y$ in G. If we start from a universal grammar G_u and S is replaced by the code $w(G_0)$ of a type-0 grammar G_0 equivalent with γ_0, then the system γ_u, associated as above with the universal grammar G_u, will simulate the work of G_u, starting from $w(G_0)$. Hence $L(\gamma'_u) = L(G'_u) = L(G_0) = L(\gamma_0)$, for $G'_u = (N_u, T, w(G_0), P_u)$.

Clearly, A_u contains only one string, hence γ_u is of the type $(\mu[1], FIN)$. □

Notice that the universal μH system γ_u furnished by the proof of Theorem 8.9 (the same assertion is true for systems with forbidding contexts, with local or global targets, with fitness mappings, and for programmed systems; in the case of evolving H systems, A_0 will be empty and here will be added the string $XB_0w(G_0)Y$, containing the "program" of γ_0, instead of the string XB_0SY in the proof of Theorem 8.6) has only one axiom. Moreover, the "program" to be run on our "computer" also consists of one string-axiom only.

The existence of universal H systems with permitting contexts provides a partial answer to the third open problem formulated after Corollary 8.1: there is n such that a result of the form $RE = EH_2([n], p[2])$ is true. This can be proved as follows. Start the construction in the proofs of Lemmas 8.2 and 8.3 from a universal type-0 grammar. We get an H system with fixed components – hence with a fixed number of axioms and a fixed set of splicing rules. In order to generate a given language $L \in RE$ we have to introduce one more axiom of the type $XBw(G)Y$, for $L = L(G), w(G)$ a code of G. The radius of rules remains unchanged, the number of axioms is bounded.

Thus, the problem can be reformulated: which is the smallest n such that $RE = EH_2([n], p[2])$?

The proof of Theorem 8.9 is effective, it constructively provides an extended μH system which is universal for the class of μH systems or, directly, for the class of type-0 Chomsky grammars or of Turing machines. Starting from a universal Turing machine, we get in this way a universal extended μH system.

Instead of presenting a universal system, let us estimate the number of splicing rules obtained if we follow the constructions on which the proof of Theorem 8.9 is based.

Consider a universal Turing machine M in a class $UTM(m, n)$ as in Sect. 3.3 and having p moves. For each move of M (given as a rewriting rule, as shown in Sect. 3.3, or in any other appropriate manner), in the proof of Lemma 8.10 we construct about $(n+m)^4$ rules of type 1 and about $(n+m)$ rules of type 2. In total, we obtain

$$N_1 = p((n+m)^4 + n + m)$$

rules (we consider only the tape symbols and the states of the Turing machine as symbols of the alphabet, although further auxiliary symbols might be necessary). Furthermore, we consider about

$$N_2 = (n+m)^5 + (n+m)^3 + (n+m)^5 + (n+m)^3$$

rules of types 3, 4, 5, 6. In total, we obtain about

$$N_3 = p((n+m)^4 + n + m) + 2(n+m)^5 + 2(n+m)^3$$

rules. For the three universal machines presented in Sect. 3.3 we get the results below:

m	n	p	N_3
7	4	26	705716
5	5	29	492290
4	6	22	222220

These figures are definitely out of the reach of any practical attempt of realizing such a universal μH system. Note, however, that when proving the results mentioned above, we were not interested in keeping small the size of the output, but rather in getting a simple proof for the correctness of the construction. Thus, it remains as a *research topic* to find small universal H systems of various types. To this aim, a direct construction will be necessary, avoiding the passing through grammars and Turing machines.

8.8 Bibliographical Notes

Extended H systems with permitting and forbidding context conditions are introduced in [62], but restricted forms of the splicing operation were considered first in [166], where only non-iterated operations are examined; such operations with respect to infinite (regular) sets of rules are investigated in [102]. Sect. 8.2 is based on [62]; part of the results are also presented in [31].

Lemma 8.4 is from [21] (where one conjectures that permitting context H systems of radius one and with one-sided contexts – in each rule $(r; C_1, C_2)$, one of C_1, C_2 is empty – characterize the linear languages; the conjecture is disproved in [160], where one proves that such systems with all sets C_2 empty can generate all context-free languages). A more detailed study of permitting context H systems of small radius is carried out in [143] and [144]: instead of the radius, one considers the *weight* of a rule $u_1\#u_2\$u_3\#u_4$ as the four-tuple $(|u_1|, |u_2|, |u_3|, |u_4|)$; the weight of a system γ is the smallest (componentwise) four-tuple (n_1, n_2, n_3, n_4) with the components maximizing the corresponding components of the weight of rules in γ. In this framework, a strenghtening of the result in Theorem 8.4 for local targets is obtained in [144]: $RE = EH_2([1], lt[1])$, which therefore is an optimal result for the number of axioms and probably also optimal for the radius (this proves the power of control by means of local target languages).

Splicing operations with target languages are considered in [166]; Sect. 8.3 is based on [157]. Programmed and evolving H systems are considered in [168]. H systems with double splicing were introduced in [162].

The use of multisets was first considered in [46], where non-recursive languages are generated in this way. A representation of recursively enumerable languages as morphic images of languages in $EH_2(\mu FIN, FIN)$ is given in [150]. The result is improved in [62] in the form of Theorem 8.8 above. Sect. 8.7 is based on [62].

Explicit universal H systems were constructed in [6] (with infinite regular sets of splicing rules, following the proof of the Basic Universality Lemma) and in [61] (of μH type; starting from Minsky's Turing machine in $UTM(7, 4)$ and modifying the construction in Lemma 8.10, in [61] one produces – by means of a computer program – a universal μH system with 56 *types* of splicing rules, each of them involving one, two, or three variables running on domains of five elements; that is, the system contains some thousand splicing rules; this is much better than the previous estimations, but still completely unpractical).

A recent attempt to produce an explicit universal H system, again with multiplicites, is done in [57], and it leads to a universal system with the number of axioms bounded by $17n^2 + 225n + 784$ and the number of splicing rules bounded by $29n^2+405n+1414$, where n is the cardinality of the terminal alphabet. These bounds are comparable with those in [61].

Chapter 9

Splicing Circular Strings

In certain circumstances – in several bacteria, for instance – the DNA molecules are present in the form of a circular sequence. More generally, we can consider situations where both linear and circular DNA sequences are present. The restriction enzymes can cut both the linear and the circular double stranded sequences, hence recombination by ligation can also appear in such a case. Many variants are possible, because a recombination can have as input two circular strings, or one circular and one linear string, and can have as output one or two circular strings, one or two linear strings, or both a circular and a linear string.

From a mathematical point of view, the study of such variants is not so elegant as the study of linear splicing in the previous chapters, but it can provide significant simplifications of some constructions above, because, for instance, we no longer need the *rotate* activity in the proofs based on the rotate-and-simulate idea.

9.1 Variants of the Splicing Operation for Circular Strings

Consider an alphabet V. A circular string over V is a sequence $x = a_1a_2\dots a_n$ for $a_i \in V, 1 \leq i \leq n$, with the assumption (convention) that a_1 follows a_n. In other words, x can be represented by any circular permutation of $a_1a_2\dots a_n$, for instance, $a_{i+1}\dots a_na_1\dots a_i$, for any $1 \leq i \leq n-1$. Thus, a circular string over V is an equivalence class of all linear strings equal to each other modulo a circular permutation. We denote by $\hat{x}$ the circular string associated to the linear string $x \in V^*$. The set of all circular strings over V is denoted by $V^\diamond$. Any subset of $V^\diamond$ is called a circular language.

To a usual language $L \subseteq V^*$ we can associate the circular language $Cir(L) = \{\hat{x} \mid x \in L\}$. (For singleton languages we also write $Cir(x) = \hat{x}$.) Conversely, to a circular language $L \subseteq V^\diamond$ we can associate the *full lineariza-*

tion $Lin(L) = \{x \mid \hat{x} \in L\}$. A language $L_1 \subseteq V^*$ is a *linearization* of a circular language $L_2 \subseteq V^\diamond$ if $Cir(L_1) = L_2$.

Having a family FL of languages, we can consider its circular counterpart, $FL^\diamond = \{Cir(L) \mid L \in FL\}$. Thus, we can speak about $FIN^\diamond, REG^\diamond, RE^\diamond$, etc.

The operations of union, intersection, intersection with regular circular languages, direct and inverse morphisms, can also be defined for circular languages (but not the operations of concatenation and Kleene closure).

Lemma 9.1. *If FL is a family of languages closed under circular permutation, then* $L \in FL^\diamond$ *if and only if* $Lin(L) \in FL$.

Proof. Let us denote by $cp(L)$ the set of all circular permutations of strings in L. Clearly, $Lin(Cir(L)) = cp(L)$, for every language L. Now, if $L_0 \in FL^\diamond$, from the definition of $FL^\diamond$ we have $L_0 = Cir(L)$ for some $L \in FL$. Because $Lin(L_0) = Lin(Cir(L)) = cp(L)$ and FL is closed under circular permutation, we have $Lin(L_0) \in FL$. Conversely, if $L \subseteq V^\diamond$ such that $Lin(L) \in FL$, because $L = Cir(Lin(L))$, we have $L \in FL^\diamond$. □

Corollary 9.1. *Let FL be a family of languages which is closed under circular permutation. If FL is closed under union, direct morphisms, inverse morphisms, intersection with regular languages, then also* $FL^\diamond$ *is closed under union, direct morphisms, inverse morphisms, intersection with regular circular languages, respectively.*

Let us note that all families *FIN, REG, CF, CS, RE* are closed under circular permutation, but *LIN* is not closed: $cp(\{a^n b^n \mid n \geq 1\}) \cap a^+b^+a^+ = \{a^n b^{n+m} a^m \mid n, m \geq 1\}$ is not a linear language.

Let us now define some natural splicing operations involving circular strings.

Consider an alphabet V and a splicing rule $r = u_1\#u_2\$u_3\#u_4$ over V. For $\hat{x}, \hat{y}, \hat{z} \in V^\diamond$ and $v, w \in V^*$, we write:

$$
\begin{aligned}
(\hat{x}, \hat{y}) \models_r^1 \hat{z} \quad \text{iff} \quad & x = x_1u_1u_2, \\
& y = y_1u_3u_4, \\
& z = x_1u_1u_4y_1u_3u_2, \\
& \text{for some } x_1, y_1 \in V^*, \\
\hat{x} \models_r^2 (\hat{y}, \hat{z}) \quad \text{iff} \quad & x = x_1u_1u_2x_2u_3u_4, \\
& y = x_1u_1u_4, \\
& z = x_2u_3u_2, \\
& \text{for some } x_1, x_2 \in V^*, \\
(\hat{x}, v) \models_r^3 w \quad \text{iff} \quad & x = x_1u_1u_2, \\
& v = v_1u_3u_4v_2, \\
& w = v_1u_3u_2x_1u_1u_4v_2,
\end{aligned}
$$

$$v \models_r^4 (\hat{x}, w) \quad \text{iff} \quad \begin{aligned} &\text{for some } x_1, v_1, v_2 \in V^*, \\ &v = v_1 u_1 u_2 v_2 u_3 u_4 v_3, \\ &x = u_2 v_2 u_3, \\ &w = v_1 u_1 u_4 v_3, \\ &\text{for some } v_1, v_2, v_3 \in V^*. \end{aligned}$$

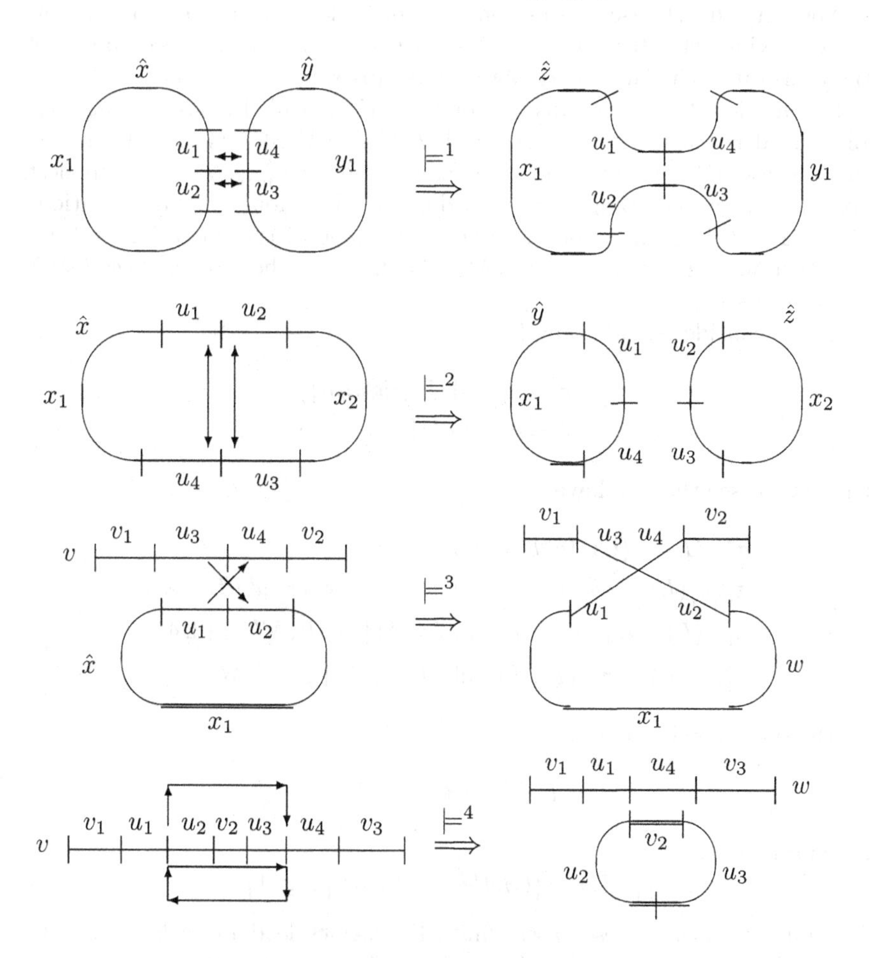

Figure 9.1: Splicing circular strings (I)

In the case of $\models^1$ two circular strings are cut at sites u_1u_2, u_3u_4, and then pasted together to form a new circular string. In the case of $\models^2$, a single circular string gives rise to two circular strings, by cutting it at two sites, u_1u_2, u_3u_4, and pasting together the ends of the two fragments. The operation $\models^3$ cuts a circular string at a site u_1u_2 and a linear string at a site u_3u_4, then the linear string obtained by cutting the circular one is linked to the two fragments obtained by cutting the linear string. In short, a specific linearization of the circular string is inserted into the linear string, as specified by the sites in the splicing rule. Finally, by $\models^4$ we pass from a single linear string, cut at two positions, at sites u_1u_2, u_3u_4, to a circular string and a linear one. Figure 9.1 illustrates these variants. One further variant will be considered in the subsequent section.

Note that in all cases we use one splicing rule only, in the same way as in the previous chapters: we cut at sites u_1u_2, u_3u_4 and we recombine the sticky ends in such a way as to obtain the substrings u_1u_4, u_3u_2.

For an alphabet V, a subset L of $V^* \cup V^\circ$ is called a *mixed* language. For a usual H scheme $\sigma = (V, R)$, with $R \subseteq V^*\#V^*\$V^*\#V^*$, and a mixed language $L \subseteq V^* \cup V^\circ$, we define the mixed language $\sigma^*_{mix}(L)$ as the smallest mixed language containing L and closed under all the four splicing operations $\models^i, i = 1, 2, 3, 4$, defined above. When only some of the operations $\models^i$ are used, then we write $\sigma^*_M(L)$, where $M \subseteq \{1, 2, 3, 4\}$ is the set of indices i such that $\models^i$ is used.

Let us consider some examples:

$$\sigma = (\{a, b\}, \{a\#b\$b\#a\}),$$
$$L = \{Cir(ab)\}.$$

It is easy to see that we have

$$\sigma^*_{\{1\}}(L) = \{Cir(a^n b^n) \mid n \geq 1\},$$
$$\sigma^*_{\{1,2\}}(L) = \{\hat{y} \mid y = a^n, \text{ or } y = b^n, \text{ or } y = a^n b^n, n \geq 1\},$$
$$\sigma^*_{\{1\}}(L) = \sigma^*_M(L), \text{ for all } M = \{1\} \cup M', M' \subseteq \{3, 4\},$$
$$\sigma^*_{\{1,2\}}(L) = \sigma^*_M(L), \text{ for all } M = \{1, 2\} \cup M', M' \subseteq \{3, 4\}.$$

For the same H scheme and

$$L' = \{ba, Cir(ba)\},$$

we obtain

$$\sigma^*_{\{3\}}(L') = \{Cir(ba)\} \cup \{b^n a^n \mid n \geq 1\}.$$

Consequently, in all these cases, finite H schemes lead finite languages to non-regular (usual or circular) languages. This contrasts the situation met for linear strings (Lemma 7.14) and makes the splicing of circular strings interesting from DNA computing point of view: non-restricted splicing with respect to finite sets of splicing rules leads to non-regular languages even when starting from finite languages.

A regularity preserving result can be obtained for the mixed splicing for a particular class of H schemes. Specifically, an H scheme $\sigma = (V, R)$ is said to be *reflexive* if whenever $u_1\#u_2\$u_3\#u_4 \in R$, then also $u_1\#u_2\$u_1\#u_2 \in R$ and $u_3\#u_4\$u_3\#u_4 \in R$.

A proof of the following counterpart of Lemma 7.15 can be found in [90]:

Theorem 9.1. *Let FL be a full AFL closed under circular permutation. If $\sigma = (V, R)$ is a reflexive H scheme with a finite set R and L is a mixed language over V such that $L \cap V^* \in FL$ and $L \cap V^\circ \in FL^\circ$, then $\sigma^*_{mix}(L) \cap V^* \in FL$ and $\sigma^*_{mix}(L) \cap V^\circ \in FL^\circ$.*

Thus, in order to obtain computability models stronger than finite automata based on non-controlled mixed splicing we have to use H schemes which are not reflexive.

Returning to the example above, if we add the rules $a\#b\$a\#b$ and $b\#a\$b\#a$ to σ, then we obtain an H scheme σ' such that

$$\sigma'^*_{\{1\}}(L) = \{\hat{y} \mid y \in \{a, b\}^+, |y|_a = |y|_b\},$$
$$\sigma'^*_{\{1,2\}}(L) = \{\hat{y} \mid y \in \{a, b\}^+\}.$$

When using only the relation $\models^1$, a non-regular circular language is still obtained, but, because $\sigma'^*_{\{1,2\}}(L) = \sigma'^*_{mix}(L)$ (no linear string is present), Theorem 9.1 can be applied, and the circular language $\sigma'^*_{\{1,2\}}(L)$ is regular.

The precise characterization of the power of (extended) H systems based on the splicing of circular strings of various types considered above, or on mixed splicing is an important *research topic.*

9.2 One Further Variant and its Power

From two directions, we get the same suggestion on how one further splicing operation involving circular strings can be defined.

One direction is the rotate-and-simulate idea in the proofs of several characterizations of *RE* in the previous chapters. In the rotation steps, we start from strings $Xw\alpha Y$, with $w \in V^*, \alpha \in V$, for a given alphabet V, and two special symbols X, Y, we remove α from the right hand end, getting XwY' (usually, $Y' = Y_\alpha$, to remember the removed symbol), then we add α to the left hand end, producing $X\alpha wY'$; finally, we return Y' to Y, obtaining $X\alpha wY$. Since $Cir(w\alpha) = Cir(\alpha w)$ we can consider that the block YX has been interchanged with α, by a substitution of the form

$$\alpha YX \to YX\alpha,$$

and not that α has been moved in $Cir(Xw\alpha Y)$. In this way we can treat in a uniform way both the rules $u \to v$ in the grammar to be simulated and the rotating steps: both of them are steps when a substring of the circular string is replaced by another substring. If we want to preserve the control block YX, perhaps replaced by one symbol only, then we need interchanging rules $\alpha YX \to YX\alpha$. In fact, if we are able to substitute substrings of a circular string by other substrings, then the rotation is no longer necessary. What we need in addition is a linearization step, or a convention of reading linear strings from circular strings.

Another motivation for considering a splicing operation with circular strings able to perform a substitution comes from the characterization of RE languages by means of an iterated gsm. Iterating a gsm means to parse a string and to pass from its last symbol to the first symbol again, continuing in this way. If the two ends of the strings were connected, as in a circular string, then we can imagine that the gsm simply continues the parsing, going along the circular string.

The operation necessary in both these cases, that of a substring substitution in a circular string, leads to the following way of using a splicing rule $r = u_1\#u_2\$u_3\#u_4$ over some alphabet V. For $\hat{x}, \hat{y} \in V^\diamond$ and $z, w \in V^*$, we write

$$\begin{aligned}(\hat{x}, z) \models_r^5 (\hat{y}, w) \quad \text{iff} \quad & x = x_1u_1u_2x_2u_3u_4, \\ & z = u_2z_1u_3, \\ & y = x_1u_1u_2z_1u_3u_4, \\ & w = u_2x_2u_3, \\ & \text{for some } x_1, x_2, z_1 \in V^*.\end{aligned}$$

The operation is illustrated in Fig. 9.2. One sees how the strings $\hat{x}$ and z interchange the substrings x_2 and z_1. The circular string is cut in two places, leaving free the subword $u_2x_2u_3$; the linear string already has ends which match the ends of the remaining part of the circular string, hence a new circular string can be produced.

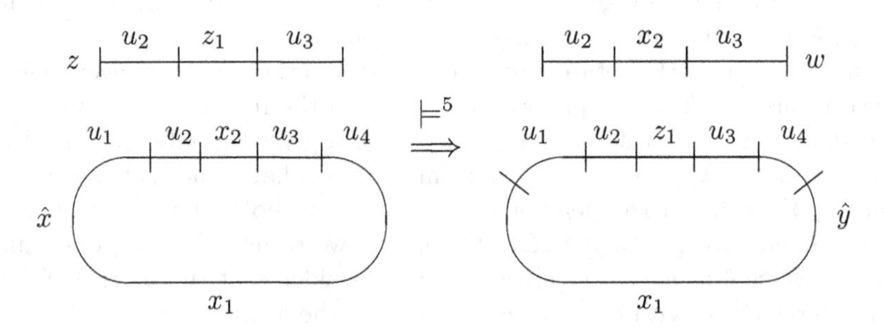

Figure 9.2: Splicing circular strings (II)

Based on this operation, we define a language generating device in the following way.

An *extended circular H system* is a construct

$$\gamma = (V, T, A_1, A_2, R_1, R_2),$$

where V is an alphabet, $T \subseteq V$ (the terminal alphabet), $A_1 \subseteq V^\circ, A_2 \subseteq V^*$ are finite sets (of axioms), and R_1, R_2 are finite sets of splicing rules over V. For such a system and for $L_1 \subseteq V^\circ, L_2 \subseteq V^*$ we define

$$\sigma_{\{5\}}(L_1, L_2) = (L_1', L_2'),$$

where

$$\begin{aligned} L_1' &= \{\hat{y} \mid (\hat{x}, z) \models_r^5 (\hat{y}, w), \\ &\quad \text{for some } \hat{x} \in L_1, z \in L_2, r \in R_1\}, \\ L_2' &= \{w \mid (\hat{x}, z) \models_r^5 (\hat{y}, w), \\ &\quad \text{for some } \hat{x} \in L_1, z \in L_2, r \in R_1\}. \end{aligned}$$

Then we define

$$\begin{aligned} &\sigma_{\{5\}}^0(L_1, L_2) = (L_1, L_2), \\ &\sigma_{\{5\}}^{i+1}(L_1, L_2) = \sigma_{\{5\}}^i(L_1, L_2) \cup \sigma_{\{5\}}(\sigma_{\{5\}}^i(L_1, L_2)), \ i \geq 0, \\ &\sigma_{\{5\}}^*(L_1, L_2) = \bigcup_{i \geq 0} \sigma_{\{5\}}^i(L_1, L_2), \end{aligned}$$

where the union is defined componentwise.

For $r = u_1\#u_2\$u_3\#u_4$ in R_2 and $\hat{x} \in V^\circ, y, z \in V^*$, we write

$$\begin{aligned} \hat{x} \models_r^6 (y, z) \quad \text{iff} \quad & x = x_1u_1u_2x_2u_3u_4, \\ & y = u_4x_1u_1, \\ & z = u_2x_2u_3, \\ & \text{for some } x_1, x_2 \in V^*. \end{aligned}$$

(The circular string is cut at the sites u_1u_2, u_3u_4, producing two linear strings.)

Then for $L \subseteq V^\circ$ we define

$$\begin{aligned} \sigma_{\{6\}}(L) = \{y \in V^* \mid \hat{x} \models_r^6 (y, z) \text{ or } \hat{x} \models_r^6 (z, y), \\ \text{for some } \hat{x} \in L \text{ and } r \in R_2\}. \end{aligned}$$

The language generated by γ is defined by

$$L(\gamma) = \sigma_{\{6\}}(L_1) \cap T^*,$$

where

$$\sigma_{\{5\}}^*(A_1, A_2) = (L_1, L_2),$$

for some $L_2 \subseteq V^*$.

Therefore, we start from two sets of axioms, we splice the circular strings with the linear ones according to the operation $\models^5$, with respect to the rules in R_1, iteratively; finally we cut the circular strings obtained in this way by an operation $\models^6$ with respect to the rules in R_2; in the generated language we keep only the strings composed of terminal symbols.

We do not know how powerful the extended circular H systems are, but a simple restriction on the splicing operations $\models^5, \models^6$ will lead to a characterization of RE, modulo a projection which erases certain markers used in the process of string generation.

We say that $\gamma = (V, T, A_1, A_2, R_1, R_2)$ is a *restricted* extended circular H system if the rules $r \in R_1 \cup R_2$ have associated strings $v_r \in V^*$ (we present the rules as pairs (r, v_r)) and the operations $\models^5, \models^6$ are defined in the following ways: for $r = u_1\#u_2\$u_3\#u_4, (r, v_r) \in R_1 \cup R_2$, and $\hat{x}, \hat{y} \in V^\circ, z, w \in V^*$, we write

$$
\begin{aligned}
(\hat{x}, z) \models_r^{5'} (\hat{y}, w) \quad \text{iff} \quad & x = x_1u_1u_2x_2u_3u_4, \\
& z = u_2z_1u_3, \\
& y = x_1u_1u_2z_1u_3u_4, \\
& w = u_2x_2u_3, \\
& \text{for some } x_1, x_2, z_1 \in V^*, \text{ such that} \\
& v_r \in Sub(x_2) \text{ and } u_1u_2x_2u_3u_4 \text{ cannot be written} \\
& \text{in the form } x_1'u_1u_2x_2'u_3u_4x_3' \text{ with} \\
& |x_2'| < |x_2|, v_r \in Sub(x_2'), \\
\hat{x} \models_r^{6'} (y, z) \quad \text{iff} \quad & x = x_1u_1u_2x_2u_3u_4, \\
& y = u_4x_1u_1, \\
& z = u_2x_2u_3, \\
& \text{for some } x_1, x_2 \in V^* \text{ such that} \\
& v_r \in Sub(x_2) \text{ and } u_1u_2x_2u_3u_4 \text{ cannot be written} \\
& \text{in the form } x_1'u_1u_2x_2'u_3u_4x_3' \text{ with} \\
& |x_2'| < |x_2|, v_r \in Sub(x_2').
\end{aligned}
$$

In words, the sites u_1u_2, u_3u_4 should be placed around a substring x_2 of x which contains at least an occurrence of the string v_r and x_2 is minimal with this property, no proper substring of x_2 containing v_r can be bracketed by the sites u_1u_2, u_3u_4. Thus, the strings v_r act as "anchors" for the splicing rules, as promoters whose influence is manifested locally, to the first occurrence of u_1u_2 to the left of them and to the first occurrence of u_3u_4 to the right of them.

Having such information on the string x_2, which is replaced by z_1 by an operation $\models^{5'}$, is a very powerful feature: using this operation we can again characterize the recursively enumerable languages.

Let $EH_2(cFIN, rFIN)$ denote the family of languages generated by restricted extended circular H systems ("c" in front of the first occurrence of FIN indicates the fact that we start with circular axioms, "r" in front of the second FIN indicates the fact that the splicings are restricted in the sense defined above).

Theorem 9.2. *Every recursively enumerable language is a projection of a language in the family* $EH_2(cFIN, rFIN)$.

Proof. Consider a language $L \in RE, L \subseteq T^*$. As we know from Theorem 3.14, we can write this language in the form $L = g^*(a_0) \cap T^*$, for a gsm $g = (K, V, V, s_0, F, P)$, $a_0 \in V$.

We construct the restricted extended circular H system

$$\gamma = (V', T', A_1, A_2, R_1, R_2),$$

as follows.

Consider a new symbol, E, and add to P all the rules of the form $s_f E \rightarrow E s_0$, for $s_f \in F$. Denote by P' the set obtained in this way. Assume that $P' = \{r_1, \ldots, r_n\}$, with

$$r_i : q_i b_i \rightarrow c_{i,1} \ldots c_{i,t_i} q_i',$$

for $t_i \geq 0, q_i, q_i' \in K, b_i \in V \cup \{E\}, c_{i,1} \in V \cup \{E\}, c_{i,j} \in V, 2 \leq j \leq t_i$, $1 \leq i \leq n$.

Consider also the new symbols $d_0, d_1, \ldots, d_n$. Then

$$\begin{aligned}
V' &= V \cup K \cup \{E, d_0, d_1, \ldots, d_n\}, \\
T' &= T \cup \{d_0, d_1, \ldots, d_n\}, \\
A_1 &= \{Cir(d_1 d_2 \ldots d_n d_0 s_0 d_0 a_0 d_1 d_2 \ldots d_n d_0 E)\}, \\
A_2 &= \{d_{i+1} \ldots d_n d_0 c_{i,1} d_1 d_2 \ldots d_n d_0 c_{i,2} d_1 d_2 \ldots d_n d_0 \\
&\quad \ldots c_{i,t_i} d_1 d_2 \ldots d_n d_0 q_i' d_0 \mid 1 \leq i \leq n\}, \\
R_1 &= \{(d_1 d_2 \ldots d_i \# d_{i+1} \ldots d_n d_0 \$ d_0 \#,\ q_i d_0 b_i) \mid 1 \leq i \leq n\}, \\
R_2 &= \{(\# d_1 d_2 \ldots d_n d_0 \$ d_0 \#,\ q_f d_0 E) \mid q_f \in F\}.
\end{aligned}$$

Let us see how this system works.

Assume that we have a circular string $\hat{x}$ for some

$$x = x_1 b d_1 d_2 \ldots d_n d_0 q d_0 a d_1 d_2 \ldots d_n d_0 c x_2,$$

for $x_1, x_2 \in (V \cup \{d_0, d_1, \ldots, d_n\})^*, a, b, c \in V \cup \{E\}, q \in K$; at the beginning we have $x_1 = \lambda, x_2 = \lambda, q = s_0, a = a_0$, and b, c are missing. If qa corresponds to a rule r_i in P', that is $q = q_i, a = b_i$, then there is a splicing rule $d_1 d_2 \ldots d_i \# d_{i+1} \ldots d_n d_0 \$ d_0 \#$ in R_1. Due to the restricted mode of applying these rules, we cut from x the substring $x_3 = d_{i+1} \ldots d_n d_0 q_i d_0 b_i d_1 \ldots d_n d_0$ (the substring $q_i d_0 b_i$ appears in x_3 and x_3 is minimal with this property). There is only one string in A_2 starting with $d_{i+1} \ldots d_n d_0$ and ending with d_0, namely, $d_{i+1} \ldots d_n d_0 c_{i,1} d_1 d_2 \ldots d_n d_0 c_{i,2} \ldots c_{i,t_i} d_1 d_2 \ldots d_n d_0 q_i' d_0$. Substituting it in x instead of the string x_3 cut above, we get a circular string $\hat{y}$ for

$$y = x_1 b d_1 \ldots d_n d_0 c_{i,1} d_1 \ldots d_n d_0 c_{i,2} \ldots c_{i,t_i} d_1 \ldots d_n d_0 q_i' d_0 c x_2.$$

Therefore, all circular strings $\hat{x}$ obtained by using the rules in R_1 have the following properties:

- only one state $q \in K$ appears in x in a substring of the form qd_0a, for $a \in V \cup \{E\}$,
- all two consecutive occurrences of symbols in $V \cup \{E\}$ are separated by a block $d_1d_2 \dots d_nd_0$,
- q is also separated from its left neighbour in $V \cup \{E\}$ by a block $d_1d_2 \dots d_nd_0$.

The previous splicing operation has simulated the use of the rule r_i on the circular string. The linear string x_3 produced at the same time can be used only for splicing an occurrence of itself, in a splicing performed by the same rule $(d_1 \dots d_i \# d_{i+1} \dots d_nd_0 \$ d_0 \#,\ q_id_0b_i)$, hence neither the circular string nor the string x_3 are changed.

Consequently, using the rules in R_1 we can simulate the work of g, starting from a_0, iteratively due to the existence of rules $q_fE \to Es_0$ in P'. When a circular string $\hat{x}$ is obtained, with x of the form

$$x = x_1ad_1d_2 \dots d_nd_0q_fd_0Ed_1d_2 \dots d_nd_0bx_2,$$

with $x_1, x_2 \in (V \cup \{d_0, d_1, \dots, d_n\})^*, a, b \in V \cup \{\lambda\}, q_0 \in F$, then the corresponding rule in R_2 can also be applied.

We obtain two linear strings

$$y = d_1d_2 \dots d_nd_0q_fd_0Ed_1d_2 \dots d_nd_0,$$
$$z = bx_2x_1a.$$

The first string above contains the nonterminal symbols q_f, E; the second one is accepted in $L(\gamma)$ when containing only symbols in $T \cup \{d_0, d_1, \dots, d_n\}$. Because we have cut the circular string $\hat{x}$ at the position indicated by E, which marks the end of the string in $g^*(a_0)$ simulated by γ, we thus obtain a string in $g^*(a_0)$ in the correct circular permutation, with the symbols separated by blocks $d_1d_2 \dots d_nd_0$.

Consider now the projection $pr_T : (T \cup \{d_0, d_1, \dots, d_n\})^* \longrightarrow T^*$ defined by

$$pr_T(a) = a, \text{ for } a \in T,$$
$$pr_T(d_i) = \lambda, \text{ for } 0 \le i \le n.$$

We obtain $L = pr_T(L(\gamma))$, which completes the proof. □

A way to implement the restricted splicing operations of the type $\models^{5'}$ in the previous system γ could be the following one.

Consider an encoding $h : (V \cup K \cup \{E\})^* \longrightarrow \{\mathrm{A, C, G, T}\}^*$ of elements of $V \cup K \cup \{E\}$ as strings over the DNA alphabet.

For each rule $r_i \in R_1$ we consider a restriction enzyme enz_i, characterized by the pattern $\begin{bmatrix} u_i z_i v_i \\ \bar{u}_i \bar{z}_i \bar{v}_i \end{bmatrix}$. (We use again the notations established in Chap. 4.) This means that a sequence containing the subsequence $\begin{bmatrix} u_i z_i v_i \\ \bar{u}_i \bar{z}_i \bar{v}_i \end{bmatrix}$ will be cut in such a way that we obtain the sticky ends $\begin{bmatrix} u_i \\ \bar{u}_i \end{bmatrix} \binom{\lambda}{\bar{z}_i}$ and $\binom{z_i}{\lambda} \begin{bmatrix} v_i \\ \bar{v}_i \end{bmatrix}$, respectively. Denote by α_i the string $u_i z_i v_i$ appearing in the upper strand of the pattern associated with $enz_i, 1 \leq i \leq n$.

Take one further restriction enzyme, enz_0, characterized by the pattern $\begin{bmatrix} u_0 z_0 v_0 \\ \bar{u}_0 \bar{z}_0 \bar{v}_0 \end{bmatrix}$; denote $\alpha_0 = u_0 z_0 v_0$. The sequences α_i correspond to the symbols $d_i, 0 \leq i \leq n$.

Then, the elements of γ are constructed as follows:

- the circular axiom:

$$\binom{circ(h(s_0)\alpha_0 h(a_0)\alpha_1\alpha_2 \ldots \alpha_n \alpha_0 h(E)\alpha_1\alpha_2 \ldots \alpha_n \alpha_0)}{\lambda},$$

- for each rule $r_i : sa \to xs' \in P$ or $r_i : s_f E \to E s_0$, $1 \leq i \leq n$, we introduce

  1. the auxiliary string

$$w_i = \binom{\lambda}{\bar{u}_i \bar{z}_i \bar{v}_i} \binom{\lambda}{\bar{u}_{i+1}} \binom{z_{i+1}}{\lambda} \binom{\lambda}{\bar{v}_{i+1}} \cdots \binom{\lambda}{\bar{u}_n} \binom{z_n}{\lambda} \binom{\lambda}{\bar{v}_n}$$
$$\binom{\lambda}{\bar{u}_0} \binom{z_0}{\lambda} \binom{\lambda}{\bar{v}_0} \binom{\lambda}{\overline{h(s)}\bar{u}_0} \binom{z_0}{\lambda} \binom{\lambda}{\bar{v}_0} \binom{\lambda}{\overline{h(a)}\bar{u}_1} \binom{z_1}{\lambda}$$
$$\binom{\lambda}{\bar{v}_1} \cdots \binom{\lambda}{\bar{u}_n} \binom{z_n}{\lambda} \binom{\lambda}{\bar{v}_n} \binom{\lambda}{\bar{u}_0 \bar{z}_0 \bar{v}_0},$$

  2. the axiom xs', for $x = b_1 b_2 \ldots b_k, b_i \in V, 1 \leq i \leq k, k \geq 1$, in the form

$$w_i' = \binom{z_i v_i}{\lambda} \binom{\alpha_{i+1} \ldots \alpha_n \alpha_0 h(b_1) \alpha_1 \ldots \alpha_n \alpha_0 h(b_2) \alpha_1}{\lambda}$$
$$\ldots \binom{\alpha_n \alpha_0 \ldots \alpha_o h(b_k) \alpha_1 \ldots \alpha_n \alpha_0 h(s')}{\lambda} \begin{bmatrix} u_0 \\ \bar{u}_0 \end{bmatrix} \binom{\lambda}{\bar{z}_0},$$

  and

$$w_i'' = \binom{z_i v_i}{\lambda} \binom{\alpha_{i+1} \ldots \alpha_n \alpha_0 h(s')}{\lambda} \begin{bmatrix} u_0 \\ \bar{u}_0 \end{bmatrix} \binom{\lambda}{\bar{z}_0},$$

  when $x = \lambda$,

- for each $s_f \in F$ we also introduce the auxiliary string

$$w(s_f) = \binom{\lambda}{\bar{u}_1\bar{z}_1\bar{v}_1}\binom{\lambda}{\bar{u}_2}\binom{z_2}{\lambda}\binom{\lambda}{\bar{v}_2}\cdots\binom{\lambda}{\bar{u}_n}\binom{z_n}{\lambda}\binom{\lambda}{\bar{v}_n}\binom{\lambda}{\bar{u}_0}$$
$$\binom{z_0}{\lambda}\binom{\lambda}{\bar{v}_0}\binom{\lambda}{\overline{h(s_f)}\bar{u}_0}\binom{z_0}{\lambda}\binom{\lambda}{\bar{v}_0}\binom{\lambda}{\overline{h(E)}\bar{u}_1}\binom{z_1}{\lambda}\binom{\lambda}{\bar{v}_1}$$
$$\cdots\binom{\lambda}{\bar{u}_{n-1}}\binom{z_{n-1}}{\lambda}\binom{\lambda}{\bar{v}_{n-1}}\binom{\lambda}{\bar{u}_n\bar{z}_n\bar{v}_n},$$

- instead of the splicing rules in $R_1 \cup R_2$ we add to the test tube the enzymes $enz_1, \ldots, enz_n, enz_0$.

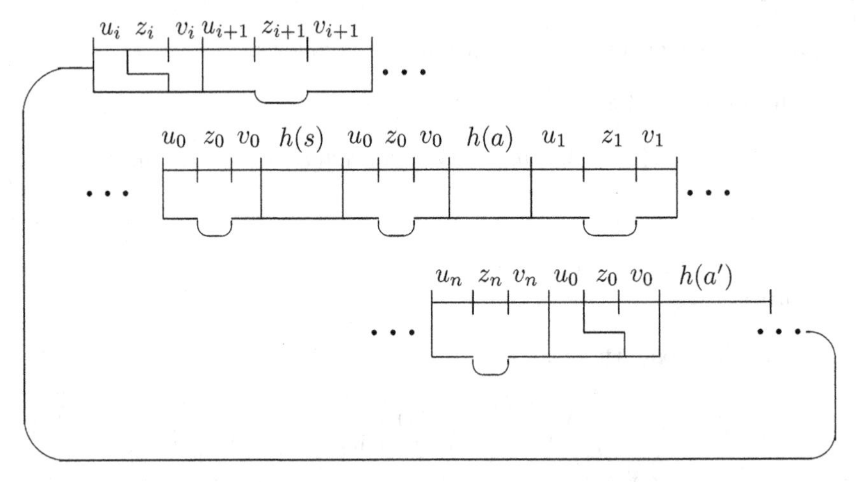

Figure 9.3: Partial (selective) annealing of the auxiliary string

Now, assume that we have a circular single stranded string, w, containing the codes $h(a)$ of several letters $a \in V$, the code of the end marker E, and the code of a state $s \in K$; each two codes $h(a), h(a'), h(E), a, a' \in V$, are separated by the block $\alpha_1\alpha_2\ldots\alpha_n\alpha_0$, whereas $h(s)$ is separated from the code of the left neighboring symbol by the same block $\alpha_1\alpha_2\ldots\alpha_n\alpha_0$ and from the code of the right neighboring symbol by α_0 only. The enzymes cannot cut single stranded sequences. If $h(s)\alpha_0 h(s)$ corresponds to the rule $r_i : sa \to xs'$ in $P \cup \{s_f E \to Es_0 \mid s_f \in F\}$, then the string w_i is also present in the test tube. It will anneal to the single stranded string w in such a way that we obtain a circular word with double stranded portions on the positions of α_i to the left of $h(s)$, of all strings u_j, v_j between this α_i and the string α_0 to the right of $h(a)$, of $h(s)$, of $h(a)$, and of α_0 to the right of $h(a)$; the portions z_i do not anneal, because they are the same in w_i as those in w (they are not complementary).

The situation is illustrated in Fig. 9.3.

Consequently, there are only two places where the restriction enzymes can recognize some patterns and cut: in the left hand portion shown in Fig. 9.3, where enz_i can cut, and in the right hand, where enz_0 can cut.

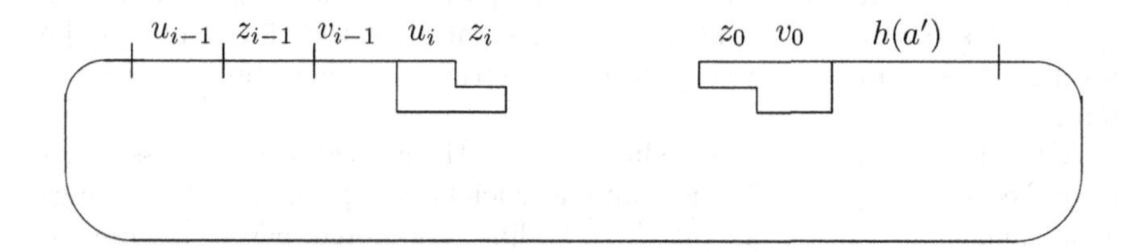

Figure 9.4: The result of the cutting operation

After cutting, we get the situation in Fig. 9.4: one end is $\begin{bmatrix} u_i \\ \bar{u}_i \end{bmatrix} \begin{pmatrix} \lambda \\ \bar{z}_i \end{pmatrix}$ and the other is $\begin{pmatrix} z_0 \\ \lambda \end{pmatrix} \begin{bmatrix} v_0 \\ \bar{v}_0 \end{bmatrix}$. These are exactly the ends corresponding to w_i' or w_i''. By ligation, we get again a circular word, of the type illustrated in Fig. 9.5 (we have not specified the parts of x, $h(b_i)$, and blocks $\alpha_1 \dots \alpha_n \alpha_0$ between them).

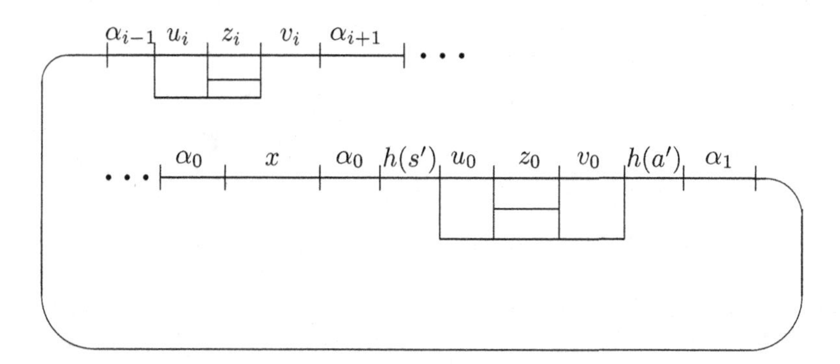

Figure 9.5: The result of the ligation

We have two double stranded portions, one corresponding to $u_i z_i$ (hence *not* to the whole α_i) and one corresponding to α_0. The enzyme enz_0 can cut again, but then the string is lost, because no terminal string can be derived from it, $h(s')$ will never be removed.

We have returned to a circular word similar to that we started with – modulo the existence of the two mentioned double stranded portions. We remove them (we do not know how this can be done without also denaturing double stranded sequences in $A_1 \cup A_2$) and we repeat the procedure. When encountering the marker E we either continue or we anneal the string $w(s_f)$.

It will produce a circular sequence similar to that in Fig. 9.3, making possible the cutting of $u_1z_1v_1$ by enz_1 at the left of the block $h(s_f)\alpha_0 h(E)$ and of $u_n z_n v_n$ by enz_n at the right of $h(s_f)\alpha_0 h(E)$. No ligation is possible, we get two linear strings which cannot enter further annealings with auxiliary symbols, hence the work of the system stops here. The strings containing only blocks $h(a), a \in T$, and $\alpha_0, \alpha_1, \ldots, \alpha_n$ are selected (filtered) and the result of the computation consists of the strings obtained by ignoring the blocks $\alpha_0, \alpha_1, \ldots, \alpha_n$.

Of course, the previous procedure is just a theoretical proposal, assuming error-free annealings and ligation, and, a crucial assumption, that two strings which anneal are linearly arranged, no folding is accepted, hence the complementary fragments are placed at the same distances in the two strands. How to get rid of this strong assumption (replacing it with unstable annealings when the matching fragments are not long enough, or by other techniques) is not an easy question.

9.3 Bibliographical Notes

The problem of splicing circular strings is formulated in [87]. Sect. 9.1 is based on [176], [177]. Circular strings are also considered in [206], [223]. Sect. 9.2 is based on [142], where a similar operation is investigated.

Chapter 10

Distributed H Systems

One of the important drawbacks of the models considered in the previous chapters is the fact that we need several splicing rules. Each rule corresponds to two restriction enzymes. However, each enzyme needs specific temperature, acidity, salinity, and other reaction conditions. This means that, in general, from using a splicing rule to using another one we have to change these reaction conditions. This operation dramatically decreases the efficiency of the computation (in terms of duration), hence should be avoided as much as possible.

One suggestion to accomplish this comes from the grammar systems area: using distributed architectures, separating parts of the model which are able to work independently, preferably in parallel, arranging a way of cooperating between these parts, and synthesizing the result of the computation from the partial results produced by the mentioned parts. We are led in this way to structures similar to parallel communicating grammar systems.

10.1 Splicing Grammar Systems

The first model we consider here is rather similar to a parallel communicating grammar system: the components are usual context-free grammars, working separately, synchronously, on their own sentential forms, and splicing their sentential forms according to a given set of splicing rules. Thus, we have a hybrid model, involving both rewriting operations and splicing operations. This, of course, is not completely implementable in biochemical terms, but this could be a more realistic approach to molecular computing than confining systems to only biochemical tools: devise hybrid systems, containing both DNA modules and operations and classic, electronic modules.

From a generative point of view, our systems, called *splicing grammar systems*, turn out to be computationally complete, even when having components of a rather simple type.

A *splicing grammar system* (of degree $n, n \geq 1$) is a construct

$$\Gamma = (N, T, (S_1, P_1), (S_2, P_2), \ldots, (S_n, P_n), R),$$

where

(i) N, T are disjoint alphabets, $S_i \in N$, and $P_i, 1 \leq i \leq n$, are finite sets of rewriting rules over $N \cup T$,

(ii) R is a finite subset of $(N \cup T)^*\#(N \cup T)^*\$(N \cup T)^*\#(N \cup T)^*$, with $\#, \$$ two distinct symbols not in $N \cup T$.

The sets P_i are called the *components* of Γ.

For two n-tuples (we call them *configurations*) $x = (x_1, x_2, \ldots, x_n)$, and $y = (y_1, y_2, \ldots, y_n)$, $x_i, y_i \in (N \cup T)^*, 1 \leq i \leq n$, we write $x \Longrightarrow y$ if and only if one of the following two conditions holds:

(i) for each $1 \leq i \leq n$, either $x_i \Longrightarrow_{P_i} y_i$, or $x_i = y_i \in T^*$,

(ii) there exist $1 \leq i, j \leq n$ such that $x_i = x_i' u_1 u_2 x_i''$, $x_j = x_j' u_3 u_4 x_j''$, and $y_i = x_i' u_1 u_4 x_j''$, $y_j = x_j' u_3 u_2 x_i''$, for $u_1 \# u_2 \$ u_3 \# u_4 \in R$; for all $k \neq i, j$, we have $y_k = x_k$.

In the above definition, point (i) defines a rewriting step, whereas point (ii) defines a splicing step, corresponding to a communication step in parallel communicating grammar systems. Note that no priority of any of these operations over the other one is assumed. In case (ii) we denote the passing from (x_i, x_j) to (y_i, y_j) by $(x_i, x_j) \models (y_i, y_j)$, as usual.

The language generated by the system Γ is the language generated by its first component, that is,

$$L(\Gamma) = \{x_1 \in T^* \mid (S_1, \ldots, S_n) \Longrightarrow^* (x_1, \ldots, x_n),\ x_j \in (N \cup T)^*, 2 \leq j \leq n\}.$$

We denote by $SGS_n(X)$ the families of languages $L(\Gamma)$, generated by splicing grammar systems of degree at most $n, n \geq 1$, with components of type X. We consider here $X \in \{REG, RL, CF\}$, where for REG we use regular rules in the restricted sense, RL indicates the use of right-linear rules, and for CF we use λ-free context-free rules. When no restriction is imposed on the number of components, then we replace the subscript n with $*$.

Note that in view of the definition of the relation $\Longrightarrow$ between configurations of a splicing grammar system, we implicitly use multiplicities of strings, because each component of the system has in every moment – both after a rewriting and a splicing step – exactly one current string; no string of arbitrary multiplicity is used.

Surprisingly enough (similar results are not known for parallel communicating grammar systems), we can characterize RE by such systems with only two λ-free context-free components.

Theorem 10.1. *$RE = SGS_n(CF) = SGS_*(CF)$, for all $n \geq 2$.*

Proof. The inclusions $SGS_n(CF) \subseteq SGS_{n+1}(CF) \subseteq SGS_*(CF) \subseteq RE$, $n \geq 1$, are obvious. We have only to prove the inclusion $RE \subseteq SGS_2(CF)$.

Consider a language $L \subseteq T^*, L \in RE$, and take a grammar $G = (N, T, S, P)$ in the Geffert normal form as specified in Theorem 3.5(1): $N = \{S, A, B, C\}$ and P contains context-free rules of the form $S \to x, x \in (\{S, A, B, C\} \cup T)^+$, as well as the rule $ABC \to \lambda$. Denote by P' the set of context-free rules in P. We construct the splicing grammar system

$$\Gamma = (N \cup \{X, Y, Z, S_2\}, T, (S, P_1), (S_2, P_2), R),$$

with

$$P_1 = P' \cup \{A \to A\},$$
$$P_2 = \{S_2 \to YXXABCZ,\ X \to XX, X \to X\},$$
$$R = \{\#ABC\$YX\#XABC,\ \#XABC\$YXABC\#\}.$$

We claim that $L(G) = L(\Gamma)$.

Let us examine the work of Γ. Since $P' \subseteq P_1$, each terminal string w such that $S \Longrightarrow^* w$ using rules in P' is in $L(\Gamma)$. (As P_2 contains the rule $X \to X$, no restriction on the length of the derivations is imposed.) Moreover, each string $w \in (N \cup T)^*$ such that $S \Longrightarrow_{P'}^* w$ can be reproduced in component P_1 of Γ. The problem remains of simulating in Γ the erasing rule $ABC \to \lambda$ in P.

Consider the first step of a derivation when a splicing operation is performed. This means that the current configuration is of the form

$$(uABCv,\ YX^iABCZ),$$

for some $i \geq 2$. Since X does not appear in $uABCv$, only the first splicing rule can be used. This is possible only when $i = 2$ (hence the rule $X \to XX$ has not been used in the second component). We get

$$(u|ABCv,\ YX|XABCZ) \models (uXABCZ,\ YXABCv).$$

Case 1: We continue by applying the second splicing rule.

Then we get:

$$(u|XABCZ,\ YXABC|v) \models (uv,\ YXABCXABCZ).$$

The substring ABC has been removed from $uABCv$. Observe that in the second component we have obtained a string with the prefix $YXABC$, hence by using the rule $X \to XX$ in P_2 we can obtain $YXXABC$ (at the same time, $A \to A$ can be used in P_1, providing that the first component has not reached a terminal string), that is the same string we have started with (continued with a suffix ended with the symbol Z). Let us consider, in general, that we have a configuration

$$(uABCv,\ YXXABCwZ),$$

with $w \in (X^+ABC)^*$; initially we have $w = \lambda$, after removing one subword ABC we have $w = X^jABC$, for some $j \geq 1$. By two splicings as above, we obtain

$$(u|ABCv,\ YX|XABCwZ) \models (u|XABCwZ,\ YXABC|v)$$
$$\models (uv,\ YXABCXABCwZ).$$

Using the rule $X \to XX$ in P_2 and $A \to A$ in P_1, we get the configuration $(uv, YXXABCXABCwZ)$. The form of the configuration is preserved, hence the operation can be iterated until removing all occurrences of ABC from the string of the first component. If some rewriting steps are mixed with the splicing steps, then the rewritings in P_1 either correspond to rewritings in G using rules in P' on the prefix u above, or they use the rule $A \to A$, hence change nothing in the first component; if the second component uses the rule $X \to X$, then nothing is changed. Using the rule $X \to XX$ is the subject of the second case and this is discussed below. Consequently, in this way we can generate each string in $L(G)$, that is $L(G) \subseteq L(\Gamma)$.

The second splicing above can also be performed for a substring $XABC$ in w:

$$(uXABCw_1|XABCw_2Z,\ YXABC|v)$$
$$\models (uXABCw_1v,\ YXABCXABCw_2Z).$$

In order to remove the symbol X from the first string (or from a string obtained after rewriting the first string by rules in P') we have to eventually use the second splicing rule, cutting the string in front of X. Assume that we perform this operation for the previous configuration (the result is similar for a configuration derived from it). We have

$$(u|XABCw_1v,\ YXABC|XABCw_2Z)$$
$$\models (uXABCw_2Z,\ YXABCXABCw_1v).$$

An occurrence of X (as well as of Z) is again present in the first component. Consequently, the nonterminal X cannot be eliminated, the derivation will never produce a terminal string.

Case 2: After using the first splicing rule, we perform a derivation step when $X \to XX$ is used in P_2.

We consider the general form of a configuration when this operation can occur:

$$(u|ABCv,\ YX|XABCwZ) \models (uXABCwZ,\ YXABCv)$$
$$\Longrightarrow (u'XABCwZ,\ YXXABCv).$$

We can have $u' = u$ when the rule $A \to A$ has been used in P_1, or $u \Longrightarrow u'$ by a rule in P'. If $X \to XX$ is used again in P_2, then a splicing can never be performed, hence no terminal string can be ever obtained.

Assume that we have performed several rewriting steps and we want now to splice again using the first rule. We have two possibilities:

(2.1.) Splicing for some occurrence of ABC in u':

$$(u_1|ABCu_2XABCwZ,\ YX|XABCv)$$
$$\models (u_1XABCv,\ YXABCu_2XABCwZ).$$

As we have seen at the end of the discussion about Case 1, the symbol X can never be eliminated from the string in the first component.

(2.2.) Splicing for some occurrence of ABC in w:

$$(u'XABCw_1X|ABCw_2Z,\ YX|XABCv)$$
$$\models (u'XABCw_1XXABCv,\ YXABCw_2Z).$$

If $w_2 \neq \lambda$, this implies that w_2 contains at least one occurrence of X and then the symbol X will never be eliminated from the string in the first component. When $w_2 = \lambda$, and we splice as

$$(u'XABCw_1X|XABCv,\ YXABC|Z)$$
$$\models (u'XABCw_1XZ,\ YXABCXABCv),$$

from now on X will be always present in the first component. The same conclusion is obtained if we splice using a site $XABC$ in w_1.

If $w_2 = \lambda$ and we remove the first occurrence of X in the first component, then we get:

$$(u'|XABCw_1XXABCv,\ YXABC|Z)$$
$$\models (u'Z,\ YXABCXABCw_1XXABCv).$$

The nonterminal Z can now be removed only using the first splicing rule. Assume that we perform

$$(u_1'|ABCu_2'Z,\ YX|XABCw'v) \models (u_1'XABCw'v,\ YXABCu_2'Z).$$

In order to remove the symbol X from the string in the first component, we will introduce again Z. The derivation will never produce a terminal string in any of the two components.

This concludes the discussion about Case 2, proving that no string outside $L(G)$ can be produced, hence $L(\Gamma) \subseteq L(G)$, too. Therefore, we obtain $L(\Gamma) = L(G)$. □

Remark 10.1. We can define a language associated to a splicing grammar system Γ as above also in a "more democratic" manner, as the union of all terminal strings generated by each component of Γ. More specifically, if $L_i(\Gamma)$ is the language generated by the ith component of Γ, then we can consider the language $L_t(\Gamma) = \bigcup_{i=1}^{n} L_i(\Gamma)$. In the proof above we have $L_1(\Gamma) = L_t(\Gamma)$, because $L_2(\Gamma) = \emptyset$ (the nonterminal Y is always present in the left hand end of the strings in the second component of the system). Therefore, also for the language $L_t(\Gamma)$, too, we get a characterization of RE as in Theorem 10.1. □

The previous result is optimal as far as it concerns the number of components. Is it also optimal as regards the type of rewriting rules used in its components? What about systems with regular or with right-linear rules? These systems are also rather powerful. Proofs of the following two inclusions can be found in [75]:

$$LIN \subseteq SGS_2(REG),$$
$$CF \subseteq SGS_3(REG).$$

Both these inclusions are proper.

Indeed, let us consider the splicing grammar system

$$\Gamma = (\{S_1, S_2, A, B, C\}, \{a, b, c, d\}, (S_1, P_1), (S_2, P_2), R),$$

with

$$P_1 = \{S_1 \to aS_1,\ S_1 \to aA,\ B \to cB,\ C \to c\},$$
$$P_2 = \{S_2 \to dB,\ B \to bB,\ A \to cA,\ A \to cC\},$$
$$R = \{a\#A\$b\#B,\ c\#B\$d\#b\}.$$

The derivations in Γ proceed as follows. After an initial rewriting phase,

$$(S_1,\ S_2) \Longrightarrow (aS_1,\ dB) \Longrightarrow^* (a^{n+1}S_1,\ db^nB) \Longrightarrow (a^{n+2}A,\ db^{n+1}B),$$

$n \geq 0$, or consisting of only one step,

$$(S_1, S_2) \Longrightarrow (aA, dB),$$

we have to perform a splicing, according to the first rule, and this is the only way to reach a terminal string, because A cannot be rewritten to a terminal string in the first component and B cannot be rewritten to a terminal string in the second component. No splicing can be performed in the second case. In the first one, we get

$$(a^{n+2}|A,\ db^{n+1}|B) \models (a^{n+2}B,\ db^{n+1}A).$$

At least one rewriting step must be performed (no splicing rule can be used now):

$$(a^{n+2}B,\ db^{n+1}A) \Longrightarrow^* (a^{n+2}c^mB,\ db^{n+1}c^mA),\ m \geq 1.$$

If we splice again, using the second rule, then we bring the symbol A to the first component and the work of the system is blocked:

$$(a^{n+2}c^m|B,\ d|b^{n+1}c^m A) \models (a^{n+2}c^m b^{n+1}c^m A,\ dB).$$

Therefore, we have to rewrite A by cC in the second component; C cannot be rewritten in P_2, hence a splicing must be immediately performed and after that C is replaced with c in the first component:

$$\begin{aligned}&(a^{n+2}c^m B,\ db^{n+1}c^m A) \Longrightarrow (a^{n+2}c^{m+1}|B,\ d|b^{n+1}c^{m+1}C)\\ &\models (a^{n+2}c^{m+1}b^{n+1}c^{m+1}C,\ dB) \Longrightarrow (a^{n+2}c^{m+1}b^{n+1}c^{m+2},\ dbB).\end{aligned}$$

Consequently,

$$L(\Gamma) = \{a^{n+1}c^m b^n c^{m+1} \mid n \geq 1, m \geq 2\}.$$

Clearly, this language is not context-free.

In fact, splicing grammar systems with three regular components "almost characterize" the recursively enumerable languages.

Theorem 10.2. *Each language $L \subseteq T^*, L \in RE$, can be written in the form $L = L' \cap T^*$, for some $L' \in SGS_3(REG)$.*

Proof. Take L as in the proof of Theorem 10.1, generated by a grammar $G = (N, T, S, P)$ in the Geffert normal form, with $N = \{S, A, B, C\}$. We construct a splicing grammar system for which the symbols of N will be considered terminals.

Consider the system

$$\Gamma = (N', T', (S_1, P_1), (S_2, P_2), (S_3, P_3), R),$$

where

$$\begin{aligned}
N' &= \{S_1, S_2, S_3, X, X', Y, Z, Z', Z''\},\\
T' &= \{S, A, B, C\} \cup T \cup \{d_0, d_1, d_2, d_3\},\\
P_1 &= \{S_1 \to SY,\ Y \to d_0Y,\ X \to d_0X',\ Z \to d_0Z'\},\\
P_2 &= \{S_2 \to d_1X,\ X \to d_3Z,\ Y \to d_0Y,\ Z' \to d_3Z,\ Z' \to d_3Z''\}\\
&\quad \cup \{X \to \alpha X \mid \alpha \in \{S, A, B, C\} \cup T\},\\
P_3 &= \{S_3 \to d_2X,\ X' \to d_2X,\ X \to d_2X\},\\
R &= \{\#S\$d_1\#xX \mid S \to x \in P\}\\
&\quad \cup \{\#d_0X'\$d_1S\#,\ d_1\#Sd_0\$d_2\#X,\ \#ABC\$d_3\#Z,\\
&\quad\quad \#d_0Z'\$d_3ABC\#,\ \#d_0d_0\$Z''\#\}.
\end{aligned}$$

The idea behind this construction is the following.

Each derivation in Γ has two phases, in the first one simulates the use of rules $S \to x$ in P, and in the second one simulates the rule $ABC \to \lambda$. The passing from the first phase to the second one is marked by replacing X with Z in the second component.

In the first phase, in a nondeterministic way, P_2 generates a string d_1xX, for some $x \in (\{S, A, B, C\} \cup T)^*$. Such a string can be used in a splicing only when $S \to x$ is a rule in P. Thus, P_2 has the role of producing the right-hand members of rules in P. By the splicing rules in R which do not contain the symbol Z and its prime variants, we can then simulate the use of the rule $S \to x$ for rewriting the only occurrence of S in the first component. Between the splicing operations, rewriting steps are necessary, just for changing the nonterminals. During such rewritings, the dummy terminal d_0 is introduced in P_1 and P_2, and d_2 in P_3. In order to iterate the process without producing parasitic strings (here, this means strings in $(L(\Gamma) \cap T^*) - L$), we need the third component. Its role is to "clean" the second component (see explanations below).

After introducing the symbol Z, we start the simulation of the rule $ABC \to \lambda$, and this is performed in a similar way as in the proof of Theorem 10.1. No rule $S \to x$ in P can be simulated during this phase; this is not losing generality because the derivation in a grammar in the Geffert normal form can be arranged in such a way to use first the context-free rules and then the erasing rule, without modifying the generated string.

Let us examine in some detail the work of Γ. We start with

$$(S_1,\ S_2,\ S_3) \Longrightarrow (SY,\ d_1X,\ d_2X).$$

No splicing can be performed (1) before producing a string x in the second component such that $S \to x \in P$, or (2) before introducing the symbol Z. For a splicing in case (2) we need at least a substring ABC of the string in the first component. Hence, we have to continue with case (1):

$$(SY,\ d_1X,\ d_2X) \Longrightarrow^* (Sd_0^nY,\ d_1xX,\ d_2^{n+1}X),\ n = |x| \geq 1.$$

The process is blocked when x cannot be continued in such a way as to obtain a string xx' with $S \to xx' \in P$; x' can be the empty string. Assume that $S \to x \in P$. Then we can perform a splicing:

$$(|Sd_0^nY,\ d_1|xX,\ d_2^{n+1}X) \models (xX,\ d_1Sd_0^nY,\ d_2^{n+1}X).$$

If we continue by

$$(xX,\ d_1|Sd_0^nY,\ d_2^{n+1}|X) \models (xX,\ d_1X,\ d_2^{n+1}Sd_0^nY),$$

then the system is blocked, no rewriting and no splicing is possible. We have, instead, to rewrite and then splice:

$$\begin{aligned}&(xX,\ d_1Sd_0^nY,\ d_2^{n+1}X) \Longrightarrow (x|d_0X',\ d_1S|d_0^{n+1}Y,\ d_2^{n+2}X)\\ &\models (xd_0^{n+1}Y,\ d_1Sd_0X',\ d_2^{n+2}X).\end{aligned}$$

The only possible continuation is a splicing again:

$$(xd_0^{n+1}Y,\ d_1|Sd_0X',\ d_2^{n+2}|X) \models (xd_0^{n+2}Y,\ d_1X,\ d_2^{n+2}Sd_0X').$$

We have returned to a configuration with the nonterminals Y, X in the first components and with the string in the second component "cleaned", equal again to d_1X. We can iterate the process.

Let us display an arbitrary step of this type. For the dummy symbols d_0, d_2 we will write d_0^*, d_2^* in order to indicate that they appear in a number of copies which is not of interest for our argument. A similar writing concerns the blocks $d_2^*Sd_0$ in the third component. Hence, we have:

$$\begin{aligned}
&(u|Svd_0^*Y,\ d_1|xX,\ (d_2^*Sd_0)^*X)\\
&\models (uxX,\ d_1Svd_0^*Y,\ (d_2^*Sd_0)^*X)\\
&\Longrightarrow (ux|d_0X',\ d_1S|vd_0^*Y,\ (d_2^*Sd_0)^*d_2X) \qquad (*)\\
&\models (uxvd_0^*Y,\ d_1|Sd_0X',\ (d_2^*Sd_0)^*d_2|X)\\
&\models (uxvd_0^*Y,\ d_1X,\ (d_2^*Sd_0)^*d_2Sd_0X')\\
&\Longrightarrow^* (uxvd_0^*Y,\ d_1yX,\ (d_2^*Sd_0)(d_2Sd_0)d_2^*X).
\end{aligned}$$

The process also runs as described above in the general case.

Note that from the configuration $(*)$ we can also splice the strings of the second and the third components, providing $v = \lambda$, but after such an operation the system will be blocked, no rewriting (because of Y in the third component) and no splicing is possible.

When X is present in the second component, we can also use the rule $X \rightarrow d_3Z$, which determines the passing to the second phase of the derivation, that where substrings ABC can be removed from the string in the first component:

$$(uABCvd_0^*Y,\ d_1wX,\ -X) \Longrightarrow (uABCvd_0^*Y,\ d_1wd_3Z,\ -X)$$

(from now on, the third component plays no role, hence we ignore its string; we only point out the nonterminal X in order to see that no splicing can involve this string, and that rewriting operations can be done for ever, no limit on the length of the derivation is imposed in this way). We have to splice

$$(u|ABCvd_0^*Y,\ d_1wd_3|Z,\ -X) \models (uZ,\ d_1wd_3ABCvd_0^*Y,\ -X).$$

No further splicing is possible, we have to rewrite,

$$\Longrightarrow (ud_0Z',\ d_1wd_3ABCvd_0^*Y,\ -X).$$

Now, we have to splice:

$$\begin{aligned}
&(u|d_0Z',\ d_1wd_3ABC|vd_0^*Y,\ -X)\\
&\models (uvd_0^*Y,\ d_1wd_3ABCd_0Z',\ -X). \qquad (**)
\end{aligned}$$

After a rewriting,

$$\Longrightarrow (uvd_0^*Y,\ d_1wd_3ABCd_0d_3Z,\ -X),$$

we have again the substring d_3Z present in the second component. If uv contains further occurrences of ABC, then we can splice again, removing them. Note that two rewritings are performed at each such iteration, hence the string in the first component is of the form uvd^iY for some $i \geq 2$.

All substrings ABC can be removed in this way. However, at any step when reaching a configuration of the form (**), the rule $Z' \to d_3Z''$ can be used in P_2. This determines the elimination of the nonterminal in the first component:

$$\begin{aligned}(uvd_0^iY,\ zZ',\ -X) &\Longrightarrow (uvd_0^j|d_0^kY,\ zd_3Z''|,\ -X)\\ &\models (uvd_0^j,\ zd_3Z''d_0^kY,\ -X).\end{aligned}$$

A terminal string (from the point of view of Γ) is obtained.

Note that the splicing rule $\#d_0d_0\$Z''\#$ cannot be used for splicing the strings in the third and the second components, because the string in the third component does not contain substrings d_0d_0 (this can be easily seen by examining the derivations discussed above).

It is now clear that if all substrings ABC were removed and if the last splicing has been done in the front of the suffix d_0^iY (hence $j = 0$ in the previous writing), then we get a string in $L(G)$. Consequently, $L(G) = L(\Gamma) \cap T^*$. □

Corollary 10.1. *For every family FL such that $FL \subset RE$ and FL is closed under intersection with regular sets, we have $SGS_3(REG) - FL \neq \emptyset$.*

Many important families in the basic Chomsky hierarchy, in the regulated rewriting area, and in the Lindenmayer area have these properties: languages generated by matrix grammars with and without appearance checking but without using λ-rules, languages generated by matrix grammars with λ-rules and without appearance checking, $ET0L$, CS, recursive, etc. $SGS_3(REG)$ contains languages outside each of them, which proves once again the power of *cooperation* in the parallel communicating grammar systems style, combined with the power of the *splicing* operation.

We do not know whether or not $RE = SGS_n(REG)$, for some n. However, if we allow chain rules, $X \to Y$, and rules of the form $X \to xY$, for X, Y nonterminal symbols and x a terminal string, then we can get a characterization of RE. The proof of the following result can be found in [75].

Theorem 10.3. $RE = SGS_4(RL)$.

We have considered splicing grammar systems of a hybrid character. Distributed H systems using only splicing operations (and not also rewriting operations) will be proposed in the following three sections.

10.2 Communicating Distributed H Systems

The model we consider in this section is the splicing counterpart of the parallel communicating grammar systems with communication by command: the components work by splicing and communicate by sending to each other strings which pass certain filters specified in advance.

A *communicating distributed H system* (of degree $n, n \geq 1$) is a construct

$$\Gamma = (V, T, (A_1, R_1, V_1), \ldots, (A_n, R_n, V_n)),$$

where V is an alphabet, $T \subseteq V$, A_i are finite languages over V, R_i are finite sets of splicing rules over V, and $V_i \subseteq V$, $1 \leq i \leq n$.

Each triple $(A_i, R_i, V_i), 1 \leq i \leq n$, is called a *component* of Γ; A_i, R_i, V_i are the set of *axioms*, the set of *splicing rules*, and the *selector* (or *filter*) of the component i, respectively; T is the terminal alphabet of the system.

We denote

$$B = V^* - \bigcup_{i=1}^{n} V_i^*.$$

The pair $\sigma^{(i)} = (V, R_i)$ is the underlying H scheme associated to the component i of the system.

An n-tuple $(L_1, \ldots, L_n), L_i \subseteq V^*, 1 \leq i \leq n$, is called a *configuration* of the system; L_i is also called the *contents* of the ith component, understanding the components as test tubes where the splicing operations are carried out.

For two configurations $(L_1, \ldots, L_n), (L'_1, \ldots, L'_n)$, we define

$$(L_1, \ldots, L_n) \Longrightarrow (L'_1, \ldots, L'_n) \text{ iff}$$
$$L'_i = \bigcup_{j=1}^{n} (\sigma_2^{(j)*}(L_j) \cap V_i^*) \cup (\sigma_2^{(i)*}(L_i) \cap B),$$
$$\text{for each } i, 1 \leq i \leq n.$$

In words, the contents of each component are spliced according to the associated set of rules (we pass from L_i to $\sigma_2^{(i)*}(L_i), 1 \leq i \leq n$), and the result is redistributed among the n components according to the selectors $V_1, \ldots, V_n$; the part which cannot be redistributed (which does not belong to some $V_i^*, 1 \leq i \leq n$) remains in the component. As we have imposed no restriction over the alphabets V_i, for example, we did not suppose that they are pairwise disjoint, when a string in $\sigma_2^{(j)*}(L_j)$ belongs to several languages V_i^*, then copies of this string will be distributed to all components i with this property.

The language generated by Γ is defined by

$$L(\Gamma) = \{w \in T^* \mid w \in L_1 \text{ for some } L_1, \ldots, L_n \subseteq V^* \text{ such that } (A_1, \ldots, A_n) \Longrightarrow^* (L_1, \ldots, L_n)\}.$$

That is, the first component of the system is designated as its *master* and the language of Γ is the set of all terminal strings generated (or collected by communications) by the master.

We denote by CDH_n the family of languages generated by communicating distributed H systems of degree at most $n, n \geq 1$. When n is not specified, we replace the subscript n with $*$.

Another possibility is to consider as the language generated by Γ, the union of all languages generated by its components, in a similar way as discussed in Remark 10.1, but we do not follow up this suggestion here.

Communicating distributed H systems characterize RE. Before proving this result, let us examine an example:

Consider the system

$$\begin{aligned}
\Gamma &= (\{a,b,c,d,e\}, \{a,b,c\}, (A_1, R_1, V_1), (A_2, R_2, V_2)),\\
A_1 &= \{cabc, ebd, dae\},\\
R_1 &= \{b\#c\$e\#bd,\ da\#e\$c\#a\},\\
V_1 &= \{a,b,c\},\\
A_2 &= \{ec, ce\},\\
R_2 &= \{b\#d\$e\#c,\ c\#e\$d\#a\},\\
V_2 &= \{a,b,d\}.
\end{aligned}$$

The only possible splicings in the first component are:

$$\begin{aligned}
(x_1b|c, e|bd) &\models (x_1bbd, ec), \quad \text{for } x_1 \in \{a,b,c,d,e\}^*,\\
(da|e, c|ax_2) &\models (daax_2, ce), \quad \text{for } x_2 \in \{a,b,c,d,e\}^*.
\end{aligned}$$

One further occurrence of a and one further occurrence of b can be added in this way to the strings x_1bc, cax_2, respectively (at the same time, c is replaced by d, in both cases). We start from $cabc$; from the second component we communicate strings over $\{a,b,c\}$. From the first component we can communicate to the second one only strings over $\{a,b,d\}$. This means that, starting from a string ca^nb^mc, $n,m \geq 1$ (initially, $n = m = 1$), in the first component we produce $da^{n+1}b^{m+1}d$, no further splicing can be done here, and the string is communicated to the second component; moreover, both rules in R_1 must be applied, otherwise a symbol c is still present in the string. In the second component we can perform:

$$\begin{aligned}
(\alpha a^ib^j|d, e|c) &\models (\alpha a^ib^jc, ed), \quad \alpha \in \{c,d\},\\
(c|e, d|a^ib^j\alpha) &\models (ca^ib^j\alpha, de), \quad \alpha \in \{c,d\}.
\end{aligned}$$

Only the string ca^ib^jc can be communicated to the first component, hence the process can be iterated.

Consequently, we obtain

$$L(\Gamma) = \{ca^nb^nc \mid n \geq 1\},$$

which is not a regular language.

Thus, $CDH_2 - REG \neq \emptyset$.

Theorem 10.4. *$RE = CDH_n = CDH_*$, for all $n \geq 3$.*

Proof. The inclusions $CDH_n \subseteq CDH_{n+1} \subseteq CDH_* \subseteq RE, n \geq 1$, are obvious. We have only to prove the inclusion $RE \subseteq CDH_3$.

Consider a type-0 grammar $G = (N, T, S, P)$, take a new symbol, B, and denote, for an easy reference,

$$N \cup T \cup \{B\} = \{D_1, D_2, \ldots, D_n\}.$$

Since $N \neq \emptyset, T \neq \emptyset$, we have $n \geq 3$. We construct the communicating distributed H system

$$\Gamma = (V, T, (A_1, R_1, V_1), (A_2, R_2, V_2), (A_3, R_3, V_3)),$$

with

$$\begin{aligned} V &= N \cup T \cup \{X, Y, X', Y', Z, B\} \\ &\cup \{X_i, Y_i \mid 0 \leq i \leq 2n\}, \end{aligned}$$

$$\begin{aligned} A_1 &= \{XBSY\} \cup \{ZvY \mid u \to v \in P\} \\ &\cup \{X_{2i}D_iZ \mid 1 \leq i \leq n\} \\ &\cup \{X_{2i}Z \mid 0 \leq i \leq n-1\} \\ &\cup \{ZY_{2i} \mid 0 \leq i \leq n\}, \\ R_1 &= \{\#uY\$Z\#vY \mid u \to v \in P\} \\ &\cup \{\#D_iY\$Z\#Y_{2i},\ X\#\$X_{2i}D_i\#Z \mid 1 \leq i \leq n\} \\ &\cup \{\#Y_{2i+1}\$Z\#Y_{2i},\ X_{2i+1}\#\$X_{2i}\#Z \mid 0 \leq i \leq n-1\}, \\ V_1 &= N \cup T \cup \{B, X, Y\} \cup \{X_{2i+1}, Y_{2i+1} \mid 0 \leq i \leq n-1\}, \end{aligned}$$

$$\begin{aligned} A_2 &= \{ZY_{2i-1}, X_{2i-1}Z \mid 1 \leq i \leq n\} \cup \{ZZ\}, \\ R_2 &= \{\#Y_{2i}\$Z\#Y_{2i-1},\ X_{2i}\#\$X_{2i-1}\#Z \mid 1 \leq i \leq n\} \\ &\cup \{X'B\#\$\#ZZ,\ \#Y'\$ZZ\#\}, \\ V_2 &= N \cup T \cup \{B, X', Y'\} \cup \{X_{2i}, Y_{2i} \mid 1 \leq i \leq n\}, \end{aligned}$$

$$\begin{aligned} A_3 &= \{ZY, XZ, ZY', ZX'\}, \\ R_3 &= \{\#Y_0\$Z\#Y,\ X_0\#\$X\#Z,\ \#Y_0\$Z\#Y',\ X_0\#\$X'\#Z\}, \\ V_3 &= N \cup T \cup \{B, X_0, Y_0\}. \end{aligned}$$

Let us examine the work of Γ. The underlying idea is again rotate-and-simulate. Starting from strings of the form XwY (the axiom $XBSY$ is of this form), the first component can simulate the rules of P in a suffix of w, by using splicing rules $\#uY\$Z\#vY$, for $u \to v \in P$, or can start rotating the string. In the first case, the string obtained is again bounded by the markers X, Y, hence the process can be iterated. When removing a symbol D_i from the right hand end of w one replaces Y with Y_{2i}:

$$(Xw_1|D_iY, Z|Y_{2i}) \models (Xw_1Y_{2i}, ZD_iY),$$

providing that $w = w_1D_i, 1 \leq i \leq n$ (observe that B can be removed like any symbol in $N \cup T$).

No string containing an occurrence of Z can be moved from a component to another one. If such strings obtained by splicing enter new splicing operations, then no terminal string can be obtained using the resulting strings: both of them contain the symbol Z and by splicing them no new string is obtained. Consider, for instance, the string ZD_iY. Using again the rule $\#D_i\$Z\#Y_{2i}$ we obtain the strings ZY_{2i}, ZD_iY. A similar result will be obtained in all cases below.

The string Xw_1Y_{2i} cannot be communicated, but a further splicing is possible in the first component:

$$(X|w_1Y_{2i}, X_{2j}D_j|Z) \models (XZ, X_{2j}D_jw_1Y_{2i}),$$

for some $1 \leq j \leq n$. The two operations can be performed in the reverse order and the result is the same.

Strings bounded by markers X_r, Y_s with even r, s cannot enter new splicings in the first component and can be communicated to the second component. Two splicings are possible here, decreasing by one the subscripts of X and Y. If only one splicing is performed, then the string cannot be communicated. Thus, we get:

$$(X_{2j}|D_jw_1Y_{2i}, X_{2j-1}|Z) \models (X_{2j}Z, X_{2j-1}D_jw_1Y_{2i}),$$
$$(X_{2j-1}D_jw_1|Y_{2i}, Z|Y_{2i-1}) \models (X_{2j-1}D_jw_1Y_{2i-1}, ZY_{2i}).$$

Again, the order of the two operations is not important.

A string with odd subscripts of the end markers can be communicated to the first component. These operations can be iterated and they must be continued, otherwise there is no way to remove the nonterminal symbols. When in the first component we obtain X_0 or Y_0, the string can no longer be communicated to the second component. If one of the end markers X, Y has the subscript 0 and the other subscript is strictly larger, then the string is "lost", it cannot be communicated and it cannot enter new splicings. If both markers have the subscript 0, then the string can be communicated to the third component.

In the third component, a string of the form X_0wY_0 can be transformed to XwY (and this string is passed to the first component, thus making possible the iteration of the whole process, of simulation of rules in P or of rotation), or to $X'wY'$, or to a string with mixed forms of the markers X, Y, with and without a prime. In the last case, the string is once again "lost", because it cannot be further processed.

A string of the form $X'wY'$ can be communicated only to the second component, where only two splicings are possible:

$$(X'B|w_1Y', |ZZ) \models (X'BZZ, w_1Y'),$$
$$(w_1|Y', ZZ|) \models (w_1, ZZY'),$$

providing that $w = Bw_1$ (which ensures that the string has the same permutation as the corresponding string produced by G). A string without end markers cannot enter new splicings. If it is a terminal one, then it can be communicated to the first component, hence it is an element of $L(\Gamma)$; otherwise it is "lost".

Therefore, the subscripts of the two markers X, Y must reach at the same time the value 0. This is possible only when they have started from the same value. In the case above, we must have $i = j$. This means that the symbol D_i which has been erased from the right end of w has been simultaneously introduced in the left end of w. In this way, the rotation phase is correctly implemented, hence all circular permutations of the string can be obtained. Consequently, all derivations in G can be simulated in Γ and, conversely, only strings in $L(G)$ can be sent as terminal strings to the first component of Γ. Thus, $L(G) = L(\Gamma)$. □

Communicating distributed H systems of degree 1 do not use communication, hence they are extended finite H systems. In view of the results in Chap. 7 we can write

$$CDH_1 = REG \subset CDH_2$$

(the properness of the second inclusion is proved by the example considered before Theorem 10.4).

It is an *open problem* whether or not the inclusion $CDH_2 \subseteq CDH_3$ is proper, hence whether or not the result in Theorem 10.4 can be strengthened, to $n = 2$. We expect a negative answer. (We *conjecture* that $CDH_2 \subset CF$.)

This problem is most interesting from a mathematical point of view, though rather less so for DNA computing: the motivation of considering distributed H systems is to decrease the number of splicing rules used in each component; a small number of components intuitively means components of large size, which conflicts with our goal.

Consider now the very problem which has motivated the definition of distributed H systems – limiting the number of splicing rules working together. For a communicating distributed H system $\Gamma = (V, T, (A_1, R_1, V_1), \ldots, (A_n, R_n, V_n))$ we denote by $tubes(\Gamma)$ the degree of Γ (the number n, of components), by $rad(\Gamma)$ the maximum radius of rules in Γ, and

$$size(\Gamma) = \max\{card(R_i) \mid 1 \leq i \leq n\}.$$

By a modification of the construction from the proof of Theorem 10.4, we can characterize the family RE by communicating distributed H systems of minimal size, namely one (of course, this is obtained at the expense of leaving the number of components unbounded).

Theorem 10.5. *For each type-0 grammar $G = (N, T, S, P)$ we can construct a communicating distributed H system Γ such that $L(G) = L(\Gamma)$ and*

$$tubes(\Gamma) = 2(card(N \cup T) + 1) + card(P) + 9,$$
$$size(\Gamma) = 1,$$
$$rad(\Gamma) = card(N \cup T) + 2.$$

Proof. For a type-0 grammar $G = (N, T, S, P)$, consider a new symbol, B, and denote, for an easy reference, $N \cup T \cup \{B\} = \{D_1, D_2, \ldots, D_n\}$. Since $N \neq \emptyset, T \neq \emptyset$, we have $n \geq 3$. Assume the rules in P are labeled in a one-to-one manner with $r_1, \ldots, r_m$, for $m = card(P)$. We construct the communicating distributed H system Γ with the alphabet

$$V = N \cup T \cup \{X, X', X'', Y, Y', Y'', Z, B, C, E\},$$

the terminal alphabet T, and with the components described below. We identify these components with elements α in the set

$$M = \{1, 2, 3, 4, 5, 6, 7, 8, 9\} \cup \{D_i, D_i' \mid 1 \leq i \leq n\} \cup \{r_i \mid 1 \leq i \leq m\}.$$

$$\begin{aligned}
\alpha = r_i : \quad & A_\alpha = \{XBSY, Zv_iY\}, \\
& R_\alpha = \{\#u_iY\$\#v_iY\}, \text{ for } r_i : u_i \to v_i \in P, \\
& V_\alpha = N \cup T \cup \{X, Y, B\}, \text{ where } i = 1, 2, \ldots, m; \\
\alpha = D_i : \quad & A_\alpha = \{ZC^iY\}, \\
& R_\alpha = \{\#D_iY\$Z\#C^iY'\}, \\
& V_\alpha = N \cup T \cup \{X, Y, B\}, \text{ where } i = 1, 2, \ldots, n; \\
\alpha = 1 : \quad & A_\alpha = \{ZY''\}, \\
& R_\alpha = \{\#CY'\$Z\#Y''\}, \\
& V_\alpha = N \cup T \cup \{X, Y', B, C\}; \\
\alpha = 2 : \quad & A_\alpha = \{X'CZ\}, \\
& R_\alpha = \{X\#\$X'C\#Z\}, \\
& V_\alpha = N \cup T \cup \{X, Y'', B, C\}; \\
\alpha = 3 : \quad & A_\alpha = \{ZY'\}, \\
& R_\alpha = \{\#Y''\$Z\#Y'\}, \\
& V_\alpha = N \cup T \cup \{X', Y'', B, C\}; \\
\alpha = 4 : \quad & A_\alpha = \{XZ\}, \\
& R_\alpha = \{X'\#\$X\#Z\},
\end{aligned}$$

$$\begin{aligned}
& & V_\alpha &= N \cup T \cup \{X', Y', B, C\};\\
&\alpha = D_i': & A_\alpha &= \{X''D_iZ\},\\
& & R_\alpha &= \{XC^i\#\$X''D_i\#Z\},\\
& & V_\alpha &= N \cup T \cup \{X, Y', B, C\}, \text{ where } i = 1, 2, \ldots, n;\\
&\alpha = 5: & A_\alpha &= \{ZY\},\\
& & R_\alpha &= \{\#Y'\$Z\#Y\},\\
& & V_\alpha &= N \cup T \cup \{X'', Y', B\};\\
&\alpha = 6: & A_\alpha &= \{XZ\},\\
& & R_\alpha &= \{X\#Z\$X''\#\},\\
& & V_\alpha &= N \cup T \cup \{X'', Y, B\};\\
&\alpha = 7: & A_\alpha &= \{ZE\},\\
& & R_\alpha &= \{\#Y\$Z\#E\},\\
& & V_\alpha &= T \cup \{X, Y, B\};\\
&\alpha = 8: & A_\alpha &= \{ZZ\},\\
& & R_\alpha &= \{\#ZZ\$XB\#\},\\
& & V_\alpha &= T \cup \{X, E, B\};\\
&\alpha = 9: & A_\alpha &= \{ZZ\},\\
& & R_\alpha &= \{\#E\$ZZ\#\},\\
& & V_\alpha &= T \cup \{E\}.
\end{aligned}$$

The work of Γ proceeds as follows.

No terminal string can enter a splicing, because all splicing rules in $R_\alpha, \alpha \in M$, contain control symbols $X, X', X'', Y, Y', Y'', Z$.

The components identified by $\alpha = r_i, 1 \le i \le m$, are used for simulating the rules of P. This is done in the right hand end of a current string of the form XwY, for $w \in (N \cup T \cup \{B\})^*$ (only such strings are accepted by the filters of these components). Such an operation is of the form

$$(Xw_1|u_iY, Z|v_iY) \models (Xw_1v_iY, Zu_iY).$$

The string Xw_1v_iY is of the same type (contains similar symbols) as the input string Xw_1u_iY, the byproduct string Zu_iY can enter a new splicing as above only if $u_i = v_i$ (hence nothing new is produced), or splicings of the form

$$(Z|u_iY, Z|v_iY) \models (Zv_iY, Zu_iY),$$

hence the strings are reproduced.

A string of the form XwY, with w containing only symbols in $N \cup T \cup \{B\}$, can also be communicated to a component identified by $\alpha = D_i, 1 \le i \le n$. Such a component replaces the symbol D_i appearing on the rightmost position of w by C^i:

$$(Xw_1|D_iY, Z|C^iY') \models (Xw_1C^iY', ZD_iY).$$

The string ZD_iY can enter only a splicing of the form

$$(Z|D_iY, Z|C^iY') \models (ZC^iY', ZD_iY),$$

hence nothing new is produced. The first string can be passed both to the component identified by $\alpha = 1$ and to any component identified by $\alpha = D_j', 1 \leq j \leq n$. No splicing is possible in the second case.

The components corresponding to $\alpha = 1, 2, 3, 4$ move symbols C from the right hand end of a string to its left hand end, as follows:

$$\begin{aligned}
&(XC^swC^t|CY', Z|Y'') \models (XC^swC^tY'', ZCY'),\\
&(X|C^swC^tY'', X'C|Z) \models (XZ, X'CC^swC^tY''),\\
&(X'C^{s+1}wC^t|Y'', Z|Y') \models (X'C^{s+1}wC^tY', ZY''),\\
&(X'|C^{s+1}wC^tY', X|Z) \models (X'Z, XC^{s+1}wC^tY').
\end{aligned}$$

One occurrence of C has been moved from the right hand end of the string $XC^swC^{t+1}Y', s, t \geq 0$, to the left hand end of this string. The operation can be iterated (the obtained string again contains symbols from $N \cup T \cup \{X, Y', B, C\}$). All the strings obtained during these steps and containing the symbol Z cannot enter splicings which produce new strings, the "main string", that marked with X, Y and primed versions of these symbols, has to circulate on the path 1, 2, 3, 4 and back to 1.

In any moment, the string $XC^swC^tY', s, t \geq 0$, produced by the component associated with $\alpha = 4$ can be passed to any component identified by $\alpha = D_i', 1 \leq i \leq n$. When $i \leq s$ we obtain

$$(XC^i|C^{s-i}wC^tY', X''D_i|Z) \models (XC^iZ, X''D_iC^{s-i}wC^tY').$$

If $s > i$ or $t \geq 1$, then the string $X''D_iC^{s-i}wC^tY'$ cannot be passed to another component, because none of them accepts strings containing at the same time occurrences of X'' and C. A further splicing in the component associated with $\alpha = D_i'$ is not possible for strings starting with X''. Therefore, we must have $s = i$ and $t = 0$, which means that the symbol D_i replaced by C^i above has to be reintroduced in the left hand end of the string.

In this way, the string is circularly permuted, making possible the simulation of rules in G on any position of the string generated by G and simulated in Γ.

Now, the components identified by $\alpha = 5, 6$ replace Y' by Y and X'' by X, thus both the simulation of rules in G and the rotation of the string can be iterated. We leave the details to the reader.

In any moment, any string containing only symbols in $T \cup \{B\}$ can be communicated to the component identified by $\alpha = 7$, which introduces the symbol E. The only continuation is the removing of the auxiliary symbols. It is important to note that we can remove B only together with X, hence only when the obtained string is in the same permutation as in G. The operations performed by the last three components of our system are:

$$(XBx|Y, Z|E) \models (XBxE, ZY),$$
$$(|ZZ, XB|xE) \models (xE, XBZZ),$$
$$(x|E, ZZ|) \models (x, ZZE).$$

Note that in all these operations we have $x \in T^*$.

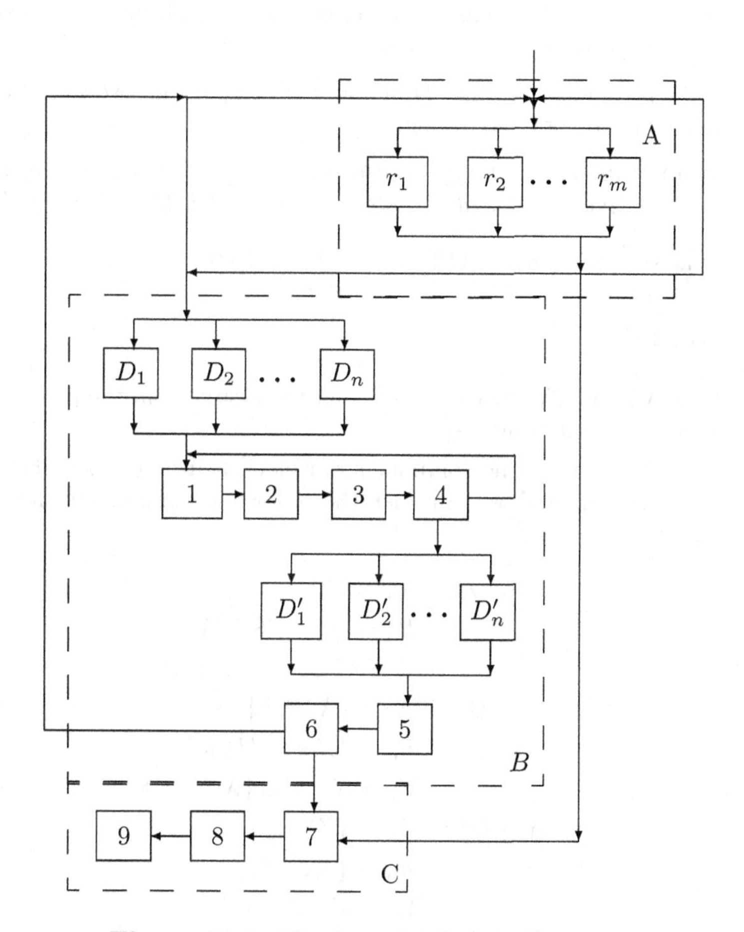

Figure 10.1: The flow of strings in Γ

The path followed by the string of the form XwY, maybe with X, Y replaced by primed versions of them, is indicated in Fig. 10.1. Block A contains the components performing the simulation of rules in P, block B contains the components performing the circular permutation of the string, whereas block C contains the components ending the process.

According to the previous discussion, one can see that each derivation in G can be simulated in Γ and, conversely, all terminal strings reaching the first component of Γ are strings in $L(G)$. Consequently, $L(G) = L(\Gamma)$.

It is easy to see that Γ contains $m + 2n + 9$ components (remember that $n = card(N \cup T \cup \{B\})$) and that $rad(\Gamma) = n + 1$ (this value is reached by R_α for $\alpha = D_i$). Each component contains only one splicing rule, that is $size(\Gamma) = 1$. □

Remark 10.2. If we also consider the parameter $ax(\Gamma)$, defined as the maximum number of axioms in the components of Γ, from the previous construction we also obtain $ax(\Gamma) = 1$. □

At the price of increasing the number of components, we can also bound the radius of the obtained system.

Theorem 10.6. *For each type-0 grammar $G = (N, T, S, P)$ we can construct a communicating distributed H system Γ such that $L(G) = L(\Gamma)$ and*

$$\begin{aligned}
&tubes(\Gamma) \leq 3(card(N \cup T) + 1) + 2 \cdot card(P) + 4, \\
&size(\Gamma) = 1, \\
&rad(\Gamma) = 2.
\end{aligned}$$

Proof. We modify the construction of the system Γ in the proof of Theorem 10.5 in the following way.

Firstly, we remove the components in block B in Fig. 10.1, those which rotate the string, and we consider the following components, for all $i = 1, 2, \ldots, n$:

$$\begin{aligned}
\alpha = D_i : \quad & A_\alpha = \{ZY_i\}, \\
& R_\alpha = \{\#D_iY\$Z\#Y_i\}, \\
& V_\alpha = N \cup T \cup \{X, Y, B\}, \\
\alpha = D_i' : \quad & A_\alpha = \{X'D_iZ\}, \\
& R_\alpha = \{X\#\$X'D_i\#Z\}, \\
& V_\alpha = N \cup T \cup \{X, Y_i, B\}; \\
\alpha = D_i'' : \quad & A_\alpha = \{ZY\}, \\
& R_\alpha = \{\#Y_i\$Z\#Y\}, \\
& V_\alpha = N \cup T \cup \{X', Y_i, B\}; \\
\alpha = 1 : \quad & A_\alpha = \{XZ\}, \\
& R_\alpha = \{X'\#\$X\#Z\}, \\
& V_\alpha = N \cup T \cup \{X', Y, B\}.
\end{aligned}$$

The rotation is now performed in the following way:

$$(Xw|D_iY, Z|Y_i) \models (XwY_i, ZD_iY),$$

$$(X|wY_i, X'D_i|Z) \models (XZ, X'D_iwY_i),$$
$$(X'D_iw|Y_i, Z|Y) \models (X'D_iwY, ZY_i),$$
$$(X'|D_iwY, X|Z) \models (X'Z, XD_iwY).$$

Secondly, we also modify the simulating components, those in block A in Fig. 10.1. Specifically, we start from a grammar G in Kuroda normal form (with the rules of the forms $C \to x, C \in N, x \in (N \cup T)^*, |x| \leq 2$, and $CD \to EF$, with $C, D, E, F \in N$); moreover, we assume the rules of G are labeled in a one-to-one manner.

For each rule $r : C \to x$ we introduce in Γ the component

$$\begin{aligned} \alpha = r: \quad & A_\alpha = \{Zx\}, \\ & R_\alpha = \{\#CY\$Z\#x\}, \\ & V_\alpha = N \cup T \cup \{X, Y, B\}, \end{aligned}$$

whereas for each rule $r : CD \to EF$ we introduce the components

$$\begin{aligned} \alpha = r: \quad & A_\alpha = \{ZY_r\}, \\ & R_\alpha = \{C\#DY\$Z\#Y_r\}, \\ & V_\alpha = N \cup T \cup \{X, Y, B\}, \\ \alpha = r': \quad & A_\alpha = \{ZEFY\}, \\ & R_\alpha = \{\#CY_r\$Z\#E\}, \\ & V_\alpha = N \cup T \cup \{X, Y_r, B\}. \end{aligned}$$

The use of symbols Y_r ensures the correct simulation of the non-context-free rules in P.

The terminating components (identified by $\alpha = 7, 8, 9$ in the proof of Theorem 10.5) remain unchanged.

As in the proof of Theorem 10.5 one can see that the obtained system is equivalent with G. Moreover, it is of radius two, of size one, but it contains at most $3n + 2m + 4$ components (where $n = card(N \cup T \cup \{B\})$ and $m = card(P)$). □

The proofs of Theorems 10.4, 10.5, and 10.6 are constructive and provide universal communicating distributed H systems when starting from a universal type-0 grammar, in the usual sense of universality. The "program" of the particular distributed H system to be simulated by the universal one is introduced in the axiom set of the second component, instead of the axiom $XBSY$: as in Sect. 8.7, the symbol S should be replaced by the code of the simulated system.

10.3 Two-Level Distributed H Systems

The distributed H systems in the previous section make essential use of a feature which makes them inefficient from a biochemical point of view, namely communication. This means the transport of many strings from a component to another one and the checking of the filter conditions. Moving (long) DNA sequences from one place to another takes time and can break the molecules, while checking the filter conditions can be done, in present laboratory conditions, only manually, which is bad from the efficiency point of view.

The model we shall define below will no longer involve communication between components in the sense considered in the communicating distributed H systems.

A *two-level distributed H system* (of degree $n, n \geq 1$), is a construct

$$\Gamma = (V, T, (w_1, A_1, I_1, E_1), \dots, (w_n, A_n, I_n, E_n)),$$

where V is an alphabet, $T \subseteq V$, $w_i \in V^*, A_i \subseteq V^*$, and $I_i, E_i \subseteq V^*\#V^*\$V^*\#V^*$, for $\#, \$$ symbols not in V. All sets $A_i, I_i, E_i, 1 \leq i \leq n$, are finite; T is the terminal alphabet, (w_i, A_i, I_i, E_i) is the ith *component* of the system; w_i is the *active axiom*, A_i is the set of not-so-active (we say *passive*) axioms, I_i is the set of *internal splicing rules*, and E_i is the set of *external splicing rules* of component i, $1 \leq i \leq n$.

One can imagine a two-level H system as consisting of n (active) DNA sequences, $z_1, \dots, z_n$ (initially they are $w_1, \dots, w_n$), with their left hand end fixed on a solid support, surrounded each by arbitrarily many copies of passive strings (initially, those in sets $A_1, \dots, A_n$, respectively), and having around both "strong" restriction enzymes which can "see" only the active strings and "weak" restriction enzimes, acting only locally, on z_i and on an associated string in the passive set. As a result of a local splicing, a string with a prefix which is also a prefix of z_i will be obtained, hence again fixed on the support, and one more string which will be added to the surrounding set of passive strings. The external splicing has priority over the internal splicing. When an external splicing is performed, according to a rule in some set E_i, then the associated string z_i is the first term of the splicing, hence a new string fixed on the solid support is obtained, having a common prefix with z_i; the second term of the splicing, some $z_j, j \neq i$, remains unchanged after this operation – one can assume that a copy of z_j has been produced and sent to component i just for participating in the splicing.

Formally, these operations are defined as follows.

The *contents* of a component $i, 1 \leq i \leq n$, is described by a pair (x_i, M_i), where $x_i \in V^*$ is the *active* string and $M_i \subseteq V^*$ is the set of *passive* strings. An n-tuple $\pi = [(x_1, M_1), \dots, (x_n, M_n)]$ is called a *configuration* of the system. For $1 \leq i \leq n$ and a given configuration π as above, we define

$$\mu(x_i, \pi) = \begin{cases} \textit{external}, & \text{if there are } r \in E_i \text{ and } x_j, j \neq i, \\ & \text{such that } (x_i, x_j) \models_r (u, v), \text{ for some } u, v \in V^*, \\ \textit{internal}, & \text{otherwise.} \end{cases}$$

Then, for two configurations $\pi = [(x_1, M_1), \dots, (x_n, M_n)]$ and $\pi' = [(x_1', M_1'), \dots, (x_n', M_n')]$, we write $\pi \Longrightarrow_{ext} \pi'$ if the following conditions hold:

1. there is $i, 1 \leq i \leq n$, such that $\mu(x_i, \pi) = external$,

2. for each $i, 1 \leq i \leq n$, with $\mu(x_i, \pi) = external$, we have $(x_i, x_j) \models_r (x'_i, z_i)$, for some $j, 1 \leq j \leq n, j \neq i$, $r \in E_i$, and $z_i \in V^*$; moreover, $M'_i = M_i \cup \{z_i\}$,

3. for each $i, 1 \leq i \leq n$, with $\mu(x_i, \pi) = internal$, we have $(x'_i, M'_i) = (x_i, M_i)$.

For two configurations π and π' as above, we write $\pi \Longrightarrow_{int} \pi'$ if the following conditions hold:

1. for all $i, 1 \leq i \leq n$, we have $\mu(x_i, \pi) = internal$,

2. for each $i, 1 \leq i \leq n$, either $(x_i, z) \models_r (x'_i, z')$, for some $z \in M_i, z' \in V^*, r \in I_i$, and $M'_i = M_i \cup \{z'\}$, or

3. no rule $r \in I_i$ can be applied to (x_i, z), for any $z \in M_i$, and then $(x'_i, M'_i) = (x_i, M_i)$.

The relation $\Longrightarrow_{ext}$ defines an external splicing, $\Longrightarrow_{int}$ defines an internal splicing. Note that in both cases all the splicing operations are performed in parallel and the components not able to use a splicing rule do not change their contents. We stress the fact that the external splicing has priority over the internal one and that all operations have as the first term an active string; the first string obtained by splicing becomes the new active string of the corresponding component, the second string becomes an element of the set of passive strings of that component.

We write $\Longrightarrow$ for both $\Longrightarrow_{ext}$ and $\Longrightarrow_{int}$. The *language generated* by a two-level distributed H system Γ is defined by

$$L(\Gamma) = \{w \in T^* \mid [(w_1, A_1), \ldots, (w_n, A_n)] \Longrightarrow^* [(x_1, M_1), \ldots, (x_n, M_n)],$$
$$\text{for } w = x_1, x_i \in V^*, 2 \leq i \leq n, \text{ and } M_i \subseteq V^*, 1 \leq i \leq n\}.$$

We denote by LDH_n the family of languages generated by two-level distributed H systems with at most n components, $n \geq 1$. When no restriction is imposed on the number of components, we write LDH_*.

Here is an example: Consider the system

$$\Gamma = (\{a, b, C, D\}, \{a, b\}, (w_1, A_1, I_1, E_1), (w_2, A_2, I_2, E_2)),$$

with

$$\begin{aligned} w_1 &= aD, \\ A_1 &= \{aD, Da\}, \\ I_1 &= \{b\#C\$\#aD,\ b\#C\$D\#a\}, \\ E_1 &= \{a\#D\$D\#b\}, \end{aligned}$$

$$\begin{aligned} w_2 &= DbC, \\ A_2 &= \{bC\}, \\ I_2 &= \{b\#C\$\#bC\}, \\ E_2 &= \emptyset. \end{aligned}$$

A computation in Γ runs as follows:

$$\begin{aligned} &[(a|D, \{aD, Da\}), (D|bC, \{bC\})] \\ \Longrightarrow_{ext}\ &[(ab|C, \{|aD, Da, DD\}), (Db|C, \{|bC\})] \\ \Longrightarrow_{int}\ &[(aba|D, \{aD, Da, DD, C\}), (D|b^2C, \{bC, C\})] \\ \Longrightarrow_{ext}\ &[(abab^2|C, \{|aD, Da, DD, C\}), (Db^2|C, \{|bC, C\})] \\ \Longrightarrow_{int}\ &[(abab^2aD, \{aD, Da, DD, C\}), (Db^3C, \{bC, C\})] \Longrightarrow^* \dots \\ \Longrightarrow_{ext}\ &[(abab^2 \dots ab^k|C, \{aD, D|a, DD, C\}), (Db^k|C, \{b|C, C\})] \\ \Longrightarrow_{int}\ &[(abab^2 \dots ab^ka, \{aD, Da, DD, C, DC\}), (Db^{k+1}C, \{bC, C\})], \end{aligned}$$

for some $k \geq 1$.

After an alternate sequence of external and internal splicings, the active string of component 1 becomes $abab^2 \dots ab^kC$, which can be turned out to a terminal string by replacing C with a. Therefore,

$$L(\Gamma) = \{abab^2 \dots ab^ka \mid k \geq 1\}.$$

This language is not semilinear, hence it is not context-free.

A variant of the model above, with the two levels better distinguished, is the following one:

A *separated two-level distributed H system* is a construct

$$\Gamma = (V, T, (w_1, A_1, I_1), \dots, (w_n, A_n, I_n), E),$$

where V is an alphabet, $T \subseteq V$ (the terminal alphabet), $w_i \in V^*$ (active axiom), $A_i \subseteq V^*$ (passive axioms), $I_i, 1 \leq i \leq n$, and E are sets of splicing rules over V; all the sets A_i, I_i, E are finite. The elements of I_i are called *internal* splicing rules, $1 \leq i \leq n$, and those of E are called *external* splicing rules; (w_i, A_i, I_i) is the ith *component* of $\Gamma, 1 \leq i \leq n$.

The language generated by Γ, denoted by $L(\Gamma)$, is equal to the language generated by the two-level H system $\Gamma' = (V, T, (w_1, A_1, I_1, E), \dots, (w_n, A_n, I_n, E))$. We denote by $SLDH_n, n \geq 1$, the family of languages generated by separated two-level H systems of degree at most n; when n is not specified, we write $SLDH_*$.

Theorem 10.7. *$RE = SLDH_n = LDH_n = SLDH_* = LDH_*$, for all $n \geq 3$.*

Proof. The inclusions $SLDH_n \subseteq LDH_n, n \geq 1$, $SLDH_* \subseteq LDH_*$, $LDH_n \subseteq LDH_{n+1}, SLDH_n \subseteq SLDH_{n+1}, n \geq 1$, follow from the definitions, $LDH_* \subseteq RE$ is obvious. Therefore, we have only to prove the inclusion $RE \subseteq SLDH_3$.

Consider a type-0 grammar $G = (N, T, S, P)$. We construct the separated two-level distributed H system

$$\Gamma = (V, T, (w_1, A_1, I_1), (w_2, A_2, I_2), (w_3, A_3, I_3), E),$$

with

$$\begin{aligned}
V &= N \cup T \cup \{X, Z, Z_s, Z_l, Z_r, Y, C_1, C_2, C_3\}, \\
w_1 &= SXXC_1, \\
A_1 &= \{ZvXZ_s \mid u \to v \in P\} \\
&\cup \{ZXX\alpha Z_l, Z\alpha XXZ_r \mid \alpha \in N \cup T\}, \\
I_1 &= \{\#uXZ\$Z\#vXZ_s \mid u \to v \in P\} \\
&\cup \{\#\alpha XZ\$Z\#XX\alpha Z_l,\ \#XZ\$Z\#\alpha XXZ_r \mid \alpha \in N \cup T\},
\end{aligned}$$

$$\begin{aligned}
w_2 &= C_2Z, \\
A_2 &= \{C_2Y\}, \\
I_2 &= \{C_2\#Y\$C_2\#Z\},
\end{aligned}$$

$$\begin{aligned}
w_3 &= C_3Z, \\
A_3 &= \{C_3Y\}, \\
I_3 &= \{C_3\#Z\$C_3\#Y,\ C_3\#Y\$C_3\#Z\},
\end{aligned}$$

$$\begin{aligned}
E = &\{C_2\#Z\$X\#X,\ X\#X\$C_3\#Z,\ X\#Z_s\$C_2\#X, \\
&C_2\#X\$C_3\#Y,\ \#XXC_1\$C_2Z\#\} \\
&\cup \{XX\alpha\#Z_l\$C_2X\#,\ \alpha XX\#Z_r\$C_2X\alpha\# \mid \alpha \in N \cup T\}.
\end{aligned}$$

The idea behind this construction is as follows. We simulate the work of G in the first two components, on their active strings. Namely, the substring XX of the active string of the first component shows the place where the simulation is done – immediately to the left of XX. To this aim, the active string of the first component is cut between the two occurrences of X, the prefix remains in the first component and the suffix is saved in the second component. This is done by external splicings performed simultaneously in the first and the second components. The simulation is performed by an internal splicing in the first component. Then the two strings are concatenated again by an external splicing, in the presence of the symbol Z_s (the subscript s stands for "simulation"). The substring XX can be moved to the left or to the right, over one symbol, in a similar way for both directions; the symbols X_l, X_r control the operation (l = left, r = right). After removing the substring XX, together with the symbol C_1, no further splicing is possible.

Let us examine in some details the work of Γ.

Consider a configuration

$$[(w_1 XXw_2C_1, M_1), (C_2Z, M_2), (C_3Z, M_3)] \qquad (*)$$

Initially we have $w_1 = S, w_2 = \lambda, M_1 = A_1, M_2 = A_2, M_3 = A_3$. We have to splice the active strings according to the rules $C_2\#Z\$X\#X$, $X\#X\$C_3\#Z$ in E and we obtain the configuration:

$$[(w_1XZ, M_1' = M_1 \cup \{C_3Xw_2C_1\}), (C_2Xw_2C_1, M_2' = M_2 \cup \{w_1XZ\}),$$
$$(C_3Z, M_3)].$$

No external splicing is possible, we perform internal splicings in the first and the third components. There are three possibilities:

1) If $w_1 = w_1'u$, for some $u \to v \in P$, then we can use the rule $\#uXZ\$Z\#vXZ_s$ in component 1 and $C_3\#Z\$C_3\#Y$ in component 3 and we get:

$$[(w_1'vXZ_s, M_1'' = M_1' \cup \{ZuXZ\}), (C_2Xw_2C_1, M_2'),$$
$$(C_3Y, M_3' = M_3 \cup \{C_3Z\})].$$

External splicings are possible, using the rules $X\#Z_s\$C_2\#X, C_2\#X\$C_3\#Y$, and leading to the configuration

$$[(w_1'vXXw_2C_1, M_1''' = M_1'' \cup \{C_2Z_s\}), (C_2Y, M_2'' = M_2' \cup \{C_3Xw_2C_1\}),$$
$$(C_3Y, M_3')].$$

No external splicing is possible and no internal splicing in the first component, but we can perform internal splicings in components 2 and 3, leading to:

$$[(w_1'vXXw_2C_1, M_1'''), (C_2Z, M_2''' = M_2'' \cup \{C_2Y\}),$$
$$(C_3Z, M_3'' = M_3' \cup \{C_3Y\})].$$

We have returned to a configuration of the form we started with, $(*)$.

The new passive strings, produced during these operations, will enter no new splicing in components 1 and 2, and they are always C_3Y, C_3Z in the third component.

2) If $w_1 = w_1'\alpha$, for some $\alpha \in N \cup T$, then we can use the rule $\#\alpha XZ\$Z\#XX\alpha Z_l$ in component 1 and the rule $C_3\#Z\$C_3\#Y$ in component 3, and we get:

$$[(w_1'XX\alpha Z_l, M_1'' = M_1' \cup \{Z\alpha XZ\}), (C_2Xw_2C_1, M_2'),$$
$$(C_3Y, M_3' = M_3 \cup \{C_3Z\})].$$

External splicings must be done, using the rules $XX\alpha\#Z_l\$C_2X\#$ and $C_2\#X\$C_3\#Y$, leading to:

$$[(w_1'XX\alpha w_2C_1, M_1''' = M_1'' \cup \{C_2XZ_l\}),$$
$$(C_2Y, M_2'' = M_2' \cup \{C_3Xw_2C_1\}), (C_3Y, M_3')].$$

No external splicing and no internal splicing is possible in the first component, but we can perform internal splicings in the other components; we obtain the configuration

$$[(w_1'XX\alpha w_2C_1, M_1'''), (C_2Z, M_2''' = M_2' \cup \{C_2Y\}),$$
$$(C_3Z, M_3'' = M_3' \cup \{C_3Y\})].$$

We have also returned to a configuration of type $(*)$.

3) If in configuration $(*)$ we have $w_2 = \alpha w_2'$, for some $\alpha \in N \cup T$, then we can proceed as follows; by using the rule $\#XZ\$Z\#\alpha XXZ_r$ in the first component and $C_3\#Z\$C_3\#Y$ in the third one, we first produce

$$[(w_1\alpha XXZ_r, M_1'' = M_1' \cup \{ZXZ\}), (C_2Xw_2C_1, M_2'),$$
$$(C_3Y, M_3' = M_3 \cup \{C_3Z\})],$$

then we have to perform external splicings, according to the rules $\alpha XX\#Z_r\$C_2X\alpha\#$, $C_2\#X\$C_3\#Y$, leading to

$$[(w_1\alpha XXw_2'C_1, M_1''' = M_1'' \cup \{C_2X\alpha Z_r\}),$$
$$(C_2Y, M_2'' = M_2' \cup \{C_3Xw_2C_1\}), (C_3Y, M_3')].$$

No external splicing is possible and no internal splicing in the first component; splicing internally in the other components, we get

$$[(w_1\alpha XXw_2'C_1, M_1'''), (C_2Z, M_2''' = M_2'' \cup \{C_2Y\}),$$
$$(C_3Z, M_3'' = M_3' \cup \{C_3Y\})].$$

We have again returned to a configuration of type $(*)$.

In case 1 we have simulated the rule $u \to v$ in the presence of XX, in case 2 we have moved XX across a symbol to the left, in case 3 we have moved XX across a symbol to the right. The operations above can be repeated an arbitrary number of times. Changing in this way the place of XX, we can simulate the rules of G in any place of the word.

When we have a configuration of the form

$$[(wXXC_1, M_1), (C_2Z, M_2), (C_3Z, M_3)],$$

then the external splicing is possible using the rules $\#XXC_1\$C_2Z\#$, $C_2\#Z\$X\#X$, leading to

$$[(w, M_1' = M_1 \cup \{C_2ZXXC_1\}), (C_2XC_1, M_2' = M_2 \cup \{wXZ\}),$$
$$(C_3Z, M_3)].$$

No further splicings, internal or external, are possible in the first component. If w is a terminal string, then it is accepted in the language generated by Γ, if not, then it will never lead to a terminal string.

Consequently, we have $L(G) = L(\Gamma)$. □

We do not know whether the threshold 3 in the previous theorem is optimal or not.

Starting the construction in the proof of Theorem 10.7 from a universal type-0 grammar, we obtain a universal separated two-level distributed H system, in the usual sense. The "program" to be executed on such a "computer" is introduced as the active axiom w_1, instead of $SXXC_1$.

10.4 Time-Varying Distributed H Systems

The distributed architecture we introduce in this section is also related to programmed and evolving H systems investigated in Sect. 8.4: at different moments we use different sets of splicing rules. The passing from a set of rules to another one is now specified in a cycle. Thus, the new model corresponds both to periodically time-varying grammars in the regulated rewriting area and to controlled tabled Lindenmayer systems. We can also interpret these systems as counterparts of cooperating distributed grammar systems with the order of enabling components controlled by a graph having the shape of a ring.

As a biochemical motivation, these models start from the assumption that the splicing rules are based on enzymes whose work essentially depends on the environment conditions. Hence, in any moment, only a subset of the set of all available rules are active. If the environment changes periodically, then the active enzymes also change periodically.

A *time-varying distributed H system* (of degree $n, n \geq 1$) is a construct

$$\Gamma = (V, T, A, R_1, R_2, \dots, R_n),$$

where V is an alphabet, $T \subseteq V$ (terminal alphabet), A is a finite subset of V^* (axioms), and R_i are finite sets of splicing rules over $V, 1 \leq i \leq n$. The sets $R_i, 1 \leq i \leq n$, are called the components of the system.

At each moment $k = n \cdot j + i$, for $j \geq 0, 1 \leq i \leq n$, the component R_i is used for splicing the currently available strings. Specifically, we define

$$\begin{aligned} L_1 &= A, \\ L_{k+1} &= \sigma_2^{(i)}(L_k), \text{ for } i \equiv k (mod\ n), k \geq 1, \end{aligned}$$

where $\sigma^{(i)} = (V, R_i), 1 \leq i \leq n$.

Therefore, from a step k to the next step, $k+1$, one passes only the result of splicing the strings in L_k according to the rules in R_i for $i \equiv k (mod\ n)$; the strings in L_k which cannot enter a splicing are removed.

The language generated by Γ is defined by

$$L(\Gamma) = (\bigcup_{k\geq 1} L_k) \cap T^*.$$

We denote by $VDH_n, n \geq 1$, the family of languages generated by time-varying distributed H systems of degree at most n, and by VDH_* the family of all languages of this type.

The way of working of time-varying H systems is surprisingly powerful. (The explanation lies in the fact that from a step to another step one passes only the result of splicing operations done at the previous step; strings produced at different "generations" cannot be spliced together.)

For example, let us consider the system (of degree 1)

$$\Gamma = (\{a, b, c\}, \{a, b, c\}, \{cab\}, \{a\#b\$c\#a\}).$$

We obtain

$$\begin{aligned} L_1 &= \{cab\}, \\ L_2 &= \{caab, cb\}, \text{ by } (ca|b, c|ab) \models (caab, cb), \\ L_3 &= \{ca^4b, cb\}, \text{ by } (caa|b, c|aab) \models (ca^4b, cb), \\ &\ldots\ldots\ldots\ldots\ldots \\ L_k &= \{ca^{2^{k-1}}b, cb\}, \; k \geq 1. \end{aligned}$$

Therefore,

$$L(\Gamma) = \{ca^{2^n}b \mid n \geq 0\} \cup \{cb\},$$

which is a non-context-free language (not even in the family MAT^λ).

Because each regular language can be generated by a time-varying H system of degree 1 (follow the same construction as in the proof of Lemma 7.18, adding splicing rules which pass the axioms from one step to the next; because the axioms are of a well specified form, this can be easily achieved), we have

Lemma 10.1. *$REG = EH_2(FIN, FIN) \subset VDH_1 \subseteq VDH_2 \subseteq \ldots \subseteq VDH_* \subseteq RE$.*

This hierarchy collapses (at most) at level 7:

Theorem 10.8. *$RE = VDH_n = VDH_*, n \geq 7$.*

Proof. Consider a type-0 grammar $G = (N, T, S, P)$ with $N \cup T = \{\alpha_1, \ldots, \alpha_{n-1}\}, n \geq 3$, and $P = \{u_i \to v_i \mid 1 \leq i \leq m\}$. Let $\alpha_n = B$ be a new symbol. We construct the time-varying distributed H system

$$\Gamma = (V, T, A, R_1, \ldots, R_7),$$

with

$$\begin{aligned} V &= N \cup T \cup \{X, Y, Y', Z, B\} \\ &\cup \{Y_i, Y_i', X_i \mid 0 \leq i \leq n\}, \\ A &= \{XBSY, ZY, ZY', ZZ, XZ\} \\ &\cup \{Zv_iY \mid 1 \leq i \leq m\} \\ &\cup \{ZY_j, ZY_j', X_jZ \mid 0 \leq j \leq n\} \\ &\cup \{X_j\alpha_jZ \mid 1 \leq j \leq n\}, \end{aligned}$$

and the following sets of splicing rules:

$$\begin{aligned} R_1 &= \{\#u_iY\$Z\#v_iY \mid 1 \leq i \leq m\} \\ &\cup \{\#Y\$Z\#Y,\ Z\#\$Z\#\} \\ &\cup \{\#Y_j\$Z\#Y_j \mid 1 \leq j \leq n\}, \\ R_2 &= \{\#\alpha_jY\$Z\#Y_j \mid 1 \leq j \leq n\} \\ &\cup \{\#Y\$Z\#Y',\ Z\#\$Z\#\} \\ &\cup \{\#Y_j\$Z\#Y_j' \mid 1 \leq j \leq n\}, \\ R_3 &= \{X\#\$X_j\alpha_j\#Z \mid 1 \leq j \leq n\} \\ &\cup \{\#Y'\$Z\#Y,\ Z\#\$Z\#\} \\ &\cup \{\#Y_j'\$Z\#Y_j \mid 1 \leq j \leq n\}, \\ R_4 &= \{\#Y_j\$Z\#Y_{j-1} \mid 1 \leq j \leq n\} \\ &\cup \{\#Y\$Z\#Y,\ Z\#\$Z\#\}, \\ R_5 &= \{X_j\#\$X_{j-1}\#Z \mid 1 \leq j \leq n\} \\ &\cup \{\#Y\$Z\#Y,\ Z\#\$Z\#\}, \\ R_6 &= \{\#Y_0\$Z\#Y,\ \#Y_0\$ZZ\#,\ \#Y\$Z\#Y',\ Z\#\$Z\#\} \\ &\cup \{\#Y_j\$Z\#Y_j' \mid 1 \leq j \leq n\}, \\ R_7 &= \{X_0\#\$X\#Z,\ X_0B\#\$\#ZZ,\ \#Y'\$Z\#Y,\ Z\#\$Z\#\} \\ &\cup \{\#Y_j'\$Z\#Y_j \mid 1 \leq j \leq n\}. \end{aligned}$$

This system works as follows.

Consider a string of the form $XwY, w \in (N \cup T \cup \{B\})^*$; for the axiom $XBSY$ we have $w = BS$.

If $w = w'u_i, 1 \leq i \leq m$, then the first component can simulate the rule $u_i \to v_i \in P$ for a suffix of w. A string XwY can also be passed to R_2 unmodified, by using the rule $\#Y\$Z\#Y$. Similarly, by using the rule $Z\#\$Z\#$, any axiom (in general, any string containing an occurrence of Z) can be passed from R_1 to R_2 – and the same assertion is true for all consecutive components).

A string XwY can enter in R_2 two splicings:

$$\begin{aligned} &(Xw'|\alpha_jY, Z|Y_j) \models (Xw'Y_j, Z\alpha_jY), \text{ for } w = w'\alpha_j, 1 \leq j \leq n, \\ &(Xw|Y, Z|Y') \models (XwY', ZY). \end{aligned}$$

The string $Xw'Y_j$ can enter only one splicing in R_3:

$$(X|w'Y_j, X_i\alpha_i|Z) \models (XZ, X_i\alpha_i w'Y_j) \qquad (*)$$

for some $i, 1 \leq i \leq n$.

A string of the form $X_i x Y_j, 1 \leq i, j \leq n$, will enter splicings in R_4, R_5 which will decrease by one each of i and j, thus producing $X_{i-1}xY_{j-1}$.

A string $X_i x Y_j, 1 \leq i, j \leq n$, will be transformed in R_6 into $X_i x Y_j'$ and this one will be transformed in R_7 into $X_i x Y_j$. R_1 will pass this string unmodified to R_2 which will again replace Y_j by Y_j'; R_3 will return to $X_i x Y_j$. The components R_4, R_5 will again decrease by one the subscripts of X and Y. Eventually, one of X, Y will get the subscript 0. We have three possibilities:

1) R_6 receives a string $X_0 x Y_j$ with $j \geq 1$. The only applicable rule is $\#Y_j\$Z\#Y_j'$; the string $X_0 x Y_j'$ is passed to R_7 which returns to $X_0 x Y_j$; again Y_j is replaced by Y_j', then R_3 returns to $X_0 x Y_j$ which reaches R_4. R_4 produces $X_0 x Y_{j-1}$. No splicing can be done in R_5 on this string, hence no terminal string is obtained in this way.

2) R_6 receives a string $X_i x Y_0$ with $j \geq i$. If Y_0 is replaced by Y, then the string $X_i x Y$ cannot be spliced in R_7. The same assertion is true if Y_0 is deleted. No terminal string can be produced in this way.

3) R_6 receives a string $X_0 x Y_0$. (This means that the string $X_i\alpha_i w'Y_j$ obtained after the splicing $(*)$ has $i = j$, hence the same symbol α_j which was deleted from the right hand end of the string has been introduced in the left hand end.) If R_6 replaces Y_0 by Y, then the only continuation in R_7 is to replace X_0 by X, hence the whole process can be iterated. If R_6 removes Y_0 and R_7 replaces X_0 by X, then the obtained string cannot pass over R_1, hence it is lost. If R_6 removes Y_0 and R_7 removes X_0B, then we get a string without markers, which cannot enter further splicings. If it is terminal, then it belongs to $L(\Gamma)$, otherwise it is lost.

Consequently, every derivation in G can be simulated in Γ by a standard simulate-and-rotate procedure, that is, $L(G) \subseteq L(\Gamma)$.

Assume now that R_2 has produced the string XwY'. If R_3 replaces Y' by Y, then the string XwY will pass unchanged through R_4, R_5, then R_6 will produce XwY' and R_7 will return to XwY, and we arrive back to R_1 with XwY.

If XwY' is spliced in R_3 by a rule $X\#\$X_j\alpha_j\#Z$, $1 \leq j \leq n$, then we get the string $X_j\alpha_j wY'$. This string is blocked by R_4, where it cannot be spliced any more.

The strings obtained by the splicings mentioned above and containing occurrences of Z can pass from a component to another one due to the rules $Z\#\$Z\#$ (and also to rules using symbols Y, Y', etc). If strings of this form enter further splicings, this will happen only together with other strings containing occurrences of Z, either axioms or by-products of other splicings. Thus, both the resulting strings will contain occurrences of Z, hence no terminal string can be produced in this way.

For instance, after a splicing in R_2 using a rule $\#\alpha_j Y\$Z\#Y_j$, $1 \leq j \leq n$, we get the string $Z\alpha_j Y$. It can pass unmodified through $R_3 - R_7$, but in R_1 we can perform

$$(Z|\alpha_j Y, Z|v_i Y) \models (Zv_i Y, Z\alpha_j Y),$$

if $\alpha_j \to v_i$ is the ith rule of P. The input strings are reproduced.

The reader can trace the development of other strings of the type of $Z\alpha_j Y$ above, and the result will be similar: no terminal string which is not in $L(G)$ can be produced. In conclusion, $L(G) = L(\Gamma)$. □

The constant 7 in the equality $RE = VDH_7$ can probably be replaced by a smaller integer. We do not persist in this direction, because of the motivation we have started with: diminishing the *size* of the components. As we have also done for communicating distributed H systems, this objective can be reached: time-varying distributed H systems with components consisting of only three splicing rules are enough.

Theorem 10.9. *Each recursively enumerable language can be generated by a time-varying distributed H system whose components contain at most three splicing rules.*

Proof. Consider a type-0 grammar $G = (N, T, S, P)$ with $N \cup T = \{\alpha_1, \ldots, \alpha_{n-1}\}, n \geq 3$, and $P = \{u_i \to v_i \mid 1 \leq i \leq m\}$. Let $\alpha_n = B$ be a new symbol. We construct the time-varying distributed H system

$$\Gamma = (V, T, A, R_1, \ldots, R_{2n+m+2}),$$

with

$$\begin{aligned} V &= N \cup T \cup \{X, Y, Y', Z, B\}, \\ A &= \{XBSY, ZY, ZY', ZZ\} \\ &\cup \{ZvY \mid u \to v \in P\} \\ &\cup \{X\alpha_i Z \mid 1 \leq i \leq n\}, \end{aligned}$$

and the following sets of splicing rules

$$\begin{aligned} R_i &= \{\#u_i Y\$Z\#v_i Y,\ \#Y\$Z\#Y,\ Z\#\$Z\#\},\ 1 \leq i \leq m, \\ R_{m+2j-1} &= \{\#\alpha_j Y\$Z\#Y,\ \#Y\$Z\#Y',\ Z\#\$Z\#\},\ 1 \leq j \leq n, \\ R_{m+2j} &= \{X\alpha_j\#Z\$X\#,\ \#Y'\$Z\#Y,\ Z\#\$Z\#\},\ 1 \leq j \leq n, \\ R_{m+2n+1} &= \{XB\#\$\#ZZ,\ \#Y\$Z\#Y',\ Z\#\$Z\#\}, \\ R_{m+2n+2} &= \{\#Y\$ZZ\#,\ \#Y'\$Z\#Y,\ Z\#\$Z\#\}. \end{aligned}$$

The idea behind this construction is again rotate-and-simulate. The components $R_i, 1 \leq i \leq m$, simulate the rules in P, in the right hand end of the strings of the form XwY produced by Γ (starting with $XBSY$). The components $R_{m+2j-1}, 1 \leq j \leq n$, remove one occurrence of the corresponding symbol α_j from the right hand end of the strings, whereas the pair of components $R_{m+2j}, 1 \leq j \leq n$, reintroduce an occurrence of α_j in the left hand end of the strings. Thus, these components $R_i, m+1 \leq i \leq m+2n$, circularly permute the strings, making possible the simulation of rules in P in any desired position. The components R_{m+2n+1}, R_{m+2n+2} remove the end markers X (only in the presence of B, hence in the right permutation) and Y. All components contain rules used just for passing the strings unmodified to the next step. These rules are $\#Y\$Z\#Y$ in $R_i, 1 \leq i \leq m$, $\#Y\$Z\#Y'$ and $\#Y'\$Z\#Y$ alternating in $R_{m+2j-1}, R_{m+2j}, 1 \leq j \leq n$, and in R_{m+2n+1}, R_{m+2n+2}, as well as the rules $Z\#\$Z\#$ present in all components, and used for passing the axioms from one step to the next. The role of Y' is to prevent wrong splicings, by introducing a symbol α_j in a string from which a symbol α_j has not been removed. This is the essential point of this construction, hence we shall examine its implementation in some detail.

Consider a string XwY and assume that the component R_1 works on it. The rule $\#Y\$Z\#Y$ changes nothing, it only passes the string to the next component. If we reach a component $R_i, 1 \leq i \leq n$, and a rule $\#u_iY\$Z\#v_iY$ can be used, then again a string bounded by X, Y is obtained.

When reaching R_{m+1}, a string of the form XwY can enter the splicing

$$(Xw|Y, Z|Y') \models (XwY', ZY)$$

or, when $w = w'\alpha_1$, the splicing

$$(Xw'|\alpha_1 Y, Z|Y) \models (Xw'Y, Z\alpha_1 Y).$$

In the first case, XwY' can also enter two different splicings in the next component, R_{m+2}:

$$(Xw|Y', Z|Y) \models (XwY, ZY'),$$
$$(X\alpha_1|Z, X|wY') \models (X\alpha_1 wY', XZ).$$

By the first splicing, we have returned to XwY, which is passed to the next component. Also, the string $X\alpha_1 wY'$ obtained in the second case is passed to the next component, but there is no splicing rule here which can be applied to this string. Thus $X\alpha_1 wY'$ is no longer present at the next step, and this prevents the production of "wrong" strings: α_1 is introduced to the left of w without first removing an occurrence of α_1 from the right hand of w.

In the second case, that when α_1 has been removed from $w'\alpha_1$, the string $Xw'Y$ reaches R_{m+2} where only one splicing is possible:

$$(X\alpha_1|Z, X|w'Y) \models (X\alpha_1 w'Y, XZ).$$

The string $X\alpha_1 w'Y$ is a correct one-step circular permutation of $Xw'\alpha_1 Y$, and it is again bounded by X, Y.

Therefore, we pass to R_{m+3} a string of the form XzY. The case of $j = 1$ is similar to the case of arbitrary $j, 1 \leq j \leq n$, hence the components

R_{m+2j-1}, R_{m+2j}, perform the desired rotations, or they pass the strings unmodified to R_{m+2n+1}.

A string of the form XwY can enter two splicings in R_{m+2n+1}:

$$(Xw|Y, Z|Y') \models (XwY', ZY),$$
$$(XB|w'Y, |ZZ) \models (XBZZ, w'Y), \text{ if } w = Bw'.$$

In the first case, there is only one possible splicing in R_{m+2n+2}:

$$(Xw|Y', Z|Y) \models (XwY, ZY'),$$

hence we pass the string XwY unchanged to R_1, resuming the cycle. In the second case, we also have only one possibility of splicing in R_{m+2n+2}:

$$(w'|Y, ZZ|) \models (w', ZZY).$$

The string w' is not marked by X, Y, hence it cannot enter new splicings if it is passed to R_1. If it is not terminal, then it is lost. Consequently, any derivation in G can be simulated in Γ, following the usual simulate-and-rotate procedure. As in the proofs in the previous sections, the "byproducts" of the splicings, strings which are not of the forms XwY, XwY', are never producing terminal strings outside $L(G)$. Thus, $L(G) = L(\Gamma)$, completing the proof. □

Note that each component of the system Γ above contains exactly three splicing rules.

10.5 Summary of Computationally Complete H Systems

We list now the classes of H systems which were proved in Chaps. 8, 9, 10 to characterize the recursively enumerable languages when using finite sets of axioms and finite sets of splicing rules. We cluster them in four categories:

Regulated rewriting-like control:

- extended H systems with permitting contexts,
- extended H systems with forbidding contexts,
- extended H systems with local targets,
- extended H systems with a global target,
- programmed extended H systems,
- extended H systems with double splicing,
- ordered H systems;

Other control mechanisms:

- extended H systems using multisets,
- evolving extended H systems,
- extended H systems with a fitness mapping;

Distributed architectures:

- splicing grammar systems,
- communicating distributed H systems,
- two-level distributed H systems,
- time-varying distributed H systems;

Using circular strings:

- extended circular H systems.

10.6 Bibliographical Notes

Splicing grammar systems as in Sect. 10.1 were introduced in [36], where one characterizes the family RE using systems with three context-free components. Theorems 10.1 and 10.2 are from [154]. Theorem 10.3 is from [75], where related results can also be found ($LIN \subseteq SGS_2(REG)$, $CF \subseteq SGS_3(REG)$).

Distributed H systems as in Sect. 10.2 were introduced in [32], in the non-extended case, with the first component used only for selecting by its filter the terminal strings produced by other components. (In this way, one further component is necessary, for instance, in Theorem 10.4.) In [32] one proves that $CDH_* = RE$. In [224] one improves this result to $CDH_9 = RE$, then in [158] one proves that $CDH_6 = RE$. The strengthening to three components (Theorem 10.4) is obtained in [180]; the proof of Theorem 10.4 is from [162] (it is slightly simpler than that in [180]). Theorems 10.5 and 10.6 are from [163].

Two-level distributed H systems in the non-separated form are introduced in [161], where a result like Theorem 10.7 is proved for them; the case of separated systems (hence the proof of Theorem 10.7) is considered in [158].

The time-varying distributed H systems are also introduced in [158], where one can find the proof of Theorem 10.9. Theorem 10.8 is from [162]. Recently, in [124] it was reported that time-varying H systems of degree two characterize the family of recursively enumerable languages, thus significantly improving the result in Theorem 10.8.

A variant of distributed H systems was introduced in [126]. They correspond to the cooperating distributed grammar systems in [28], [29] and use the 1-splicing relation $\vdash$ instead of $\models$. At each step one splices the string obtained at the previous step with an axiom. The components are enabled in a nondeterministic manner and they work in the maximal mode: when active, a component works as much as it can (this is the t mode of derivation in grammar systems, [28], [29]). Extended distributed H systems of this type with three components characterize the recursively enumerable languages, but it not known whether or not a similar result holds true for systems with two components.

Chapter 11

Splicing Revisited

Besides the variants of the splicing operation discussed in the previous chapters (especially in Chap. 8) and of the generative mechanisms based on them, several others were already investigated in the literature, mainly from a mathematical point of view, without directing the research toward (universal) computability models. In this chapter we present some of these variants, as a challenge to the reader for building further computability models.

11.1 Restricted Splicing; The Non-Iterated Case

In Chap. 8 we have considered certain regulated variants of the splicing operation, using permitting or forbidding contexts, target languages, fitness mappings, and order restrictions. Several other variants are possible. They are important in view of the fact that the unrestricted splicing operation (with respect to a finite set of splicing rules) cannot produce languages which are not regular.

We introduce here some further regulations on the splicing operation, namely for the operation $\vdash$. The reason to consider 1-splicing are, on the one hand, the fact that some of these variants do not look adequate for the 2-splicing, $\models$, on the other hand, the fact that we remain at the mathematical level, investigating only the non-iterated operations, as in Sect. 7.2, hence we prefer to work in the most general mathematical framework. The extent to which these restrictions are able to lead to controlled H systems which characterize the family of recursively enumerable languages remains to be investigated.

Consider an alphabet V and a splicing rule $r = u_1\#u_2\$u_3\#u_4$ over V. For $x, y \in V^*$, we define:

$$(x, y) \vdash_r^{pr} z \quad \text{iff} \quad (x, y) \vdash_r z \text{ and } x \in Pref(z), x \neq z,$$

$$\begin{aligned}
(x,y) \vdash_r^{in} z \quad &\text{iff} \quad (x,y) \vdash_r z \text{ and } |z| > \max\{|x|,|y|\},\\
(x,y) \vdash_r^{mi} z \quad &\text{iff} \quad (x,y) \vdash_r z \text{ and } |z| \geq |z'| \text{ for all}\\
&\qquad z' \in V^* \text{ such that } (x,y) \vdash_r z'.
\end{aligned}$$

In the first case the result of the splicing should be a proper continuation of the first term of the splicing, in the second one we must obtain a string which is strictly longer than each of the two terms of the splicing, in the third case we must splice so as to obtain one of the longest possible outputs. The indications pr, in, mi stand for "prefix", "increasing", "most increasing".

These restrictions are defined at the level of the splicing operation. We can also define restricted splicing schemes, with the operation being the subject of conditions formulated with respect to the whole set of splicing rules or with respect to the terms of the splicing. Of the first type is the ordered restriction discussed at the end of Sect. 8.2. We introduce here a class of restrictions of the second type, allowing splicing only among *similar* strings; various degrees of similarity can be considered.

In all cases below, we work with H schemes with a finite set of rules.

An H scheme with *clusters* is a triple $\sigma = (V, R, C)$, where (V, R) is a splicing scheme and C is a partition of V^*. For $r \in R$ and $x, y, z \in V^*$, we write

$$(x,y) \vdash_r^{cl} z \text{ iff } (x,y) \vdash_r z \text{ and } x, y \text{ belong to the same class of } C.$$

When C is a finite set of regular languages we write $\vdash_r^{rc}$ (rc for "regular clusters") instead of $\vdash_r^{cl}$. When C consists of singleton classes, $\{x\}, x \in V^*$, then we write $\vdash_r^{sf}$ (from "self-splicing"), and when C consists of classes $C_i = \{x \in V^* \mid |x| = i\}, i \geq 0$, then we write $\vdash_r^{sl}$ (from "same-length" splicing).

We denote by D the set $\{pr, in, mi, rc, sf, sl\}$, identifying the variants of the splicing operation defined above. For a splicing scheme σ with the alphabet V and the set of rules R, and for $L \subseteq V^*$, $g \in D$, we define

$$\sigma_g(L) = \{z \in V^* \mid (x,y) \vdash_r^g z, \text{ for } x, y \in L, r \in R\}.$$

As we have proceeded in Sect. 7.2 with $\vdash$ (the *free* splicing), we investigate now the relations between the operations $\vdash^g, g \in D$, and usual operations with languages. The aim is to settle the closure properties of families in the Chomsky hierarchy under the new operations of restricted splicing.

Lemma 11.1. *If a family FL of languages is closed under union with singleton languages, concatenation with symbols, left derivative, shuffle with symbols, and any splicing operation belonging to $D - \{in\}$, then FL is closed under the operation Suf of taking the suffixes.*

Proof. Take $L \in FL, L \subseteq V^*$, and consider the symbols a, b not in V. For the splicing scheme

$$\sigma = (V \cup \{a, b\}, \{a\#\$b\#\}),$$

we have

$$Suf(L) = \partial_a^l(\sigma_g(L_0)),$$

where

$$L_0 = \{a\} \cup (L \sqcup\!\sqcup \{b\}),$$

for both $g \in \{pr, mi\}$.

Indeed, the only possible splicing must involve the string a and a string xby for $xy \in L, x, y \in V^*$. There is only one place to apply the rule of σ, hence the result is unique, ay.

If we consider σ as having the total cluster $(V \cup \{a, b\})^*$, then g above can also be equal to rc.

For the sf, sl cases, we replace L_0 above with

$$L_0' = \{a\}L \sqcup\!\sqcup \{b\}.$$

Splicing $axby$ with itself, in the only possible way, we get ay; in general, for $axby$ and $ax'by'$ we have only one possibility of splicing, obtaining ay'. Therefore, we can obtain all elements of $Suf(L)$ and only strings in this set. □

Lemma 11.2. *If a family FL of languages is closed under union, concatenation with symbols, weak codings, and any splicing operation* $\vdash^g, g \in D - \{sf\}$, *then FL is closed under concatenation.*

Proof. Take two languages $L_1, L_2 \in FL, L_1, L_2 \subseteq V^*$, consider the symbols a, b not in V and the splicing scheme

$$\sigma = (V \cup \{a, b\}, \{a\#\$\#b\}).$$

We have

$$L_1L_2 = h(\sigma_g(L_0)),$$

where h is the weak coding defined by $h(a) = h(b) = \lambda$ and $h(c) = c$ for all $c \in V$,

$$L_0 = L_1\{a\} \cup \{b\}L_2,$$

and $g \in \{pr, in, mi\}$. As in the proof of Lemma 11.1, it is easy to cover also the case $g = rc$.

For $g = sl$ we replace L_0 above with

$$L_0' = L_1\{a\}^+ \cup \{b\}^+L_2.$$

Starting from strings xa^n, b^my with $n = |y|$ and $m = |x|$, we obtain $|xa^n| = |b^my|$ and $(xa^i|a^{n-i}, b^{m-j}|b^jy) \vdash^{sl} xa^ib^jy$, for all $i, j \geq 1$. Thus, $L_1L_2 = h(\sigma_{sl}(L_0'))$. □

Lemma 11.3. *If a family FL of languages is closed under concatenation with symbols, weak codings, and self-splicing, then FL is closed under doubling (the operation leading from L to* $d(L) = \{xx \mid x \in L\}$*).*

Proof. For $L \in FL, L \subseteq V^*$, and $a, b \notin V$, consider the splicing scheme

$$\sigma = (V \cup \{a\}, \{b\#\$\#a\}).$$

We obtain

$$d(L) = h(\sigma_{sf}(\{a\}L\{b\})),$$

for h being defined by $h(a) = h(b) = \lambda$ and $h(c) = c$ for all $c \in V$.

The only possible splicing is $(axb|, |axb) \vdash_r^{sf} axbaxb$, hence the equality holds. □

Lemma 11.4. *If a family FL of languages is closed under union, concatenation with symbols, right and left derivatives, and prefix splicing, then FL is closed under intersection.*

Proof. Take $L_1, L_2 \in FL, L_1, L_2 \subseteq V^*$, and consider the symbols a, b, c not in V. For the splicing scheme

$$\sigma = (V \cup \{a, b, c\}, \{a\#\$b\#\})$$

we have

$$L_1 \cap L_2 = \partial_a^l(\partial_{cc}^r(\sigma_{pr}(L_0))),$$

where

$$L_0 = \{a\}L_1\{c\} \cup \{b\}L_2\{cc\}.$$

Indeed, from the form of the splicing rule in σ, we can only splice a string axc with a string $bycc$, hence with $x \in L_1, y \in L_2$. Because the result must be a prolongation to the right of axc, we must have $x = y$. Consequently, $\sigma_{pr}(L_0) = \{a\}(L_1 \cap L_2)\{cc\}$. □

Lemma 11.5. *If a family FL of languages is closed under shuffle, concatenation with regular languages, non-erasing gsm mappings, and restricted morphisms, then FL is closed under length-increasing splicing.*

Proof. Take a language $L \in FL, L \subseteq V^*$, and a symbol $c \notin V$. For every $a \in V$ consider also the new symbols a', a'', as well as the coding h_1 defined by $h_1(a) = a', a \in V$. Take also a splicing scheme $\sigma = (V, R)$. For each rule $r \in R$, take two associated symbols, d_r, d'_r. Consider the regular languages

$$L_1 = \{xu_1d_ru_2y \mid x, y \in V^*, r = u_1\#u_2\$u_3\#u_4 \in R\},$$
$$L_2 = \{xu_3d'_ru_4y \mid x, y \in V^*, r = u_1\#u_2\$u_3\#u_4 \in R\}.$$

Take also

$$L'_1 = L_1 \cap (\bigcup_{r \in R}(L \sqcup\!\sqcup \{d_r\})),$$
$$L'_2 = h_1(L_2 \cap (\bigcup_{r \in R}(L \sqcup\!\sqcup \{d'_r\}))),$$

where h_1 is extended by $h_1(d'_r) = d'_r$. Both these languages are in FL, because the coding h_1 as well as the intersection with regular languages can be obtained by using non-erasing gsm mappings. Consider now the language

$$L_3 = (L'_1 c^* \sqcup\!\sqcup c^* L'_2) \cap L_4,$$

where

$$L_4 = \{ac \mid a \in V\}^+ \{ab' \mid a, b \in V\}^* \{d_r d'_r \mid r \in R\} \{ab' \mid a, b \in V\}^* \\ \{ca' \mid a \in V\}^+.$$

Also L_3 is in FL. The intersection with L_4 selects from $L'_1 c^* \sqcup\!\sqcup c^* L'_2$ the strings of the form

$$a_1 c \dots a_k c a_{k+1} b'_1 \dots a_{k+j} b'_j d_r d'_r a_{k+j+1} b'_{j+1} \dots \\ \dots a_{k+j+i} b'_{j+i} c b'_{j+i+1} \dots c b'_{j+i+l}, \qquad (*)$$

for $k \geq 1, j \geq 0, i \geq 0, l \geq 1$, corresponding to the strings

$$x = a_1 a_2 \dots a_{k+j+i} \in L, \quad y = b_1 b_2 \dots b_{j+i+l} \in L.$$

We have

$$(x, y) \vdash_r^{in} a_1 \dots a_{k+j} b_{j+1} \dots b_{j+i+l} = z.$$

The obtained string has the length $k + j + i + l$, which is strictly greater than both $|x|$ and $|y|$.

We can easily construct a gsm g which can parse a string w of the form $(*)$, performing the following operations:

- leave unchanged the prefix $a_1 c \dots a_k c$,
- from the last occurrence of the symbol c in the prefix, until d_r, leave unchanged the symbols $a_{k+1}, \dots, a_{k+j}$, and replace each symbol $b'_1, \dots, b'_j$ with an occurrence of c,
- replace $d_r d'_r$ with cc,
- from d'_r to the next occurrence of c, replace each $a_{k+j+1}, \dots, a_{k+j+i}$ with an occurrence of c and each $b'_{j+1}, \dots, b'_{j+i}$ with $b_{j+1}, \dots, b_{j+i}$, respectively,
- in the suffix $cb'_{j+i+1} \dots cb'_{j+i+l}$, replace the primed symbols with their non-primed variants.

In this way, we obtain the above string z, shuffled with $|z| + 2$ occurrences of the symbol c. By a restricted morphism h_2 we can erase c, leaving unchanged the other symbols. Therefore, $\sigma_{in}(L) = h_2(g(L_3))$, which means that $\sigma_{in}(L) \in FL$. □

Lemma 11.6. *If a family FL of languages is closed under concatenation and arbitrary gsm mappings, then FL is closed under most-increasing splicing.*

Proof. We have $\sigma_{mi}(L) = g(L\{c\}L)$ for a gsm g which checks the presence in strings of $L\{c\}L$ of substrings u_1u_2, u_3u_4 from the rules of σ, removes the parts to be removed, and checks at the same time the fact that u_1u_2 is considered on the rightmost possible position before c and u_3u_4 is considered on the leftmost possible position after c, thus producing the largest result. The reader can construct such a gsm with full details. □

Lemma 11.7. *If a family of languages is closed under doubling and arbitrary gsm mappings, then it is closed under self-splicing.*

Proof. For $L \subseteq V^*, c \notin V$, and $\sigma = (V, R)$, we have

$$\sigma_{sf}(L) = g(d(L\{c\})),$$

where g is a gsm which simulates the application of rules in R as in the previous proof. □

Lemma 11.8. *If a family FL of languages is closed under shuffle and arbitrary gsm mappings, then FL is closed under prefix-splicing.*

Proof. Take $\sigma = (V, R), L \subseteq V^*$, $L \in FL$, and the coding $h : V \longrightarrow \{a' \mid a \in V\}$ defined by $h(a) = a', a \in V$. Take the language

$$L' = L \sqcup\!\sqcup\, h(L)$$

and construct a gsm g performing the following operations:

- scan a prefix $x_1 \in V^*$ and leave it unchanged,
- choose a rule $r = u_1\#u_2\$u_3\#u_4$ in R, $u_1, u_2, u_3, u_4 \in V^*$; parse the string u_1 and leave it unchanged,
- parse the substring $h(y_1u_3)$, for $y_1 \in V^*$, and remove it,
- check that a string follows of the form $a_1a'_1a_2a'_2 \ldots a_ka'_k$, for $a_i \in V$, $1 \leq i \leq k$, such that $a_1a_2 \ldots a_k = u_2x_2$ for some $x_2 \in V^*$; at the same time check whether or not $a_1a'_1a_2a'_2 \ldots a_ka'_k$ is followed by $a'_{k+1} \ldots a'_{k+s}$, $s \geq 1$, such that $a'_1a'_2 \ldots a'_ka'_{k+1} \ldots a'_{k+s} = h(u_3y_2)$ for some $y_2 \in V^*$; during this phase, all symbols a_i are erased and all symbols a'_i are replaced by the associated symbols a_i.

The construction of g is straightforward.

The equality $\sigma_{pr}(L) = g(L')$ is obvious. Because the coding h can be realized by a gsm, the proof is complete. □

Lemma 11.9. *If a family FL of languages is closed under union, intersection with regular sets, and free splicing, then it is also closed under regular clustered splicing with finitely many regular sets in the partition.*

Proof. Consider a splicing scheme $\sigma = (V, R, C)$, with a partition $C = \{C_1, \ldots, C_n\}$ of V^* with $C_i \in REG, 1 \leq i \leq n$. Then, for each $L \subseteq V^*$, $L \in FL$, we obtain

$$\sigma_{rc}(L) = \bigcup_{i=1}^{n} \sigma_1(L \cap C_i).$$

As $L \cap C_i \in FL$, we have $\sigma_1(L \cap C_i) \in FL$ for all i, hence $\sigma_{rc}(L) \in FL$. □

We are now ready to settle the closure properties of families in Chomsky hierarchy under the operations $\vdash^g, g \in D$.

Theorem 11.1. *The closure properties in Table 11.1 hold, where for each pair $(g, FL), g \in D$, at the intersection of the row of g and the column of FL we have written the smallest family FL' (among the five families considered here) such that $\sigma_g(L) \in FL'$ for all $L \in FL$. (When $FL' = FL$, this means that the family FL is closed under the type g of splicing, and, conversely, $FL \neq FL'$ indicates the nonclosure.)*

Table 11.1. Closure properties of families in the Chomsky hierarchy

Splicing variant	REG	LIN	CF	CS	RE
pr	REG	RE	RE	RE	RE
in	REG	CS	CS	CS	RE
mi	REG	CF	CF	RE	RE
rc	REG	CF	CF	RE	RE
sf	CS	RE	RE	RE	RE
sl	LIN	RE	RE	RE	RE

Proof. The positive closure properties of REG are obtained from: pr = Lemma 11.8, in = Lemma 11.5, mi = Lemma 11.6, rc = Lemma 11.9.

The positive closure properties of CF follow from: mi = Lemma 11.6, rc = Lemma 11.9.

The unique positive closure property of CS, namely the closure under length-increasing splicing, follows from Lemma 11.5.

The closure of RE under all operations above follows either from Church–Turing Thesis, or from lemmas above.

The fact that there are $L \in REG$ such that $\sigma_{sf}(L) \notin CF$ is proved by Lemma 11.3 (for instance, $d(\{a^n b^m \mid n, m \geq 1\}) = \{a^n b^m a^n b^m \mid n, m \geq 1\} \notin CF$). On the other hand, if $\sigma = (V, R)$ and $L \in REG$, then $\sigma_{sf}(L) \in CS$. This can be seen as follows. If $L \in REG$, then $d(L\{c\})$ is a right-linear simple matrix language (see [40]). This family is closed under arbitrary gsm mappings and is strictly included in CS, hence $\sigma_{sf}(L) \in CS$.

If a language $L \subseteq V^*$ has the property that $card(V^n \cap L) \leq 1$ for all $n \geq 0$ (such a language is called *thin*, see [169]) and σ is a splicing scheme, then $\sigma_{sf}(L) = \sigma_{sl}(L)$, hence from Lemma 11.3 we find that the closure of REG under sl splicing would imply the fact that $d(L) \in REG$ for each thin regular language L. However, this does not hold: $d(\{a^n b \mid n \geq 1\}) = \{a^n b a^n b \mid n \geq 1\} \notin REG$; note that $\{a^n b \mid n \geq 1\}$ is thin. However, if $L \in REG, L \subseteq V^*$, and $\sigma = (V, R)$ is a splicing scheme, then $\sigma_{sl}(L) \in LIN$. Indeed, take $a, c \notin V$ and consider the language

$$L_1 = (L\{c\}L \sqcup\!\sqcup \{a^n c a^n \mid n \geq 1\}) \cap (V\{a\})^+\{cc\}(V\{a\})^+.$$

Because $\{a^n c a^n \mid n \geq 1\}$ is linear, $L\{c\}L$ is regular and LIN is closed under shuffle with regular sets (as well as under intersection with regular sets), it follows that $L_1 \in LIN$. All its strings are of the form

$$b_1 a b_2 a \ldots b_n a c c d_1 a d_2 a \ldots d_n a,$$

for $x = b_1 b_2 \ldots b_n \in L, y = d_1 d_2 \ldots d_n \in L$. By a gsm g we can now simulate the splicing of strings $x, y \in L$ as selected by L_1, that is $g(L_1) = \sigma_{sl}(L)$. Consequently, $\sigma_{sl}(L) \in LIN$.

The fact that there are linear languages L such that $\sigma_g(L) \notin CF$ for $g \in \{pr, in, sf, sl\}$ is proved as follows: pr = Lemma 11.4 (there are $L_1, L_2 \in LIN$ such that $L_1 \cap L_2 \notin CF$), sf = Lemma 11.3, sl = Lemma 11.3, used for thin linear languages (in the same way as above for thin regular languages: $L = \{a^n b^n \mid n \geq 1\}$ is thin and linear, but $d(L) \notin CF$). For in we consider a new example:

For $\sigma = (\{a, b, c\}, \{b\#c\$a\#c\})$ and

$$L = \{a^n b^n c \mid n \geq 1\} \cup \{a^n c^n \mid n \geq 1\},$$

we get

$$\sigma_{in}(L) = \{a^n b^n c^m \mid 2 \leq m < 2n\}.$$

Indeed, the only possible splicing is of the type

$$(a^n b^n | c, a^m | c^m) \vdash_r a^n b^n c^m.$$

For the length-increasing mode we must have $2n + m > 2n + 1$ (hence $m \geq 2$) as well as $2n + m > 2m$ (hence $m < 2n$). This language is not context-free.

All the non-closure properties of CS follow from Lemma 11.1.

Thus, all the assertions presented in Table 11.1 are proved. □

Several remarks about the results in Table 11.1 are worth mentioning:

- two variants of splicing "break" the regularity barrier and five of them break the context-freeness barrier,
- no splicing operation (as considered above) preserves linear languages; this is due mainly to the fact that by splicing (aided by the semi-AFL operations, under which LIN is closed) we can simulate concatenation,
- none of the previous operations – except the length-increasing splicing – preserves the context-sensitive languages; this is due mainly to the fact that by splicing we can perform the erasing of arbitrarily long prefixes and suffixes.

It is an *open problem* whether or not the assertions corresponding to families LIN, CF on the rows of pr, sf, and sl can be improved. More exactly, given a splicing scheme σ and a language $L \in LIN$ (or $L \in CF$), the question remains whether or not the relations $\sigma_g(L) \in CS, g \in \{pr, sf, sl\}$, are true, or the strongest assertion here is $\sigma_g(L) \in RE$, as in Table 11.1. Anyway, the families $S_g(LIN, FIN) = \{\sigma_g(L) \mid L \in LIN\}, g \in \{pr, sf, sl\}$, seem to be very large.

Theorem 11.2. *Every language $L \in RE$ is the morphic image of a language in $S_{pr}(LIN, FIN)$.*

Proof. From Corollary 3.3 we know that each recursively enumerable language $L \subseteq V^*$ can be written in the form $L = h_1(L_1 \cap L_2)$, where h_1 is a morphism and $L_1, L_2 \in LIN$. Consider the symbols a, b, c not in V. As in the proof of Lemma 11.4, take

$$\sigma = (V \cup \{a, b, c\}, \{a\#\$b\#\}),$$
$$L_0 = \{a\}L_1\{c\} \cup \{b\}L_2\{cc\}.$$

For $L' = \sigma_{pr}(L_0)$ and $h : (V \cup \{a, c\})^* \longrightarrow V^*$ defined by $h(a) = h(c) = \lambda$ and $h(d) = h_1(d)$ for $d \in V$, we obtain $L = h(L')$. □

Corollary 11.1. *For every family FL of languages which is strictly included in RE and is closed under arbitrary morphisms we have $S_{pr}(LIN, FIN) - FL \neq \emptyset$.*

It is a research topic to consider extended H systems based on the operations $\vdash^g$, $g \in D$, defined above. Of particular interest is the case $g = rc$, which can have some "realistic" interpretations. For instance, the clusters can be defined by a specific similarity relation favoring certain enzymes to work on the strings in the same class.

11.2 Replication Systems

In Sect. 9.1 we have considered the following operation, leading a circular string $\hat{x}$ and a linear one v to a linear string w:

$$(\hat{x}, v) \models_r^3 w \quad \text{iff} \quad x = x_1u_1u_2,$$
$$v = v_1u_3u_4v_2,$$
$$w = v_1u_3u_2x_1u_1u_4v_2,$$
$$\text{for some } x_1, v_1, v_2 \in V^*,$$

for $r = u_1\#u_2\$u_3\#u_4$ a splicing rule over V. Thus, we have inserted the string obtained by cutting $\hat{x}$ between u_1 and u_2 in v, namely between u_3, u_4. This was possible, *because we know that u_1u_2, u_3u_4 describe matching sticky ends.* This operation can also be considered starting from two linear strings, hence with the string to be inserted being a substring of a linear string; thus, we need two cutting places in order to produce it. In this way we are led to the following general definition.

A *replication system* of degree $n, n \geq 1$, is a construct

$$\gamma = (V, A, C_1, C_2, \ldots, C_n),$$

where V is an alphabet, A is a finite language over V, and $C_i, 1 \leq i \leq n$, are finite subsets of $V^*\#V^*$. The elements of A are called *axioms*, those of $C_i, 1 \leq i \leq n$, are *cutting rules.*

For $x, y, z \in V^*$ and $1 \leq i \leq n$, we write

$$\begin{aligned}(x, y) \rhd_i z \quad \text{iff} \quad & x = x_1u_1v_1x_2, \\ & y = y_1u_2v_2y_2u_3v_3y_3, \\ & z = x_1u_1v_2y_2u_3v_1x_2, \\ & \text{for } u_1\#v_1, u_2\#v_2, u_3\#v_3 \in C_i, \\ & \text{and } x_1, x_2, y_1, y_2, y_3 \in V^*.\end{aligned}$$

Thus, we insert the substring $v_2y_2u_3$ of y between u_1, v_1 in x, knowing that the patterns u_1v_1, u_2v_2, u_3v_3 correspond to (restriction enzymes producing) matching sticky ends (this is encoded in the fact that the three cutting rules belong to the same set C_i).

Now, we can proceed as in the case of the splicing operation. For a replication system $\gamma = (V, A, C_1, C_2, \ldots, C_n)$ and a language $L \subseteq V^*$ we define

$$\begin{aligned}\sigma(L) &= \{z \in V^* \mid (x, y) \rhd_i z, \text{ for } x, y \in L, 1 \leq i \leq n\}, \\ \sigma^0(L) &= L, \\ \sigma^{i+1}(L) &= \sigma^i(L) \cup \sigma(\sigma^i(L)), i \geq 0, \\ \sigma^*(L) &= \bigcup_{i \geq 0} \sigma^i(L).\end{aligned}$$

Then, the language generated by γ is defined by

$$L(\gamma) = \sigma^*(A)$$

(we start from the axioms in A and we iteratively grow strings by replication according to the cutting rules in $C_1, \ldots, C_n$).

A cutting rule $u\#v$ is said to be of *radius* p if $p = \max\{|u|, |v|\}$. A replication system is said to be of radius p if p is the maximum radius of its cutting rules.

We denote by $REP_n([m])$ the family of languages generated by replication systems with at most n components, $n \geq 1$, and of radius at most $m, m \geq 1$;

when the number of components is not bounded, the subscript n is replaced by $*$; when the radius is not bounded, we replace $[m]$ with FIN. Thus,

$$REP_*(FIN) = \bigcup_{n\geq 1}\bigcup_{m\geq 1} REP_n([m]) = \bigcup_{n\geq 1} REP_n(FIN).$$

Because we always increase the length of the strings, we have

Lemma 11.10. $REP_*(FIN) \subseteq CS$.

The replication systems $\gamma = (V, A, C_1, \ldots, C_n)$ such that $card(C_i) = 1$, $1 \leq i \leq n$, are said to be *uniform*. (When defining $(x, y) \triangleright_i z$, we use the same unique cutting rule both in x and y: $x = x_1uvx_2, y = y_1uvy_2uvy_3$ and $z = x_1uvy_2uvx_2$.)

In the definition of the relation $\triangleright$ we have two important choices: (1) of the string to be inserted, and (2) of the place where the insertion is performed. Several variants can be considered in each case. We leave as a *research topic* the study of these variants for the replication systems in the general form defined above and we investigate here only some possibilities occurring when answering the first question above, for a particular type of replication system. Namely, we consider *simple replication systems*, that is systems which are uniform, of radius one, of degree one, and with one axiom only.

Therefore, a simple replication system is a construct

$$\gamma = (V, w, a\#b),$$

where V is an alphabet, $w \in V^+$ (axiom), and $a, b \in V$. Moreover, we define the replication only for identical strings (self-replication; anyway, in the first step of the system's working, because only w is available, we have to perform an operation of the type $(w, w) \triangleright z$).

Specifically, for a simple replication system $\gamma = (V, w, a\#b)$ and $x, y \in V^*$ we write

$$\begin{array}{lll} x \rightsquigarrow y & \text{iff} & (1)\ \ x = x_1abx_2,\ x_1, x_2 \in V^*, \\ & & (2)\ \ y = x_1azbx_2,\ \text{for } z = bz' = z''a,\ z', z'' \in V^*, \\ & & (3)\ \ z \in Sub(x). \end{array}$$

No restriction is imposed about the position of the substring ab in x (condition (1)) or on the way z was selected from $Sub(x)$ (condition (3)). We discuss here ten possibilities arising from considering restrictions on condition (3).

Condition (3) can be replaced by more restrictive ones as follows (in all cases, $z = bz' = z''a$):

1. $z = x$ (*total*; t),

2. $z \in Pref(x)$ (*arbitrary prefix; ap*),
3. $z \in Pref(x)$ and z is maximal (if $x = z_1u_1, z_1 = bz_1' = z_1''a$, then $|z| \geq |z_1|$) (*maximal prefix; Mp*),
4. $z \in Pref(x)$ and z is minimal (if $x = z_1u_1, z_1 = bz_1' = z_1''a$, then $|z| \leq |z_1|$) (*minimal prefix; mp*),
5. $z \in Sub(x)$ and z is leftmost (if $x = u_1zu_2, x = u_1'z_1u_2', z_1 = bz_1' = z_1''a$, then $|u_1| \leq |u_1'|$) (*arbitrary leftmost; al*),
6. $z \in Sub(x)$, z is leftmost and maximal ($x = u_1zu_2$ and if $x = u_1'z_1u_2', z_1 = bz_1' = z_1''a$, then $|u_1| \leq |u_1'|$; moreover, if $x = u_1zu_2 = u_1z_1u_2', z_1 = bz_1' = z_1''a$, then $|z| \geq |z_1|$) (*maximal leftmost; Ml*),
7. $z \in Sub(x)$, z is leftmost and minimal ($x = u_1zu_2$ and if $x = u_1'z_1u_2', z_1 = bz_1' = z_1''a$, then $|u_1| \leq |u_1'|$; moreover, if $x = u_1zu_2 = u_1z_1u_2', z_1 = bz_1' = z_1''a$, then $|z| \leq |z_1|$) (*minimal leftmost; ml*),
8. $z \in Sub(x)$ and z is maximal (if $x = u_1zu_2$ and $x = u_1'z_1u_2', z_1 = bz_1' = z_1''a, |u_1'| \leq |u_1|, |u_2'| \leq |u_2|$, then $z = z_1$) (*arbitrary maximal; aM*),
9. $z \in Sub(x)$ and z is minimal (if $x = u_1zu_2$ and $x = u_1'z_1u_2', z_1 = bz_1' = z_1''a, |u_1'| \geq |u_1|, |u_2'| \geq |u_2|$, then $z = z_1$) (*arbitrary minimal; am*).

The case in the definition of $\rightsquigarrow$ corresponds to

10. $z \in Sub(x)$ (*any* subword, *free; af*).

We denote by D the set $\{t, ap, Mp, mp, al, Ml, ml, af, aM, am\}$, of previously defined modes of choosing the inserted string. We stress the fact that in all cases 1 – 10, z is a string of the form $z = bz' = z''a$. (When $a = b$, we can have $z = a$, but for $a \neq b$ we have $|z| \geq 2$.)

For a replication choosing the inserted string in the mode $g \in D$, we write $x \rightsquigarrow_g y$.

We denote by $\rightsquigarrow_g^*$ the reflexive and transitive closure of the relation $\rightsquigarrow_g$, $g \in D$. (We call it a *replication chain.*) Then the *language generated* by the replication system $\gamma = (V, w, a\#b)$ in the mode $g \in D$ is defined by

$$L_g(\gamma) = \{z \in V^* \mid w \rightsquigarrow_g^* z\}.$$

We denote by $SREP(g)$ the family of languages of the form $L_g(\gamma)$, where $g \in D$ and γ is a simple replication system.

Ten families of languages are obtained. We briefly investigate their interrelationships and relationships with Chomsky families. We start by considering two examples.

Example 11.1. Consider the replication system

$$\gamma_1 = (\{a, b\}, baba, a\#b).$$

In all cases, the obtained strings are in $(ba)^+$: we start from $(ba)^2$; when $(ba)^n \rightsquigarrow_g y$, we insert a string $b(ab)^i a, i \geq 0$, between two symbols ab, hence the string y is of the form $(ba)^m, m > n$.

We have

$$L_g(\gamma_1) = (ba)^+ ba, \text{ for } g \in \{ap, mp, al, ml, af, am\}.$$

Indeed, in any minimal mode (prefix, leftmost, arbitrary), we have to insert the string ba. In all free modes we *can* insert ba. Therefore $(ba)^+ ba \subseteq L_g(\gamma_1)$. The reverse inclusion has been pointed out above. Moreover,

$$L_g(\gamma_1) = \{(ba)^{2^n} \mid n \geq 1\}, \text{ for } g \in \{Mp, Ml, aM, t\}.$$

The maximal prefix (leftmost string, substring) of the form $z = bz'a$ of a string $(ba)^n$ is $(ba)^n$ itself, hence $x \rightsquigarrow_g y$ for g as above means $y = x^2$, which proves the equality.

Example 11.2. Consider also the system

$$\gamma_2 = (\{a, b\}, abb, b\#b).$$

We obtain

$$L_g(\gamma_2) = \begin{cases} ab^+b, & \text{for } g \in \{al,\ ml,\ af,\ am\}, \\ abb, & \text{for } g \in \{ap, mp, Mp, t\}, \\ \{ab^{2^n} \mid n \geq 1\}, & \text{for } g \in \{aM,\ Ml\}. \end{cases}$$

In order to obtain counterexamples, we consider a series of necessary conditions.

Lemma 11.11. (i) *Any replication language is either a singleton or it is infinite.* (ii) *For any replication language $L \subseteq V^*$ we have $card(Pref(L) \cap V) = 1$, $card(Suf(L) \cap V) = 1$.*

Proof. (i) Consider a replication system $\gamma = (V, w, a\#b)$. If w cannot be replicated (either no substring ab is present in w, or w contains no suitable substring $z = bz' = z''a$), then $L_g(\gamma) = \{w\}$.

If we have $w \rightsquigarrow_g w'$, then $w = w_1abw_2$, there is $z = bz' = z''a$ in $Sub(w)$ fulfilling the condition associated with g, and $w' = w_1azbw_2$.

Clearly, w' contains again the substring ab, as well as substrings starting with b and ending with a (z is such an example), therefore we can replicate again. This is obvious for all modes $g \in \{al,\ ml,\ Ml,\ af,\ am,\ aM\}$.

If $z \in Pref(w)$, hence $w = zw_3$, then $w' = w_1azbw_2$, and there is $x \in Pref(w')$ such that $x = bx' = x''a$. Indeed, because $z = bz' \in Pref(w)$ and $w_1a \in Pref(w)$, we have $b \in Pref(w_1a) \subseteq Pref(w')$; because $z = z''a$ and $w' = w_1az''abw_2$, the prefix $w_1az''a$ ends with a. Consequently, we can also replicate again in any of the modes $g \in \{ap,\ mp,\ Mp\}$.

For the t case, we have $w = z$ and $w' = w_1awbw_2$. As above, we obtain $w' = bw''a$, hence we can replicate again in this case, too.

The process can be iterated in all modes, hence we obtain an infinite language.

(ii) Directly from the definitions of replication operations $\rightsquigarrow_g$ we see that the first and the last symbol of the string is never modified, they remain the same as those of the axiom. □

Lemma 11.12. *If $L \in SREP(g)$, for $g \in \{ap, mp, Mp, t\}$, is an infinite language with $b = Pref(L) \cap V$, then for each $n \geq 1$, there is $x \in L$ such that $|x|_b \geq n$.*

Proof. Take $\gamma = (V, w, a\#b)$ with $L_g(\gamma) = L$ infinite, for g as specified above. Because L is infinite, there is w' such that $w \rightsquigarrow_g w'$. It follows that w starts with b, hence $b = Pref(L) \cap V$. Moreover, the replication can continue an arbitrarily large number of times, at each step using a prefix of the current string (the whole string in the case of $g = t$). At each replication, a further copy of the symbol b is introduced: if $x \rightsquigarrow_g y, x = x_1abx_2$, then $y = x_1azbx_2$ for $z \in Pref(x), z = bz'$. Therefore, an arbitrarily large number of occurrences of b can be produced. □

Lemma 11.13. *If $L \in SREP(t)$ is an infinite language, then for both $a = Suf(L) \cap V$ and $b = Pref(L) \cap V$ and for each $n \geq 1$, there is $x \in L$ such that $|x|_a \geq n, |x|_b \geq n$.*

Proof. We proceed exactly as in the previous proof, taking into consideration both the first symbol of w and the last symbol of w (they are preserved on these positions by replication, and the number of their occurrences in the current string increases unboundedly). □

Lemma 11.14. *Any infinite language $L \in SREP(t)$, with $V = alph(L)$, has $\Psi_V(L) = \{\pi \cdot 2^n \mid n \geq 0\}$, for some $\pi \in \mathbf{N}^k, k = card(V)$.*

Proof. If $L = L_t(\gamma)$ for some $\gamma = (V, w, a\#b)$ and L is infinite, then there are arbitrarily long replication chains

$$\rho : w = w_1 \rightsquigarrow_t w_2 \rightsquigarrow_t w_3 \rightsquigarrow_t \cdots$$

Clearly, $\Psi_V(w_{i+1}) = 2 \cdot \Psi_V(w_i), i \geq 1$, hence $\Psi_V(w_i) = 2^{i-1} \cdot \Psi_V(w_1), i \geq 1$, and this is true for all chains ρ. Consequently, $\Psi_V(L) = \{\Psi_V(w) \cdot 2^n \mid n \geq 0\}$, that is, π in the lemma is $\Psi_V(w)$. □

Of course, Lemma 11.14 implies Lemma 11.13.

Lemma 11.15. *For any infinite language $L \in SREP(g)$, $g \in \{Mp, Ml, aM\}$, with $V = alph(L)$, there are $\pi_1, \pi_2 \in \mathbf{N}^k, k = card(V)$, such that $\Psi_V(L) = \{\pi_1 + \pi_2 \cdot 2^n \mid n \geq 0\}$.*

Proof. Take $\gamma = (V, w, a\#b)$ such that $L = L_g(\gamma)$, and examine an arbitrarily long replication chain

$$\rho : w = w_1 \rightsquigarrow_g w_2 \rightsquigarrow_g w_3 \rightsquigarrow_g \cdots$$

Assume that $w_1 = x_1abx_2, w_2 = x_1az_1bx_2$, for z_1 maximal in w_1 of the form $z_1 = bz_1' = z_1''a$. This means that when writing $w_1 = y_1z_1y_2$, z_1 starts with the leftmost occurrence of b in w_1 and ends with the rightmost occurrence of a in w_1.

If $w_1 = y_1by_1'$ and this is the leftmost occurrence of b in w_1, then $w_2 = y_1by_1''$ again: if the insertion place is to the right hand of this occurrence of b, then the prefix y_1b of w_1 appears unmodified in w_2; if the insertion place is to the left hand of this occurrence of b, then it must imply this b (being the leftmost), hence y_1 ends with a and we get $w_2 = y_1z_1bx_2 = y_1bz_1'bx_2$, that is, again the prefix y_1b is preserved.

Similarly, if a in $w_1 = y_2'ay_2$ is the rightmost occurrence of a in w_1, then the suffix ay_2 appears in w_2, too.

Therefore, $w_1 = y_1z_1y_2, w_2 = y_1z_2y_2$, where z_2 starts with b, ends with a, and it is maximal in w_2 with these properties. Moreover,

$$\begin{aligned} |z_2| &= |w_2| - |y_1y_2| = |w_1z_1| - |y_1y_2| \\ &= |y_1z_1y_2| + |z_1| - |y_1y_2| = 2|z_1|. \end{aligned}$$

Consequently, at every step of ρ we get a string of the form $y_1z_iy_2$ with $\Psi_V(z_{i+1}) = 2 \cdot \Psi_V(z_i), i \geq 1$. For $\pi_1 = \Psi_V(y_1y_2)$ and $\pi_2 = \Psi_V(z_1)$, we have the assertion in the lemma (clearly, if $z_i \in Pref(w_i)$, then $z_{i+1} \in Pref(w_{i+1})$). □

In contrast to the preceding two lemmas, we have

Lemma 11.16. *For all $g \in \{af, am, ap, mp, al, ml\}$, if $L \in SREP(g)$ is an infinite language, then there is a constant k such that whenever $x \in L$, there exists $y \in L$, with $x \leadsto_g y$, such that $|x| < |y| \leq |x| + k$.*

Proof. Take $\gamma = (V, w, a\#b)$ and $z \in Sub(w)$ such that $w \leadsto_g w'$, with $w = w_1abw_2$, $w' = w_1azbw_2$. For $g = af$ we can insert z again. For $g = am$ we can insert z again if z is minimum, or a shorter string otherwise.

If $z \in Pref(w)$, then $z \in Pref(w_1az) \subseteq Pref(w')$, hence for $g = ap$ we can use z again, whereas for $g = mp$ we can use z or a shorter prefix.

The same assertions hold for the leftmost cases of choosing the string to be inserted.

The argument can be iterated, hence at every step $x \leadsto_g y$ with $w \leadsto_g^* x$, we have $|x| < |y| \leq |x| + |w|$. Taking $k = |w|$, we have the lemma. □

The reader can easily find languages not satisfying these necessary conditions.

From Example 11.1 (each family $SREP(g)$ contains infinite languages) and Lemma 11.11 (there are finite languages not in $SREP(g)$, for any $g \in D$), we get

Theorem 11.3. *The family FIN is incomparable with each family $SREP(g), g \in D$.*

From Example 11.1 we also know that all families $SREP(g), g \in \{Mp, Ml, aM, t\}$, contain non-context-free languages. In fact, a stronger result is true.

Theorem 11.4. *Each family $SREP(g), g \in \{Mp, Ml, aM, t\}$, is incomparable with each subfamily of MAT^λ containing at least a non-singleton finite language.*

Proof. For the system $\gamma = (\{a\}, aa, a\#a)$ we obtain

$$L_g(\gamma) = \{a^{2^n} \mid n \geq 1\},$$

for each $g \in \{Mp, Ml, aM, t\}$. According to [85], this language is not in the family MAT^λ.

On the other hand, there are finite languages not in $SREP(g)$, for all $g \in D$. □

Some other replication families contain only regular languages:

Theorem 11.5. $SREP(g) \subseteq REG,\ g \in \{am, mp\}$.

Proof. Take a replication system $\gamma = (V, w, a\#b)$. If $w = a_1a_2\ldots a_n$, with $a_i \in V, 1 \leq i \leq n, n \geq 1$, then we consider the string

$$w' = s_0a_1s_1a_2\ldots s_{n-1}a_ns_n,$$

where $s_0, s_1, \ldots, s_n$ are new symbols.

Identify in w all the positions of the substring ab. Let $1 \leq i_1 < i_2 < \ldots < i_r < n$ be these positions:

$$a_{i_q} = a,\ a_{i_q+1} = b,\ 1 \leq q \leq r.$$

All these positions are insertion places.

Identify now by pairs $(j_1, k_1), \ldots, (j_t, k_t)$ all minimal substrings of w of the form $z_l = bz_l' = z_l''a$. More specifically,

$$z_l = a_{j_l}a_{j_l+1}\ldots a_{k_l},\ 1 \leq l \leq t.$$

For each such a string z_l and for each insertion position i_q specified above, consider the string

$$z_{l,q} = s_{j_l-1}^{(l,q)}a_{j_l}s_{j_l}^{(l,q)}a_{j_l+1}\ldots s_{k_l-1}^{(l,q)}a_{k_l}s_{k_l}^{(l,q)},$$

where $s_h^{(l,q)}$ are new symbols, associated with h, z_l and i_q.

Denote by K the set of all symbols $s_i, 0 \leq i \leq n$, and $s_h^{(l,q)}, 1 \leq l \leq t, j_l-1 \leq h \leq k_l, 1 \leq q \leq r$. Construct the finite nondeterministic automaton (with λ-moves)

$$M = (K, V, s_0, \{s_n\}, \delta),$$

where δ is defined as suggested in Fig. 11.1, such that it covers all links in w', in strings z_l, as well as the links called for by insertions of strings z_l in w' in the corresponding positions:

$$s_i = \delta(s_{i-1}, a_i), \text{ for } 1 \leq i \leq n,$$
$$s_h^{(l,q)} \in \delta(s_{h-1}^{(l,q)}, a_h), \text{ for } j_l \leq h \leq k_l, 1 \leq l \leq t, 1 \leq q \leq r,$$
$$s_{j_l-1}^{(l,q)} \in \delta(s_{i_q}, \lambda),$$
$$s_{i_q} \in \delta(s_{k_l}^{(l,q)}, \lambda), \text{ for } 1 \leq l \leq t, 1 \leq q \leq r.$$

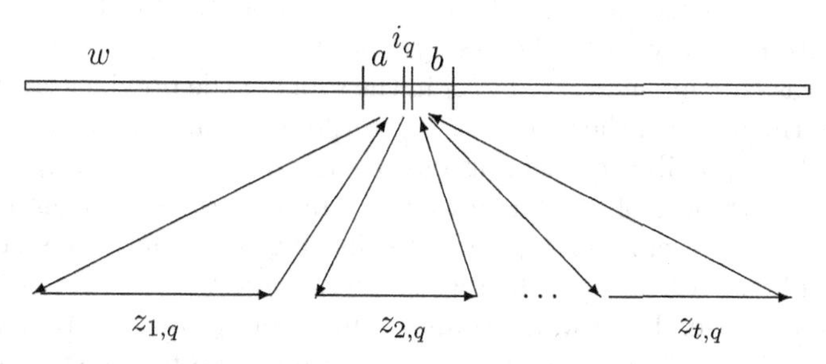

Figure 11.1: The work of the finite automaton M from the proof of Theorem 11.5

We obtain $L_{am}(\gamma) = L(M)$.

The inclusion $L(M) \subseteq L_{am}(\gamma)$ is obvious: we start from $w \in L_{am}(\gamma)$ and insert, in correct places, correct substrings.

Conversely, consider a string $x \in L_{am}(\gamma)$. If $x = w$, then $x \in L(M)$. If $w \leadsto_{am} x$, then $w = w_1abw_2, x = w_1aybw_2$, for some minimal $y \in Sub(w)$, $y = by' = y''a$. From the definition of δ, we have $x \in L(M)$. Being minimal, y does not contain a substring ab, hence the insertion places in x are either insertion places in w_1, w_2, or the two new ones obtained by inserting y, because $ayb = aby'b = ay''ab$. But each of these two subwords ab in ayb involves one symbol already existing in w. If a string $z = bz' = z''a$ is to be inserted in x in one of these places, this is possible also in the automaton A (see again Fig. 11.1). Consequently, by induction on the length of the replication chain we find that in each string x such that $w \leadsto_{am}^* x$, the insertion places correspond to the insertion places in w, hence when $x \leadsto_{am} x'$, we have $x' \in L(M)$ if $x \in L(M)$. In conclusion, $L_{am}(\gamma) \subseteq L(M)$.

The above discussion covers also the case $g = mp$: in this case we have $t = 1$ (only one string to be inserted), but the rest of the argument is identical. □

Observe that "minimal loops" can be used in the construction of the automaton M, because of the minimality of the insertions. However, as will be seen in the next theorem, a modification of the construction works also in the free mode.

Theorem 11.6. *$SREP(g) \subseteq REG$, $g \in \{ml, af\}$.*

Proof. We use the same construction as in the proof of the previous theorem, with the following differences.

For the case $g = ml$ we have to consider the (unique) minimal leftmost substring z of w of the form $z = bz' = z''a$, instead of the strings $z_1, \ldots, z_t$ considered above. As the replication mode is minimal leftmost, this string is the only one which can be inserted in any word of $L_{lm}(\gamma)$. (Note that this string z may not be minimal in the am sense: for example, consider $w = cbcbab$, for the insertion context (a, b); in the leftmost minimal mode we have to take $z = bcba$, whereas in the am mode this string is not minimal, because it contains the substring ba. This is the reason why the above proof does not automatically cover the case $g = ml$.)

For the case $g = af$, let $z_1, z_2, \ldots, z_t$ in the proof of the previous theorem be all the substrings of w of the form $z_l = bz'_l = z''_l a$, not containing the substring ab, but with z'_l possibly containing occurrences of b, and also z''_l possibly containing occurrences of a. (These z's and t may be different from the ones of the previous theorem.) Let y be any substring of w that is insertable at some stage of the derivation. Then we can write $y = z_{i_1} z_{i_2} \ldots z_{i_m}$, where z_{i_j} are strings as above. If we want to insert the string y in a position i_q, then from the definition of δ we see that from s_{i_q} we can go to the state in front of each $z_{i_j}, 1 \leq j \leq m$, and back to s_{i_q}, hence we have a path through the states of M covering the string $z_{i_1} z_{i_2} \ldots z_{i_m} = y$. We obtain the inclusion $L_{af}(\gamma) \subseteq L(M)$, for this case as well. □

In contrast to the previous theorems, we have

Theorem 11.7. *The families $SREP(ap)$ and $SREP(al)$ contain non-context-free languages.*

Proof. Let us consider the replication system

$$\gamma = (\{a, b, c, d\}, bcabda, a\#b).$$

In both modes ap and al, all the generated strings start with the symbol b, hence the two modes coincide, $L_{ap}(\gamma) = L_{al}(\gamma)$.

Consider the regular language

$$R = (bca)^+(bda)^+.$$

Denote $V = \{a, b, c, d\}$ (with this ordering of symbols). If $L_{ap}(\gamma)$ is a context-free language, then the language

$$L = L_{ap}(\gamma) \cap R$$

is context-free, too, hence $\Psi_V(L)$ is a semilinear set. Take the semilinear set

$$T = \{(n, m, p, p) \mid n, m, p \geq 1\}.$$

The family of semilinear sets of vectors is closed under intersection. Therefore, $L_{ap}(\gamma) \in CF$ implies the fact that $\Psi_V(L) \cap T$ is a semilinear set.

However, we claim that

$$\Psi_V(L) \cap T = \{(2^{n+1}, 2^{n+1}, 2^n, 2^n) \mid n \geq 1\}, \quad (*)$$

which, clearly, is not a semilinear set.

Let us prove the equality (*).

If we have generated a string $x = (bca)^i(bda)^i$ (initially we have $i = 1$) and we use exactly x as the string to be inserted between a and b in the middle of x, then we obtain $y = (bca)^{2i}(bda)^{2i}$. The process can be iterated, hence each string of the form $z = (bca)^{2^j}(bda)^{2^j}, j \geq 0$, is in $L_{ap}(\gamma)$. Such strings are also in R, hence they are in L. As $\Psi_V(z) = (2^{j+1}, 2^{j+1}, 2^j, 2^j)$, we have the inclusion $\supseteq$.

Conversely, let us examine the strings $y \in L$ for which $\Psi_V(y) \in T$. Being in L, y is of the form $y = (bca)^i(bda)^j, i, j \geq 1$. Because $\Psi_V(y) \in T$, we must have $i = j$. Therefore we have to consider the strings in $L_{ap}(\gamma)$ of the form $y = (bca)^i(bda)^i, i \geq 1$.

Because we work in the ap mode, when replicating a string $(bca)^k(bda)^k, k \geq 1$, we have to choose a string of the form $z_1 = (bca)^r, r \leq k$, or of the form $z_2 = (bca)^k(bda)^s, s \leq k$. Therefore, in every moment, what we obtain is a string y' with $|y'|_c \geq |y'|_d$. If we get a string y' with $|y'|_c > |y'|_d$, then no string y'' such that $y' \leadsto_{ap}^* y''$ will have $|y''|_c = |y''|_d$.

If we have a string y' which is not an element of $(bca)^+(bda)^+$, that is a string of the form $u_1du_2cu_3$, then no string y'' such that $y' \leadsto_{ap}^* y''$ will be in $(bca)^+(bda)^+$.

Consequently, in order to generate strings of the form $y = (bca)^i(bda)^i$ we have to use only strings of the same form, and at every replication step the inserted string is the whole current string. This implies that each replication doubles the whole string, hence i above is of the form $i = 2^n, n \geq 0$. This proves the inclusion $\subseteq$ in relation (*). □

Corollary 11.2. *The families $SREP(ap)$ and $SREP(al)$ are incomparable with each subfamily of CF containing at least one non-singleton finite language.*

Because MAT contains non-semilinear languages, we cannot infer from the previous theorem that $SREP(ap)$ and $SREP(al)$ contain languages not in the family MAT or MAT^λ, as in Theorem 11.4.

The difference between the modes *ap*, *al* and the modes *af*, *am*, *ml*, *mp* is quite unexpected (*af* and *am* behave in the same way, but *ap* and *mp*, or *al* and *ml* do not).

From Lemmas 11.14 and 11.15 we find that for the system γ_1 in Example 11.1, the languages $L_g(\gamma_1)$, for $g \in \{ap, mp, al, ml, af, am\}$, are not in $SREP(g')$, for $g' \in \{Mp, Ml, aM, t\}$, hence

Lemma 11.17. *$SREP(g) - SREP(g') \neq \emptyset$, for all $g \in \{ap, mp, al, ml, af, am\}$, $g' \in \{Mp, Ml, aM, t\}$.*

From Lemma 11.16 we find that $L_g(\gamma_1) \notin SREP(g')$, for $g \in \{Mp, Ml, aM, t\}$, $g' \in \{af, am, al, ml, ap, mp\}$, hence

Lemma 11.18. *$SREP(g) - SREP(g') \neq \emptyset$, for all $g \in \{Mp, Ml, aM, t\}$, $g' \in \{af, am, al, ml, ap, mp\}$.*

From Example 11.2 and Lemma 11.12, we obtain

Lemma 11.19. *$SREP(g) - SREP(g') \neq \emptyset$, for all $g \in \{al, ml, af, am, aM, Ml\}$, $g' \in \{ap, mp, Mp, t\}$.*

Lemma 11.20. *$SREP(Ml) = SREP(aM)$.*

Proof. Take $\gamma = (V, w, a\#b)$ and a replication step $z \rightsquigarrow_{aM} z'$. We have $z = z_1abz_2$ and $z' = z_1axbz_2$, for $x \in Sub(z), z = x_1xx_2$, $x = bx' = x''a$, and x is maximal in z, no superstring y of x can be written as $y = by' = y''a$. If x is not leftmost in z, then $z = x_1'bx_2'$ such that $|x_1'| < |x_1|$. Then x is not maximal: $z = x_1'yx_2$ for $y = by' = y''a, |y| > |x|$, hence $x \in Sub(y)$, a contradiction. Therefore $z \rightsquigarrow_{Ml} z'$, too, that is $L_{aM}(\gamma) \subseteq L_{Ml}(\gamma)$.

Conversely, it is easy to see that $z \rightsquigarrow_{Ml} z'$ implies $z \rightsquigarrow_{aM} z'$, hence $L_{Ml}(\gamma) \subseteq L_{aM}(\gamma)$. In conclusion, $L_{Ml}(\gamma) = L_{aM}(\gamma)$, hence $SREP(aM) \subseteq SREP(Ml)$.

The reverse inclusion can be obtained in the same way. □

Lemma 11.21. *$SREP(t) \subset SREP(g)$, $g \in \{Ml, Mp, aM\}$.*

Proof. Take a replication system $\gamma = (V, w, a\#b)$. If $L_t(\gamma) = \{w\}$, then $L_t(\gamma) = L_g(\gamma')$, for $\gamma' = (V \cup \{c\}, w, c\#c)$, with $c \notin V$, for all g as above.

If $L_t(\gamma) \neq \{w\}$, then $w = bw'a$ and all z such that $w \rightsquigarrow_g^* z$ are of the form $z = bz'a$ for all g. Consequently, at every moment of a replication chain the obtained string is at the same time the maximal prefix, maximal leftmost, and arbitrary maximal of the form bya. Therefore, $L_t(\gamma) = L_g(\gamma)$ for all $g \in \{Ml, Mp, aM\}$. We obtain $SREP(t) \subseteq SREP(g)$, for g as in the lemma.

From Lemma 11.17 we know that these inclusions are proper for $g \in \{aM, Ml\}$. For $g = Mp$ we consider the system

$$\gamma = (\{a, b\}, aab, a\#a).$$

We obtain

$$L_{Mp}(\gamma) = \{a^{2^n}b \mid n \geq 1\}.$$

According to Lemmas 11.13 and 11.14, this language is not in the family $SREP(t)$. □

Lemma 11.22. *$SREP(g) - SREP(g') \neq \emptyset$, $g \in \{am, ml, mp\}$, $g' \in \{aM, t, Mp, ap, al, Ml, af\}$.*

Proof. Consider the replication system

$$\gamma = (\{a, b\}, aabaa, a\#a).$$

For g as above, the string to be inserted is always a, hence

$$L_g(\gamma) = a^+aba^+a.$$

Take a system $\gamma' = (V, aabaa, c\#d)$ such that $L_g(\gamma) = L_{g'}(\gamma')$, for any g' as above. (The shortest string in a language is always the axiom of the replication system generating it.) Clearly, $c, d \in \{a, b\}$. If one of c, d is equal with b, then in each $z \leadsto_{g'} z'$ one introduces one new occurrence of b, which leads to parasitic strings. Therefore, $(c, d) = (a, a)$. For all modes g', we can choose for insertion the whole current string (this is mandatory for $g' \in \{aM, t, Mp, Ml\}$) and this again doubles the number of occurrences of the symbol b. The equality $L_g(\gamma) = L_{g'}(\gamma')$ is not possible. □

Lemma 11.23. *$SREP(g) - SREP(g') \neq \emptyset$, for all $g \in \{af, ap, al\}$, $g' \in \{am, mp, ml\}$.*

Proof. Consider again the system

$$\gamma = (\{a, b\}, aabaa, a\#a).$$

For all modes g as above, we can replicate

$$aabaa \leadsto_g a(aabaa)abaa = z$$

(we insert the whole string, between the first two occurrences of a).

Assume that $L_g(\gamma) = L_{g'}(\gamma')$ for some $\gamma' = (V, aabaa, c\#d)$. If $(c, d) = (a, a)$, then we have to insert a (this is the minimal substring of the form $x = ax' = x''a$), hence the number of occurrences of b is not increased, and the string z above cannot be obtained. If $(c, d) = (a, b)$, then the string to be inserted is ba, hence we obtain

$$L_{g'}(\gamma') = a(ab)^+aa.$$

If $(c, d) = (b, a)$, then the string to be inserted is ab, and we obtain

$$L_{g'}(\gamma') = aa(ba)^+a.$$

In none of the possible cases can we obtain the string z above, hence the equality $L_g(\gamma) = L_{g'}(\gamma')$ is not possible. □

Lemma 11.24. *$SREP(Mp) \subset SREP(Ml)$.*

Proof. If $\gamma = (V, w, a\#b)$ is a replication system and $L_{Mp}(\gamma) = \{w\}$, then we obviously have $L_{Mp}(\gamma) \in SREP(Ml)$.

If $L_{Mp}(\gamma)$ is infinite, then w is of the form $w = zw_1, z = bz' = z''a$, and all strings in $L_{Mp}(\gamma)$ start with the symbol b. Consequently, the leftmost string to be inserted is always a prefix of the current string, hence $L_{Mp}(\gamma) = L_{Ml}(\gamma)$, that is $SREP(Mp) \subseteq SREP(Ml)$.

This inclusion is proper, because $L = \{ba^{2^n} \mid n \geq 1\}$ is not in the family $SREP(Mp)$ (Lemma 11.12), but $L = L_{Ml}(\gamma)$, for $\gamma = (\{a, b\}, baa, a\#a)$. □

Lemma 11.25. *$SREP(mp) \subset SREP(ml)$, $SREP(ap) \subset SREP(al)$.*

Proof. The inclusions follow as in the previous proof: if $\gamma = (V, w, a\#b)$, then $L_{mp}(\gamma) = L_{ml}(\gamma)$ and $L_{ap}(\gamma) = L_{al}(\gamma)$.

The strictness follows from Lemma 11.19. □

Lemma 11.26. *$SREP(g) - SREP(am) \neq \emptyset$, $g \in \{mp, ml\}$.*

Proof. Consider the system

$$\gamma = (\{a, b, c\}, bcabacbca, a\#b).$$

For both $g = mp$ and $g = ml$, the string to be inserted is always bca. This is obvious for the axiom of γ, $bcabacbca$. We get

$$bcabacbca \leadsto_g (bca)^2 bacbca,$$

hence after n insertions we generate the string $(bca)^n bacbca$. Therefore, for each string x in $L_g(\gamma)$ we have $|x|_a = |x|_b = |x|_c$.

Assume now that $L_g(\gamma) = L_{am}(\gamma')$ for some $\gamma' = (V, w, d\#e)$. Being the shortest string of $L_g(\gamma)$, $bcabacbca = w$ is the axiom of γ'.

Assume that $d \neq e$. Whichever d, e are among a, b, c, because all substrings ed appear in w, we can choose such a minimal substring to be inserted, and we can obtain a string w' such that $|w'|_d = |w'|_e = |w'|_f + 1$, for $\{a, b, c\} = \{d, e, f\}$. Such a string is not in $L_g(\gamma)$, a contradiction. If $d = e$, then the string to be inserted is $d = e$, which only increases the number of occurrences of $d = e$, again leading to parasitic strings. □

Lemma 11.27. *$SREP(g) - SREP(af) \neq \emptyset$, $g \in \{ap, al\}$.*

Proof. Take again the system

$$\gamma = (\{a, b, c\}, bcabacbca, a\#b),$$

in the previous proof. For both $g = ap$ and $g = al$, the string to be inserted starts with bca, the prefix of the axiom. Moreover, in any moment of a replication chain, the prefix bca is preserved, hence it must be considered for insertion. Consequently, there are no constant k and no symbols $d, e \in \{a, b, c\}$ such that $L_g(\gamma)$ contains strings $w_1, w_2, \ldots$, such that $|w_i|_d$ is strictly increasing with i whereas $|w_i|_e \leq k$.

Assume now that $L_g(\gamma) = L_{af}(\gamma')$, for some $\gamma' = (V, w, d\#e)$. Clearly, $w = bcabacbca$. Irrespective of which the symbols d, e are, the substring ed appears in w (and in all strings produced by replication). Therefore, $L_{af}(\gamma')$ contains strings with an arbitrarily large number of occurrences of d, e and a bounded number of occurrences of $f = \{a, b, c\} - \{d, e\}$. Such strings are not in $L_g(\gamma)$, a contradiction. □

Lemma 11.28. *$SREP(af) - SREP(al) \neq \emptyset$, $SREP(am) - SREP(ml) \neq \emptyset$.*

Proof. Consider the system $\gamma = (\{a, b, c\}, w, a\#b)$ with $w = bcacbab$. In both modes *af* and *am* of choosing the string to be inserted, we can choose the substring ba of w, hence we can obtain strings containing arbitrarily many occurrences of a and b, but only a bounded number of occurrences of c. In order to generate such strings by a system $\gamma' = (V, w, d\#e)$ in one of the modes *al, ml*, we must have $\{d, e\} = \{a, b\}$. If $(d, e) = (a, b)$, then the string to be inserted will start with the prefix bca. If $(d, e) = (b, a)$, then the string to be inserted starts at the leftmost occurrence of a (on the third position of each generated string) and contains the neighbouring occurrence of c. In both cases, the number of a, b, c occurrences increases simultaneously, hence we cannot generate the above mentioned strings of $L_{af}(\gamma), L_{am}(\gamma)$. □

Summarizing these lemmas, we get

Theorem 11.8. *All of the families $SREP(g), g \in D$, are pairwise incomparable, except for the following strict inclusions and equality:*

$$SREP(mp) \subset SREP(ml), \ SREP(ap) \subset SREP(al),$$
$$SREP(t) \subset SREP(Mp) \subset SREP(Ml) = SREP(aM).$$

Corollary 11.3. *All families $SREP(g), g \in D$, are strictly included in CS.*

Lemmas 11.14 – 11.16 above provide necessary conditions for a language $L \subseteq V^*$ to be in a family $SREP(g)$ in terms of the properties of the Parikh set associated with L, $\Psi_V(L)$. This suggests considering more systematically this set for replication languages, as well as the length set, $length(L) = \{|x| \mid x \in L\}$. When L is generated in a deterministic way, along a sequence

$$w_0 \rightsquigarrow w_1 \rightsquigarrow w_2 \rightsquigarrow \dots$$

for some $\gamma = (V, w_0, a\#b)$ (hence $L = L(\gamma) = \{w_0, w_1, w_2, \dots\}$), we can also consider the growth function of L, $growth_L(n) = |w_n|, n \geq 0$.

In the case of deterministic replication systems, the growth function study is also interesting from a non-mathematical point of view, because it can characterize the power of the systems.

We list here, without proofs, some results about growth functions or length sets of replication languages of various types; proofs and further details can be found in [136].

1. A sequence $u(n)$ of nonnegative integers is the growth function for a replication system γ with respect to the t mode of replication if and only if $u(n)$ is a geometric progression with ratio 2 and with the initial element not equal to 1.

2. A sequence $u(n)$ of nonnegative integers is the growth function for a replication system γ with respect to the g mode of replication, where $g \in \{Mp, Ml, aM\}$, if and only if $u(n)$ is of the form $u(n) = l + 2^n k$, where $k \geq 2, l \geq 0$.

3. A sequence $u(n)$ of nonnegative integers is the growth function for a replication system γ with respect to the g mode of replication, where $g \in \{mp, ml\}$, if and only if $u(n)$ is an arithmetical progression $u(n) = l + nk$, with $1 \leq k < l$.

4. A set of nonnegative integers $N \subseteq \mathbf{N}$ is the length set for a replication system γ with respect to the am mode of replication, if and only if either

   (1) there exist nonnegative integers l, r, and $k_1, k_2, \ldots k_r \geq 2$, with $l \geq \sum_{i=1}^{r} k_i$, such that $N = \{l + c_1 k_1 + c_2 k_2 + \ldots + c_r k_r \mid c_1, \ldots, c_r \in \mathbf{N}\}$, or

   (2) there exists $l \in \mathbf{N}, l \geq 2$, such that $N = \{n \mid n \geq l\}$.

5. A linear set $H = \{v_0 + c_1 v_1 + \ldots + c_r v_r \mid c_i \in \mathbf{N}\}$, where $v_i \in \mathbf{N}^p$, for any $i, 0 \leq i \leq r$, and $p \geq 1$, is the Parikh set of a replication system with respect to the arbitrary minimal mode of replication if and only if $v_0 \geq \sum_{i=1}^{r} v_i$, and there exist an $s \in \{1, 2\}$ and $j_1, \ldots, j_s, 1 \leq j_1 < \ldots < j_s \leq p$, such that for any $i, 1 \leq i \leq r, v_i(j_1) = \ldots = v_i(j_s) = 1$ and, in addition, if $s = 1$, then $v_0(j_1) \geq 2$.

6. Let $\gamma = (V, w, a\#b)$ be a replication system. Then the Parikh set of the language L generated by γ with respect to the af mode of replication is linear, that is, $\Psi_V(L) = \{v_0 + c_1 v_1 + \ldots + c_r v_r \mid c_i \in \mathbf{N}, 1 \leq i \leq r\}$, for an $r \geq 1$, and $v_0, \ldots, v_r \in \mathbf{N}^p$, where $p = card(V)$.

Consequently, the growth functions (respectively, length sets, Parikh sets) of the simple replication systems studied above are either exponential or linear. Nothing lies in between. It would be of interest to point out models with a polynomial nonlinear growth.

11.3 Bibliographical Notes

Restricted variants of the splicing operation (including permitting and forbidding contexts, target languages, priority conditions) are considered in [166] (for finite sets of splicing rules, as in Sect. 11.1 above) and then are investigated in [102] (where regular sets of rules are also considered); in both these papers only the non-iterated splicing is examined.

Self-splicing H systems of the form $\gamma = (V, \{w\}, R)$, with one axiom only and a finite set of splicing rules, are also investigated in [38], where one proves, for instance, that such systems generate a family of languages which is an anti-AFL incomparable with REG, CF, and with several families of L languages; moreover, it is shown that it is not decidable whether or not a context-free language can be generated by a self-splicing H system as above.

Replication systems in the simple form are introduced and investigated in [134]. Section 11.2 is mainly based on this paper. Further results can be found in [135] (uniform replication systems with finitely many axioms, of degree one, of radius one; one also investigates the replication operation as an operation with formal languages) and [136] (generative power, growth functions).

There are several other papers dealing with the splicing operation, H systems, or related notions. For instance, [56] gives characterizations of REG, CF, RE starting from Post Tag systems and using certain related classes of extended H systems.

Splicing on graphs is considered in [58].

A variant of H systems, closer to the initial form of the operation introduced by T. Head and using cutting rules as in replication systems, is considered in [59], [67]: one gives cutting rules associated with *markers*, which identify pairs of cutting patterns producing matching ends; two strings are first cut, then the fragments are recombined, if this is possible, as indicated by markers.

The problem of learning H systems is investigated in [105], [210]. In [106] one gives a proof of the Regularity Preserving Lemma, carrying out an excessively detailed construction in terms of finite automata.

An operation with DNA sequences, related to the splicing, is the *crossovering*: given two strings x, y, we pass from x to y and back to x, and so on, a number of times (which is prescribed or not), combining subwords of x and y as specified by the positions of jumping from one string to the other. The 1-splicing is a particular case, with only one jump, from x to y. When the number of jumps is not specified, a crossovering operation can be simulated by an iterated splicing. An interesting case appears when the number of jumps is fixed in advance (and the places where they are done are ordered in the used strings, hence we combine substrings appearing in a sequence, from the left to the right, in x and y). Results about such an operation appear in [10], [96], [139].

The recombination of strings also appears as a basic operation in the "evolutionary grammars" from [37].

Surveys of results and bibliographical information about the splicing operation, H systems, and computing by splicing can be found in [90], [152], [155], [170], [171].

Bibliography

[1] L. M. Adleman: Molecular computation of solutions to combinatorial problems. *Science*, 226 (November 1994), 1021–1024

[2] L. M. Adleman: On constructing a molecular computer. [117], 1–22

[3] L. M. Adleman, P. W. K. Rothemund, S. Roweiss, E. Winfree: On applying molecular computation to the Data Encryption Standard. [13], 28–48

[4] A. V. Aho, J. D. Ullman: *The Theory of Parsing, Translation, and Compiling.* Prentice Hall, Englewood Cliffs, N.J., Vol. I: 1971, Vol. II: 1973

[5] B. Alberts, D. Bray, J. Lewis, M. Raff, K. Roberts, J. D. Watson: *Molecular Biology of the Cell.* 3rd ed., Garland Publishing, New York, 1994

[6] G. Alford: An explicit construction of a universal extended H system. *Unconventional Models of Computation* (C. S. Calude, J. Casti, M. L. Dinneen, eds.), Springer, Berlin, 1998, 108–117

[7] M. Amos: *DNA Computation.* PhD Thesis, Univ. of Warwick, Dept. of Computer Sci., 1997

[8] M. Amos, S. Wilson, D. A. Hodgson, G. Owenson, A. Gibbons: Practical implementation of DNA computations. *Unconventional Methods of Computation* (C. S. Calude, J. Casti, M. J. Dinneen, eds.), Springer, Berlin, 1998, 1–18

[9] M. Arita, M. Hagiya, A. Suyama: Joining and rotating data with molecules. *IEEE Intern. Conf. on Evolutionary Computing*, Indianapolis, 1997, 243–248

[10] A. Atanasiu: On the free crossovering operation. *Ann. Univ. Buc., Matem.-Inform. Series*, 45, 1 (1996), 3–8

[11] B. Baker, R. Book: Reversal-bounded multipushdown machines. *J. Computer System Sci.*, 8 (1974), 315–332

[12] E. B. Baum: A DNA associative memory potentially larger than the brain. [117], 23–28

[13] E. Baum, D. Boneh, P. Kaplan, R. Lipton, J. Reif, N. Seeman (eds.): *DNA Based Computers.* Proc. of the Second Annual Meeting, Princeton, 1996

[14] R. Beene (ed.): *RNA-Editing: The Alteration of Protein Coding Sequences of RNA.* Ellis Horwood, Chichester, UK, 1993

[15] A. J. Blumberg: Parallel computation on a DNA substrate. *Proc. of the Third DIMACS Workshop on DNA Based Computers*, Philadelphia, 1997, 275–289

[16] D. Boneh, C. Dunworth, R. J. Lipton: Breaking DES using a molecular computing. [117], 37–66

[17] D. Boneh, C. Dunworth, R. J. Lipton, J. Sgall: On the computational power of DNA. *Discrete Appl. Math.*, 71 (1996), 79–94

[18] F. J. Brandenburg: Representations of language families by homomorphic equality operations and generalized equality sets. *Theoretical Computer Sci.*, 55 (1987), 183–263

[19] C. Calude, J. Hromkovic: Complexity: A language-theoretic point of view. [193], Vol. 2, 1–60

[20] C. Calude, Gh. Păun: Global syntax and semantics for recursively enumerable languages. *Fundamenta Informaticae*, 4, 2 (1981), 245–254

[21] V. T. Chakaravarthy, K. Krithivasan: A note on extended H systems with permitting/forbidding contexts of radius one. *Bulletin of the EATCS*, 62 (June 1997), 208–213

[22] J. Chen, D. H. Wood: A new DNA separation technique with low error rate. *Proc. of the Third DIMACS Workshop on DNA Based Computers*, Philadelphia, 1997, 43–56

[23] J. Collado-Vides: The search for a grammatical theory of gene regulation is formally justified by showing the inadequacy of context-free grammars. *CABIOS*, 7 (1991), 321–326

[24] M. Conrad: Information processing in molecular systems. *Currents in Modern Biology*, 5 (1972), 1–14

[25] M. Conrad: On design principles for a molecular computer, *Comm. of the ACM*, 28 (1985), 464–480

[26] M. Conrad: The price of programmability. *The Universal Turing Machine: A Half-Century Survey* (R. Herken, ed.), Kammerer and Unverzagt, Hamburg, 1988, 285–307

[27] M. Conrad, K.-P. Zauner: Design for a DNA conformational processor. *Proc. of the Third DIMACS Workshop on DNA Based Computers*, Philadelphia, 1997, 290–295

[28] E. Csuhaj-Varju, J. Dassow: On cooperating distributed grammar systems. *J. Inf. Process. Cybern., EIK*, 26, 1–2 (1990), 49–63

[29] E. Csuhaj-Varju, J. Dassow, J. Kelemen, Gh. Păun: *Grammar Systems. A Grammatical Approach to Distribution and Cooperation.* Gordon and Breach, London, 1994

[30] E. Csuhaj-Varju, R. Freund, F. Wachtler: Test tube systems with cutting/recombination operation. *Second Pacific Conf. on Biocomputing*, Hawaii, 1997

[31] E. Csuhaj-Varju, R. Freund, L. Kari, Gh. Păun: DNA computing based on splicing: universality results. *Proc. First Annual Pacific Symp. on Biocomputing*, Hawaii, 1996 (L. Hunter, T. E. Klein, eds.), World Scientific, Singapore, 1996, 179–190

[32] E. Csuhaj-Varju, L. Kari, Gh. Păun: Test tube distributed systems based on splicing. *Computers and AI*, 15, 2–3 (1996), 211–232

[33] E. Csuhaj-Varju, J. Kelemen, Gh. Păun: Grammar systems with WAVE-like communication. *Computers and AI*, 15, 5 (1996), 419–436

[34] K. Culik II: A purely homomorphic characterization of recursively enumerable sets. *Journal of the ACM*, 26 (1979), 345–350

[35] K. Culik II, T. Harju: Splicing semigroups of dominoes and DNA. *Discrete Appl. Math.*, 31 (1991), 261–277

[36] J. Dassow, V. Mitrana: Splicing grammar systems. *Computers and AI*, 15, 2–3, (1996), 109–122

[37] J. Dassow, V. Mitrana: Evolutionary grammars. *Proc. of German Conf. on Bioinformatics*, Leipzig, 1996 (R. Hofestadt, T. Lengauer, M. Loffler, D. Schomburg, eds.), *Lect. Notes in Computer Sci.* 1278, Springer, Berlin, 1997

[38] J. Dassow, V. Mitrana: Self cross-over systems. *Computing with Bio-Molecules. Theory and Experiments* (Gh. Păun, ed.), Springer, Berlin, 1998, 283–294

[39] J. Dassow, V. Mitrana: On some operations suggested by genome evolution. *Proc. Second Pacific Conf. on Biocomputing*, Hawaii, 1997 (R. B. Altman et al., eds.), World Scientific, Singapore, 1997, 97–108

[40] J. Dassow, Gh. Păun: *Regulated Rewriting in Formal Language Theory*. Springer-Verlag, Berlin, 1989

[41] J. Dassow, Gh. Păun, A. Salomaa: Grammars with controlled derivation. [193], Vol. 2, 101–154

[42] M. D. Davis, E. J. Weyuker: *Computability, Complexity, and Languages*. Academic Press, New York, 1983

[43] M. T. Dawson, R. Powell, F. Gannon: *Gene Technology*. BIOS Scientific Publishers, Oxford, 1996

[44] R. Deaton, R. C. Murphy, J. A. Rose, M. Garzon, D. R. Franceschetti, S. E. Stevens, Jr.: A DNA based implementation of an evolutionary search for good encodings for DNA computation. *IEEE Intern. Conf. on Evolutionary Computing*, Indianapolis, 1997, 267–271

[45] A. De Luca, A. Restivo: A characterization of strictly locally testable languages and its applications to subsemigroups of a free semigroup. *Inform. Control*, 44 (1980), 300–319

[46] K. L. Denninghoff, R. W. Gatterdam: On the undecidability of splicing systems. *Intern. J. Computer Math.*, 27 (1989), 133–145

[47] V. Diekert, M. Kudlek: Small deterministic Turing machines. *Papers on Automata and Languages*, Dept. of Math., Univ. of Economics, Budapest, 188-4 (1989), 77–87

[48] K. Drlica: *Understanding DNA and Gene Clonig. A Guide for the CURIOUS*. John Wiley and Sons, New York, 1992

[49] P. Duris, J. Hromkovic: One-way simple multihead finite automata are not closed under concatenation. *Theoretical Computer Sci.*, 27 (1983), 121–125

[50] A. Ehrenfeucht, Gh. Păun, G. Rozenberg: On representing recursively enumerable languages by internal contextual languages. *Theoretical Computer Sci.*, 205, 1–2 (1998), 61–83

[51] S. Eilenberg: *Automata, Languages, and Machines*. Academic Press, New York, Vol. A: 1974, Vol. B: 1976

[52] T. L. Eng: Linear DNA self-assembly with hairpins generate the equivalent of linear context-free grammars. *Proc. of the Third DIMACS Workshop on DNA Based Computers*, Philadelphia, 1997, 296–301

[53] J. Engelfriet: Reverse twin-shuffles. *Bulletin of the EATCS*, 60 (1996), 144

[54] J. Engelfriet, G. Rozenberg: Fixed point languages, equality languages, and representations of recursively enumerable languages. *Journal of the ACM*, 27 (1980), 499–518

[55] R. P. Feynman, In D. H. Gilbert (ed.): *Miniaturization.* Reinhold, New York, 1961, 282–296

[56] C. Ferretti, S. Kobayashi, T. Yokomori: DNA splicing systems and Post systems. *First Annual Pacific Conference on Biocomputing*, Hawaii, 1996 (L. Hunter, T. E. Klein, eds.), World Scientific, Singapore, 1996, 288–299

[57] C. Ferretti, G. Mauri, S. Kobayashi, T. Yokomori: On the universality of Post and splicing systems. *Theoretical Computer Sci.*, 231, 2 (2000), 157–170.

[58] R. Freund: Splicing systems on graphs. *IEEE Conf. on Intelligence in Neural and Biologic Systems*, Herndon-Washington, 1995, 189–195

[59] R. Freund, F. Freund: Test tube systems or "how to bake a DNA cake." *Acta Cybern.*, 12, 4 (1997), 445–459

[60] R. Freund, F. Freund: Test tube systems with controlled applications of rules. *IEEE Intern. Conf. on Evolutionary Computing*, Indianapolis, 1997, 237–242

[61] R. Freund, F. Freund, M. Oswald: Universal H systems using multisets. manuscript, 1997

[62] R. Freund, L. Kari, Gh. Păun: DNA computing based on splicing. The existence of universal computers. *Technical Report 185-2/FR-2/95*, Technical Univ. Wien, 1995, and *Theory of Computing Systems*, 32 (1999), 69–112

[63] R. Freund, V. Mihalache: Molecular computations on circular and linear strings. *Unconventional Models of Computation* (C. S. Calude, J. Casti, M. J. Dinneen, eds.), Springer, Berlin, 1998, 201–217

[64] R. Freund, Gh. Păun, G. Rozenberg, A. Salomaa: Bidirectional sticker systems. *Third Annual Pacific Conf. on Biocomputing*, Hawaii, 1998 (R. B. Altman, A. K. Dunker, L. Hunter, T. E. Klein, eds.), World Scientific, Singapore, 1998, 535–546

[65] R. Freund, Gh. Păun, G. Rozenberg, A. Salomaa: Watson-Crick finite automata. *Proc of the Third Annual DIMACS Symp. on DNA Based Computers*, Philadelphia, 1997, 305–317

[66] R. Freund, Gh. Păun, G. Rozenberg, A. Salomaa: Watson-Crick automata. *Techn. Report 97-13*, Dept. of Computer Sci., Leiden Univ., 1997

[67] R. Freund, F. Wachtler: Universal systems with operations related to splicing. *Computers and AI*, 15, 4 (1996), 273–293

[68] B. Fu, R. Beigel: On molecular approximation algorithms for NP-optimization problems. *Proc. of the Third DIMACS Workshop on DNA Based Computers*, Philadelphia, 1997, 93–101

[69] B. S. Galiukschov: Semicontextual grammars (in Russian). *Mat. logica i mat. ling.*, Talinin Univ., 1981, 38–50

[70] Y. Gao, M. Garzon, R. C. Murphy, J. A. Rose, R. Deaton, D. R. Franceschetti, S. E. Stevens, Jr.: DNA implementation of nondeterminism. *Proc. of the Third DIMACS Workshop on DNA Based Computers*, Philadelphia, 1997, 204–211

[71] R. W. Gatterdam: Splicing systems and regularity. *Intern. J. Computer Math.*, 31 (1989), 63–67

[72] R. W. Gatterdam: Algorithms for splicing systems. *SIAM J. Comput.*, 21, 3 (1992), 507–520

[73] R. W. Gatterdam: DNA and twist free splicing systems. *Words, Languages, Combinatorics, II* (M. Ito, H. Jürgensen, eds.), World Scientific, Singapore, 1994, 170–178

[74] V. Geffert: Normal forms for phrase-structure grammars. *RAIRO. Th. Inform. and Appl.*, 25 (1991), 473–496

[75] G. Georgescu: On the generative capacity of splicing grammar systems. *New Trends in Formal Languages. Control, Cooperation, and Combinatorics* (Gh. Păun, A. Salomaa, eds.), *Lect. Notes in Computer Sci.* 1218, Springer, Berlin, 1997, 330–345

[76] A. Gibbons, M. Amos, D. Hodgson: Models of DNA computing. *Proc. of 21st MFCS Conf.*, 1996, Cracow, *Lect. Notes in Computer Sci.* 1113, Springer, Berlin, 1996, 18–36

[77] S. Ginsburg: *The Mathematical Theory of Context-free Languages*, McGraw Hill, New York, 1966

[78] S. Ginsburg: *Algebraic and Automata-Theoretic Properties of Formal Languages*. North-Holland, Amsterdam, 1975

[79] J. Gruska: Descriptional complexity of context-free languages. *Proc. MFCS Symp.*, 1973, 71–83

[80] F. Guarnieri, M. Fliss, C. Bancroft: Making DNA add. *Science*, 273 (July 1996), 220–223

[81] V. Gupta, S. Parthasarathy, M. J. Zaki: Arithmetic and logic operations with DNA. *Proc. of the Third DIMACS Workshop on DNA Based Computers*, Philadelphia, 1997, 212–220

[82] M. Harrison: *Introduction to Formal Language Theory*. Addison-Wesley, Reading, Mass., 1978

[83] J. Hartmanis: About the nature of computer science. *Bulletin of the EATCS*, 53 (June 1994), 170–190

[84] J. Hartmanis: On the weight of computation. *Bulletin of the EATCS*, 55 (February 1995), 136–138

[85] D. Hauschild, M. Jantzen: Petri nets algorithms in the theory of matrix grammars. *Acta Informatica*, 31 (1994), 719–728

[86] T. Head: Formal language theory and DNA: An analysis of the generative capacity of specific recombinant behaviors. *Bulletin of Mathematical Biology*, 49 (1987), 737–759

[87] T. Head: Splicing schemes and DNA. in *Lindenmayer Systems; Impacts on Theoretical Computer Science and Developmental Biology* (G. Rozenberg, A. Salomaa, eds.), Springer, Berlin, 1992, 371–383

[88] T. Head: Splicing representations of strictly locally testable languages. *Discrete Appl. Math.*, 87 (1998), 139–147

[89] T. Head: Splicing systems and molecular processes. *IEEE Intern. Conf. on Evolutionary Computing*, Indianapolis, 1997, 203–205

[90] T. Head, Gh. Păun, D. Pixton: Language theory and molecular genetics. Generative mechanisms suggested by DNA recombination. [193], Vol. 2, 295–360

[91] G. T. Herman, G. Rozenberg: *Developmental Systems and Languages*. North-Holland, Amsterdam, 1975

[92] J. Hertz, A. Krogh, R. G. Palmer: *Introduction to the Theory of Neural Computation*. Addison-Wesley, Reading, Mass., 1991

[93] J. E. Hopcroft, J. D. Ullman: *Introduction to Automata Theory, Languages and Computation*. Addison-Wesley, Reading, Mass., 1979

[94] J. Hromkovic: One-way multihead deterministic finite automata. *Acta Informatica*, 19, 4 (1983), 377–384

[95] L. Hunter: Molecular biology for computer scientists. *Artificial Intelligence and Molecular Biology* (L. Hunter, ed.), AAAI Press/MIT Press, Menlo Park, Calif., 1993, 1–46

[96] L. Ilie, V. Mitrana: Crossing-over on languages. A formal representation of the chromosome recombination. *Proc. of First German Conf. on Bioinformatics*, Leipzig, 1996 (R. Hofestadt, T. Lengauer, M. Loffler, D. Schomburg, eds.), *Lect. Notes in Computer Sci.* 1278, Springer, Berlin, 1997, 202–204

[97] O. H. Ibarra, C. E. Kim: On 3-head versus 2-head finite automata. *Inform. Control*, 4 (1975), 193–200

[98] N. Jonoska, S. A. Karl: Ligation experiments in computing with DNA. *IEEE Intern. Conf. on Evolutionary Computing*, Indianapolis, 1997, 261–265

[99] L. Kari: *On Insertion and Deletion in Formal Languages.* Ph.D. Thesis, University of Turku, 1991

[100] L. Kari: DNA computing: tomorrow's reality. *Bulletin of the EATCS*, 59 (June 1996), 256–266

[101] L. Kari, Gh. Păun, G. Rozenberg, A. Salomaa, S. Yu: DNA computing, sticker systems, and universality. *Acta Informatica*, 35, 5 (1998), 401–420

[102] L. Kari, Gh. Păun, A. Salomaa: The power of restricted splicing with rules from a regular set. *J. Universal Computer Sci.*, 2, 4 (1996), 224–240

[103] L. Kari, Gh. Păun, G. Thierrin, S. Yu: At the crossroads of DNA computing and formal languages: Characterizing RE using insertion-deletion systems. *Proc. 3rd DIMACS Workshop on DNA Based Computing*, Philadelphia, 1997, 318–333

[104] L. Kari, G. Thierrin: Contextual insertion/deletion and computability. *Information and Computation*, 131, 1 (1996), 47–61

[105] S. M. Kim: Identifying genetically spliced languages. *IEEE Intern. Conf. on Evolutionary Computing*, Indianapolis, 1997, 231–235

[106] S. M. Kim: Computational modeling for genetic splicing systems. *SIAM J. Computing*, 26, 5 (1997), 1284–1309

[107] S. Kobayashi, T. Yokomory, G. Sampei, K. Mizobuchi: DNA implementation of simple Horn clause computation. *IEEE Intern. Conf. on Evolutionary Computing*, Indianapolis, 1997, 213–217

[108] J. H. Koza, J. P. Rice: *Genetic Algorithms: The Movie.* MIT Press, Cambridge, Mass., 1992

[109] K. Krithivasan, V. T. Chakaravarthy, R. Rama: Array splicing systems. in *New Trends in Formal Languages. Control, Cooperation, and Combinatorics* (Gh. Păun, A. Salomaa, eds.), *Lect. Notes in Computer Sci.* 1218, Springer, Berlin, 1997, 346–365

[110] M. Latteux, B. Leguy, B. Ratoandromanana: The family of one-counter languages is closed under quotient. *Acta Informatica*, 22 (1985), 579 – 588

[111] L. F. Landweber, R. J. Lipton: DNA to DNA: A potential "killer app"?. *Proc. of the Third DIMACS Workshop on DNA Based Computers*, Philadelphia, 1997, 59–68

[112] E. Laun, K. J. Reddy: Wet splicing systems. *Proc. of the Third DIMACS Workshop on DNA Based Computers*, Philadelphia, 1997, 115–126

[113] T. H. Leete, J. P. Klein, H. Rubin: Bit operations using a DNA template. *Proc. of the Third DIMACS Workshop on DNA Based Computers*, Philadelphia, 1997, 159–171

[114] W.-H. Li, D. Graur: *Fundamentals of Molecular Evolution.* Sinauer Ass., Sunderland, Mass., 1991

[115] R. J. Lipton: Using DNA to solve NP-complete problems. *Science*, 268 (April 1995), 542–545

[116] R. J. Lipton: Speeding up computations via molecular biology. [117], 67–74

[117] R. J. Lipton, E. B. Baum, eds.: *DNA Based Computers.* Proc. of a DIMACS Workshop, Princeton, 1995, Amer. Math. Soc., 1996

[118] M. Lothaire: *Combinatorics on Words.* Addison-Wesley, Reading, Mass., 1983

[119] M. Maliţa, Gh. Ştefan: DNA computing with the connex memory. *IEEE Intern. Conf. on Evolutionary Computing*, Indianapolis, 1997, 225–229

[120] G. Manganaro, J. P. deGyvez: DNA computing based on chaos. *IEEE Intern. Conf. on Evolutionary Computing*, Indianapolis, 1997, 255–260

[121] S. Marcus: Contextual grammars. *Rev. Roum. Math. Pures Appl.*, 14 (1969), 1525–1534

[122] S. Marcus: Linguistic structures and generative devices in molecular genetics. *Cah. Ling. Th. Appl.*, 11, 2 (1974), 77–104

[123] S. Marcus: Language at the crossroad of computation and biology. *Computing with Bio-Molecules. Theory and Experiments* (Gh. Păun, ed.), Springer, Berlin, 1998, 1–35

[124] M. Margenstern, Y. Rogozhin: Time-varying distributed H systems of degree 2 generate all RE languages. *MFCS '98 Workshop on Frontiers of Universality*, Brno, 1998

[125] A. F. Markham, et al.: Solid phase phosphotriester synthesis of large oligodeoxyribonucleotides on a polyamide support. *Nucleic Acids Research*, 8, 22 (1980), 5193–5205

[126] C. Martin-Vide, Gh. Păun: Cooperating distributed splicing systems. *Journal of Automata, Languages and Combinatorics*, 4, 1 (1999), 3–16

[127] C. Martin-Vide, Gh. Păun, G. Rozenberg, A. Salomaa: Universality results for finite H systems and for Watson–Crick finite automata. *Computing with Bio-Molecules. Theory and Experiments* (Gh. Păun, ed.), Springer, Berlin, 1998, 200–220

[128] C. Martin-Vide, Gh. Păun, A. Salomaa: Characterizations of recursively enumerable languages by means of insertion grammars. *Theoretical Computer Sci.*, 205, 1–2 (1998), 195–205

[129] A. Mateescu: Splicing on routes: A framewok for DNA computation. *Unconventional Models of Computation* (C. S. Calude, J. Casti, M. J. Dinneen, eds.), Springer, Berlin, 1998, 273–285

[130] A. Mateescu, Gh. Păun, G. Rozenberg, A. Salomaa: Simple splicing systems. *Discrete Appl. Math.*, 84 (1998), 145–163

[131] A. Mateescu, Gh. Păun, A. Salomaa: Variants of twin-shuffle languages. Manuscript, 1998

[132] R. McNaughton, S. Papert: *Counter-Free Automata*. MIT Press, Cambridge, Mass., 1971

[133] V. Mihalache: Prolog approach to DNA computing. *IEEE Intern. Conf. on Evolutionary Computing*, Indianapolis, 1997, 249–254

[134] V. Mihalache, Gh. Păun, G. Rozenberg, A. Salomaa: Generating strings by replication: a simple case. *Report TUCS* No. 17, Turku, Finland, 1996

[135] V. Mihalache, A. Salomaa: Language-theoretic aspects of string replication. *Intern. J. Computer Math.*, 64, 1-2 (1996)

[136] V. Mihalache, A. Salomaa: Growth functions and length sets of replicating systems. *Acta Cybern.*, 12, 3 (1996), 235–247

[137] V. Mihalache, A. Salomaa: Lindenmayer and DNA: Watson–Crick D0L systems. *Bulletin of the EATCS*, 62 (June 1997), 160–175

[138] M. Minsky: *Computation. Finite and Infinite Machines.* Prentice Hall, Englewood Cliffs, N.J., 1967

[139] V. Mitrana: On the interdependence between shuffle and crossing-over operations. *Acta Informatica*, 34, 4 (1997), 257–266

[140] M. Ogihara, A. Ray: DNA-based parallel computation by "counting". *Proc. of the Third DIMACS Workshop on DNA Based Computers*, Philadelphia, 1997, 265–274

[141] M. Ogihara, A. Ray: The minimum DNA computation model and its computational power. *Unconventional Models of Computation* (C. S. Calude, J. Casti, M. J. Dinneen, eds.), Springer, Berlin, 1998, 309–322

[142] J. Parkkinen, S. Parkkinen, Gh. Păun: Computing by splicing: gsm's working on circular words. Manuscript, 1996

[143] A. Păun: Extended H systems with permitting contexts of small radius. *Fundamenta Informaticae*, 31, 2 (1997), 185–193

[144] A. Păun, M. Păun: Controlled and distributed H systems of small radius. *Computing with Bio-Molecules. Theory and Experiments* (Gh. Păun, ed.), Springer, Berlin, 1998, 239–254

[145] Gh. Păun: On the iteration of gsm mappings. *Revue Roum. Math. Pures Appl.*, 23, 4 (1978), 921–937

[146] Gh. Păun: On semicontextual grammars. *Bull. Math. Soc. Sci. Math. Roumanie*, 28(76) (1984), 63–68

[147] Gh. Păun: Two theorems about Galiukschov semicontextual grammars. *Kybernetika*, 21 (1985), 360–365

[148] Gh. Păun: A characterization of recursively enumerable languages. *Bulletin of the EATCS*, 45 (October 1991), 218– 222

[149] Gh. Păun: On the splicing operation. *Discrete Appl. Math.*, 70 (1996), 57–79

[150] Gh. Păun: On the power of the splicing operation. *Intern. J. Computer Math.*, 59 (1995), 27–35

[151] Gh. Păun: The splicing as an operation on formal languages. *Proc. of IEEE Conf. on Intelligence in Neural and Biologic Systems*, Herndon-Washington, 1995, 176–180

[152] Gh. Păun: Splicing. A challenge to formal language theorists. *Bulletin of the EATCS*, 57 (October 1995), 183–194

[153] Gh. Păun: Regular extended H systems are computationally universal. *J. Automata, Languages, Combinatorics*, 1, 1 (1996), 27–36

[154] Gh. Păun: On the power of splicing grammar systems. *Ann. Univ. Buc., Matem.-Inform. Series*, 45, 1 (1996), 93–106

[155] Gh. Păun: Five (plus two) universal DNA computing models based on the splicing operation. *Proc. of the Second DIMACS Workshop on DNA Based Computers*, Princeton, June 1996, 67–86

[156] Gh. Păun: Computing by splicing: How simple rules?. *Bulletin of the EATCS*, 60 (October 1996), 145–150

[157] Gh. Păun: Splicing systems with targets are computationally complete. *Inform. Processing Letters*, 59 (1996), 129–133

[158] Gh. Păun: DNA computing; Distributed splicing systems. *Structures in Logic and Computer Science. A Selection of Essays in Honor of A. Ehrenfeucht* (J. Mycielski, G. Rozenberg, A. Salomaa, eds.), *Lect. Notes in Computer Sci.* 1261, Springer, Berlin, 1997, 351–370

[159] Gh. Păun: *Marcus Contextual Grammars.* Kluwer Academic Publ., Boston, 1997

[160] Gh. Păun: Controlled H systems and Chomsky hierarchy. *Fundamenta Informaticae*, 30, 1 (1997), 45–57

[161] Gh. Păun: Two-level distributed H systems. *Proc. of the Third Conf. on Developments in Language Theory*, Thessaloniki, 1997 (S. Bozapalidis, ed.), Aristotle Univ. of Thessaloniki, 1997, 309–327

[162] Gh. Păun: DNA computing based on splicing: universality results. *Proc. of Second Intern. Colloq. Universal Machines and Computations*, Metz, 1998, Vol. I, 67–91

[163] Gh. Păun: Distributed architectures in DNA computing based on splicing: Limiting the size of components. *Unconventional Models of Computation* (C. S. Calude, J. Casti, M. J. Dinneen, eds.), Springer, Berlin, 1998, 323–335

[164] Gh. Păun: *(DNA) Computing by carving.* Research Report CTS-97-17, Center for Theoretical Study of the Czech Academy of Sciences, Prague, 1997, and *Soft Computing*, 3, 1 (1999), 30–36

[165] Gh. Păun, G. Rozenberg: Sticker systems. *Theoretical Computer Sci.*, 204 (1998), 183–203

[166] Gh. Păun, G. Rozenberg, A. Salomaa: Restricted use of the splicing operation. *Intern. J. Computer Math.*, 60 (1996), 17–32

[167] Gh. Păun, G. Rozenberg, A. Salomaa: Computing by splicing. *Theoretical Computer Sci.*, 168, 2 (1996), 321–336

[168] Gh. Păun, G. Rozenberg, A. Salomaa: Computing by splicing. Programmed and evolving splicing systems. *IEEE Intern. Conf. on Evolutionary Computing*, Indianapolis, 1997, 273–277

[169] Gh. Păun, A. Salomaa: Thin and slender languages. *Discrete Appl. Math.*, 61 (1995), 257–270

[170] Gh. Păun, A. Salomaa: From DNA recombination to DNA computing. *Proc. of First German Conf. on Bioinformatics*, Leipzig, 1996 (R. Hofestadt, T. Lengauer, M. Loffler, D. Schomburg, eds.), *Lect. Notes in Computer Sci.* 1278, Springer, Berlin, 1997, 210–220

[171] Gh. Păun, A. Salomaa: DNA computing based on the splicing operation. *Mathematica Japonica*, 43, 3 (1996), 607–632

[172] Gh. Păun, L. Sântean: Parallel communicating grammar systems: the regular case. *Ann. Univ. Buc., Series Matem.-Inform.*, 38 (1989), 55–63

[173] M. Penttonen: One-sided and two-sided contexts in phrase structure grammars. *Inform. Control*, 25, 4 (1974), 371–392

[174] N. Pisanti: DNA computing: a survey. *Bulletin of the EATCS*, 64 (February 1998), 188–216

[175] D. Pixton: Regular splicing systems. Manuscript, 1995

[176] D. Pixton: Linear and circular splicing systems. *IEEE Conf. on Intelligence in Neural and Biological Systems*, Herndon-Washington, 1995, 181–188

[177] D. Pixton: Regularity of splicing languages. *Discrete Appl. Math.*, 69 (1996), 101–124

[178] D. Pixton: Splicing in abstract families of languages. *Technical Report of SUNY Univ. at Binghamton, New York*, 1997, and *Theoretical Computer Science*, 234 (2000), 135–166.

[179] L. Priese: A. Note on nondeterministic reversible computations. *Technical Report 26/97*, Univ. Koblenz-Landau, Institut für Informatik, 1997

[180] L. Priese, Y. Rogozhin, M. Margenstern: Finite H systems with 3 tubes are not predictable. *Pacific Symp. on Biocomputing*, Hawaii, 1998 (R. B. Altman, A. K. Dunker, L. Hunter, T. E. Klein, eds.), World Sci., Singapore, 1998, 547–558

[181] P. Pudlak: Complexity theory and genetics. *Proc. 9th Conf. on Structure in Complexity Theory*, 1994, 183–195

[182] J. H. Reif: Local parallel biomolecular computation. *Proc. of the Third DIMACS Workshop on DNA Based Computers*, Philadelphia, 1997, 243–264

[183] J. H. Reif: Paradigms for biomolecular computation. *Unconventional Models of Computation* (C. S. Calude, J. Casti, M. J. Dinneen, eds.), Springer, Berlin, 1998, 72–93

[184] M. P. Robertson, A. D. Ellington: New directions in nucleic acid computing: Selected robozymes that can implement re-write rules. *Proc. of the Third DIMACS Workshop on DNA Based Computers*, Philadelphia, 1997, 69–73

[185] Y. Rogozhin: Small universal Turing machines. *Theoretical Computer Sci.*, 168 (1996), 215–240

[186] A. L. Rosenberg: On multihead finite automata. *IBM J. R. and D.*, 10 (1966), 388–394

[187] P. W. K. Rothemund: A DNA and restriction enzyme implementation of Turing machines. [117], 75–120

[188] B. Rovan: A framework for studying grammars. *Proc. MFCS 81, Lect. Notes in Computer Sci.* 118, Springer, Berlin, 1981, 473–482

[189] S. Roweis, E. Winfree, R. Burgoyne, N. Chelyapov, M. Goodman, P. Rothemund, L. Adleman: A sticker based architecture for DNA computation. [13], 1–27

[190] G. Rozenberg, A. Salomaa: *The Mathematical Theory of L Systems.* Academic Press, New York, 1980

[191] G. Rozenberg, A. Salomaa: *Cornerstones of Undecidability.* Prentice Hall, New York, 1994

[192] G. Rozenberg, A. Salomaa: Watson-Crick complementarity, universal computations and genetic engineering. *Techn. Report* 96-28, Dept. of Computer Science, Leiden Univ., Oct. 1996

[193] G. Rozenberg, A. Salomaa, eds.: *Handbook of Formal Languages.* 3 volumes, Springer-Verlag, Berlin, 1997

[194] Y. Sakakibara, C. Ferretti: Splicing on tree-like structures. *Proc. of the Third DIMACS Workshop on DNA Based Computers*, Philadelphia, 1997, 348–358

[195] A. Salomaa: *Theory of Automata.* Pergamon, Oxford, 1969

[196] A. Salomaa: On the index of context-free grammars and languages. *Inform. Control*, 14 (1969), 474–477

[197] A. Salomaa: *Formal Languages*. Academic Press, New York, 1973

[198] A. Salomaa: *Computation and Automata*. Cambridge Univ. Press, Cambridge, 1985

[199] A. Salomaa: Equality sets for homomorphisms of free monoids. *Acta Cybernetica*, 4 (1978), 127–139

[200] A. Salomaa: *Jewels of Formal Language Theory*. Computer Science Press, Rockville, Md., 1981

[201] A. Salomaa: *Public-Key Cryptography*. 2nd enlarged edition, Springer, Berlin, 1996

[202] A. Salomaa: Turing, Watson–Crick, and Lindenmayer. Aspects of DNA complementarity. *Unconventional Models of Computation* (C. S. Calude, J. Casti, M. J. Dinneen, eds.), Springer, Berlin, 1998, 94–107

[203] M. P. Schützenberger: On finite monoids having only trivial subgroups. *Inform. Control*, 8 (1965), 190–194

[204] D. B. Searls: The linguistics of DNA. *American Scientist*, 80 (1992), 579–591

[205] C. E. Shannon: A universal Turing machine with two internal states. *Automata Studies, Annals of Mathematical Studies*, 34, Princeton Univ. Press, 1956, 157–165

[206] R. Siromoney, K. G. Subramanian, V. R. Dare: Circular DNA and splicing systems. *Parallel Image Analysis, Lect. Notes in Computer Sci.* 654, Springer, Berlin, 1992, 260–273

[207] W. Smith: DNA computers in vitro and in vivo. [117], 121–186

[208] W. M. Sofer: *Introduction to Genetic Engineering*. Butterworth-Heinemann, Boston, 1991

[209] C. C. Squier: Semicontextual grammars: an example. *Bull. Math. Soc. Sci. Math. Roumanie*, 32(80) (1988), 167–170

[210] Y. Takada, R. Siromoney: On identifying DNA splicing systems from examples. *Proc. of AII'92, Lect. Notes in AI* 642, Springer, Berlin, 1992, 305–319

[211] P. Turakainen: A unified approach to characterizations of recursively enumerable languages. *Bulletin of the EATCS*, 45 (October 1991), 223–228

[212] A. M. Turing: On computable numbers, with an application to the Entscheidungsproblem. *Proc. London Math. Soc.*, Ser. 2, 42 (1936), 230–265; a correction, 43 (1936), 544–546

[213] A. M. Turing: Computing machinery and intelligence. *Mind*, 59 (1950), 433–460

[214] M. R. Walker, R. Rapley: *Route Maps in Gene Technology.* Blackwell Science, Oxford, 1997

[215] E. Winfree: Complexity of restricted and unrestricted models of molecular computability. [117], 187–198

[216] E. Winfree: On the computational power of DNA annealing and ligation. [117], 199–210

[217] E. Winfree, X. Yang, N. Seeman: Universal computation via self-assembly of DNA; some theory and experiments. [13], 172–190

[218] D. Wood: Iterated a-NGSM maps and Γ-systems. *Inform. Control*, 32 (1976), 1–26

[219] D. Wood: *Theory of Computation.* Harper and Row, New York, 1987

[220] T. Yokomori, M. Ishida, S. Kobayashi: Learning local languages and its application to protein α-chain identification. *Proc. 27th Hawaii Intern. Conf. System Sci.*, Vol. V, 1994, 113–122

[221] T. Yokomori, S. Kobayashi: DNA evolutionary linguistics and RNA structure modeling: a computational approach. *IEEE Conf. on Intelligence in Neural and Biological Systems*, Herndon-Washington, 1995, 38–45

[222] T. Yokomori, S. Kobayashi: DNA-EC: A model of DNA computing based on equality checking. *Proc. of the Third DIMACS Workshop on DNA Based Computers*, Philadelphia, 1997, 334–347

[223] T. Yokomori, S. Kobayashi, C. Ferretti: On the power of circular splicing systems and DNA computability. *Report CSIM 95-01*, Univ. of Electro-Comm., Chofu, Tokyo, 1995, and *IEEE Intern. Conf. on Evolutionary Computing*, Indianapolis, 1997, 219–224

[224] C. Zandron, C. Ferretti, G. Mauri: A reduced distributed splicing system for RE languages. *New Trends in Formal Languages. Control, Cooperation, Combinatorics* (Gh. Păun, A. Salomaa, eds.), *Lect. Notes in Computer Sci.* 1218, Springer, Berlin, 1997, 346–366

Index

Monographs in Theoretical Computer Science • An EATCS Series

K. Jensen
Coloured Petri Nets
Basic Concepts, Analysis Methods and Practical Use, Vol. 1
2nd ed.

K. Jensen
Coloured Petri Nets
Basic Concepts, *Analysis Methods* and Practical Use, Vol. 2

K. Jensen
Coloured Petri Nets
Basic Concepts, Analysis Methods and *Practical Use,* Vol. 3

A. Nait Abdallah
The Logic of Partial Information

Z. Fülöp, H. Vogler
Syntax-Directed Semantics
Formal Models Based on Tree Transducers

A. de Luca, S. Varricchio
Finiteness and Regularity in Semigroups and Formal Languages

E. Best, R. Devillers, M. Koutny
Petri Net Algebra

S. P. Demri, E. S. Orlowska
Incomplete Information: Structure, Inference, Complexity

J. C. M. Baeten, C. A. Middelburg
Process Algebra with Timing

L. A. Hemaspaandra, L. Torenvliet
Theory of Semi-Feasible Algorithms

E. Fink, D. Wood
Restricted-Orientation Convexity

Zhou Chaochen, M. R. Hansen
Duration Calculus
A Formal Approach to Real-Time Systems

M. Große-Rhode
Semantic Integration of Heterogeneous Software Specifications

Texts in Theoretical Computer Science • An EATCS Series

J. L. Balcázar, J. Díaz, J. Gabarró
Structural Complexity I

M. Garzon
Models of Massive Parallelism
Analysis of Cellular Automata
and Neural Networks

J. Hromkovič
Communication Complexity
and Parallel Computing

A. Leitsch
The Resolution Calculus

A. Salomaa
Public-Key Cryptography
2nd ed.

K. Sikkel
Parsing Schemata
A Framework for Specification
and Analysis of Parsing Algorithms

H. Vollmer
Introduction to Circuit Complexity
A Uniform Approach

W. Fokkink
Introduction to Process Algebra

K. Weihrauch
Computable Analysis
An Introduction

J. Hromkovič
Algorithmics for Hard Problems
Introduction to Combinatorial
Optimization, Randomization,
Approximation, and Heuristics
2nd ed.

S. Jukna
Extremal Combinatorics
With Applications
in Computer Science

P. Clote, E. Kranakis
Boolean Functions
and Computation Models

L. A. Hemaspaandra, M. Ogihara
The Complexity Theory Companion

C. S. Calude
Information and Randomness.
An Algorithmic Perspective
2nd ed.

J. Hromkovič
Theoretical Computer Science
Introduction to Automata,
Computability, Complexity,
Algorithmics, Randomization,
Communication and Cryptography

A. Schneider
Verification of Reactive Systems
Formal Methods and Algorithms

S. Ronchi Della Rocca, L. Paolini
The Parametric Lambda Calculus
A Metamodel for Computation

Y. Bertot, P. Castéran
Interactive Theorem Proving
and Program Development
Coq'Art: The Calculus
of Inductive Constructions

L. Libkin
Elements of Finite Model Theory

M. Hutter
Universal Artificial Intelligence
Sequential Decisions
Based on Algorithmic Probability

G. Păun, G. Rozenberg, A. Salomaa
DNA Computing
New Computing Paradigms
2nd corr. printing

W. Kluge
Abstract Computing Machines
A Lambda Calculus Perspective

J. Hromkovič
Dissemination of Information
in Communication Networks
Broadcasting, Gossiping, Leader
Election, and Fault Tolerance